The Interpretation of Quantum Mechanics

PRINCETON SERIES IN PHYSICS
Edited by Philip W. Anderson, Arthur S. Wightman, and Sam B. Treiman (published since 1976)

Studies in Mathematical Physics: Essays in Honor of Valentine Bargmann *edited by Elliott H. Lieb, B. Simon, and A. S. Wightman*

Convexity in the Theory of Lattice Gasses *by Robert B. Israel*

Works on the Foundations of Statistical Physics *by N. S. Krylov*

Surprises in Theoretical Physics *by Rudolf Peierls*

The Large-Scale Structure of the Universe *by P.J.E. Peebles*

Statistical Physics and the Atomic Theory of Matter, From Boyle and Newton to Landau and Onsager *by Stephen G. Brush*

Quantum Theory and Measurement *edited by John Archibald Wheeler and Wojciech Hubert Zurek*

Current Algebra and Anomalies *by Sam B. Treiman, Roman Jackiw, Bruno Zumino, and Edward Witten*

Quantum Fluctuations *by E. Nelson*

Spin Glasses and Other Frustrated Systems *by Debashish Chowdhury*
(*Spin Glasses and Other Frustrated Systems* is published in co-operation with World Scientific Publishing Co. Pte. Ltd., Singapore.)

Weak Interactions in Nuclei *by Barry R. Holstein*

Large-Scale Motions in the Universe: A Vatican Study Week *edited by Vera C. Rubin and George V. Coyne, S.J.*

Instabilities and Fronts in Extended Systems *by Pierre Collet and Jean-Pierre Eckmann*

More Surprises in Theoretical Physics *by Rudolf Peierls*

From Perturbative to Constructive Renormalization *by Vincent Rivasseau*

Supersymmetry and Supergravity (2d ed.) *by Julius Wess and Jonathan Bagger*

Maxwell's Demon: Entropy, Information, Computing *edited by Harvey S. Leff and Andrew F. Rex*

Introduction to Algebraic and Constructive Quantum Field Theory *by John C. Baez, Irving E. Segal, and Zhengfang Zhou*

Principles of Physical Cosmology *by P.J.E. Peebles*

Scattering in Quantum Field Theories: The Axiomatic and Constructive Approaches *by Daniel Iagolnitzer*

QED and the Men Who Made It: Dyson, Feynman, Schwinger, and Tomonaga *by Silvan S. Schweber*

The Interpretation of Quantum Mechanics *by Roland Omnès*

The Interpretation
of Quantum Mechanics

ROLAND OMNÈS

Princeton Series in Physics

PRINCETON UNIVERSITY PRESS · PRINCETON, NEW JERSEY

Copyright©1994 by Princeton University Press
Published by Princeton University Press, 41 William Street,
Princeton, New Jersey 08540
In the United Kingdom: Princeton University Press,
Chichester, West Sussex

All Rights Reserved

Library of Congress Cataloging-in-Publication Data

Omnès, Roland
The interpretation of quantum mechanics / Roland Omnès.
p. cm. — (Princeton series in physics)
Includes index.
ISBN 0-691-03336-6 (cl.)—ISBN 0-691-03669-1 (pbk.)
1. Quantum theory. I. Title. II. Series.
QC174.12.046 1994
530.1'2—dc20   93-47445

This book has been composed in Times Roman

Princeton University Press books are printed on acid-free paper and meet the
guidelines for permanence and durability of the Committee on Production
Guidelines for Book Longevity of the Council on Library Resources

Printed in the United States of America

1 3 5 7 9 10 8 6 4 2

(Pbk.)

Contents

PREFACE	xi
HOW TO READ THIS BOOK	xv

1. Elementary Quantum Mechanics — 3

The Beginnings of Quantum Mechanics	4
1. The Mechanics of Waves, Matrices, and Quanta	4
Mathematical Formalism	16
2. Quantum Mechanics in a Hilbert Space	16
3. The Spectral Theorem	24
Feynman Histories	31
4. Feynman's Formulation of Quantum Dynamics	31
Probabilities and States	35
5. Quantum Probabilities	35
6. The Density Operator	38
7. Dynamics	41
8. How to Describe a Complex System	44
Reference Frames	46
9. How to Construct the Physical Hilbert Space	46
10. Relativistic Invariance	51
Appendix: The Uncertainty Relation for Energy	56
Problems	58

2. The Problems of Measurement Theory — 60

Experimental Devices	60
1. What Is a Measurement?	60
2. Some Examples of Measurements	62
3. The Stern–Gerlach Experiment	68
Elementary Measurement Theory	72
4. Von Neumann's Formal Theory of Measurements	72
5. The Problem of Macroscopic Interferences	77

The Copenhagen Interpretation	81
6. The Prescriptions of Measurement Theory	81
7. The Copenhagen Epistemology	85
8. Schrödinger's Cat and Wigner's Friend	91
What an Interpretation Should Be	92
9. Why Does Physics Need an Interpretation?	93
10. The Criteria of an Interpretation. Consistency and Completeness	98
Problems	100

Interlude   103

Five Ideas for Constructing a Consistent Interpretation. Properties, Histories, Logic, Classical Approximations, and Decoherence   103

3. Foundations and Properties   110

The Basic Principles	111
1. The Mathematical Framework	111
2. Dynamics	113
The Properties of a System	114
3. Properties	114
4. Probabilities and States	117
5. Gleason's Theorem	119
6. Taking Time into Account	120
Appendix: Gleason's Theorem	121

4. Histories   122

The Notion of History	123
1. Experiments and Histories	123
2. Definition of Histories	125
The Probabilities of Histories	126
3. Logical Criteria for the Probabilities	126
4. Connection with Feynman Histories*	130
Consistency Conditions	132
5. The Consistency Conditions for Additive Probabilities	132
Appendix A: General Form of the Consistency Conditions	137
Appendix B: The Uniqueness of Probabilities	140
Problems	143

5. The Logical Framework of Quantum Mechanics — 144

About Logic — 144
    1. Why Does One Need Logic? — 144
    2. What Is Logic? — 148
Quantum Logic — 154
    3. Defining a Quantum Logic — 155
    4. The Complementarity "Principle" — 159
A Foundation for Interpretation — 162
    5. The Universal Rule of Interpretation — 162
Applications — 165
    6. Interferences — 165
    7. The Straight-Line Motion of a Particle — 171
    8. Decays — 176
    9. Approximations in Logic — 179
Appendix A: Formal Logic — 183
Appendix B: Formal Logic and Consistency — 186
Appendix C: The No-Contradiction Theorem — 188
Appendix D: Logic and Time Reversal — 193
Problems — 198

6. Recovering Classical Physics — 201

Objects — 204
    1. Orientation — 204
    2. Quantum Mechanics of Collective Observables — 207
    3. Classical Variables — 208
Classical Properties — 213
    4. A Return to Classical Logic — 213
    5. The Quantum Form of a Classical Property — 216
    6. The Construction of Quasi-Projectors — 226
Dynamics — 227
    7. The Dynamical Correspondence — 227
Justifying Common Sense — 234
    8. Recovering Common Sense — 234
Appendix A: Elements of Microlocal Analysis — 238
Appendix B: Semiclassical Theorems — 248
Appendix C: Consistency of Classical Logic — 259
Appendix D: A Criterion for the Existence of Collective Observables — 261
Problems — 262

7. Decoherence — 268

Orientation — 270
1. An Intuitive Approach — 273
Solvable Models — 279
2. A Simple Model — 279
3. Another Example: The Pendulum — 282
More General Models — 291
4. The General Theory — 291
5. Decoherence by the External Environment — 297
6. Back to Schrödinger's Cat — 303
Can One Circumvent Decoherence? — 304
7. A Criticism of Decoherence — 304
8. One Cannot Circumvent Decoherence — 307
9. Justifying the Assumptions* — 309
10. The Direction of Time — 315
Appendix: Decoherence from an External Environment — 319
Problem — 322

8. Measurement Theory — 324

1. Reality and Theory. Facts and Phenomena — 324
2. An Introduction to Measurement Theory — 325
Measurement of a Single Observable — 328
3. What Is a Measurement? — 329
4. The Main Theorems — 334
Wave Function Reduction — 337
5. Two Successive Measurements — 337
Actual Facts — 341
6. Actual Facts and the Present Time — 342
7. Everett's Answer — 345
8. A Law of Physics Different from All Others — 350
The Notion of Truth — 353
9. The Criteria of Truth — 353
10. Up to What Point Can One Know the State? — 359
11. Explicit States — 364
Appendix A: The Theorems of Measurement Theory — 369
Appendix B: The Density Operator and Information Theory — 375

9. Questioning Quantum Mechanics — 378

The Einstein–Podolsky–Rosen Experiment — 378
1. The Background — 380

2. Analyzing the EPR Experiment	383
3. Bohm's Version of the EPR Experiment	387
4. Truth and Reliability in the EPR Experiment	391
5. The Relativistic Case	397
6. Separability	398
Hidden Variables	400
7. Hidden Variable Theories	400
8. Bell's Inequalities	403
Problem	407

10. Nonclassical Macroscopic Systems — 409

Nonclassical Superconductors	410
1. Superconductors	410
2. The Aharonov–Bohm Effect	418
3. The Basis of Leggett's Experiments	423
Chaotic Systems	426
4. Classical and Quantum Statistics of Chaotic Systems	426

11. Experiments — 433

Experiments Showing Histories	434
1. The Decay of a Particle	434
2. Repeated Measurements	436
3. The Zeno Effect	438
4. Observing a Unique Atom	443
Interferences	448
5. The Badurek–Rauch–Tuppinger Experiment	448
6. Delayed-Choice Experiments	456
Leggett's Experiments	463
7. The Experiments with SQUIDs	463

12. Summary and Outlook — 467

The Rules of Interpretation	467
1. The Basic Principles	467
2. Properties	469
3. The Logical Framework	471
4. The Foundations of Classical Physics	476
5. Classical Logic	482
6. Decoherence	484
7. Can One Circumvent Decoherence?	486
8. The Theory of Phenomena	488

    9. Measurement Theory   489
  10. Actual Facts   492
  11. Truth Criteria   494
  12. The State of a System   497
  13. The Relation with Bohr's Interpretation   498
  14. Gell-Mann and Hartle's Approach   502
  15. Perspectives   503
Philosophical Aspects   506
  16. Twenty Theses   506
  17. Is the Theory Objective?   509
  18. Is the Theory Realistic?   512
  19. About the Foundations of Science   516
  20. Total Realism   527

NOTES   533

INDEX   545

Preface

Almost all of physics now relies upon quantum mechanics. This theory was discovered around the beginning of this century. Since then, it has known a progress with no analogue in the history of science, finally reaching a status of universal applicability.

The radical novelty of quantum mechanics almost immediately brought a conflict with the previously admitted corpus of classical physics, and this went as far as rejecting the age-old representation of physical reality by visual intuition and common sense. The abstract formalism of the theory had almost no direct counterpart in the ordinary features of the world around us, as, for instance, nobody will ever see a wave function when looking at a car or a chair. An ever-present randomness also came to contradict classical determinism. This was really troublesome because determinism not only was a dominant philosophy among scientists but also is something whose existence is necessary for our understanding of the world. Such trivial and basic features as remembering the past from memory or recalling it from a record or a photograph, as well as the fact that a laboratory apparatus or a mechanical device is going to work in a predictable fashion, are altogether essential features of our understanding of reality and our relation to it. They are fundamentally linked with determinism, and as such, one could wonder whether they hold in a world obeying quantum laws.

When a theory is so strange that it must be interpreted, whether it be relativity or quantum mechanics, the aim of this interpretation is to reconcile the fundamental, outrageously abstract concepts with plain empirism: the kind of ordinary situation met in a laboratory or anywhere else. Interpretation must also reconcile the theory with the cornerstones of physics—namely, the existence of facts and their agreement with common sense together with its everyday determinism, at least as far as large-scale phenomena are concerned. This endeavor could define the main purpose of this book.

The following text does not consist of a commentary upon the writings of Bohr and Heisenberg and what they said, as was often the case for interpretative books in the past. Neither is it a new, formal, axiomatic approach, nor is it pleading for forthcoming incomplete theories purporting to replace quantum mechanics. It is essentially a fresh approach to the older interpretation we all owe to Bohr, though putting it upon new and firmer foundations.

It was found on several occasions that the traditional Copenhagen views, though essentially correct for the vast majority of experimental or natural conditions, should be modified in order to obtain better consistency. This does not mean that it was fundamentally wrong, but it needed some trimming after more than seventy years.

This renewal of the conventional interpretation is the outcome of a recent change of emphasis in the general strategy of research in this domain, which has rapidly given many new and significant results. It came after a rather long episode during which the main interest was the question of existence or nonexistence of hidden variables, following the discovery of a corresponding test by John Bell. A new effort aiming, on the contrary, at a clarification and a justification of Bohr's interpretation has now followed, though some of its actors may elicit strong criticisms of various excesses in the nebula known as the Copenhagen interpretation.

This change began in the period 1975–1982 with the discovery and understanding of the decoherence effect, which is responsible for breaking quantum linear superpositions of different macroscopic properties. The existence of these superpositions, which seemed to follow directly from the basic theory, had been plaguing interpretation since the beginning, because it meant that different classical data could not be understood as clearly distinct events. This is also known as the problem of Schrödinger's cat. Soon after this breakthrough, one learned (in 1984) how to describe correctly the properties of a quantum system by narrating its history in detail, and this was to lead, in 1988, to a sound logical description of the behavior of a quantum system. This might look like a much more modest achievement, but quantum mechanics has always defied logic—stating, for instance, that a unique photon must follow both arms of an interferometer. It became possible to firmly control such apparent aberrations, to make clear what makes sense and what does not in quantum mechanics, and to make this soundness the basis and the unique assumption of the whole of interpretation. The fruitfulness of this new approach became clear when it allowed a complete derivation of classical physics, explaining why and when determinism can hold under large-scale conditions, though the basic theory underlying it is purely probabilistic. It became at last possible to understand quantum mechanics by resorting to a basic inversion in the process of understanding: rather than starting from common sense to try fitting it with a physical theory obviously foreign to it, the trick was to understand what principles of a logical nature stand at the bottom of physics itself and to derive from them when and how common sense is valid.

It became clear only recently that the various pieces of the puzzle that had been discovered more or less independently could be fitted together to make clear that no inconsistency had been left aside, no paradox unsolved, no well-known difficulty untouched. The overall result is an interpretation possessing the two major properties of consistency and completeness: consistent as being explicitly free from any logical self-contradiction or paradox, complete as providing a definite prediction for every experimental situation.

Quite a few of Bohr's views had to be modified accordingly, though these are only minor changes as far as the practice of physics is concerned. The modifications of the underlying conceptual framework are much more significant, but they also often provide a more straightforward justification of some rules or notions that had been put forward by Bohr with no obvious or completely convincing argument. As an example of these changes, one may mention that the usual axioms of measurement theory can be reduced to a unique logical principle, that there exist macroscopic systems behaving in a quantum mechanical way and showing macroscopic tunnel effects; wave packet reduction, though undoubtedly very useful for practical purposes, is not a necessary assumption and in any case not a physical effect undergone by the measured object; old paradoxes such as that by Einstein, Podolsky, and Rosen, become almost trivial and in any case easily untied; many conventional or more philosophical ideas concerning objectivity and reality need serious reconsideration.

A satisfactory feature of the new approach is that the interpretation of quantum mechanics has now recovered the standard of a mature science, i.e., a deductive construction relying upon clear-cut principles. One no longer needs to justify it by relying upon the word of past authorities, because its logical consistency as well as its agreement with experiment are now obvious. Another convenient feature is that interpretation is now easier to understand. These were also my two motives for writing the present book.

This is intended to be a book by which one can learn the interpretation of quantum mechanics, not an interpretation. A reader who is not particularly interested in the most recent fashions and the latest results and who wants only to learn what is the conventional Copenhagen interpretation will find this book useful, as will the reader who wants to keep in touch with research in this domain. This is because the older interpretation is given and also justified; rather than glossing over its difficulties as usual, I show how they can be avoided. Quite a few rather obscure recommendations by Niels Bohr become clear because they now follow from a consistent construction.

The interpretation of quantum mechanics is developed here by starting from what one can learn in a first-year course on the subject, and most of the book has been kept at that level. This is obtained by restricting the discussion to the main ideas, giving only the simplest proofs and the most important examples. When some theoretical results of a higher level were found to be necessary, they were explained, as far as possible, in an intuitive way rather than with elaborate proofs or difficult calculations. The proofs and the calculations, when they are really necessary for a deeper understanding or for practical applications, have been provided in a few technical appendixes.

The content of the present book is an extended version of lectures given at Collège de France in Paris during the winter of 1991, and I thank the Administrator of this prestigious place and my friend, Marcel Froissart, for offering me this opportunity. I also taught a simplified version of this text in five one-hour lessons to the students of a first-year course in quantum mechanics at Orsay. The synthesis it offers comes essentially from two previous works, one by Murray Gell-Mann and Jim Hartle, the other my own.[1] Since it is a synthesis, it relies upon many different contributions, to be described in due place in the book. One should, however, mention from the beginning the contributions by Robert Griffiths on histories and by Wojciech Zurek on decoherence.

During the years when my own works were being completed and this book was written, I benefited from many useful suggestions and helpful criticisms by various people. I wish to thank particularly those who have left a mark in the present book: Alain Aspect, Roger Balian, Claude Cohen-Tanoudji, Alain Connes, Bernard d'Espagnat, Michel Devoret, Murray Gell-Mann, Jean Ginibre, Robert Griffiths, Jim Hartle, Lars Hörmander, Philippe Nozières, Serge Reynaud, Abner Shimony, Walter Thirring, and Wojciech Zurek, not failing to mention the late John Bell and Léon van Hove. I also express my gratitude to Jean-Louis Basdevant, who read carefully an initial French version of the present book and made many useful remarks. My warmest thanks go to Arthur Wightman who, after proposing to include this book in the Princeton series, went carefully through the manuscript and improved it considerably, though of course any remaining mistakes can only be mine. I also thank the staff of Princeton University Press and the copyeditor for their excellent work on this book. Finally, I thank you, Liliane, for your infinite patience.

How To Read This Book

It has become customary to tell the reader how to read a book, as if he were not able to browse through it by himself. I rarely understand this kind of introduction, except after having read most of the book. This is why I shall only mention here what should not be read.

If you already know quantum mechanics, don't read the first chapter. It contains the usual preliminaries.

If you want to know what is new in the present approach, go immediately to the first part of the last chapter. After that you may wish to read what comes before. By all means, don't read the second part of the last chapter. It is more or less philosophical in nature and would lead you to believe that the rest of the book is like it. That would be misleading.

If, on the contrary, you are inclined toward the theory of knowledge and what quantum mechanics can add to it, read the second part of the last chapter and browse through the text containing no equations, particularly the end of Chapter 2, the interlude, and some parts of Chapter 8.

If you were looking for a novel and you found only this book in the library, you didn't make a complete mistake as, under another title such as "The murder of Professor Schrödinger's cat," it contains an enigma with a solution at the end to exercise little grey cells. If you want to understand quantum mechanics, I beg you to read a significant part of the text, and, from my heart, I hope you will be satisfied. If there is at least one such person, many years of work will not have been useless.

Finally, although the main text has been kept at a rather uniform level, it may be mentioned that the appendices are more mathematical. This is also true of a few sections, which are marked by an asterisk; I have tried my best that they might be skipped without breaking the line of argument. Similarly, a problem marked by an asterisk indicates that it is on the border of research or that it involves nontrivial calculations.

The Interpretation of Quantum Mechanics

1

Elementary Quantum Mechanics

The present chapter contains a brief introduction to quantum mechanics covering more or less the content of an elementary textbook. It remains at an elementary level, though for later convenience a few other questions have been added or developed in more detail. They include considerations about Hilbert spaces and some results about group theory, density operators, and Feynman path integrals.

For a long time in the history of physics, its framework could be essentially identified with classical mechanics as it describes the motion of a planet or the working of an engine: a science of concrete objects that was thought to reach certainty. In classical mechanics, every part of a physical object is described by some coordinates, taking into account its shape, its position, and its orientation in space. There is no difficulty in choosing these coordinates since they represent the object as one sees it. Their change in time is described by the corresponding velocities, i.e., their time derivatives. Both coordinates and velocities can vary in a continuous way, and together they define the instantaneous state of motion. The later behavior of the system can be derived from the knowledge of the initial state by using the basic principles of the theory as set forth by Newton.

Classical mechanics nevertheless had its mysteries. It did not say much about the matter constituting the objects until the advent of atomism. Another even deeper mystery was put forward in the early days: How does it happen that the material world can be so well described by purely mathematical concepts? This correspondence between reality and theory was first attributed to some divine design, and it may be noticed that, though one does not refer usually to that explanation anymore, the problem is still there and it has become more striking than ever with the progress of science. The obvious question of the relationship between mathematics and physics is too often forgotten by sheer habit, and it will be useful to keep it in mind in our study of quantum mechanics.

The comfortable framework of classical mechanics was shattered when the new laws of physics governing atoms and elementary particles were discovered. The subtle difficulties already present in classical physics were not all alleviated, and new ones were added. The mathematical framework ceased to be clear and intuitive, becoming instead very strange, often more a hindrance than a help. Classical determinism, which had sometimes been considered too coercive, was replaced by its exact opposite—pure indeterminism, which was even more difficult to cope with. A striking dichotomy had appeared between the purely probabilistic laws governing the microscopic world and the certainty we attribute to the working of experimental instruments and to ordinary facts.

This radical change occurred during the first quarter of the present century. Its outcome was a new and precise form of the laws of physics with its own well-defined concepts, however difficult. Some time has since elapsed, but there has been no essential change in these principles. Their interpretation has, however, become somewhat clearer recently, and this, including the foundations of quantum mechanics—how to understand them, or, more properly, how to relate them with common sense, even if not in so simple a way as during the classical Golden Age—will be the main subject of this book.

The subject is rather wide, and it could be approached at various levels or from rather different standpoints. Since a book must begin with a first encounter with the subject and by a chapter number one, I have chosen to follow more or less the course of history.[1]

THE BEGINNINGS OF QUANTUM MECHANICS

1. The Mechanics of Waves, Matrices, and Quanta

Quantum is a Latin word meaning some quantity or some definite amount of something. It is used in physics with the same meaning as the word *discrete* is in mathematics, i.e., some quantity or variable that can take only sharply defined values, as opposed to a continuously varying quantity. In the initial stages of development of quantum theory, one felt that the quantum character of some physical quantities, such as the internal energy of an atom or an angular momentum component, was an essential part of the new physics, and the name has remained. One still speaks of quantum mechanics, though *quantum physics* would be a more appropriate name because it provides a general framework for the whole of physics rather than dealing with a special aspect of mechanics.

The Discovery of Quanta

Quanta were born in 1900 with a theory of black-body radiation put forward by Max Planck. Black-body radiation is the electromagnetic radiation emitted by a body in thermodynamical equilibrium, or the radiation contained in a cavity when its walls are at a uniform temperature. By letting radiation escape through a small aperture, one can measure its spectrum and its energy density.

This spectrum had been studied experimentally and also predicted by thermodynamics. According to this theory, the radiation intensity ΔI_ν

emitted in a small frequency interval $\Delta\nu$ should be proportional to the cube of the frequency ν. This was obviously a troublesome result since integrating over all frequencies gives an infinite total intensity. The experimental data agreed with this prediction at small frequencies, but were at variance with it at higher frequencies: rather than increasing like ν^3, ΔI_ν decreases exponentially. There was therefore a sharp contradiction between theory and experiment, although no obvious flaw could be found in the theory despite its clearly impossible result. The solution of this problem called for a basic revision of our physical concepts.

Max Planck assumed as a new postulate that radiation is not emitted continuously but by discrete amounts, each bunch of light carrying away a discrete amount of energy $h\nu$, where h is a fundamental constant. In order for $h\nu$ to be an energy, h has to be an action, with dimensions ML^2T^{-1}. It can also be considered as the product ET of an energy and a time, or the product MVL of a momentum by a length. When Planck introduced this assumption in thermodynamics, he obtained a spectrum in good agreement with the data.

Planck believed that the emission of quanta was due to the mechanism of radiation emission by the atoms and was consequently a property of these atoms. In 1905, however, Albert Einstein showed that the empirical properties of the photoelectric effect could be explained by assuming that light consists of particles, or photons, each one having an energy $h\nu$ and moving at the velocity of light. According to him, the existence of quanta was not (or not only) due to the emission process but was a property of light itself. The best evidence for this idea came much later, when Compton found that the scattering of gamma rays by electrons has the kinematical characteristics of the collision between two particles.

At the same time, around the beginning of the century, many physicists were also very interested in the spectrum of the light emitted or absorbed by atoms. It had been known for some time that the light emitted or absorbed by a given species of atoms has a discrete line spectrum. One was soon to learn more about the atoms when, in 1911, the scattering of alpha particles allowed Ernest Rutherford to establish that an atom consists of electrons located around a small positively charged heavy nucleus.

Classical mechanics was apparently well suited to describe such an atom: the Coulomb force acting between the nucleus and an electron is formally the same as a newtonian gravity force, so the behavior of a hydrogen atom could be reduced to the Kepler problem of a planet revolving around a star. The predictions one could draw from this classical treatment were unfortunately full of inconsistencies: when applying Maxwell's theory of radiation to the emission of light by an atom, one did not find any hint of its quantum character and, even worse, one predicted that the atoms were not stable because they were expected to lose continuously some energy by radiation, the electrons falling rapidly upon the nucleus.

Bohr's Atomic Model

In 1908, Walter Ritz discovered a remarkable property of the atomic line spectra: all the frequencies in the spectrum of a given atom can be obtained from the simple formula $\nu = \nu_n - \nu_m$, where the frequencies ν_n ($n = 1, 2, 3, \ldots$) characterize the atom. Niels Bohr suggested that, by taking into account the quantum character of light, this is an indication of a change in the atom energy by a quantity $h\nu = E_n - E_m$, the energies E_n being quantized values of the atom's internal energy and therefore more basic characteristics of the atom than the frequencies.

To develop this idea, in 1913 Bohr modified the planetary model of the atom by assuming a new constraint unknown to classical physics. He assumed that, among all the possible electronic orbits in a hydrogen atom, the only ones allowed must satisfy the condition $\int p\,dx = nh$, where the integral is taken along a Kepler elliptic trajectory and n is an integer introducing quantization. It is then easy to compute the energy levels for the hydrogen atom to get

$$E_n = -\frac{m}{2n^2}\frac{e^4}{\hbar^2}$$

where m and e are the mass and the electric charge of the electron, and the quantity $\hbar$ is the ratio $h/2\pi$. This result agreed quite well with experimental data. It looked, therefore, as if classical physics was still valid in the case of atoms, except for the necessity of adding further constraints. From there followed a rather long period of slow improvements, ending finally in an accumulation of difficulties that can be summarized as a compulsory quantization of classical physics and its failure.

A Host of Mechanics

The next significant step was made by Louis de Broglie in 1923. It was decisive insofar as it threw away for the first time the whole framework of classical concepts. The idea was no longer to correct and supplement classical physics by quantum constraints but to reconsider the conceptual representation of physical objects. Drawing an analogy with light and its dual characters as wave and particle, de Broglie proposed that a wave should be associated with every kind of particle and particularly with the electron. This radically new assumption did not lead immediately to any experimental consequence, and one had to wait until 1927 for its almost accidental confirmation by Davisson and Germer, who observed the diffraction of electrons by a crystal and thereby showed the wave character of electrons.

At about the same time, Werner Heisenberg was also reconsidering the foundations of physics, though starting from quite different considerations. His leading idea was that physics should only use observable quantities.

This meant that one should not use nor even mention classical orbits, since no conceivable experiment can show their actual existence. He tested his ideas against the problem consisting in computing the relative intensities of the spectral lines.

To do so, he assumed the existence of quantized atomic energy levels E_n, as Bohr had done. Taking a hint from classical electromagnetism, where the radiated intensity depends upon the electric dipole of the atom, which is given essentially by the position of the electron with respect to the nucleus, he had to introduce this position. But the position cannot be measured while emission takes place, and one therefore should not refer directly to it. Heisenberg therefore postulated that, when the atom transits from one energy level n to another level m, the position is completely characterized by a number x_{nm} representing how the position can manifest itself during the transition. When pushing it further, after noticing that the classical intensity in fact involves the electron acceleration, Heisenberg also introduced similar quantities v_{nm} and γ_{nm} to represent velocity and acceleration.

When trying to find the relations between the quantities x_{nm}, v_{nm}, γ_{nm}, and E_n, he was led to reformulate the laws of dynamics in terms of them only. He relied for that purpose upon a correspondence principle that had been put forward by Bohr, according to which the laws of quantum physics should approach those of classical physics when the conditions are such that Planck's constant is relatively small. In such a case, the relations between the discrete quantities Heisenberg had introduced had to reproduce classical mechanics as well as the classical theory of radiation. This led him to postulate some simple dynamical relations between these strange mathematical objects and to learn more about their algebraic nature. In this way he found quite unexpected and subtle rules of calculation; in particular, that multiplication was not commutative. Nonetheless, in 1925 he obtained, for the first time, the relative intensities of spectral lines in good agreement with experiment.

Max Born, Pascual Jordan, and Paul Dirac soon realized that the use of noncommutative quantities to replace position and momentum is an essential feature of Heisenberg's theory, and they recognized in their algebra the rules of matrix calculus. They accordingly left aside the philosophical setup that had motivated Heisenberg and concentrated upon the foundation of a new kind of mechanics where the dynamical variables are not ordinary numbers but new, noncommutative mathematical objects.

Max Born, in collaboration with Pascual Jordan, soon developed an already complete formulation of this new mechanics, using infinite matrices to represent the basic physical quantities (1925). Dirac, in the same year, did not try to specify so precisely these quantities, considering them only as abstract mathematical objects, say Q_j for the position coordinates and P_j for the momentum coordinates. The multiplication rules of these quantities were supposed to follow from new dynamical principles. Dirac found once again the hint for these principles in the correspondence principle: the new kind of mechanics should reduce to the classical kind when Planck's constant is relatively negligible. This led him to postulate a general form for the commutators between two quantum dynamical variables, which amounts to replacing them by the Poisson brackets occurring in classical mechanics, up to an imaginary factor. In the special case of position and momentum coordinates, this gives the following expressions:

$$[Q_j, Q_k] = 0,$$
$$[P_j, P_k] = 0,$$
$$[Q_j, P_k] = i\hbar \delta_{jk},$$

where the commutator of two noncommuting quantities A and B is the difference $AB - BA$ and is written as $[A, B]$, and the symbol δ_{jk} being equal to 1 when $j = k$ and to 0 otherwise. The same commutation relations were also satisfied by the matrices used by Born and Jordan.

Soon afterward, in 1926, Dirac on the one hand and Born, Heisenberg, and Jordan on the other obtained a complete formulation of quantum dynamics that could be applied to any physical system. They derived many consequences from it and particularly the properties of angular momentum, from which they could explain many features of the atomic line multiplets.

In the same year (1926), de Broglie's wave hypothesis was given a precise formulation by Erwin Schrödinger. His essential progress was to discover an explicit equation for defining the wave function $\Psi(q)$ and its time evolution, somewhat similar to Maxwell's equations for an electromagnetic wave. In the case of a stationary state of the atom with energy E_n, where Bohr's model suggests the existence of a standing wave, Schrödinger proposed the following equation for the corresponding wave function $\Psi_n(q)$:

$$H\left(q, \frac{\hbar}{i}\frac{\partial}{\partial q}\right)\Psi_n(q) = E_n \Psi_n(q).$$

The function $H(q, p)$ is the classical energy of the atom when expressed in terms of position and momentum coordinates (the Hamilton function), but the momentum variable p occurring in it is replaced by the differential operator $(\hbar/i)(\partial/\partial q)$. So, for instance, the hydrogen atom has the Hamilton function

$$H(x, p) = \frac{p^2}{2m} - \frac{e^2}{r},$$

from which one obtains a differential operator

$$H\left(x, \frac{\hbar}{i}\frac{\partial}{\partial x}\right) = -\frac{\hbar^2}{2m}\Delta - \frac{e^2}{r},$$

ELEMENTARY QUANTUM MECHANICS

and the wave function is a solution of the partial differential equation

$$-\frac{\hbar^2}{2m}\Delta\Psi_n(x) - \frac{e^2}{r}\Psi_n(x) = E_n\Psi_n(x). \tag{1.1}$$

This kind of equation had already been studied in mathematics and was rather easy to solve. Assuming the wave function vanishes at infinity (because the electron has to remain in the vicinity of the nucleus) one can compute the possible energies E_n explicitly. One gets the values already obtained by Bohr, together with many other consequences in good agreement with more detailed spectral data.

More generally, Schrödinger assumed that dynamics, i.e., the time evolution of the wave function, is governed by another partial differential equation:

$$i\hbar\frac{\partial\Psi(q)}{\partial t} = H\left(q, \frac{\hbar}{i}\frac{\partial}{\partial q}\right)\Psi(q). \tag{1.2}$$

The explicit occurrence of the complex number i in this equation implies that the wave function is generally complex. In particular, by comparing equations (1.1) and (1.2), one finds that the time evolution of the wave function of a stationary state is given by

$$\Psi(q,t) = e^{-(iE_nt/\hbar)}\Psi_n(q).$$

Schrödinger was soon to realize that the wave function of a many-electron atom is not defined in ordinary, physical, three-dimensional space, as Louis de Broglie would have it, but in a much more abstract configuration space, i.e., the space described by the position coordinates of all the electrons. The wave is therefore very different from a simple electromagnetic wave and it is a far more abstract kind of mathematical field, much more difficult to understand since it takes place in a formal space and its values are complex. This last feature was something yet unknown to physics. How to interpret the wave function, what physical meaning should be given to it, became of immediate concern.

The Probabilistic Meaning of the Wave Function

The year 1926 had not reached an end when Max Born discovered the empirical meaning of the wave function. He analyzed in detail what happens when an electron scatters upon an obstacle, for instance, when it collides with an atom before being detected by a Geiger counter. He translated the properties of the wave function in terms of the behavior of the electron, using analogies with the diffraction of X-rays. This gave him the meaning of the wave function: the electron can be anywhere in a place where the wave function is different from zero, and there is no way of saying exactly its solutions, their sum $\Psi_1 + \Psi_2$ is another solution. The corresponding probability is proportional to $|\Psi_1 + \Psi_2|^2$ and it can show interference effects because it has the same form as the intensity of interfering electromagnetic waves when they are formally represented by complex quantities. The first example of interferences was found in 1927 by Davisson and Germer, when electrons were seen to be diffracted by the atomic lattice of a crystal.

However, the probabilistic significance of the wave function raised a very difficult issue. One can compute the probability of a specific event, such as an electron going in some direction after being diffracted, but the theory offers no hint, no mechanism, and no cause to determine what actual direction an individual electron will choose to follow. One can never say where a particle actually is, but only give the chances of finding it here or there. If nothing more can be added to the knowledge of the wave function in order to increase the predictability of the theory, one must recognize it as intrinsically probabilistic, strictly confined to the prediction of probabilities and nothing else.

The status of probabilities in quantum mechanics is therefore very different from what it is in classical physics, not so much when it comes to their use but rather when one considers their conceptual nature. In classical physics, probabilities only express some ignorance, some lack of information concerning the fine details of a given situation. When, for instance, a bullet is fired by a gun and hits a target apparently at random, this is only because one does not know in detail how much a hand is shaking, what kind of fuzziness there is in the eye of the person holding the gun, or if the sighting mark is centered. Many such uncontrolled causes are deemed responsible for the occurrence of randomness, the point being that actual causes exist and, in principle, if one knew them better the predictions would also be better.

ELEMENTARY QUANTUM MECHANICS

Quantum probabilities are very different because they assume that a more precise knowledge is impossible at the atomic level, as a matter of principle. As time went on, it became even clearer through the accumulation of many experiments that such an ultimate limitation could not be avoided. Nevertheless, one knows that this was never completely accepted by Einstein who, in his own words, could not admit that "God is playing dice."

The Uncertainty Relations

A most remarkable consequence of these probabilistic considerations is given by the famous *uncertainty relations*, which were discovered by Heisenberg. Knowing the probability distribution for a position coordinate x, one can easily get its dispersion or uncertainty Δx. One can show that, by using the fact that the wave function associated with a particle having a specific momentum p is an exponential $\exp(ip \cdot x/\hbar)$, the probability amplitude for the values of the momentum is the Fourier transform of the wave function. One can then compute the uncertainty Δp_x for the x coordinate of momentum. These two quantities are constrained by the inequality

$$\Delta x \, \Delta p_x \geq \hbar/2, \tag{1.4}$$

which is an uncertainty relation. It shows that, as a matter of principle, one cannot know with perfect precision both position and momentum. Any attempt to get a better knowledge of one of them unavoidably spoils the precision of the other.

This was not a complete novelty, because it amounts to a familiar property of Fourier transforms, which was already known in other physical applications. When, for instance, the transverse extent of an optical wave is chopped off by letting it cross a hole or a slit, this entails a spreading among the values of the wave-number vector, more pronounced when the slit is narrower. The resulting phenomenon is diffraction and, in that sense, the uncertainty relations are only witnesses to the wave character of the underlying physics.

What is new is that a precise knowledge of such basic physical notions as position and momentum is forbidden, in spite of our tendency to see them intuitively as ordinary properties of a little chunk of matter. This is the first place where quantum mechanics clearly conflicts with common sense. It compels us to revise drastically our traditional way of looking at the world and not to attribute blindly a universal validity to what is suggested by our visual imagination and representation.

Heisenberg pushed the consequences of these ideas further by showing that one cannot devise an apparatus allowing a simultaneous, precise measurement of both position and momentum. This was based upon thought (Gedanken) experiments, and we shall only describe one of them, the so-called Heisenberg microscope: A particle is located in front of a microscope having its axis along the z direction, and the approximate distance D between the particle and the microscope's front lens is equal to the focal length. The particle is illuminated and it scatters light just as any speck of dust would. It can therefore be seen through the microscope so that one can know its x and y coordinates. This position is not perfectly known because the microscope has a finite resolving power: if the light wavelength is λ and the radius of the front lens is a, the uncertainty Δx is of the order of $\lambda D/a$ according to elementary optics. One should therefore use a small enough wavelength to get a sharp knowledge of position.

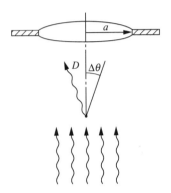

1.1. *Heisenberg's microscope*

Now comes the rub. Light is made of photons, and the particle can only be seen if at least one scattered photon enters the microscope. The momentum p of a photon is equal to h/λ. If light comes from a source far away on the axis, the rays entering the front lens after scattering will make an angle θ with the z axis up to a value $\Delta\theta = a/D$. This means that the x component of the photon momentum has itself an uncertainty $\Delta p = p\Delta\theta$. But this momentum must have been exchanged with the particle during scattering so that, when one measures the coordinate x, the momentum of the particle has just suffered an uncontrolled change, which is at least of order Δp, since many photons are needed to get an image of the particle in order to know well enough its position. It is then easy to show that the product $\Delta x \, \Delta p$ is at least of the order of h, in agreement with the uncertainty relations.

It may be noticed that, because one cannot simultaneously measure x and p, the quantities Δx and Δp occurring in the inequality (1.4) must have the following empirical meaning: A particle is assumed to be prepared experimentally in a well-defined state, i.e., with a wave function that is well defined, even if not exactly known. The same preparing process is used many times to produce many samples of the particle in the same state, and one measures the position x on a subsample and the x component of momentum upon another, which is how one can obtain both Δx and Δp_x.

ELEMENTARY QUANTUM MECHANICS

The Bohr–Einstein Controversy

One cannot leave the subject of the uncertainty relations without mentioning one of the most dramatic moments in science. In Brussels in 1927, during a Solvay meeting, an intense discussion had taken place, mostly between Bohr and Einstein, about the ideal experiments giving support to the uncertainty relations. All such experiments turned out to sustain the validity of these relations. During another Solvay meeting in 1930, Einstein returned to the same point, this time by invoking special relativity. He was questioning the uncertainty relation concerning the measurement of an energy during a time Δt, in which case quantum mechanics leads to the relation (see the appendix)

$$\Delta E \, \Delta t \geq \hbar/2. \tag{1.5}$$

The well-known relation between mass and energy, $E = mc^2$, can be used in principle to replace the measurement of an energy by a mass measurement, which can be done by using a balance. Einstein therefore proposed the following experiment: A box has an aperture on its side, which can be kept open or closed by a lid. The opening of the lid is controlled by a clock inside the box, and the box itself stands on a weighing balance. Initially, the box contains a fixed number of photons. At some given time, the clockwork opens the lid during a time Δt that can be made as short as wanted. By weighing the box before and after this event, one can in principle know with any desired precision what amount of energy has escaped, thus violating the inequality (1.5).

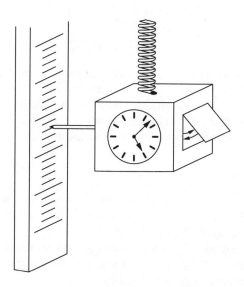

1.2. *Einstein's ideal experiment*

Another courteous but very tense discussion followed, and Bohr, when reporting it,[2] acknowledged Einstein's contribution to its clarification, though it is said that he spent a sleepless night over it. The answer was astonishing, since it came down to turning Einstein's theory of gravitation against its discoverer. It is in fact necessary to take into account all consequences of the equality between the inertial mass (entering in the value of the energy) and the gravitational mass (actually measured by the balance). These consequences are given by the theory of general relativity. One of them shows that the rate of the clock is related to the motion of the box supporting it. Let it be assumed for simplicity that one is using a spring balance where the length of the spring gives the value of the weight hanging on it.

One can measure the length of the spring by measuring the position z of a pointer, which is a part of the box and indicates a position on a vertical rule standing in the laboratory. This is a measurement of the position of the pointer and it must be associated with an uncertainty Δp in the vertical component of the box momentum, according to the ordinary uncertainty relations (1.4). This uncertainty must obviously be smaller than the total momentum that is communicated by gravity to the box during the time T while the weighting process takes place, because otherwise it would spoil the weight measurement. The momentum imparted to a mass Δm by gravity during a time T is given according to elementary mechanics by $p = Tg\,\Delta m$, where g is the acceleration of gravity. This is compensated by the force exerted by the spring, but if a momentum Δp is added when the pointer is observed at time T, one cannot be sure about the value of p and therefore about the value of Δm. The least one can ask is to have

$$\Delta p \approx h/\Delta z < Tg\,\Delta m. \tag{1.6}$$

To get a better reading precision, which means a smaller Δz, for a given precision Δm, one will have to increase the time T.

Now comes general relativity. According to this theory, when a clock is displaced vertically by a distance Δz in a gravity field, the time it shows differs from the time indicated by a fixed clock at rest in the laboratory by a quantity

$$\Delta T = gT\,\Delta z/c^2. \tag{1.7}$$

Comparing equations (1.6) and (1.7), one sees that the knowledge of the time at which the event has taken place suffers from an uncertainty $\Delta T \approx h/(\Delta m c^2)$, giving again $\Delta E\,\Delta T \approx h$.

Einstein thereby gave up his criticism, at least on these grounds. I will discuss in Chapter 9 how he came back in 1935, together with Podolsky and Rosen, with another famous ideal experiment.[3]

ELEMENTARY QUANTUM MECHANICS

Only One Theory under Various Guises

We left the history of quantum physics in the situation where, in 1926, two apparently very different formulations of the theory existed, one using matrices, the other waves. However, they always gave the same results when they were applied to the same problem. At the end of 1926, Schrödinger and Dirac succeeded in explaining this coincidence by showing an exact mathematical equivalence between the two approaches. It went as follows:

Let $\Psi_n(x)$ denote the wave function of a stationary state. Heisenberg's matrices representing position and momentum can then be explicitly obtained from these wave functions by

$$x_{nm} = \int \Psi_n^*(x) x \Psi_m(x)\, dx, \qquad p_{nm} = \int \Psi_n^*(x) \frac{\hbar}{i} \frac{\partial \Psi_m(x)}{\partial x}\, dx. \qquad (1.8)$$

Similarly, Dirac's abstract variables Q and P can also be identified with some operators acting upon the wave functions. For instance, as suggested by equations (1.8), the operator X that is associated with a position coordinate consists in multiplying a wave function $\Psi(x)$ by its argument x to give a function $x\Psi(x)$. The operator P that is associated with momentum transforms a function $\Psi(x)$ into another one that is essentially its derivative, or more precisely $(\hbar/i)(\partial \Psi(x)/\partial x)$. The outcome of this mathematical equivalence was therefore the existence of one and the same theory. This discovery did much to show the consistency of quantum mechanics, but it did not reduce its abstractness by any sizable amount.

Equations (1.8) seem to reduce altogether Heisenberg's matrix formulation and Dirac's algebraic one to the Schrödinger wave functions. One can naturally wonder whether this correspondence also goes the other way: Is it possible to consider the wave function as some sort of a vector, the operations associated with position and momentum acting like matrices upon this vector? This question might look at first sight like the product of a mind driven mad by mathematics, but on the contrary, this is precisely the most frequently used approach in the present-day formulation of quantum mechanics. The reason is to be found in the existence of quite realistic and very important physical problems, which are most conveniently expressed in that form and do not allow a description in terms of wave functions, or at least would become very burdensome if one tried to do so. The best examples are provided by the spin of a particle and by the creation or absorption of photons. Moreover, the vectorial approach is the most convenient one for getting the simplest proofs of many general results that are necessary for a complete understanding of the theory. This is why the corresponding formulation of quantum mechanics will now be described, together with its most important mathematical tool, the notion of Hilbert spaces.

MATHEMATICAL FORMALISM

2. Quantum Mechanics in a Hilbert Space

Hilbert Spaces

The years following 1927 saw a reformulation of quantum mechanics within the geometric framework of Hilbert spaces, mainly by Johann von Neumann.[4] This mathematical notion had been introduced at the turn of the century by David Hilbert, for the investigation of linear integral equations.

A *Hilbert space*[5] is a vector space, i.e., a set made of elements that can be added together or multiplied by numbers. In the case of quantum mechanics, these multiplying numbers are complex, reminding one of the complex values of wave functions. Given two vectors in the space, one can define their scalar product. This means that, given two vectors α and β, one can define their scalar product as a complex number (α, β) satisfying the following properties:

$$(\alpha, \beta) = (\beta, \alpha)^*;$$
$$(\alpha, c\beta) = c(\alpha, \beta), c \text{ being a complex number}; \qquad (1.9)$$
$$(\alpha, \beta + \beta') = (\alpha, \beta) + (\alpha, \beta');$$
$$(\alpha, \alpha) \geq 0.$$

The first condition implies that a scalar product of a vector by itself (α, α) is necessarily real. The last one says that it cannot be negative. One furthermore assumes that it is 0 only when the vector α is the zero vector.

These properties are very similar to those of a scalar product in euclidean space, except for the occurrence of complex numbers. One could therefore think of a complex euclidean space as an analogue of a Hilbert space. This is made clearer if one defines a basis of n independent vectors $(\gamma_1, \ldots, \gamma_n)$ in the n-dimensional euclidean complex vector space, considering the vectors in it as the linear combinations

$$\alpha = a_1 \gamma_1 + \ldots + a_n \gamma_n, \qquad (1.10)$$

where the coefficients $a_1, \ldots, a_n$ are complex numbers to be called the components of the vector α. One can then define the scalar product of two vectors α and β in terms of their components $\{a_j\}$ and $\{b_j\}$ by

$$(\alpha, \beta) = \sum_{j=0}^{n} a_j^* b_j,$$

an expression showing the strong analogy between a euclidean scalar product and its Hilbert brother.

ELEMENTARY QUANTUM MECHANICS

The same kind of structure can be defined in some sets of functions. Let us consider complex-valued functions $\alpha(x)$ whose argument is a real variable x. Such a function is said to be *square integrable* when the integral

$$\int_{-\infty}^{+\infty} |\alpha(x)|^2 \, dx$$

is finite. This is supposed to be a Lebesgue integral, which is not a source of any special difficulty when one considers two functions to be equal (equivalent) when their values differ only on a set of zero measure. The set consisting of all these functions is denoted by L^2, where L stands for Lebesgue and the exponent 2 means that the squares of these functions are integrable.

One can of course multiply such a function by a complex number to obtain another square-integrable function. When two functions belonging to L^2 are added, their sum is also square integrable, though this is not so obvious. This means that L^2 is a vector space, which allows one to consider square-integrable functions as vectors in the abstract space L^2. One can see how this geometric way of looking at wave functions agrees well with the idea of inverting the translation of matrix mechanics in terms of wave mechanics, since now a wave function $\alpha(x)$ is treated as a vector upon which matrices act, as will soon be seen. This structure of vector space is also well suited to the description of interference effects, which come from adding two functions, i.e., adding vectors. Of course, the vector space L^2 is not finite dimensional as in the previous example, because in order to define a function $\alpha(x)$, one must obviously introduce an infinite number of parameters, for instance, its values at every point x.

There exists a scalar product in the space L^2 enjoying the properties (1.9). Given two square-integrable functions $\alpha(x)$ and $\beta(x)$, one can define their scalar product as the quantity

$$(\alpha, \beta) = \int_{-\infty}^{+\infty} \alpha^*(x) \beta(x) \, dx.$$

One can also extend many geometric notions belonging to a euclidean space to this set of functions. For instance, two functions will be said to be orthogonal when their scalar product is 0. The norm of such a function (analogous to the length of a vector) is a real positive number $\|\alpha\|$ whose square is equal to (α, α), i.e.,

$$\|\alpha\|^2 = \int_{-\infty}^{+\infty} |\alpha(x)|^2 \, dx.$$

A vector α is said to be normalized when its norm is 1, analogously to a unit euclidean vector. A normalized function satisfies the condition (1.3) for the probability normalization of a wave function. Accordingly, the wave functions allowing a probabilistic interpretation are precisely the normalized vectors in L^2.

Bases

In a complex finite-dimensional euclidean space with dimension n, one can always find n independent orthogonal normalized vectors $\gamma_1, \gamma_2, \ldots, \gamma_n$, constituting a basis in the vector space. Their orthonormality can be summarized by the set of conditions

$$(\gamma_j, \gamma_k) = \delta_{jk}. \qquad (1.11)$$

Every vector α can be written as a linear combination of these vectors as in equation (1.10). The scalar product is still given as previously in terms of the vector components. These components are themselves given by a scalar product $a_j = (\gamma_j, \alpha)$, as one can see by taking the scalar product of the two sides of equation (1.10) by the basis vector γ_j.

It is also possible to find orthonormal bases in a space such as L^2. In that case, a basis is an infinite sequence of functions $\gamma_1(x), \gamma_2(x), \ldots, \gamma_n(x), \ldots$ satisfying the orthonormality conditions (1.11), the indices j and k now ranging to infinity. The basis must be complete, which means that every square-integrable function $\alpha(x)$ can be written as a series:

$$\alpha(x) = \sum_{j=1}^{\infty} a_j \gamma_j(x),$$

where

$$a_j = (\gamma_j, \alpha) = \int_{-\infty}^{+\infty} \gamma_j^*(x) \alpha(x)\, dx \quad \text{and} \quad (\alpha, \beta) = \sum_{j=1}^{\infty} a_j^* b_j.$$

When there exists such a countable basis, the Hilbert space is said to be separable. There are nonseparable Hilbert spaces in which one can only introduce uncountable bases $\{\gamma_\lambda\}$, where the index λ can take a continuous set of values. Some nonseparable Hilbert spaces are of interest in physics, for instance, in quantum electrodynamics. However, they will not be needed in this book and it will be better to consider only the case where the space is separable.

Operators

One can give an explicit meaning to Dirac's abstract dynamical variables as well as to the infinite matrices considered by Heisenberg and Born in terms of linear operators in a Hilbert space. Let us begin with the finite-dimensional case. A linear operator T transforms a vector α into another vector $T(\alpha)$. Linearity means that the vector $T(\alpha + \beta)$ is the sum of the two vectors $T(\alpha)$ and $T(\beta)$, whereas $T(c\alpha) = cT(\alpha)$, c being an arbitrary complex number.

If n denotes the dimension of the complex euclidean space, it is easy to show that this transformation can be written in terms of an $n \times n$ matrix whose elements T_{jk} depend upon the basis. Putting $\beta = T(\alpha)$ and letting a_j and b_j denote the coordinates of the two vectors α and β in a basis $\{\gamma_j\}$, one has by linearity

$$\beta = \sum_{j=1}^{n} a_j T(\gamma_j).$$

Hence

$$b_k = \sum_{j=1}^{n} T_{kj} a_j \quad \text{with} \quad T_{kj} = (\gamma_k, T(\gamma_j)). \tag{1.12}$$

The relation (1.12) is obviously in matrix form.

One may extend these formulas to the Hilbert space L^2, taking some care with convergence. We have already met some linear operators acting upon a function $\alpha(x)$: one can, for instance, multiply it by a given function $f(x)$, differentiate it, and also take a primitive or form an integral such as $\int K(x,y)\alpha(y)\,dy$, where a given function $K(x,y)$ is used as a kernel, and so on. Any such operator will be simply denoted by $\beta = T\alpha$, in place of $\beta = T(\alpha)$.

The problems arising from the convergence of the series or the integrals unfortunately require some caution. To see how these problems occur, consider the operation consisting in multiplying a square-integrable function $\alpha(x)$ by the variable x. This can be viewed as an operator X such that the vector notation $X\alpha$ is a shorthand for the function $x\alpha(x)$. It is clear that this operation can give a function that is not square integrable, as shown by the case where $\alpha(x)$ is the function $(1+x^2)^{-1/2}$. Similarly, the momentum operator $P = (\hbar/i)(\partial/\partial x)$ can only act upon a function having a derivative, and moreover, this derivative should also be square integrable.

The wave functions for a particle in three-dimensional space belong to the space $L^2(\mathbf{R}^3)$ of square-integrable functions in three variables x,y,z. The abstract dynamical variables for position, as considered by Dirac, are the three operators X,Y, and Z, multiplying the wave function by x,y, or z, respectively. The momentum variables and the three components of the derivative operator $(\hbar/i)\nabla$. The energy of a hydrogen atom is then given by the operator $H = P^2/2m + V(X)$, and the z component of the angular momentum is the operator $XP_y - YP_x$.

The operators just mentioned share the property of being associated with a physical quantity. They also share a common mathematical property: they are self-adjoint. A self-adjoint operator A is defined as acting so that $(\beta, A\alpha) = (A\beta, \alpha)$, whenever these scalar products exist.

The analogue of a self-adjoint operator in a finite-dimensional Hilbert space is an $n \times n$ matrix whose elements satisfy the property $A_{jk} = A^*_{kj}$, i.e., the matrix A is hermitian.

The Spectrum of an Operator

In a finite-dimensional Hilbert space with dimension n, a self-adjoint operator A is associated with a definite matrix, once a basis has been selected. This self-adjoint matrix has n orthonormal eigenvectors α_j associated with real eigenvalues a_j such that

$$A\alpha_j = a_j \alpha_j. \tag{1.13}$$

In view of the importance of this result, it may be worth recalling its proof briefly. When equation (1.13) is written explicitly in terms of the coordinates of the vector α_j, it becomes a system of n linear algebraic homogeneous equations. This system can only have a nonzero solution when its determinant vanishes, meaning that an eigenvalue a_j should be a root of the nth degree polynomial in a variable s: $\det(A - sI)$, I being the identity matrix. There are n such roots. When one of them is degenerate of order q (i.e., q roots coincide), it can be shown that q linearly independent vectors satisfy the corresponding eigenvalue equation and one can choose q such orthonormal vectors. When a root is nondegenerate, it corresponds to a unique eigenvector.

The roots are real because of self-adjointness: Taking the scalar product of both sides of equation (1.13) by the vector α_j, one gets

$$(\alpha_j, A\alpha_j) = a_j \|\alpha_j\|^2. \tag{1.14}$$

Using the properties of the scalar product and self-adjointness,

$$(\alpha_j, A\alpha_j)^* = (A\alpha_j, \alpha_j) = (\alpha_j, A\alpha_j);$$

from which it follows easily that a_j is real.

Two eigenvectors α_j and α_k associated with two different eigenvalues a_j and a_k are necessarily orthogonal. To prove this, write

$$(\alpha_k, A\alpha_j) = a_j(\alpha_k, \alpha_j) = (A\alpha_k, \alpha_j) = a_k^*(\alpha_k, \alpha_j) = a_k(\alpha_k, \alpha_j).$$

So since the eigenvalues are real, $(a_k - a_j)(\alpha_k, \alpha_j) = 0$. The eigenvalues a_k and a_j are different, hence the two eigenvectors α_k and α_j are necessarily orthogonal.

Finally, it is always possible to choose q independent eigenvectors associated with a degenerate root of order q to be orthogonal. Taking these vectors to be normalized, one thus gets an orthonormal basis of the finite-dimensional Hilbert space consisting of eigenvectors of the operator A. In this basis, the matrix representing A is diagonal and its diagonal elements are the eigenvalues.

ELEMENTARY QUANTUM MECHANICS

The set of real numbers consisting of the eigenvalues of A is called its *spectrum*. It is a finite set containing at most n numbers. One can also use the fact that the matrix of the operator A is diagonal in a basis consisting of its eigenvectors to get a less direct definition of the spectrum, which can, however, be extended to more general cases. Letting z be a complex number, one notices that the matrix $A - zI$ is also diagonal, with its diagonal matrix elements given by the numbers $a_j - z$. Therefore, when z does not coincide with one of the eigenvalues, the determinant of the matrix $A - zI$ is nonzero and the inverse matrix $(a - zI)^{-1}$ exists. Such a value for z is said to be regular, and one can define the spectrum in a roundabout way as the complement of the regular set consisting of all the regular values. Then, for any operator T, one will define a regular set as containing all the regular values for which the operator $(T - zI)^{-1}$ exists, the spectrum of T being the complement of the regular set in the complex plane.

The properties of the spectrum and the eigenvectors of a self-adjoint matrix can be extended more or less to any self-adjoint operator in any Hilbert space. This result, which is due to von Neumann, is one of the tools essential to the complete formulation of quantum mechanics, and its precise form will be given after introducing a last tool, the trace of an operator.

Traces

The *trace* of a finite-dimensional matrix T is defined as the sum of its diagonal elements:

$$\text{Tr}(T) = \sum_{j=1}^{n} T_{jj}.$$

For an operator T acting on an infinite-dimensional Hilbert space, the definition is practically the same: one selects an orthonormal basis $\{\gamma_j\}$ and writes

$$\text{Tr}\,T = \sum_{j=1}^{\infty} (\gamma_j, T\gamma_j). \tag{1.15}$$

A sufficient condition for the trace to be well defined is that the series converges absolutely in some basis. When this happens, the series (1.15) converges in every basis and the operator T is said to belong to the *trace class*.

The trace is an invariant, i.e., it does not depend upon the choice of a basis. This is easily shown in the finite-dimensional case. Letting $\{\beta_k\}$ be another basis, one has

$$\sum_{j}(\gamma_j, T\gamma_j) = \sum_{jkm}(\gamma_j, \beta_k)(\beta_k, T\beta_m)(\beta_m, \gamma_j).$$

The orthonormality of the second basis gives

$$\sum_j (\beta_m, \gamma_j)(\gamma_j, \beta_k) = (\beta_m, \beta_k) = \delta_{mk},$$

hence $\sum_j (\gamma_j, T\gamma_j) = \sum_k (\beta_k, T\beta_k).$

One may notice some useful properties of the trace. Let A and B be two operators such that their product AB is of trace class. The trace does not depend on the ordering of the product, namely,

$$\mathrm{Tr}(AB) = \mathrm{Tr}(BA). \tag{1.16}$$

This is obvious in the finite-dimensional case in view of the explicit expression

$$\mathrm{Tr}(AB) = \sum_{jk} (\gamma_j, A\gamma_k)(\gamma_k, B\gamma_j).$$

The trace of a product is also invariant under a circular permutation of the factors:

$$\mathrm{Tr}(A_1 A_2 \ldots A_r) = \mathrm{Tr}(A_2 \ldots A_r A_1).$$

This is an immediate consequence of equation (1.16), obtained by taking $A = A_1$ and $B = A_2 \ldots A_r$. Finally, the trace is obviously linear:

$$\mathrm{Tr}(A + B) = \mathrm{Tr}(A) + \mathrm{Tr}(B), \qquad \mathrm{Tr}(cA) = c\mathrm{Tr}(A).$$

It may also be noticed that the trace of a finite self-adjoint matrix is the sum of its eigenvalues.

Hilbert–Schmidt Operators

A self-adjoint operator A belongs to the Hilbert–Schmidt class when the trace of A^2 is finite:

$$\sum_{j=1}^{\infty} (A\gamma_j, A\gamma_j) < \infty. \tag{1.17}$$

Hilbert and Schmidt met this kind of operator when dealing with linear integral equations, and they found that their spectral properties are very similar to those of a finite self-adjoint matrix: There exists an orthonormal basis in the Hilbert space consisting in eigenvectors of A. The spectrum of A is discrete, consisting of discrete real eigenvalues $a_1, \ldots, a_n, \ldots$, all of them finite and their only possible limit being zero. These eigenvalues can be degenerate so that several eigenvectors may be associated with the same eigenvalue. It is convenient to take this degeneracy into account in the notation by calling α_{nr} the eigenvectors that are associated with an eigenvalue a_n, r being a so-called degeneracy index counting the eigenvectors.

ELEMENTARY QUANTUM MECHANICS

Orthonormality of the basis of eigenvectors can then be expressed by the relations $(\alpha_{jr}, \alpha_{ks}) = \delta_{jk}\delta_{rs}$. Every vector β in Hilbert space can then be written as a linear combination of eigenvectors:

$$\beta = \sum_{j=1}\sum_r b_{jr}\alpha_{jr},$$

the coefficients b_{jr} being the corresponding coordinates of the vector β. They are explicitly given as usual by $b_{jr} = (\alpha_{jr}, \beta)$.

Clearly, except for the infinite number of eigenvectors and the possibly infinite number of eigenvalues, self-adjoint Hilbert–Schmidt operators have the same spectral properties as self-adjoint finite matrices, showing how close operators can be to matrices and how significant is this geometric approach to a set of square-integrable functions.

Projectors

Another category of self-adjoint operators having a discrete spectrum plays an essential role in the interpretation of quantum mechanics. These are the *projection operators*, or, more briefly, *projectors*.

Let us begin once again with the case of a finite-dimensional Hilbert space **H** with dimension N. A subspace **M** with dimension n is a linear subspace, consisting of all linear combinations of n independent vectors $\alpha_1, \ldots, \alpha_n$. It will be convenient to assume that these vectors are orthonormal so that they constitute an orthonormal basis of **M**. One can then complete them by $N - n$ other vectors $\alpha'_1, \ldots, \alpha'_{N-n}$ to construct an orthonormal basis in the full Hilbert space **H**. Every vector β belonging to **H** can then be written as a linear combination of the vectors α_j and α'_k. If β_1 is the linear combination of the vectors α_j in this expression, and β_2 the combination of vectors α'_k, then the vector β can be written as $\beta = \beta_1 + \beta_2$, where β_1 belongs to **M** and β_2 is orthogonal to all vectors in **M**.

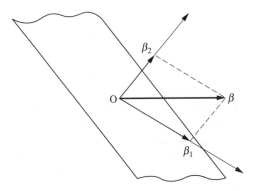

1.3. *A projector acting upon a vector*

One can go from the vector β to the vector β_1, which is called its *projection*, by replacing the last $N - n$ coordinates of β by 0. This is obviously a linear operation, so one can consider it to be the effect of a linear operator E. In the present basis, E is represented by a diagonal matrix, where the n first diagonal entries are equal to 1 and the last $N - n$ entries are 0. This implies that E is self-adjoint with the eigenvalues *1* (n times degenerate) and *0* ($N - n$ times degenerate). Multiplying E by itself, one sees that

$$E^2 = E. \tag{1.18}$$

On the other hand, let us assume that a self-adjoint operator E in the finite-dimensional Hilbert space satisfies equation (1.18). Because it is self-adjoint, it has real eigenvalues. Equation (1.18) implies that these eigenvalues a satisfy the equation $a^2 = a$, so the eigenvalues can only be equal to 0 or 1. Denoting by **M** the subspace that is generated by the eigenvectors belonging to the eigenvalue 1, it is then clear that E is the projector associated with that subspace.

From the standpoint of geometry, one can say that β_1 is the orthogonal projection of the vector β onto the subspace **M**. Equation (1.18) means that, because β_1 already belongs to **M**, its projection must coincide with itself.

It can be shown that these properties are still valid in the case of an infinite-dimensional Hilbert space, up to a topological subtlety; namely, that **M** must be a closed subspace. This will always be the case in the applications, so we need not elaborate upon this point. One can therefore say that every (closed) subspace of a Hilbert space is associated with a unique projector, i.e., a self-adjoint operator E satisfying the (idempotency) condition (1.18). When acting upon a vector β, E gives its projection upon **M**. Its eigenvalues are all equal to either 1 or 0. Every vector belonging to **M** is an eigenvector of E with eigenvalue 1, and every vector orthogonal to **M** is also an eigenvector with eigenvalue 0. On the other hand, every self-adjoint operator satisfying condition (1.18) can be associated with a closed subspace of the Hilbert space as its projector.

3. The Spectral Theorem

In the case of a finite-dimensional Hilbert space and a self-adjoint operator A, the spectral theorem says that there exists an orthonormal basis consisting of eigenvectors of A and that the eigenvalues of A are real. Its extension to a general self-adjoint operator in an infinite-dimensional Hilbert space plays a central role in quantum mechanics, and this is what will be considered now.

It will be convenient to reexpress the theorem in the finite-dimensional case in an equivalent form. To do so, let us consider the eigenvectors associated with a given eigenvalue a_j of a self-adjoint operator A. With all the possible linear combinations of these eigenvectors, one can generate a subspace $\mathbf{M}_j$. The dimension of the subspace is equal to n if the eigenvalue

a_j is n times degenerate. Let us denote by E_j the projector onto this subspace. When acting on an arbitrary vector β, it extracts its projection upon $\mathbf{M}_j$; the operator A, acting on this projection β_j, multiplies it by a_j. One can therefore consider the action of A as being decomposed into two successive actions: first extracting the projections and then multiplying them by the corresponding eigenvalues. This is expressed by the equation

$$A = \sum_j a_j E_j. \tag{1.19}$$

The completeness of the eigenvector basis means that the sum of all the projections of a vector reproduces it. This can also be written as an equation, which reads

$$I = \sum_j E_j. \tag{1.20}$$

The two relations (1.19) and (1.20) are just other expressions of the spectral theorem in the case of a finite-dimensional Hilbert space, and their form can be extended, when suitably modified, to the general case.

As an example, a Hilbert–Schmidt self-adjoint operator, though the underlying Hilbert space is infinite dimensional, satisfies these properties, as was first shown by Hilbert and Schmidt. The only difference is that the finite sum over the index j now may become infinite and the projectors may also project onto an infinite-dimensional subspace (for the eigenvalue 0).

As another simple case, when A is a projector E, equation (1.19) reduces to $E = E$, whereas equation (1.20) just means that the operator $\bar{E} = I - E$ is also a projector (onto the subspace orthogonal to the subspace associated with E).

The Difficulties of the General Case

To find out what happens more generally, it will be convenient to consider the case of the operators associated with position and momentum in the Hilbert space L^2 for the wave functions of a particle in a one-dimensional space. These are square-integrable functions depending upon a real variable x in a domain ranging from $-\infty$ to $+\infty$. The operators will be denoted as usual by X and P, X transforming a function $\alpha(x)$ into $x\alpha(x)$, and P into $(\hbar/i)\, d\alpha(x)/dx$.

It has been noticed already that these operators do not necessarily transform a square-integrable function into another one. So, let us denote by D the set of square-integrable functions $\alpha(x)$ for which $x\alpha(x)$ is also square integrable. This will be called the *domain* of the operator X. More generally, the domain of an operator A in a Hilbert space is defined as the set of vectors α such that $A\alpha$ is a well-defined vector in the Hilbert space. One will consider only operators having a dense domain, which means that every vector in the Hilbert space can be written as a limit of vectors belonging to D. It is not always easy to check that this property is satisfied, and we shall leave aside this kind of proof, being content to mention that the operators of interest in physics satisfy this condition.

The spectrum of X coincides with the set **R** of real numbers. This can be shown by going back to the general definition of the spectrum and looking for the regular values of this operator. It has been said that a complex number z is a regular value for X if the operator $(X - zI)$ has an inverse. This inverse should act upon a wave function $\alpha(x)$ to give the function

$$\beta(x) = \alpha(x)[x - z]^{-1}.$$

This is clearly square integrable for any function $\alpha(x)$ in L^2 as long as the imaginary part of z is nonzero. Furthermore, when z is a real number a, one can consider a function $\alpha(x)$ that is continuous and nonzero for $x = a$ to get a transformed function $\beta(x)$ that is not square integrable because $(x - a)^{-1}$ is not square integrable in a small domain around $x = a$. One can therefore conclude that the set of regular values for X is identical to the set of complex numbers having a nonzero imaginary part. The spectrum of X, which is its complement, is the whole real set. One can prove in a similar way that the spectrum of the operator P is also the whole real set, although the proof is a bit longer.

Because every real number a belongs to the spectrum of the operator X, one may wonder whether a can be an eigenvalue of X. If this were true, it would mean that there exists a square-integrable function $\alpha(x)$ such that $x\alpha(x) = a\alpha(x)$, whatever the value of x. This is clearly impossible since $\alpha(x)$ would have to be 0 except for $x = a$, and such a function is equivalent to zero from the standpoint of Lebesgue integration. So, there are no eigenvectors of the operator X.

The case of the operator P is similar. The eigenvalue equation would then read

$$\frac{\hbar}{i}\alpha'(x) = a\alpha(x),$$

from which one gets $\alpha(x) = \exp(iax/\hbar)$. Although $\alpha(x)$ is perfectly well defined, it is clearly not square integrable.

So, one must take into account the fact that the spectrum cannot in general be defined as a set of eigenvalues. The spectral theorem should therefore be more subtle than in the finite-dimensional case. Since this kind of difficulty always occurs when the spectrum contains a continuous part, it is necessary to face the problems arising in such a case.

The Spectral Theorem for the Position Operator

Von Neumann (1929) extended the spectral theorem to self-adjoint operators having a continuous spectrum. This is a good example of a deep mathematical result that is needed to give a firm foundation to physics. We shall now try to show what it means by considering the special case of the position operator.

The main idea consists in extending the relations (1.19) and (1.20) expressing a self-adjoint operator as a sum of projectors multiplied by numbers, which are its eigenvalues in this simple case. Because it is impossible to associate in general an eigenvector with a number belonging to a continuous spectrum, one will try instead to build up a projector that is associated with a full interval $J = [a, b]$ rather than with a discrete value. This is quite natural from the standpoint of physics, since it means that one is not trying to give exactly the value of the position but only to assert it within some fixed bounds.

Consider therefore the following operation: Starting from a square-integrable function $\alpha(x)$, let us construct another function $\beta(x)$ coinciding with $\alpha(x)$ inside the interval J and vanishing outside. It might be called the *procustean operation*, from a famous giant in Greek mythology, Procustes, whose main game consisted in cutting off the head and feet of his prisoners when their size exceeded the length of his bed. This is obviously a linear operation and it can be associated with an operator $E(J)$: $\beta = E(J)\alpha$, which is self-adjoint because

$$(\gamma, E(J)\alpha) = \int_{-\infty}^{+\infty} \gamma^*(x)\beta(x)\,dx = \int_a^b \gamma^*(x)\alpha(x)\,dx = (E(J)\gamma, \alpha).$$

It is a projector since, when iterated, it chops off $\beta(x)$ outside J, whereas this function was already 0 in that domain. One therefore has $E^2(J) = E(J)$. The eigenfunctions associated with the eigenvalue 1 are all the functions vanishing outside J, and the eigenfunctions associated with the eigenvalue 0 are those vanishing inside J.

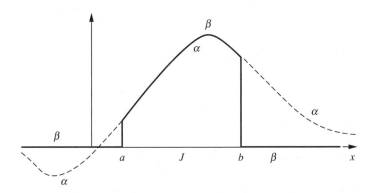

1.4. *A special projector*. Acting upon a wave function $\alpha(x)$, it gives a wave function $\beta(x)$ coinciding with $a(x)$ in the interval $J = [a, b]$ and vanishing outside it.

Let us now split the real line into an infinite set of contiguous intervals $D_j = [x_{j-1}, x_j]$, and let us denote by E_j the projector that is associated with the interval D_j. It preserves a function $\alpha(x)$ in that interval and puts it equal to zero outside. Since the intervals cover the whole real line, one has

$$I = \sum_j E_j, \qquad (1.21)$$

which means that one gets back $\alpha(x)$ by gluing all its pieces together.

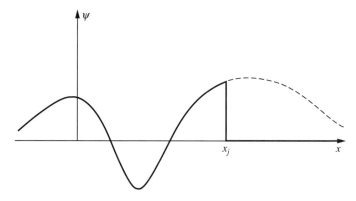

1.5. *A projector* $E(x_j)$. When acting upon a wave function $\psi(x)$, this operator leaves it unchanged for values of x smaller than x_j and gives 0 for x larger than x_j.

If the interval D_j is small enough, there is not much difference between the functions $x\alpha(x)$ and $x_j\alpha(x)$ inside that interval. This means that one makes a small error when writing that the action of the X is essentially given by

$$X \approx \sum_j x_j E_j, \qquad (1.22)$$

which is very similar to equation (1.19). One can now make a slight change in notation by introducing the projector $E(x_j)$ that is associated with the semi-infinite interval $(-\infty, x_j]$. Its effect is to cut off the graph of a function $\alpha(x)$ to the right of the point x_j. Notice that $E_j = E(x_j) - E(x_{j-1})$, which allows us to rewrite equations (1.21) and (1.22) as

$$I = \sum_j [E(x_j) - E(x_{j-1})], \qquad X \approx \sum_j x_j [E(x_j) - E(x_{j-1})].$$

ELEMENTARY QUANTUM MECHANICS

Each of these formulas looks like a Riemann sum defining an integral, and this remark leads one to anticipate that, in the limit when the length of each interval becomes infinitesimally small, one has

$$I = \int_{-\infty}^{+\infty} dE(x), \qquad X = \int_{-\infty}^{+\infty} x \, dE(x).$$

A complete explication and justification of these formulas requires more refined mathematical arguments, as one is dealing here with integrals in which a projector is replacing the infinitesimal element.[6] There are fortunately other convenient ways of using the result, and one need only know that these formulas make sense to be able to associate a well-defined projector with any given set of real numbers. They express the spectral theorem in the special case of the position operator, and their meaning is rather simple: When acting upon a function $\alpha(y)$, the operator $E(x)$ truncates it to the right of x.

In its general form, a self-adjoint operator A has a spectrum consisting of a discrete part containing eigenvalues $\{\alpha_j\}$ and a continuous part σ_c. With each discrete eigenvalue, one can associate some orthogonal eigenvectors spanning a subspace of the Hilbert space with a projector E_j. As far as the continuous spectrum is concerned, one may proceed as in the case of the position operator. One defines a projector $E(a)$ corresponding to the values belonging to the continuous spectrum and smaller than a. In most cases,[7] the spectral theorem is expressed by a simultaneous decomposition of the operator A and the identity I:

$$A = \sum_j a_j E_j + \int_{\sigma_c} a \, dE(a), \qquad I = \sum_j E_j + \int_{\sigma_c} dE(a).$$

The orthogonality of eigenvectors, as it holds in the finite-dimensional case, remains true for the eigenvectors associated with discrete eigenvalues. For the continuous spectrum, this is replaced by the properties

$$E(x)E(y) = E(\min(x, y)), \qquad E_j E(x) = E(x) E_j = 0.$$

Every vector α can be written in terms of its components along the subspaces that are associated with the discrete and continuous spectra as

$$\alpha = \sum_j \alpha_j + \int_{\sigma_c} d\alpha(a),$$

where $\alpha_j = E_j \alpha$, $d\alpha(a) = dE(a)\alpha$, whereas

$$A\alpha = \sum_j a_j \alpha_j + \int_{\sigma_c} a \, d\alpha(a).$$

These results can also be used to define a function of an operator. If $f(a)$ is a function of the real variable a, one defines the operator $f(A)$ as

$$f(A) = \sum_j f(a_j) E_j + \int_{\sigma_c} f(a) \, dE(a).$$

Dirac's Notation

Physicists often prefer to use a notation more akin to the one used for a finite-dimensional Hilbert space. It is somewhat automatic and avoids asking too many questions, which is better in practice because it usually works perfectly well. It can be rigorously justified according to a theorem by Godement,[8] but the introduction of the necessary preliminaries would mean paying too high a price for what is after all a mere convenience.

The recipe consists of acting as if a self-adjoint operator A having a continuous part in its spectrum had eigenvectors α_a satisfying the formal eigenvalue equation $A\alpha_a = a\alpha_a$. Such "vectors" do not generally belong to the Hilbert space, and it is often difficult to identify what they really are. However, in the case of the position operator, one can see that the "wave function"

$$\alpha_a(x) = \delta(x - a) \tag{1.23}$$

can do it. This kind of formal eigenvector cannot be normalized. However, it is possible to choose it "orthonormalized" in an extended sense by having $(\alpha_a, \alpha_b) = \delta(b-a)$, a relation that is effectively satisfied by equation (1.23).

In the case of the momentum operator P in L^2, one can take

$$\alpha_p(x) = (2\pi\hbar)^{-1/2} \exp(ipx/\hbar),$$

and the spectral theorem expresses in that case the basic properties of Fourier transforms.

To decompose in general a vector β into its components in a convenient "basis," one writes

$$\beta = \sum_j b_j \alpha_j + \int_{\sigma_c} b(a) \alpha_a \, da,$$

the coordinates being given by

$$b_j = (\alpha_j, \beta), \qquad \beta(a) = (\alpha_a, \beta),$$

and the scalar product by

$$(\beta, \gamma) = \sum_j b_j^* c_j + \int b(a)^* c(a) \, da.$$

ELEMENTARY QUANTUM MECHANICS

Dirac also introduced another notation for the vectors. In place of simply denoting a vector in a Hilbert space by a letter α, he wrote it $|\alpha\rangle$. The apparent clumsiness of this notation is more than compensated for when one deals with an eigenvector α_a. With the index a recalling the eigenvalue, one can simply write α_a as $|a\rangle$, so in fact the notation becomes simpler. The scalar product of two vectors α and β is then written as $\langle\alpha|\beta\rangle$, and the scalar product of the vector β by the vector $A\alpha$ as $\langle\beta|A\alpha\rangle$.

A convenient aspect of Dirac's notation is how it represents projectors, in a way extending to a complex Hilbert space Gibbs's dyadic notation. The projector along a normalized vector α is written as $|\alpha\rangle\langle\alpha|$. This corresponds to a convention in which

$$(|\alpha\rangle\langle\alpha|)|\beta\rangle = |\alpha\rangle(\langle\alpha|\beta\rangle) \equiv |\alpha\rangle\langle\alpha|\beta\rangle,$$

where the bracket $(|\alpha\rangle\langle\alpha|)$ in the first term represents an operator, whereas in the second term one has a sum over a product of the basis vectors by the coordinates of $|\beta\rangle$. Dirac found it convenient to suppress all parentheses and this is the meaning of the third term, which represents indifferently these two equivalent expressions.

FEYNMAN HISTORIES

4. Feynman's Formulation of Quantum Dynamics

The solution of Schrödinger's equation can be written in a form where interference effects are very clearly seen and the classical limit is more easily handled.[9] It was published in 1948 by Richard Feynman, who had previously used it as a foundation for quantum electrodynamics. It consists in assuming that a particle, for instance an electron, may have many different kinds of histories, with the meaning one usually gives to a history, namely, that the particle has some position and some momentum at each time, although of course these quantities do not obey classical physics. The probabilistic character of quantum mechanics appears through the fact that all conceivable histories are possible, with each one having some probability amplitude.

Feynman's formulation is equivalent to the previous ones by Heisenberg, Dirac, and Schrödinger, at least as far as continuous degrees of freedom are concerned, in the same way that these formulations are equivalent to each other. It will be convenient to show how it comes naturally from this background, though restricting our considerations for simplicity to the case of a unique particle in a one-dimensional physical space. Let us therefore consider a particle with the hamiltonian

$$H = \frac{P^2}{2m} + V(x) = K + V,$$

K denoting the kinetic energy and V the potential. Schrödinger's equation then reads

$$i\hbar \frac{\partial \psi}{\partial t} = -\frac{\hbar^2}{2m}\frac{\partial^2 \psi}{\partial x^2} + V(x)\psi.$$

The wave function at some time t'' depends linearly upon its value at some previous time t', as the Schrödinger equation is linear and of first order in time. It should then be possible to express this linearity by

$$\psi(x',t'') = \int G(x'',t'';x',t')\psi(x',t')\,dx',$$

where the *Green's function* G is a matrix element of the evolution operator $U(t) = \exp(-iHt/\hbar)$, to be discussed later in more detail. These matrix elements are defined according to Dirac's notation by

$$G(x'',t'';x',t') = \langle x''|U(t''-t')|x'\rangle.$$

This can be written in a more suggestive way by splitting the time interval of length $t = t'' - t'$ into a large number N of small intervals, each of length $\Delta\tau = t/N$, according to a sequence $t_0\,(=t'), t_1, t_2, \ldots, t_N\,(=t'')$. The exponential character of the evolution operator gives

$$U(t) = \exp(-iHt/\hbar) = \exp(-iNH\,\Delta\tau/\hbar) = U(\Delta\tau)^N, \qquad (1.24)$$

One has $U(\Delta\tau) = \exp(-i(K+V)\Delta\tau/\hbar)$. Since the time $\Delta\tau$ is very short, one can write

$$U(\Delta\tau) \approx I - (K+V)\,\Delta\tau/\hbar \approx (I - K\,\Delta\tau/\hbar)(I - V\,\Delta\tau/\hbar)$$
$$\approx \exp(-iK\,\Delta\tau/\hbar)\exp(-iV\,\Delta\tau/\hbar),$$

suggesting

$$U(t) = \lim_{N\to 0}\{\exp(-iKt/N\hbar)\exp(-iVt/N\hbar)\}^N. \qquad (1.25)$$

In this last formula, one discards the difficulties arising from the non-commutativity of K and V by using directly a product of exponentials with a small time interval for both the effect of the kinetic energy and the potential energy. This can be proved rigorously and one can insert equation (1.25) into the matrix element (1.24). One can also replace in proper places the identity operator by one of its two spectral decompositions

$$I = \int |x\rangle\langle x|\,dx, \qquad I = \int |p\rangle\langle p|\,dp.$$

One thus gets

$$G(x'',t''; x',t') = \langle x''| \int |p_1\rangle \, dp_1 \langle p_1| \exp(-iKt/N\hbar)|x_1\rangle \cdots$$
$$\times \int |x_{N-2}\rangle \, dx_{N-2} \langle x_{N-2}| \exp(-iVt/N\hbar)$$
$$\times \int |p_{N-1}\rangle \, dp_{N-1} \langle p_{N-1}| \exp(-iKt/N\hbar)$$
$$\times \int |x_{N-1}\rangle \, dx_{N-1} \langle x_{N-1}| \exp(-iVt/N\hbar)|x'\rangle.$$

This apparently complicated formula is much simplified if one uses the relations

$$\langle x|p\rangle = h^{-1/2} \exp(ipx/\hbar),$$
$$\langle x''| \exp(-iVt/N\hbar)|x'\rangle = \delta(x'-x'') \exp(-iV(x')t/N\hbar),$$
$$\langle p''| \exp(-iVt/N\hbar)|p'\rangle = \delta(p'-p'') \exp(-ip'^2 t/2mN\hbar),$$

and if one brings all the exponentials together. One then gets

$$G(x'',t''; x',t') = \lim \int \frac{dx_1 \, dp_1}{h} \cdots \frac{dx_{N-1} \, dp_{N-1}}{h} dp_N \exp(-iS/\hbar), \quad (1.26)$$

where the limit is taken when N tends to infinity and the exponent S is the sum

$$S = \sum_{k=1}^{N} p_k[x_k - x_{k-1}] - \sum_{k=1}^{N-1} \left\{ \frac{[p_{k+1}-p_k]^2 \Delta\tau}{2m} + V(x_{k+1})\Delta\tau \right\}. \quad (1.27)$$

This last sum can be recognized as the action of a classical system having the Hamilton function

$$H(x,p) = \frac{p^2}{2m} + V(x),$$

the action being written as

$$S = \int_{t'}^{t''} p \, dx - H(x,p) \, dt. \quad (1.28)$$

One can see that, if this integral were replaced by a Riemann sum by splitting the interval (t', t'') into N small intervals, one would obtain precisely the sum (1.27). It is convenient to write equation (1.26) in a symbolic form:

$$G(x'',t''; x',t') = \int \prod_{t=t'}^{t''} \frac{dx_t \, dp_t}{h} \exp(iS/\hbar).$$

It can be understood as a sum over all the possible continuous histories of the particle starting from position x' at time t' to reach position x'' at time t''. The exponent S is the classical action for that history. A history therefore corresponds to well-defined values of $\{x(t), p(t)\}$ for each value of the time t. This is also the kind of test history one uses in variational calculus when defining a classical history as giving a stationary value for the action integral. Here one must sum over all possible histories, and how this is to be performed is made explicit by equation (1.24) and letting N tend to infinity. Still more sketchily, one can also write

$$\langle x'', t'' | x', t' \rangle = \sum_{\text{histories}} \exp(-iS/\hbar). \tag{1.29}$$

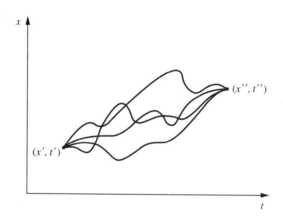

1.6. *Feynman's sum over paths*

It may be noticed that the Feynman sum gives, at least intuitively, a simple explanation of why the principle of least action applies in classical mechanics: When the beginning and the end of the history are such that the minimal action is large compared with Planck's constant, the sum of all the complex terms with unit modulus in equation (1.29) gives rise to strong interference effects. The histories that are near enough to the classical one interfere with it constructively, since the action is stationary, so there are many different histories giving rise to essentially the same value for the action. On the contrary, other histories farther away interfere destructively, so the essential contribution to the probability amplitude comes from histories differing little from the classical one.

The Feynman formulation is helpful in some aspects of the interpretation of quantum mechanics. However, we shall mainly use it for qualitative considerations, while not entering into finer details. This is why the techniques that are necessary for computing explicitly Feynman sums will not be needed in this book, and we shall limit ourselves to a few remarks. First, notice that it is easy to generalize what has been obtained to any number of

ELEMENTARY QUANTUM MECHANICS

degrees of freedom, since one should just sum over all the possible values of the coordinates $\{x(t), p(t)\}$, the denominator h being replaced by its nth power if n is the number of degrees of freedom.

One can also perform explicitly the integration upon the momentum variables p_k in equation (1.26), because the corresponding integrands are gaussian. One then obtains

$$G(x'', t''; x', t') = \int \prod_{t=t'}^{t''} \frac{dx_t}{A} \exp(-iS/\hbar),$$

where A is a complex normalization coefficient one seldom needs explicitly because it can be fixed by normalizing the total probability. One is now dealing with histories where only the position is given at each time t and the momentum does not enter; one is therefore working in configuration space rather than in phase space. The action S takes in that case the more familiar form

$$S = \int_{t'}^{t''} L(x, v)\, dt,$$

where L is the classical Lagrange function $K - V$.

One can also express the matrix elements of a time-dependent Heisenberg operator $B(t)$ as a Feynman sum, and this will be useful later on. The matrix elements can be written as

$$\langle x'', t'' | B(T) | x', t' \rangle = \int \prod_{t=t'}^{t''} \frac{dx_t}{A} \exp(-iS(x'', t''; x, T)/\hbar)$$

$$\times \langle x | B | y \rangle \exp(-iS(y, T; x', t')/\hbar),$$

where $B(T) = U^{-1}(T) B U(T)$, $t' \leq T \leq t''$. We shall need this result when B is a projector $E(J)$ associated with the values of the position X in an interval J. Its matrix elements are then given by

$$\langle x | E(J) | y \rangle = \delta(x - y) \text{ for } x \text{ inside } J, 0 \text{ otherwise.}$$

It can be seen that one can either introduce explicitly the projector, stating that the position is in J at time T, or just restrict the Feynman sum to the histories crossing the interval J at time T; the two procedures are equivalent. This remark will prove useful later on.

PROBABILITIES AND STATES

5. Quantum Probabilities

One can use the spectral theorem, either in the form given by von Neumann or as expressed by Dirac's formal calculus, to express quantum probabilities in a general form.

We have already met Born's rule expressing the probability of the position of a particle being in some region of space. However, as far as the possibility of a measurement is concerned, one might expect that measuring a position is not essentially different from measuring any other physical quantity. This suggests looking for a generalization of Born's rule to apply to any observable quantity. However, before considering this issue, it will be useful to find out what are these observable quantities, at least in a first approach.

At the beginning of quantum mechanics, there was a strong tendency to identify the physical notion of an observable quantity and the mathematical notion of a self-adjoint operator. This was particularly stressed by both Dirac and von Neumann. They even coined the word *observable* to designate both notions. The developments to be discussed later will give many examples of this correspondence between physical quantities and self-adjoint operators, and they will also show what kind of restrictions one should impose on such a correspondence. Obviously, when considering energy, momentum, angular momentum, or electric current, the correspondence is valid and, provisionally, we shall accept it as general.

An observable A, when being considered as an operator, has in general both a discrete spectrum $\{a_n\}$ and a continuous one. In the usual approach to quantum mechanics, one assumes that the result of a measurement of this observable must necessarily be a number belonging to the spectrum. One also assumes that the state of the system is characterized by a given wave function or a normalized vector ψ. The probability p_n for A to have the discrete value a_n is then defined as follows: Introducing the projector E_n associated with the eigenvalue a_n, one assumes as a postulate that $p_n = \|E_n\psi\|^2$. The probability for the observable to have a value belonging to a finite interval J belonging to the continuous spectrum is defined in the same way: One uses the projector E_J that is associated with this interval, and puts

$$p_J = \|E_J\psi\|^2. \tag{1.30}$$

In Dirac's notation, one often uses a summation sign to denote either a discrete sum or an integral over a continuous domain, so that both kinds of probabilities take the same form:

$$p_J = \sum_{a \in J; r} |\langle a, r|\psi\rangle|^2,$$

r being a degeneracy index.

One also assumes that two (or more) distinct observables A and B can be measured simultaneously when the corresponding operators commute. Mathematics tells us that, when two self-adjoint operators A and B commute, they can be simultaneously diagonalized. This means that there exists an orthonormal basis (at least in the sense of Dirac) consisting of vectors

ELEMENTARY QUANTUM MECHANICS

$|a, b, r\rangle$, where a is in the spectrum of A, b is in the spectrum of B, and r is a degeneracy index. The probability for finding the values (a, b) when simultaneously measuring A and B is then written as

$$p(a, b) = \sum_r |\langle a; b; r|\psi\rangle|^2 \quad \text{or} \quad p(a, b) = \|E(a, b)\psi\|^2,$$

where

$$E(a, b) = \sum_r |a; b; r\rangle\langle a; b; r|.$$

Examples

These rules reiterate what we knew already. In the simple case of the coordinate position X of a particle in a one-dimensional space, the projector E_J associated with the property of finding the particle in an interval J is given by

$$E_J = \int_J |x\rangle\langle x|, \tag{1.31}$$

so that equation (1.29) reads

$$p_J = \int_{-\infty}^{\infty} \int_J \langle \psi|x\rangle\langle x|\psi\rangle dx = \int_J |\psi(x)|^2 \, dx,$$

where the wave function $\psi(x)$ in Dirac's notation is $\langle x|\psi\rangle$. The last formula is identical with Born's probability rule.

Under the same conditions, the probability of finding the momentum of the particle in some domain K is

$$p_K = \int_K \langle\psi|p\rangle\langle p|\psi\rangle \, dp,$$

where

$$\langle p|y\rangle = \int \langle p|x\rangle\langle x|\psi\rangle \, dx = \int h^{-1/2} \exp(-ipx/\hbar)\psi(x) \, dx.$$

The momentum probability distribution turns out to be the squared modulus of the Fourier transform of the wave function, as was expected.

Average Values and Uncertainties

Knowing the probabilities, one can derive the average value (or mean value) of the observable A in the state ψ. In probability calculus, the average value of a random variable A, taking the values a_n with probabilities p_n, is defined as the quantity

$$\langle A \rangle = \sum_n p_n a_n.$$

Using Dirac's notation, this gives in the present case

$$\langle A \rangle = \sum_n p_n a_n + \int_\sigma dp(a) a$$
$$= \sum_{nr} |\langle a_n r | \psi \rangle|^2 + \sum_r \int_\sigma a |\langle a r | \psi \rangle|^2$$
$$= \langle \psi | \left\{ \sum_{nr} |a_n r\rangle a_n \langle a_n r| + \sum_r \int_\sigma |ar\rangle a \langle ar| \, da \right\} |\psi\rangle$$
$$= \langle \psi | A | \psi \rangle.$$

The last equality was obtained by recognizing the spectral decomposition of the operator A inside the brackets in the third line. The final expression is very simple.

Knowing the average value of a random variable, one can define another variable with zero average value by subtracting from it the average denoted by $\langle A \rangle$, namely, $A' = A - \langle A \rangle I$. The uncertainty Δa for A is defined in probability calculus by

$$(\Delta a)^2 = \langle \psi | A'^2 | \psi \rangle = \langle \psi | A^2 | \psi \rangle - \langle \psi | A | \psi \rangle^2.$$

The Heisenberg uncertainty relations in general form involve two non-commuting observables A and B. Their commutator is $[A, B] = iC$, with C a self-adjoint operator, as is easily checked. The centered variables A' and B' have the same commutator, and the uncertainties satisfy the inequality

$$\Delta a \, \Delta b \geq \frac{1}{2} |\langle \psi | C | \psi \rangle|$$

(see problem 2). In the case of the position and momentum operators, one has $A = X$, $B = P$, and $C = \hbar I$, again giving $\Delta x \, \Delta p \geq \hbar/2$.

6. The Density Operator

The pioneering versions of quantum mechanics assumed the state of a system to be always specified by a wave function, or, more generally, by a Hilbert space vector. This restriction does not correspond to many realistic situations, and it was relaxed by von Neumann and Landau.

Consider for the sake of definiteness a pendulum. Let us take its initial situation as being in a position of coordinate a with zero velocity. Because of the uncertainty relations, the position cannot be strictly equal to a, nor

ELEMENTARY QUANTUM MECHANICS

the momentum strictly zero. The best one can get is an initial situation with average position a and average momentum 0, the product of their uncertainties being minimal. Let this be so and let α_a or $|a\rangle$ be the state vector in the Hilbert space of wave functions.

In practice, it is very difficult to make sure that all the preparing processes are strictly identical when repeated a large number of times. The device initially holding the pendulum has thermal motion, it is slightly sensitive to the air motion around it, and it may feel the effect of vibrations coming from external sources, such as, for instance, a truck in a neighboring street. As a result, the starting position of the pendulum, the exact form of its initial wave function, and the initial velocity all fluctuate. If one takes only the first effect into account for simplicity, the starting position a should be considered as a random variable having a probability distribution $w(a)$ such that $\int w(a)\,da = 1$. One may wonder what happens to the statistical distribution of a series of measurements for an observable B belonging to the pendulum under these conditions.

Let us assume for definiteness that B has only nondegenerate eigenvalues $b_1, b_2, \ldots$. One may use the rule of compound probabilities to say that the probability p_n for B having the value b_n is the sum of the probabilities for a to have some given value times the probability to find b_n under this initial condition. As a result, one gets

$$p_n = \int w(a) p_n(a)\,da.$$

Quantum mechanics gives $p_n(a)$ as

$$p_n(a) = |\langle b_n | a \rangle|^2,$$

so p_n is obtained from equation (1.30). The average value of B is given by

$$\langle B \rangle = \sum_n p_n b_n = \sum_n \int w(a) |\langle b_n | a \rangle|^2 b_n \, da.$$

The spectral theorem can be used to replace this by

$$\langle B \rangle = \int w(a) \left[\sum_n \langle b_n | B | a \rangle \langle a | b_n \rangle \right] da$$

$$= \int w(a)\, \mathrm{Tr}(B E_a)\, da,$$

where the projector E_a onto the state vector $|a\rangle$ has been introduced: $E_a = |a\rangle\langle a|$.

Finally a more compact expression of the result is obtained by introducing the *density operator*

$$\rho = \int w(a) E_a \, da. \qquad (1.32)$$

This operator depends only upon the preparation process, and the average value of B is simply

$$\langle B \rangle = \text{Tr}(B\rho), \tag{1.33}$$

whatever the observable B. The restrictive assumptions made upon the spectrum of B and the form of the initial randomness do not change this last result, which is quite general.

The probability p_n for B to have the value b_n is similarly expressed. It is given by

$$p_n = \int w(a)|\langle b_n|a\rangle|^2 \, da = \text{Tr}(|b_n\rangle\langle b_n|\rho) = \text{Tr}(E_n\rho), \tag{1.34}$$

E_n being the projector onto $|b_n\rangle$. When the eigenvalue b_n is degenerate, one obtains the same result by taking E_n to be the projector onto the subspace of all the eigenvectors associated with the value b_n. This can also be extended to a continuous spectrum: the probability of finding the value of B in an interval J in the continuous spectrum is given by equation (1.34), after replacing E_n by E_J. It may be noticed that equation (1.34) has the same form as equation (1.33): the probability of the value b_n is also the average value of the observable E_n.

The density operator is also called the *density matrix* (in the case of a finite-dimensional Hilbert space), the *state operator*, or the *density functional*. We shall sometimes use the first two terms. It has a number of important properties:

1. Its trace is 1:

$$\text{Tr}\,\rho = \int w(a)\,\text{Tr}\,E_a \, da = \int w(a)\, da = 1,$$

   the second equality coming from the fact that the trace of the projector E_a upon a unique vector $|a\rangle$ is 1.
2. It is self-adjoint. This comes from the self-adjointness of the projectors E_a, from which one gets ρ as a sum with real coefficients.
3. It is, moreover, a positive operator; i.e., for any vector γ, $(\gamma, \rho\gamma) \geq 0$. This is a direct consequence of the positive character of the probability $w(a)$ together with $(\gamma, E_a\gamma) = |\langle\gamma|a\rangle|^2 \geq 0$.

Equation (1.32) for ρ shows that the matrix element $(\gamma, \rho\gamma)$ is always less than 1. The spectrum of ρ belongs therefore to the interval $[0, 1]$. All these properties remain valid when the values of A are continuous or degenerate.

A state that is described by a state vector $|a\rangle$ is called a *pure state*. The corresponding density operator is given by $\rho = |a\rangle\langle a|$, and it has only one nonzero eigenvalue, equal to 1 with eigenvector $|a\rangle$. When this simple form of ρ does not hold, the state is said to be *mixed*.

ELEMENTARY QUANTUM MECHANICS

7. Dynamics

The Schrödinger Equation

Dynamics, as expressed by the Schrödinger equation, is easily generalized to the Hilbert space framework. One assumes that an isolated system is described at a time t by a state vector $\alpha(t)$ and that there exists a self-adjoint hamiltonian operator H that is the observable of energy. The time evolution of a state vector is then governed by the equation

$$i\hbar \frac{d\alpha(t)}{dt} = H\alpha(t),$$

which is still called the Schrödinger equation.

It is often convenient to solve the Schrödinger equation formally. To do so, define an *evolution operator* $U(t)$ by

$$U(t) = \exp(-iHt/\hbar). \quad (1.35)$$

If the state vector at time 0 is $\alpha(0)$, one easily verifies that its value at time t is given by $\alpha(t) = U(t)\alpha(0)$. The evolution operator is unitary, i.e.,

$$U(t)U^\dagger(t) = U^\dagger(t)U(t) = I$$

($U^\dagger$ being the operator adjoint to U), so that it conserves the scalar products: $(U(t)\beta, U(t)\alpha) = (\beta, \alpha)$. As a consequence, the scalar product of $\alpha(t)$ with itself is conserved so that the norm $\|\alpha(t)\|$ is a constant. This is what one expects, as the normalization of probabilities should be conserved.

In this formulation of dynamics following the Schrödinger approach, the state vector evolves in time, whereas the observables are fixed operators. The time evolution of average values is given by

$$\langle B \rangle(t) = \langle \alpha(t)|B|\alpha(t)\rangle = \langle \alpha(0)|U^\dagger(t)BU(t)|\alpha(0)\rangle.$$

In classical physics, one prefers to think of the physical observables themselves, such as position or momentum, as varying with time. This can also be the case in quantum mechanics if one follows more closely Heinsenberg's first approach by using the so-called *Heisenberg representation*. The state vector in that case is considered to be independent of time, having, for instance, the fixed value $\alpha(0)$, and time-dependent observables are defined by $B(t) = U^\dagger(t)BU(t) = U^{-1}(t)BU(t)$. This is strictly equivalent from a physical standpoint because $\langle B(t) \rangle = \langle \alpha(0)|B(t)|\alpha(0)\rangle = \langle B \rangle(t)$. The time evolution of a time-dependent Heisenberg operator is easily obtained from the explicit form (1.35) of $U(t)$, giving

$$\frac{dB(t)}{dt} = \frac{i}{\hbar}[H, B(t)].$$

Dynamics and Group Properties

The exponential form (1.35) for the evolution operator shows that one has $U(t)U(t') = U(t')U(t) = U(t+t')$ for all values of t and t'. Furthermore, the operator $U(-t)$ is the inverse of $U(t)$. This shows that the family of operators $\{U(t)\}$ is an abelian (i.e., commutative) group for which the unit element is the identity operator I.

According to a theorem by M. H. Stone, every one-parameter strongly continuous abelian group of unitary operators is generated in the same way. Letting $V(a)$ be the group elements, there exists a self-adjoint operator A such that $V(a) = \exp(iAa/\hbar)$, strong continuity meaning that $\lim_{a\to 0}\|V(a) - I\| \to 0$. This self-adjoint operator A is an observable called the *group generator*. One sees that the hamiltonian is the generator of the time evolution or, if one prefers, of the group of translations for the origin of time. This is true when H itself does not depend upon time, so the choice for the origin of time is irrelevant for physics.

Another example of the same property is given by space translations. Consider again the particle in a one-dimensional space. Translating the origin by a distance a corresponds to a change of the wave function from $\psi(x)$ to $\psi(x + a)$. This is a unitary transformation because it conserves scalar products as is shown by

$$\int_{-\infty}^{+\infty} \phi^*(x+a)\psi(x+a)\,dx = \int_{-\infty}^{+\infty} \phi^*(x)\psi(x)\,dx.$$

The family of all these transformations is obviously an abelian group of operators $V(a)$, and we shall admit that it is continuous. It turns out that the corresponding generator is the momentum. This is easily shown in a special case where the wave function is analytic (i.e., well described by a Taylor series), because it then follows that

$$\psi(x+a) = \sum_{n=0}^{\infty} \frac{a^n}{n!}\left(\frac{d}{dx}\right)^n \psi(x) = \sum_{n=0}^{\infty} \frac{(iaP/\hbar)^n}{n!}\psi(x)$$
$$= \exp(iaP/\hbar)\psi(x).$$

This result can be extended to any square-integrable function by using Fourier transforms. In the case of a three-dimensional space, whatever the number of particles in the system, the group of translations has three parameters and there are three generators, which are the three components of the total momentum.

It may be noticed that this implies that any function $f(\mathbf{x})$ depending upon the position of a center of mass $\mathbf{X}$ also obeys the property

$$\exp(-i\mathbf{P}\cdot\mathbf{a}/\hbar)f(\mathbf{X})\exp(i\mathbf{P}\cdot\mathbf{a}/\hbar) = f(\mathbf{X}+\mathbf{a}),$$

ELEMENTARY QUANTUM MECHANICS 43

a being a translation vector. Taking the derivatives, one gets

$$\frac{i}{\hbar}[\mathbf{P}, f(\mathbf{X})] = \nabla f(\mathbf{X}), \qquad (1.36)$$

a property we shall use in a moment.

Ehrenfest's Theorem

In a later part of this book, much space will be given to the detailed correspondence between classical and quantum mechanics. It will be useful nevertheless to recall how it was first shown by Paul Ehrenfest in 1927.

Let us restrict ourselves for simplicity to the case of a particle in a one-dimensional space with the hamiltonian

$$H = \frac{P^2}{Em} + V(X).$$

Bohr's correspondence principle asserts that one should recover approximately the results of classical physics when considering states for which Planck's constant is relatively small with respect to other quantities of action. Once quantum mechanics had become a full-fledged theory, Ehrenfest tried to replace this heuristic principle with a theorem. This can be done in the following way:

Let us use Heisenberg's representation so that the state of the particle is some given fixed vector α in the Hilbert space of square-integrable functions. The position and momentum observables $X(t)$ and $P(t)$ depend upon time, and one looks for their dynamical relations. One first notices that the canonical commutation relations are valid for arbitrary time t because the evolution operators are unitary so that $[X(t), P(t)] = i\hbar I$. Also note that $U(t)$, as a function of H, commutes with H. So one gets the time derivative of $X(t)$ from equation (1.28) as

$$\frac{dX(t)}{dt} = \frac{i}{\hbar}[H, X(t)] = \frac{P(t)}{m}, \qquad (1.37)$$

where the last equality follows from the explicit form of H and the canonical commutation relations.

To get the time derivative of $P(t)$, one uses the fact that P is the generator of translations so that equation (1.36) gives $i/\hbar[P, V(X)] = V'(X)$, $V'(x)$ being the derivative of the function $V(x)$. Then one gets

$$\frac{dP(t)}{dt} = \frac{i}{\hbar}[H, P(t)] = \frac{i}{\hbar}[V(X(t)), P(t)] = -V'(X(t)). \qquad (1.38)$$

It is clear that equations (1.37) and (1.38), relating the time-dependent operators for position and momentum, are formally identical with the Hamilton equations for classical mechanics.

If one takes the average value of these equations in a given state, one gets

$$\frac{d\langle X(t)\rangle}{dt} = \frac{\langle P(t)\rangle}{m}, \quad \frac{d\langle P(t)\rangle}{dt} = -\langle V'(X(t))\rangle. \tag{1.39}$$

This would be identical to the classical Hamilton equations with position and momentum replaced by their average values, if one had the quantity $-V'(\langle X(t)\rangle)$ in the right-hand side of the second equation. But one doesn't. However, by localizing the wave function in some region of space where the force $-V'(x)$ varies slowly, one can make the necessary replacement up to a small error. Equations (1.39) then show that, in practice, the particle will have a classical motion.

These questions will be reconsidered later in much more detail, so it is not necessary to be more general or more rigorous at present. One may, however, notice the new light shed upon the correspondence principle, as compared to what it was at the time of Bohr's first model. It looked at that time as if the quantum effects came from supplementary conditions superimposed on classical physics, which remained the basic reference. Here one sees the opposite situation: quantum mechanics is taken as the basic form of physics, and classical physics is only the appearance it takes on a large scale in an approximate and nonuniversal way.

8. How to Describe a Complex System

It was easy to describe a system made of several particles in Schrödinger's wave formulation of quantum mechanics, at least as far as writing down the corresponding equations. One had only to consider a wave function depending upon all the coordinates of these particles. It will be useful to also know how to do it in the more formal Hilbert space approach.

Let us start from the notion of a *degree of freedom* for a system. A particle in ordinary space has three degrees of freedom, which may be taken to be its three coordinates in some reference system. Considering only two coordinates for simplicity, one can expand a wave function $\psi(x, y)$ in any basis consisting of orthonormal functions of x and y. But one can also proceed as follows: First, one expands $\psi(x, y)$ for every fixed value of y in an orthonormal set of functions depending only upon x and therefore providing a basis for a Hilbert space associated with only one degree of freedom. The corresponding coefficients are functions of y and they may be expanded in an orthonormal set of functions of y. So, one sees that the Hilbert space $L^2(\mathbf{R}^2)$ is in some sense built up from two Hilbert spaces $L^2(\mathbf{R})$. The underlying construction is called the *tensor product* of two Hilbert spaces and it is quite general and most useful.

ELEMENTARY QUANTUM MECHANICS 45

Tensor Products

Consider two Hilbert spaces **E** and **F**. One can build from them a third Hilbert space **G** to be called their tensor product. One starts from the direct product **E** × **F** consisting of all pairs of vectors $\{\alpha, \beta\}$, where the first one is in **E** and the second in **F**. For every complex number c, one considers two pairs $\{\alpha, \beta\}$ and $\{c\alpha, c^{-1}\beta\}$ to be equivalent. This is similar to the example we saw when considering products of two functions of one variable, since the pairs of functions $\{\alpha(x), \beta(y)\}$ and $\{c\alpha(x), c^{-1}\beta(y)\}$ are equivalent as far as their product $\alpha(x) \cdot \beta(y)$ is concerned. A class of equivalent pairs is called the tensor product of the two vectors α and β, and it is written as $\alpha \otimes \beta$. The tensor product **G**, written as **E** ⊗ **F**, consists of all possible linear combinations of the tensor products of vectors. This may be viewed as inspired by the construction of a function of x and y as a linear combination of functions of x times a function of y, each of them being taken in some basis of a Hilbert space of functions of one variable. So, despite its somewhat tricky construction, the tensor product of two spaces appears quite natural.

The elements of **G** are called tensors and have the general form

$$\sum_{nm} c_{nm} \varepsilon_n \otimes \phi_m, \tag{1.40}$$

where the sets of vectors $\{\varepsilon_n\}$ and $\{\phi_m\}$ are bases for **E** and **F**, respectively. When **E** and **F** have finite dimensions N and N', **G** has the dimension NN'. In Dirac's notation, one simply writes $|\alpha\rangle|\beta\rangle$ in place of $|\alpha\rangle \otimes |\beta\rangle$, as a reminder that this is only a product in the case of two functions.

The scalar product in **G** is defined by the following rules, which are again obvious when one thinks of the product of functions: $(\alpha \otimes \beta, \alpha' \otimes \beta') = (\alpha, \alpha') \cdot (\beta, \beta')$. When two operators A and B act in the spaces **E** and **F**, respectively, one defines their tensor product $A \otimes B$ as an operator in **G** to be given by $(A \otimes B)(\alpha \otimes \beta) = (A\alpha) \otimes (B\beta)$. This defines its action on a basis $\varepsilon_n \otimes \phi_m$ of the space **G**, and therefrom on every element of **G**. As a special case, one can extend an operator A on **E** to become an operator in **G** by considering the tensor product $A \otimes I'$, where I' is the identity in **F**.

Tensor coordinates, as they are given by equation (1.40), behave in an interesting way when one performs a change of bases in **E** and **F**. We shall consider only the case of finite-dimensional spaces. Let $U_{n'n}$ be the matrix for a change of coordinates in **E**, and let $V_{m'm}$ be the matrix for a change of coordinates in **F**. The new coordinates of a tensor then become

$$c'_{n'm'} = \sum_{nm} U_{n'n} V_{m'm} c_{nm}.$$

One can also consider a tensor as given by its coordinates in some basis, together with this rule for a change of bases. This point of view is often preferred when one uses tensors in relativity, hydrodynamics, elasticity, or classical electromagnetism.

The formalism allows one to consider a physical system as being built up from its various degrees of freedom, including spin. For instance, the pure states of a spin-1/2 particle can be considered as tensors, i.e., linear combinations of tensor products such as $\psi \otimes \alpha$ of a wave function $\psi(x)$ and a spinor α representing a spin state in a two-dimensional Hilbert space.

Noninteracting Systems

The tensor formalism is also convenient to give a mathematical characterization of two noninteracting physical systems S and S' (for instance, two noninteracting particles). Denoting by **H** and **H**$'$ their respective Hilbert spaces and by H and H' their hamiltonians, the special case of wave functions is easily translated into the following general form: the Hilbert space of the compound system $S + S'$ is $\mathbf{H} \otimes \mathbf{H}'$, and its hamiltonian is the sum $H'' = H \otimes I' + I \otimes H'$. Most often, this is simply written as $H + H'$, the tensor products with an identity operator being left understood.

When $S + S'$ is initially in a pure state $\phi = \psi \otimes \psi'$, the Schrödinger equation for the compound system

$$i\hbar \frac{d\phi}{dt} = (H + H')\phi$$

has the solution $\phi(t) = \psi(t) \otimes \psi'(t)$, where $\psi(t)$ and $\psi'(t)$ obey the separate Schrödinger equations

$$i\hbar \frac{d\psi}{dt} = H\psi, \qquad i\hbar \frac{d\psi'}{dt} = H'\psi'.$$

REFERENCE FRAMES

9. How to Construct the Physical Hilbert Space

When it became clear that the foundations of physics lay in quantum mechanics, one could not feel confident any more that its building blocks, such as the exact form of the Hilbert space or the exact hamiltonian, could reliably be obtained from classical mechanics by some sort of "quantization" rule. Quantization had been used, for instance, when Schrödinger wrote down the hamiltonian of an atom. An objection against that procedure was that one should not derive something fundamental, in this case, the hamiltonian, from a limiting case such as classical physics. Furthermore, this extrapolation is questionable since it cannot reach terms in the hamiltonian of higher order in Planck's constant if they exist. When spin was discovered, it was added "by hand" to what one knew before, and this too was rather unsatisfactory.

The case of relativistic particles was also perplexing. In 1928, Dirac proposed describing a relativistic electron by a four-component wave function rather than the two-component one occurring in the nonrelativistic case, where the two components represent the spin $1/2$ of the electron. Dirac's achievement was a tour de force, but it was not at all clear why one had to proceed that way and not otherwise; in other words, the underlying principles behind Dirac's formulation remained ill understood. Moreover, the hamiltonian he had introduced allowed for negative energy states that were difficult to deal with, even if they led to predicting the existence of the positron.

So, one felt it necessary to lay relativistic quantum mechanics upon clearer foundations. As a matter of fact, Galilean relativity and the Lorentz–Einstein relativity can be developed in parallel in the context of quantum mechanics. This was mainly achieved by Wigner.[10] He found that many things could be made clearer by considering systematically the constraints imposed on the quantum description of a system by requiring relativistic invariance. He could explain the occurrence of negative energy states and also how to avoid them, as well as how to get the Dirac equation naturally. But one of his most startling results was to show that, as a consequence of axiomatic quantum mechanics and special relativity, a free particle must have a well-defined mass and spin. This is an example of the hidden fecundity of quantum mechanics, where oftentimes some empirical data that one had grown to accept without too much questioning (for instance mass), or that one had grown to accept as coming from nowhere (for instance spin), can be found to be necessary ingredients from the standpoint of the theory. Other examples of these surprising rediscoveries will be introduced in later chapters.

It will be useful to describe sketchily how Wigner proceeded, if only to show that the Hilbert space of a system of particles is well defined and not obtained by a questionable induction from classical physics. This question is important from the standpoint of interpretation because it will turn out later on that one can derive all the aspects of classical physics from quantum mechanics, and this would give a circular argument if the Hilbert space one started with had been induced from classical considerations.

Wigner relied upon *relativistic invariance,* which means looking at what happens when one changes the reference frame. A key feature of the problem is that the family of all changes of reference frames is a group, which will lead us to mention a few results concerning the application of group theory to quantum physics. This might appear to take us away from our main theme and it should accordingly be mentioned that what follows through the end of this chapter will not be necessary in the later parts of the book.

Active and Passive Transformations

Consider an experiment where a preparation device, for instance an accelerator, prepares a particle in a well-defined state $|\alpha\rangle$, and a measuring device, for instance a counter, measures its probability for being in a state $|\beta\rangle$. This probability is $|\langle\beta|\alpha\rangle|^2$.

Let us assume that the whole laboratory stands upon a horizontal platform, which is turned by some angle. Clearly, the accelerator does not produce the same state as before, but it still works in exactly the same way. One can say more formally that the rules of physics do not depend upon the reference system. Let $|\alpha'\rangle$ be the new state that is prepared. Similarly, the counter now measures the probability for being in a new state $|\beta'\rangle$. Nothing significant is changed, so one expects that the results of a series of measurements will be the same, the probabilities being unchanged, i.e., $|\langle\beta|\alpha\rangle|^2 = |\langle\beta'|\alpha'\rangle|^2$.

The invariance of probabilities must be valid, whatever the states to be prepared or measured. One might then be tempted to assume that the scalar products themselves are unchanged, which would mean that there exists a unitary transformation U such that

$$|\alpha'\rangle = U|\alpha\rangle. \tag{1.41}$$

That this is true was proved by Wigner, though his proof shows that the transformation could in principle be antiunitary rather than unitary. An example of an antiunitary transformation will be given in Chapter 4 in the case of time reversal. However, in the present case, the rotation belongs to a continuous group, or, said otherwise, one might realize it by a series of small rotations. When this is the case, namely, when the change of a reference frame belongs to a continuous group, one must use equation (1.41) with a unitary operator U.

The rotation of the laboratory is called an *active transformation*. This is because it represents an act of the experimentalist to turn the whole system by a rotation R. There is no real difference, however, between this and the case where one chooses to refer the initial experiment to a new system of coordinates rotated by R^{-1} with respect to the initial one. Then one says that it is a *passive transformation*.

Transformation Groups

Rotations form a group, and the same is true for all the changes of reference frames one may think of, from the translations of the origin of space to a relativistic change of inertial frame. If one denotes such a group by G, one can define the product of two operations g and g' by gg', and this product is also a change of reference frame.

ELEMENTARY QUANTUM MECHANICS

If one excludes an inversion of the space axes as well as time reversal, every transformation belonging to the group can be reached continuously from the identity and therefore there exists a family of unitary operators $U(g)$ acting in the Hilbert space of the system. They do not, however, form a group by themselves in general, because a state vector can be multiplied by a complex number of modulus 1 without anything physical changing. Wigner proved that what one has in general is

$$U(gg') = e^{i\phi(g,g')}U(g)U(g'),$$

where $\phi(g, g')$ is a phase depending upon the transformations g and g'. In the cases of the rotation group and the Poincaré group (to be defined shortly), one can get rid of the phase by a suitable change of phase in the operators $U(g)$ themselves while going from the group itself to its so-called *universal covering*.[11] An interesting exception is the Galilei group, which includes the changes of inertial reference frames in nonrelativistic physics so that, curiously enough, the mathematical investigation of nonrelativistic physics is in some sense more difficult than in the relativistic case. Because it can be recovered from the relativistic case by letting the velocity of light become large, we shall leave it aside. So, in all cases to be considered, one will have

$$U(gg') = U(g)U(g'), \qquad (1.42)$$

with a conveniently extended group.

Generators

Let us first consider the case of the rotation group. Continuity implies that every rotation R can be obtained by iterating a large number of times a very small rotation r, which means a rotation close to the identity ($R = r^n$, with n large).

Rotations depend upon three parameters—for instance, the three Euler angles—which can vary in a continuous manner. For example, under a rotation of an angle θ around the z axis, the coordinates of a point in space change according to the matrix

$$R = \begin{bmatrix} \cos\theta & -\sin\theta & 0 \\ \sin\theta & \cos\theta & 0 \\ 0 & 0 & 1 \end{bmatrix}.$$

For θ very small, it becomes, leaving out second-order terms,

$$R = I + \theta \begin{bmatrix} 0 & -1 & 0 \\ 1 & 0 & 0 \\ 0 & 0 & 0 \end{bmatrix} = I + M_z\theta. \qquad (1.43)$$

The matrix M_z represents the action of an infinitesimal rotation around the z axis. Similarly, one can write similar matrices M_x and M_y for infinitesimal rotations around the other axes. An explicit calculation shows that their commutation relations are given by $[M_x, M_y] = M_z$, together with similar relations one can obtain by a circular permutation of the indices. Basically, the fact that the commutator of two generators of infinitesimal rotations is another generator is due to the group property.

The same considerations apply to other groups of changes in reference frame, except that the number of their generators and their commutator algebras are different.

Infinitesimal Transformations

The unitary operator $U(g)$ must be very near the identity when the change of coordinates g is infinitesimal. For instance, when g is the infinitesimal rotation with the matrix (1.43), $U(g)$ should have the form $U(g) = I + i\theta J_z/\hbar$, where I is now the identity operator in Hilbert space, and Planck's constant has been introduced for convenience. The pure imaginary number i has also been introduced for the following reason: Because $U(g)$ is unitary, it is easy to show that the operator J_z is self-adjoint. It is therefore an observable generating the rotations around the z axis.

One can define in a similar way two other operators J_x and J_y generating rotations around the x and y axes. The group property (1.42) can be used to compute the commutators of these generators. To do so, one considers a rotation R around the x axis with a very small angle α, and a rotation R' around the y axis with a small angle β. One then constructs the rotation $R'' = RR'R^{-1}R'^{-1}$. Performing the computation directly with the matrices, one finds that R'' is a rotation around the z axis with an angle $\alpha\beta$. As a consequence, the product $U(R)U(R')U^{-1}(R)U^{-1}(R')$ is equal to $U(R'')$. Expressing all these infinitesimal operators in terms of their generators, one gets

$$[J_x, J_y] = i\hbar J_z, \tag{1.44}$$

as well as two other relations that may be obtained by circular permutations upon the indices x, y, z.

Angular Momentum

The commutation relations (1.44) are known to characterize angular momentum in elementary quantum mechanics. Their consequences are well known and we shall only recall them briefly: There exist Hilbert spaces where the component J_z of angular momentum and its square $J^2 = J_x^2 + J_y^2 + J_z^2$ are a complete set of commuting observables. There is a basis in such a space whose vectors are eigenvectors of the two operators. One says

ELEMENTARY QUANTUM MECHANICS

that these spaces provide an *irreducible representation* of the rotation group. Every space where these operators can act has a basis consisting of their eigenvectors up to degeneracy indices.

The possible eigenvalues of J^2 are $j(j+1)\hbar^2$, where j is a nonnegative number, an integer, or a half-integer. The eigenvalues of J_z are $m\hbar$, where m, like j, is an integer or half-integer, and can take the values between $-j$ and $+j$.

This example shows the power of group theory when applied to a change of reference frame: the angular momentum has been found to be the set of generators for rotations, just as the hamiltonian has already been shown to be the generator of time translations and the momentum for space translations. The possibility of half-integer values of j is a strong theoretical indication for the possible existence of spin. It will now be shown that one can do much better.

10. Relativistic Invariance

The Principle of Relativity

One will assume that all kinds of accelerators and detectors can be used and described similarly in any inertial reference frame, and, furthermore, that the physical data for a given experiment as described in two frames Σ and Σ' are related by the Lorentz transformation bringing Σ onto Σ'. The Lorentz transformations are therefore treated as passive.

It should be stressed that this principle refers only to the results of straightforward measurements, i.e., to quantities that are unambiguous and unaffected by a problem of interpretation. It does not say anything about quantum mechanics as such, like stating what should be the behavior of the operator associated with an observable under a change of frame. Nevertheless, by using Wigner's theorem, one knows that this change of frame will be represented by the action on the Hilbert space of a unitary operator U belonging to a known group.

The Homogeneous Lorentz Group

Homogeneous Lorentz transformations are defined as linear changes of coordinates in space-time

$$x'_m = \sum_{n=1}^{4} \Lambda_{mn} x_n.$$

The spatial coordinates (x_1, x_2, x_3) represent a vector associated with some measurement in the reference frame Σ. If this is a position, one also introduces a time coordinate $x_0 = ct$, or an energy together with a momentum, and so on. The matrix Λ representing a change of frame in space-time with no translation of the origin for coordinates should conserve the space-time distance $s^2 = x_0^2 - x_1^2 - x_2^2 - x_3^2$. This is conveniently written as

$$s^2 = \sum_{mn} g_{mn} x_m x_n,$$

where the matrix g is given by

$$g = \begin{bmatrix} 1 & 0 & 0 & 0 \\ 0 & -1 & 0 & 0 \\ 0 & 0 & -1 & 0 \\ 0 & 0 & 0 & -1 \end{bmatrix}.$$

The conservation law $s'^2 = s^2$ can also be written as

$$\sum_{mnkr} g_{mn} \Lambda_{mk} \Lambda_{nr} x_k x_r = \sum_{mn} g_{mn} x_m x_n.$$

This can in turn be expressed by the following relation between the matrices Λ and g: $\Lambda^t g \Lambda = g$, where Λ^t denotes the transposed matrix of Λ.

This matrix equation gives as many conditions on the matrix Λ as there are independent elements in it. Because a 4×4 symmetric matrix has only 10 independent elements and there are 16 matrix entries in Λ, there remain 6 free parameters. One can therefore consider that the homogeneous Lorentz transformations constitute a continuous group with 6 parameters.

The meaning of these parameters can be found by considering a few special cases. For instance, a pure Lorentz transformation with its velocity v along the x axis has the matrix

$$L_x(v) = \begin{bmatrix} \gamma & \gamma\beta & 0 & 0 \\ \gamma\beta & \gamma & 0 & 0 \\ 0 & 0 & 1 & 0 \\ 0 & 0 & 0 & 1 \end{bmatrix},$$

where $\beta = v/c$ and $\gamma = (1 - \beta^2)^{-1/2}$. Similarly, a rotation of the space axes around the z axis by an angle θ is described by

$$R_z(\theta) = \begin{bmatrix} 1 & 0 & 0 & 0 \\ 0 & \cos\theta & -\sin\theta & 0 \\ 0 & \sin\theta & \cos\theta & 0 \\ 0 & 0 & 0 & 1 \end{bmatrix}.$$

From a physical standpoint, it is clear that a general Lorentz transformation can be obtained by a pure Lorentz transformation whose velocity v is given by the velocity of the second reference frame as seen in the first one (giving three parameters), followed by a rotation bringing the space axes along the same directions (three parameters again). The existence of six parameters is therefore rather obvious.

Infinitesimal Lorentz Transformations

Proceeding with the Lorentz group (as with the rotation group), one first considers pure Lorentz transformations with a very small velocity δv along the x axis. This gives to first order in δv,

$$L_x(\delta v) = I + \delta v/c \begin{bmatrix} 0 & 1 & 0 & 0 \\ 1 & 0 & 0 & 0 \\ 0 & 0 & 0 & 0 \\ 0 & 0 & 0 & 0 \end{bmatrix},$$

and the corresponding unitary transformation in Hilbert space has the form $[I + i(\delta v/c)K_x/\hbar]$, where K_x is a self-adjoint operator.

Similarly, a rotation by a very small angle $\delta\theta$ around the z axis corresponds to the matrix

$$R_z(\delta\theta) = I + \delta\theta \begin{bmatrix} 0 & 0 & 0 & 0 \\ 0 & 0 & -1 & 0 \\ 0 & 1 & 0 & 0 \\ 0 & 0 & 0 & 0 \end{bmatrix}$$

and to an operator $I + i\delta\theta J_z/\hbar$.

One can compute the commutators of the infinitesimal matrix generators to find that, once again, they are linear combinations of the generators. The correspondence with unitary operators gives the commutation rules between the self-adjoint generators in Hilbert space. Denoting by J_k and K_j ($j,k = x,y,z$) the generators in Hilbert space, one finds

$$\begin{aligned}[J_x, J_y] &= i\hbar J_z, \\ [J_x, K_x] &= 0, \\ [J_x, K_y] &= i\hbar K_z, \\ [K_x, K_y] &= -i\hbar J_z,\end{aligned} \quad (1.45)$$

with other commutation relations being obtained by circular permutations of the indices.

The Poincaré Group

The most general change of frame consists of a general Lorentz transformation together with a translation of the origin of space-time. It can be written as

$$x'_m = \sum_{n=0}^{3} \Lambda_{mn} x_n + a_n, \quad (1.46)$$

where Λ is a Lorentz matrix and a is a four-vector. Their family is called the Poincaré group and it obviously has 10 parameters (6 for Λ and 4 for a). One may give a matrix form of equation (1.47) by introducing a dummy fifth coordinate x_4, which is unaffected by the transformation, so that

$$x'_m = \sum_{n=0}^{4} \Pi_{mn} x_n,$$

where $\Pi_{mn} = \Lambda_{mn}$ for $m \neq 4$ and $n \neq 4$, $\Pi_{4n} = a_n$ ($n \neq 4$), and $\Pi_{44} = 1$.

This explicit matrix form can be used as before. There are still generators J_k and K_j for the rotations and the pure Lorentz transformations, and their commutation relations are still given by equation (1.45). There are also four generators for the translations along every space-time axis, and the commutation relations where these translation generators enter are found to be

$$[P_m, P_n] = 0,$$
$$[J_x, P_y] = i\hbar P_z,$$
$$[J_x, P_x] = 0,$$
$$[J_x, P_0] = 0, \qquad (1.47)$$
$$[K_x, P_0] = i\hbar P_x,$$
$$[J_x, P_x] = i\hbar P_0,$$
$$[J_x, P_y] = 0.$$

Relativistic Form of the Commutation Relations

One can choose not to separate the space and time axes and to work directly in a four-dimensional framework. To do so, consider the couple of space vectors (J_k, K_k) as being the components of a unique antisymmetric four-dimensional tensor, in analogy with the electric and magnetic fields giving an antisymmetric tensor in space-time. Define a family of self-adjoint operators M_{mn}, antisymmetric in their indices, by $M_{xy} = J_z$, $M_{0x} = K_x$, the other components being given by a permutation of the indices.

The various commutation relations can then be given the following covariant form:

$$[M_{mn}, M_{kr}] = -i\hbar(g_{mk}M_{nr} - g_{mr}M_{nk} + g_{nr}M_{mk} - g_{nk}M_{mr}),$$
$$[M_{mn}, P_r] = -i\hbar(g_{mr}P_n - g_{nr}P_m).$$

Invariants

From these commutation relations, one easily finds that the operator $P^2 = P_0^2 - P_x^2 - P_y^2 - P_z^2$ commutes with all the generators, meaning that it remains unchanged under a change of frame. What was found previously for the generators of time and space translations shows that the operator P_0 must be the hamiltonian, and P_x, for instance, the x component of the total momentum. One can then recognize in the invariant P^2 the square of the mass M, up to a factor, or more exactly the quantity M^2c^2.

An interesting feature of this approach is to show that, among all the observables characterizing an arbitrary physical system, one of them is an invariant and its eigenvalues are the possible masses of the system. In other words, *the existence of the inertial mass need not be postulated but is a necessary consequence of quantum mechanics together with the relativistic invariance of physical experiments.*

This remarkable result suggests looking for other invariants. There is only one, which can be constructed as follows: Introduce a four-vector W_m by $W_m = \frac{1}{2}\varepsilon_{mnkr}P_n M_{kr}$, where ε_{mnkr} is the familiar completely antisymmetric tensor and where the Einstein convention for summing upon repeated indices has been used. The time and space components of this four-vector are

$$W_0 = -\mathbf{P} \cdot \mathbf{J}, \quad \mathbf{W} = -P_0 \mathbf{J} + \mathbf{P} \wedge \mathbf{K}. \tag{1.48}$$

One can then compute the commutators of the quantity $W^2 = W_0^2 - W_x^2 - W_y^2 - W_z^2$ with all the generators, to find that it commutes with them. It is therefore an invariant.

Representations of the Poincaré Group

To get a better handle on the physical meaning of the invariants, it is convenient to choose a special basis in the Hilbert space. Noticing that the operators P_n commute, one chooses a basis where the four of them are diagonal, say $|p, r\rangle$, where p is a four-vector for the eigenvalues of the four operators, and r is a degeneracy index. One then has $P_n|p, r\rangle = p_n|p, r\rangle$. The quantity p^2 is an invariant under a change of inertial frame.

The simplest physical system one can think of is a unique free particle. It seems attractive to associate it with one of the simplest possible Hilbert spaces, where the invariant P^2 has only one eigenvalue m^2c^2 and the degeneracy indices will be as few as possible. This is what is called in mathematics an irreducible unitary representation of the Poincaré group.

Let us restrict ourselves to the case where the mass m is nonzero. The commutation relations (1.47) can be used to show that some Lorentz transformation can bring an eigenstate of four-momentum with eigenvalues p upon any other one having the same value of p^2. One of them therefore has the four-momentum $(mc, \mathbf{0})$, representing a system at rest in its own reference frame. In this frame, equations (1.48) show that

$$W_0 = 0, \quad \mathbf{W} = -mc^2 \mathbf{J}.$$

Up to an unessential factor, the nonzero components of $\mathbf{W}$ are given by the vector $\mathbf{J}$. They commute with P_0, so the transformations they generate do not change the rest frame. They therefore affect only the degeneracy indices. The commutation relations for the components of **J** are those of an angular momentum, which is enough to make sure that the possible eigenvalues of W^2 can be written as $m^2 c^4 s(s+1) \hbar^2$, where s is a non-negative integer or half-integer. This is the spin, and there are no other indices than the eigenvalues of J_z in an irreducible representation. One has therefore found that the possible existence of spin is a direct consequence of the Hilbert space axioms of quantum mechanics together with relativistic invariance.

Wigner and Bargmann have investigated in detail other consequences of relativistic invariance. The case of a spin $1/2$ directly leads to the Dirac equation. For a system of many particles, one can also construct localized field observables, at least for integer values of the spin. For a zero-mass particle with spin 1, the relations existing between these fields turn out to be Maxwell's equations. We shall not need these results and leave them aside accordingly.

Summary

When combined with relativistic invariance, the formal Hilbert space approach to quantum mechanics demands the existence of two invariants for a particle: its mass and spin. By an explicit deductive construction, one finds the Hilbert space of a particle and the basic operators representing four-momentum and angular momentum. From this one may proceed, at least in the nonrelativistic limit, toward a construction of an operator canonically conjugate to momentum, which is position. One then recovers the wave function for a particle.[12] Considering together several noninteracting particles, their overall Hilbert space is obtained from the previous assumptions about noninteracting systems: it is the tensor product of the Hilbert spaces of the constitutive particles. Of course, this procedure does not give the exact form of the interaction hamiltonian. It also has some difficulties when considering the case of quarks, which are never found as free particles.

APPENDIX: THE UNCERTAINTY RELATION FOR ENERGY

The uncertainty relation for energy is usually stated as giving a bound or an estimate for the error ΔE in the energy that is found when measuring it in an experiment lasting at most a time Δt. It is then written as

$$\Delta E \, \Delta t \geq \hbar/2. \tag{1A.1}$$

This is strongly suggested by the connection between the energy and the frequency of a wave function, and as such it is the well-known property of Fourier transforms according to which a signal with duration Δt necessarily has a spread in frequency $\Delta \omega$ of the order of $1/\Delta t$ or larger.

There is, however—because of the significance of Δt—a strong difference between the uncertainty relation (1A.1) and the ones involving position and momentum. Time is not an observable represented by an operator, but rather a parameter. This is reflected in the fact that one cannot find a wave function violating the position and momentum uncertainty relations, whereas it is quite legitimate to write down an eigenstate of energy at a well-defined time. It is therefore important to state clearly and correctly what the inequality (1A.1) means.

Many textbooks make use of first-order perturbation calculus, from which the following results are found: Consider a system having a continuous or very dense energy spectrum. It is perturbed during a time Δt by a weak perturbation, which is constant during that time interval. After this interaction, an initial eigenfunction of the energy $|\psi\rangle$ with energy E becomes to first order in perturbation theory

$$A|\psi\rangle + |\delta\psi\rangle, \qquad (1A.2)$$

the coefficient A differing from unity only to second order. The perturbed part $|\delta\psi\rangle$ can be projected onto the various eigenstates of energy, and the probability for an energy E' is found to be proportional to

$$\sin^2\{(E' - E)\Delta t/2\hbar\}(E' - E)^{-2}. \qquad (1A.3)$$

Hence, as Landau and Lifschitz said,[13] "we see that the most probable value of the difference $E' - E$ is of the order of $\hbar/\Delta t$."

This is absolutely correct, because the maximum of the function (1A.3) is obtained for this value of the energy difference, but it may be noticed that the probability distribution has no second moment; in other terms, the mean value of $(E'-E)^2$ is infinite. Therefore, though extremely suggestive, this argument is somewhat doubtful when one considers the perturbation as representing an interaction with a measuring apparatus. It can give a finite value for $\langle (E'-E)^2 \rangle$, as also noticed by Landau and Lifschitz, if one assumes that the interaction does not abruptly rise to a constant value at time zero, to abruptly vanish later at time Δt, but it is a smoother function of time. The physical idea behind this assumption is that one cannot prepare a measuring system acting so harshly. What we discover is that the energy uncertainty relation is consistent in the following way: if it is satisfied by the interaction with the apparatus, then it is satisfied for the measured system after their interaction. But this does not prove that one cannot prepare an apparatus with an initial and a final zero uncertainty in energy.

As a matter of fact, there exists a formal counterexample, which is given as an exercise (Problem 2) in Chapter 2: One can formally write down a strictly zero hamiltonian for a measuring apparatus M and a coupling between this apparatus and a measured system acting only during a finite time Δt. The apparatus cannot be submitted to the uncertainty relation in energy, since its energy is always strictly zero as long as it does not interact. What one finds in that case is that, if the initial wave function of the measured system is an eigenstate of energy, it is not changed at all after the measurement, and furthermore, the reading of the apparatus giving the value of the energy includes an arbitrarily small error.

Accordingly, one should be careful when stating what the inequality (1A.1) is supposed to mean. One may rely for that upon a theorem by Walter Thirring.[14] It assumes that two systems Q and M both have a continuous spectrum of energy and interact during a time Δt, which can depend upon the initial state (this may represent, for instance, the time spent by an atom in the field of a Stern–Gerlach device). Then, after a separation of the interacting and the noninteracting parts of the total wave function, as in equation (1A.2), though without assuming a perturbation expansion, one can show that the spread in energy for both Q and M satisfies the inequality

$$\Delta E \, \Delta t \geq \frac{\hbar}{8\pi}. \tag{1A.4}$$

The number $1/8\pi$ comes from a mathematical argument, and one cannot assert that it is the smallest possible bound. Most probably, it should be replaced by $1/2$ in a more precise estimate.

One can then go back to the counterexample mentioned above to try understanding why it violates the inequality (1A.4). This is because the hamiltonian of the measuring apparatus has no continuous spectrum. When this is corrected, by introducing, for instance, a kinetic energy for the apparatus, one gets a spread ΔE in the result shown by the pointer satisfying the uncertainty relation.

PROBLEMS

1. *Joint Probability Distributions*[15]

Let (Q, P) be two canonically conjugate observables ($[Q, P] = i\,\hbar I$). Let $|\psi(q)|^2$ be the probability distribution for the values of Q, and $|\phi(p)|^2$ for the values of P. One wants to construct a joint probability distribution $f(q, p)$ such that its integral over p (respectively, over q) is the quantum probability distribution for q (respectively, for p).

Show that this can be done by introducing the variables

$$u(q) = \int_{-\infty}^{q} |\psi(q')|^2 \, dq', \qquad v(p) = \int_{-\infty}^{p} |\phi(p')|^2 \, dp',$$

and choosing

$$f(q, p) = |\psi(q)|^2 |\phi(p)|^2 (1 - h(u, v)),$$

where the function $h(u, v)$ is less than 1, whatever u and v are, and such that

$$\int_{-\infty}^{+\infty} h(u, v)\, du = 0, \qquad \int_{-\infty}^{+\infty} h(u, v)\, dv = 0.$$

2. Uncertainty Relations

Consider two observables A and B with $[A, B] = iC$. Show that C is also an observable. Let ψ be the state vector of a pure state with average values $\langle A \rangle$ and $\langle B \rangle$. Define the new observables $A' = A - \langle A \rangle I$ and $B' = B - \langle B \rangle I$ and the uncertainties Δa and Δb for A and B in that state.

(i) Let λ be a real variable. Define the function $f(\lambda) = \|(A' + i\lambda B')\|^2$ and write it in terms of Δa, Δb, $\langle C \rangle$, and λ.

(ii) Show that the positivity of $f(\lambda)$ yields the general form of the uncertainty relation.

(iii) Consider the case when A and B are canonically conjugate, and show that the minimum value of the product $\Delta a \Delta b$ is realized when the wave function is gaussian.

3. A system is in an eigenstate (j, m) of angular momentum. What are the uncertainty relations for J_x and J_y? What happens when $m = j$? When j becomes large?

4. The spectrum of an observable A is continuous. Let E_J be the projector associated with the values of A in a subset J of the spectrum. Show that

$$E_J E_K = E_K E_J = E_{J \cap K}; \qquad E_J + E_K = E_{J \cap K} + E_{J \cup K}.$$

2

The Problems of Measurement Theory

This chapter is intended to provide a first encounter with *measurement theory,* which is the turning point of quantum mechanics.[1] We will first consider how some typical measurements are actually performed. Then, following von Neumann, we will see how a measurement can be described by the theory, in an idealized case, and what the main difficulties are. We will then briefly review how the resulting troublesome questions have been dealt with traditionally. The most commonly used answer will be codified according to the rules of the Copenhagen school. Finally, we shall open the way toward a more systematic approach by considering, in more detail, why quantum mechanics needs an interpretation and what it should be like.

EXPERIMENTAL DEVICES

1. What Is a Measurement?

A measurement always involves some man-made or natural devices, which are macroscopic objects consisting of a large number of atoms. Because all that happens at the atomic scale is too tenuous to be known directly, except in very exceptional cases, one needs amplifying devices to bring them up to our own scale.

The basic data of a measurement are therefore always macroscopic phenomena. We shall often use this word *phenomenon.* It comes from the Greek, where it means something apparent enough to reach perception. One can then say that measuring a characteristic of a microscopic object consists first of all in making it generate a phenomenon.[2]

A phenomenon, as it occurs in physics, is most often expressed by numbers. This is the case for the position of a pointer on a voltmeter dial or, equivalently, the numerical value of the corresponding voltage, which can also be shown by an electronic device in a digitalized form. The data are generally registered on a record in order to be used more conveniently, by either writing them down, photographing them, keeping them in a computer memory, or other means.

Physics translates these primary data into information concerning the measured object. For instance, a click of a Geiger counter is understood to mean that a charged particle has crossed the counter, though of course this assumes a great deal of knowledge about the instrument, how it is made, and how it works. It is worth noticing that, most often, no mention of quantum mechanics is made, and the statement of a datum, as well as our understanding of the apparatus, relies mainly upon classical considerations. So, when it comes to describing the conditions under which an experiment takes place, how one understands it and how a physicist acts when performing it, *everything is expressed in a phenomenological framework relying only upon common sense as it is codified by classical physics.*

A Gross Classification of Measurements

It will be convenient to distinguish between various types of measurements, if only to put some order among our ideas. We shall first distinguish between what may be called *structural* and *circumstantial measurements*.

A particle, as for instance a proton, has some structural properties that are the same for all protons: its mass, spin, electric charge, magnetic moment, charge radius, and so on. Similarly, an atom in its ground state has a large number of structural properties such as its binding energy, spin, magnetic moment, and the wave function describing its internal electrons.[3] The difference in energy between the various stationary states of an atom is also a structural property. We shall say that an experiment aimed at obtaining these quantities is a structural measurement. It is different from a circumstantial measurement, which yields a value of a quantity depending on the preparation process or, more generally, upon the past history of the system. Such is the case, for instance, when one measures the position or the momentum of a proton, or a component of its angular momentum.

This distinction is convenient, though there is no tight barrier between the two categories of measurements. Many structural measurements, the results of which now appear in catalogues, have been obtained in practice through a series of circumstantial measurements. One can approach, for instance, the knowledge of the spin of a nucleus by studying the scattering of other particles on it and measuring how the momenta of these bombarding particles are distributed. The main point of making this distinction is in fact to distinguish between what can be considered to be already known and what is brought forth by a specific experiment.

The probabilistic character of the theory should also lead us to make another distinction between the measurement of a probability and an individual measurement. Quantum mechanics can predict some precise values for specific quantities, such as the difference between the energy levels in the hydrogen atom. To check this kind of prediction is a matter of structural measurements. In many cases, however, the theory only predicts a probability. It gives, for instance, a precise prediction for the probability of Rutherford scattering of an alpha particle on an atom. One also frequently goes the other way round, the measurement of a probability allowing one to learn some feature, some fine detail, about an atom or a nucleus; for instance, about a hamiltonian. Whatever it may be, every probability measurement is necessarily the outcome of a large number of *individual* measurements, each one of them being made upon one and only one atom, one particle, one physical system, though one tries of course to realize all these measurements under identical conditions. The final outcome of the series of individual measurements is a set of empirical frequencies, which give a good estimate for the probabilities if the statistics are good enough.

2. Some Examples of Measurements

The variety of experimental devices existing is by now so large and the ingenuity of the experimentalists using them so great that one cannot attempt to review them seriously. Only a few brief indications will therefore be given here, their main interest being to make sure that later we will be talking of actual physics and not only of the dreams of theorists.

Photons and Radiation

Photons are the particles of electromagnetic radiation. In most cases, radiation is produced with a sizable intensity so that, taking into account the very small energy of photons, radiation involves a large number of them. When the intensity of radiation is greatly reduced, for instance after crossing an absorbing medium or when it extends over a large region, the average number of photons inside a detector may become of the order of unity or less. There is nevertheless some probability for the occurrence of no photon, one photon, two photons, and so on, and one should not simplify too much by assuming that radiation is in a one-photon state. It is now possible to produce systematically exact one-photon states but they are very seldom used.

One knows how to measure the direction and frequency of radiation. For instance, in the visible spectrum, one can measure a frequency by using a prism or a diffraction grating. The theory of these devices is well understood and one could even take into account the atomic content of these objects, though this does not add much. The knowledge of both the frequency and the direction of a photon allows one to get at its momentum, since one knows the velocity of light and Planck's constant. This is an example of how structural data can be used to get more complete information in the case of a circumstantial measurement.

Photons are also often detected one by one and we shall begin our discussion of measurements by considering a few photon detectors. We shall particularly stress the amplification process through which this is transformed into a manifest phenomenon. The most commonly used photon detector is the eye retina. Anatomy shows that it contains many neurons and also some cells full of pigments having the shape of a rod or a cone. The pigments are organic molecules that are able to absorb a photon in a specific frequency range and transfer its energy into molecular vibration. The molecules are contained in lamellar structures piled up one above another and this allows the vibration to be transmitted from one lamella to the next. If the vibration is strong enough (and the detection of a few photons is enough for this) its energy is converted into chemical energy by an interaction with some energy-catching molecules, which can then act upon a protein located in the wall of a neuron. The protein reacts by turning like some sort of door, opening an entry in the neuron wall through which a macroscopic number of sodium ions can pass. The result is a nervous signal that is transferred after many relays and information treatments to the brain. An eye therefore possesses the main characters of a quantum detector, namely an elementary reaction at the atomic level, an amplification process leading to a macroscopic phenomenon and, finally, a record in the brain.

Photographic Detection

The active components of a photographic emulsion are silver bromide microcrystals and the global reaction occurring at the atomic level is usually summarized by the photodissociation process $\gamma + Ag\,Br \rightarrow Ag + Br$. However, this simple explanation does not show the amplification mechanism allowing a microcrystal to behave as a measuring device for the photon. It is, as a matter of fact, a very efficient mechanism: a small number of photons (three at least, apparently) can generate the accumulation of billions of silver atoms, something obviously not explained by the simple chemical reaction just written.

The catalytic amplification effect is quite interesting and is worth describing briefly.[4] A perfect silver bromide crystal is made of ions Br^- and Ag^+ in a cubic lattice. An emulsion grain is an imperfect microcrystal containing dislocations and this lack of perfection plays an important part in the photographic process. Some defects consist in the existence of interstitial ions Ag^+, i.e., ions that are not located at normal sites in the lattice but somewhere in between. They can easily migrate from place to place while remaining in interstitial locations. Another important type of defect is provided by dislocations of the lattice. Some of them are located on the outer surface of the grain where they look like a wall standing above the surface by one atom unit, with a corner where a particularly high local electric field is produced by neighboring ions. This electric field can act as an electron trap so as to capture an electron passing by.

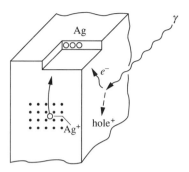

2.1. *Photographic detection.* Each incoming photon produces an electron and a hole. Charged diffusing particles (electrons and interstitial silver ions) can meet together so as to generate a silver atom in an electron trap located at the corner of a dislocation. When three such silver atoms have thus been brought alongside, any number of them become energetically favored.

When a photon enters a grain, its first effect is to produce an electron-hole so that an electron is extracted from the valence band and sent into the conduction band where it can travel easily, while a hole is also produced. The hole is a site in the crystal where a unit positive charge is present or, if one prefers, an electron is lacking and we shall leave aside its destiny, which plays no important role.

The freely moving electron will pass, sooner or later, near a trap and will get captured in the corner. Some diffusing interstitial ion Ag^+ will also arrive at the same corner, sooner or later. The ion can then capture the electron so as to turn into a neutral silver atom, which remains along the dislocation wall because of the local electric field. An important feature of the dislocation is that the electron trap is reconstituted after this neutralization and it can capture other electrons. The process can therefore go on if new photons arrive, one more silver atom and still another one being aligned along the wall.

This is where the catalytic process enters. It begins surreptitiously: when two silver atoms are neighbors in the surrounding field of the dislocation, they slightly bind as a kind of molecule. The same thing happens for three silver atoms. However, as soon as the magic threshold three is passed, there is a phase transition: it is energetically favorable for the silver interstitial ions to catch an electron in the valence band and join the other silver atoms, the three initial atoms being enough to start an accumulative process. This is a very subtle effect and the fact that the grain is in a solvent (gelatine) is essential. We shall not of course discuss it in detail, but the most interesting point concerning photographic photon detection should now be clear: an aggregate of silver atoms can grow at the expense of the population of interstitial silver ions. Finally, this is the amplification mechanism allowing the expression of a quantum event (three photons entering the grain) into a macroscopic phenomenon (billions of neutral silver atoms accumulating on the surface of the grain).

Photomultipliers

Photomultipliers are much simpler photon detectors. They rely upon a few simple physical effects—namely, the photoelectric effect, the acceleration of a charge in an electric field, and the production of radiation by an accelerated charge.

When a photon reaches the surface of a metal, it can extract an electron by the photoelectric effect. The electron is accelerated by an intense electric field so that it radiates other photons, which in turn produce more electrons through other photoelectric effects and so on, so that finally a macroscopic electric current is produced. This is of course a very schematic description of this kind of apparatus and there are several variants. The main point is that here again a macroscopic phenomenon is the signature of a quantum event.

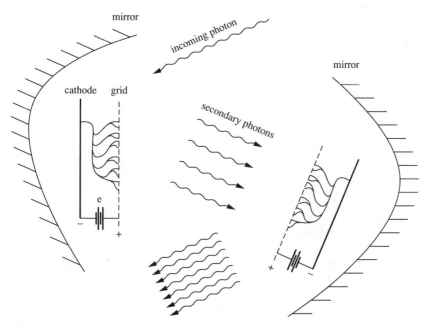

2.2. *Photomultiplier.* An incoming photon (arriving from the right) produces an electron from a cathode by a photoelectric effect. The electron is accelerated by an electric field generated by a grid and it accordingly emits new photons. A parabolic mirror directs these photons towards a second cathode so that, after a few steps, the number of electrons produced is large enough to give rise to a macroscopic signal.

The Detection of Charged Particles

Charged particles are particularly easy to produce: Electrons are extracted from a metal by a photoelectric or a thermoionic effect, the latter occurring in a heated cathode submitted to an electric field. One can also saturate matter with energy—by heating it, hitting it with a laser beam, or any other means which produces a plasma consisting of free electrons and atomic ions. These charged particles can be directed, accelerated, or decelerated by the action of electric and magnetic fields.

The detection of charged particles is usually via their ionizing properties. When a charged particle crosses a piece of matter with a sufficiently high velocity, it brings with it a moving electric field acting in a transient way upon the electrons in the nearby atoms. The effect can be strong enough to extract some electrons along the particle trajectory (ionization). The ionization rate depends upon the atoms, their density, and the velocity of the charged particle. It has been studied in great detail, both experimentally and theoretically. Each ionization draws its energy from the incoming particle, so it slows down and finally stops if the piece of matter is thick enough. The length of its track before stopping (its range) gives precise information about its initial momentum and this is by far the most commonly used method for performing a momentum measurement.

The elementary ionizations must of course be made conspicuous by means of a macroscopic phenomenon. In the case of a Geiger counter, this is realized with the help of an external electric field generated by a capacity. The counter contains a gas or a liquid in an electric field with an intensity just below electric breakdown. When a charged particle crosses the counter, it produces free electrons and ions by ionization. The electrons are accelerated by the electric field and they acquire thereby a large enough velocity to produce further ionizations. A cascade effect then occurs, through which the number of free electrons grows very rapidly (exponentially with time) and a plasma is generated in the immediate vicinity of the particle track. This plasma is highly conductive and the field produces a rapid current in it, a spark creating a signal that can be registered. The spark current discharges the capacity so that the electric field disappears for a while, the ions can recombine, the capacity is therefore regenerated, and the counter is ready for another detection.

Ionization along the track of a charged particle can also be revealed by provoking a phase transition in the surrounding medium, as in a bubble chamber. It contains hydrogen initially in a liquid state. If the pressure is slightly lowered, the liquid becomes metastable and would tend to vaporize. The chamber is however very clean and the thermal fluctuations at the temperature of liquid hydrogen are not large enough to produce the nucleating bubbles (they can produce very small bubbles consisting of a few atoms having some more space at their disposal but the surface tension of

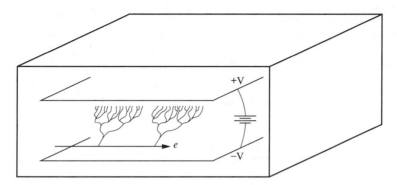

2.3. *A Geiger counter.* A charged particle ionizes a gas in a capacitor. A spark is produced by electric breakdown in the ionized medium.

the neighboring liquid gives an energy cost larger than the volume free energy of the small bubble and the bubble collapses). So, the liquid remains metastable. However, when a charged particle crosses the chamber, it will ionize the medium along its track. The resulting free electrons are able to catalyze the creation of some bubbles already big enough to overcome surface tension so that they grow in size very rapidly. One can then see a line of bubbles along the track of the particle. The process can be stopped by reestablishing the initial pressure, after taking a photograph of the pattern of bubbles. There are many variants of these methods, each one having its own technical interest, but they need not concern us.

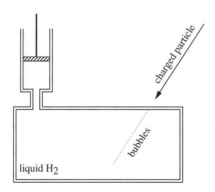

2.4. *Schematic view of a bubblechamber*

It may be noticed that this kind of detection provides a gross measurement of the particle position together with its momentum. This is not in contradiction with the uncertainty relations because the product of the errors obtained is far above the theoretical limit.

3. The Stern–Gerlach Experiment

Its Principle

Finally we mention that neutral particles can be detected by letting them produce photons or charged secondaries. One can for instance obtain a neutron detector by using a medium where some nuclei absorb the neutrons and emit positrons, which are detected as energetic charged particles.

In the Middle Ages, teachers often made use of examples that were repeated again and again. This was called a pons asinorum and it is now called a paradigm, but the idea is the same. In measurement theory, an experiment first made by Stern and Gerlach to measure a component of an atomic magnetic moment is such a paradigm. It played a crucial role in confirming the existence of spin and it still remains one of the best and clearest examples in measurement theory.

An atom with nonzero spin generally has a magnetic moment directed along its spin direction, which means that the matrix elements of the three components of these two vectors are proportional. This proportionality follows from a theorem by Wigner and Eckhart in group theory. One can also say more simply that both the spin and the magnetic moments are pseudovectors and the spin defines the only privileged direction along which the magnetic moment could have a nonzero average value. Whatever it may be, if one denotes by (M_x, M_y, M_z) the components of the magnetic moment **M** and by **S** the spin, one has $\mathbf{M} = g\mathbf{S}$, where the coefficient g is called the gyromagnetic ratio. It is accordingly the same thing to measure M_z as to measure S_z, as long as one knows g from previous structural measurements.

Let the atom be submitted to a magnetic field **B** directed along the z axis. The part of the hamiltonian depending upon the magnetic moment is given by

$$H_{\text{magn}} = -\mathbf{M} \cdot \mathbf{B} = -g\mathbf{S} \cdot \mathbf{B} = -gBS_z. \tag{2.1}$$

Since the hamiltonian for the free atom, which remains in its ground state, does not depend upon the spin direction, it will be enough to work with the hamiltonian (2.1) in order to find the motion of the atom. Using the Heisenberg representation where the operators vary with time, we have

$$\frac{d\mathbf{S}}{dt} = i\hbar[H_{\text{magn}}, \mathbf{S}(t)].$$

The commutator is 0 for the observable S_z, which is therefore a constant of motion.

THE PROBLEMS OF MEASUREMENT THEORY

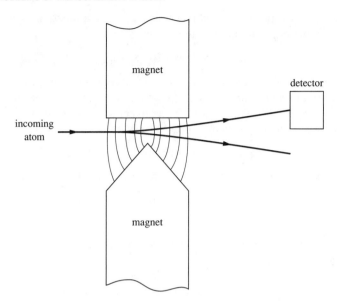

2.5. *Schematic view of the Stern–Gerlach experiment*

The Stern–Gerlach device uses a magnet with asymmetric polar caps between which the atom is sent. One cap is planar and the other one is wedge-shaped. This configuration results in a magnetic field that is normal to the planar cap (say, for definiteness, along the z axis); its only nonzero component B_z depends on the coordinate z and very little on x and y. Under these conditions, the hamiltonian (2.1) also depends on the z coordinate of the atom, through the value of the magnetic field. This gives rise to a force, which according to Ehrenfest's theorem is given by

$$\mathbf{F} = -\nabla \langle gB(z)S_z \rangle = -g\frac{\partial B}{\partial z}\langle S_z \rangle \mathbf{e}_z,$$

where $\mathbf{e}_z$ denotes a unit vector in the z direction. This force is directed along the z axis and it depends upon the z component of the spin. The average $\langle S_z \rangle$ is taken in the state of the atom and we have seen that it is a constant of motion.

Let us consider for definiteness the simplest case where the value of the spin is $1/2$. Also assume that the initial spin state is a pure state as a result of some preparing process. The atom's internal degrees of freedom remain unchanged during the experiment and the atom remains in its ground state. One can therefore represent this atom by a wave function describing only its center-of-mass position **x** and its spin state, without mentioning its internal wave function. The initial state is given by

$$|\psi\rangle = \phi(\mathbf{x})\exp(i\mathbf{p}\cdot\mathbf{x}/\hbar)|\gamma\rangle, \qquad (2.2)$$

where $\phi(\mathbf{x})$ represents a localized wave packet, **p** is the average initial momentum of the atom and $|\gamma\rangle$ is a spinor one can always write as $|\gamma\rangle = a|+1/2\rangle + b|-1/2\rangle$ using a basis of eigenvectors $|\pm 1/2\rangle$ for S_z. It will be convenient to write the wave function (2.2) as $|\psi\rangle = a|\psi_+\rangle + b|\psi_-\rangle$, where

$$|\psi_+\rangle = \phi(\mathbf{x})\exp(i\mathbf{p}\cdot\mathbf{x}/\hbar)|+1/2\rangle = \psi_+(\mathbf{x})|+1/2\rangle.$$

The Schrödinger equation for the atom is given by the hamiltonian

$$H = -\frac{\hbar^2}{2m}\Delta - gB(z)S_z,$$

so that it can be easily separated into two different equations for the two components of spin, which gives

$$i\hbar\frac{\partial}{\partial t}\psi_\pm(x,t) = -\frac{\hbar^2}{2m}\Delta\psi_\pm \pm (-gB(z)\,\hbar/2)\psi_\pm. \qquad (2.3)$$

One can apply Ehrenfest's theorem to each one of these Schrödinger equations. The first one gives a classical motion where the atom feels a force directed upwards (towards the positive values of z) when it crosses the magnetic field and, when the atom leaves the magnet, it follows a straight-line trajectory directed upwards. The second equation (2.3) gives the same results, except that the force and the final motion are now directed downwards. So, after crossing the magnet, the complete wave function has become a sum of two parts associated with the two different spin states, each one of them being localized in the vicinity of a different trajectory and their respective probability amplitudes being still given by a and b.

If some detectors are located along each trajectory, one expects that a detector located along the upper one will only be sensitive to the component ψ_+ of the wave function. This is due to the fact that the part ψ_- vanishes in the region covered by this detector, so that it cannot give rise to an interaction with the atoms in it. The probabilities for an interaction with each detector are given by $|a|^2$ and $|b|^2$ and, when the upper detector reacts, one says that the individual measurement has shown that S_z is equal to $+\hbar/2$ (or $+1/2$, Planck's constant being often not written in this context).

This is of course not quite a trivial statement. Does it mean that S_z is equal to $+\hbar/2$ when the atom is entering the detector, or already before entering the magnet? Bohr and Heisenberg considered that this was a meaningless question. One can state the value of the spin at the time where the measurement occurs and one should refrain from any statement about what happened before. Even this is not without ambiguity, since the measurement

THE PROBLEMS OF MEASUREMENT THEORY

takes some amount of time. We shall, however, not let this kind of difficulty stand in the way and we shall accept the statement that a measurement of spin has taken place without further ado (at least for the time being). We shall also take it for granted that S_z is still equal to $+\hbar/2$ when, for instance, the atom leaves the apparatus in the absence of detectors but the fact that the upper trajectory has been followed is ensured by the interaction of the atom with some other particle located along this track. The Stern–Gerlach apparatus is therefore used in that case to produce an atom in a given spin state. There is accordingly no fundamental difference between a preparing process and a measurement, at least in this case.

The Necessity of a Detector

One may wonder what happens when there is no detector or no reaction to reveal which way the atom went. Could it not be said that the splitting of the wave function is enough by itself to make sure that a measurement has taken place, even though one does not know the result? After all, the two components of the wave function are a very large distance apart somewhere behind the magnet.

This is not at all enough. Following Wigner,[5] let us assume that the initial state of the atom is a pure state with $S_x = \hbar/2$ and it crosses a Stern–Gerlach magnet with an axis along the z direction with no detector around. We also assume that a well-devised magnetic field can bring the two trajectories together, merging them into a unique trajectory as shown in Figure 2.6. Place a second Stern–Gerlach magnet along the final common part of the trajectory, with its magnetic field along the x direction (i.e., the initial direction of the spin), a detector this time being located behind. It is easy to solve the Schrödinger equation for this problem since the atom interacts

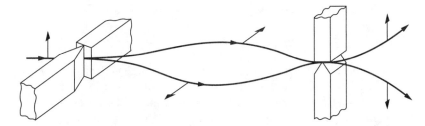

2.6. *Wigner's ideal experiment.* A spin-1/2 atom has initially its spin oriented along the x direction. After crossing a Stern–Gerlach device with a magnetic field in the z direction, the two paths along which the atom can travel are brought back together by a suitable external field. The resulting final unique path crosses a second Stern–Gerlach device directed along the x direction.

only with an external magnetic field (the result is also rather obvious without any calculation): the atom must be again finally in the pure state where S_z is equal to $+\hbar/2$ with probability 1 at the time of detection. The Schrödinger equation therefore tells us that the excursion of the atom along two well-separated trajectories had no consequence. This result unfortunately has to be accepted without an experimental check because the experiment is too difficult to perform.

On the contrary, when detectors are located behind the first magnet in Wigner's device, one usually considers that the atoms are prepared by the first measurement for the second one. This preparation is supposed to be similar to the kind of random process we used in Chapter 1 to introduce the density operator. The population of atoms is then described by a statistical mixed state with a density matrix

$$\frac{1}{2}\{|S_z = \hbar/2\rangle\langle S_z = \hbar/2| + |S_z = -\hbar/2\rangle\langle S_z = -\hbar/2|\} \qquad (2.4)$$

The probability of getting the result $S_x = +1/2$ in the second measurement is equal in that case to $1/2$. There is therefore apparently a fundamental difference whether or not a macroscopic detection phenomenon has taken place before a measurement. This conclusion might be criticized by saying that our justification of the density matrix (2.4) is rather shallow. As a matter of fact, equation (2.4) is more properly a consequence of wave packet reduction, and one has therefore already entered the domain of interpretation.

ELEMENTARY MEASUREMENT THEORY

4. Von Neumann's Formal Theory of Measurements

An elementary account of a measurement will now be given following an approach due to von Neumann.[6] The measurement is considered as an interaction between two systems, both obeying quantum mechanics.

A Measurement as an Interaction

It may be considered that a measurement proceeds schematically as follows: A physical system Q is initially in a given quantum state. This is the system to be measured and one is interested in the value of an observable A to be obtained from an individual measurement. The measurement is made by another physical system M, the measurement device, which involves some macroscopic elements such as electronic equipment, detectors, and registers.

The measuring apparatus is a macroscopic object, which is supposed to be well understood according to classical physics. One knows however that it is made of atoms and, as such, it is also basically described by quantum mechanics. We shall admit that its apparent classical behavior is only a large-scale manifestation of its quantum dynamics according to Ehrenfest's theorem. The measurement therefore appears as the result of an interaction between two physical systems Q and M, the measured one and the measuring one, both obeying quantum mechanics.

As long as it remains isolated, the measured system Q is described by a Hilbert space $\mathbf{H}_Q$ and a state vector ψ (or a density operator ρ), and its dynamics is governed by a hamiltonian H_Q. The measuring apparatus is described by a Hilbert space $\mathbf{H}_M$, a state vector ϕ (or a density operator), and a hamiltonian H_M. The observable A is associated with a self-adjoint operator acting on the Hilbert space $\mathbf{H}_Q$, and it will be assumed for convenience that it has only nondegenerate discrete eigenvalues $\{a_n\}$.

The two systems should be treated together as a larger system, since the measurement must result from their interaction. One knows how to do this by using for the system $Q + M$ the Hilbert space $\mathbf{H} = \mathbf{H}_Q \otimes \mathbf{H}_M$. As long as the two systems Q and M do not interact, their total hamiltonian is given by $H_0 = H_Q + H_M$. Their interaction should come from a coupling term in the hamiltonian and one will therefore assume the complete hamiltonian to be $H = H_0 + H_C$, where the second term H_C represents the interaction between Q and M.

What Interaction for a Measurement?

One may wonder what peculiarities of the coupling are responsible for the fact that M is effectively a measuring device for the observable A. This is the question we will consider first, though not aiming at a general theory of measurement but only as an academic example showing the main features of the problem.

Assume that the apparatus involves, among other elements, a pointer that can move on a dial, its final position providing the experimental datum. The position of the pointer will be denoted by X. This is an observable belonging to M, among many others. One assumes that the pointer is at rest as long as no interaction takes place. This means that the operator X commutes with the hamiltonian H_M and X is then a constant of motion. Before the measurement, the pointer is in its neutral position. This property can be described by taking the initial wave function of M to be a product $\phi(x, y') = \eta(x)\chi(y')$, where y' denotes variables commuting with X and where the pointer wave function $\eta(x)$ is a rather sharp wave packet centered around the average value $x = 0$. Since the variables y' describing the detailed internal working of M play only an accessory role in the present discussion, it will be enough to use the shorter notation $\phi(x)$ to represent the complete wave function of the apparatus.

Frequently the beginning of the interaction between Q and M is due to the motion of the Q system when it enters the active region of M. For instance, in a Stern–Gerlach experiment, the motion of the atom is responsible for its entering the region where the magnetic field is acting, resulting in an interaction between this field and the atom. The atom's position and its spin variables are decoupled before that interaction and the experimentalist can control the arrival of the atom in the field without disturbing the spin state. One tries to model this kind of situation by assuming a coupling $H_C = -g(t)PA$, where P is an observable canonically conjugate to the position X of the pointer, i.e., $[X, P] = \hbar I$. One knows that the operator P is the generator of pointer translations and the coupling is chosen so that this displacement is controlled by the value of A. The function $g(t)$ represents an activation factor for the coupling. It will be convenient to assume that it is 0 outside the time interval $[-\varepsilon, +\varepsilon]$.

It should be stressed that one does not presume that any realistic measuring device is well described by such a simple representation. We shall have other occasions to reconsider this question in later chapters. The present model has no other goal than to make explicit the main features and problems of measurement theory.

How the Apparatus Works

It will now be shown that this idealized model correctly represents a measuring apparatus, in so far as the final position of the pointer is effectively a good indication for the value of the observable A.

Let us assume, to makes things even simpler, that A is a constant of motion for Q. Then consider the case when the initial state of Q is an eigenstate of A with the eigenvalue a_n, so that it can be written in Dirac's notation as $|a_n, r\rangle$, r being an index representing the variables not occurring directly in the interaction. As a matter of fact, one can leave aside such parameters, which only bring notational complexity without shedding light upon the problem at hand. Let us then see what occurs:

Before the interaction, the global state is simply $|\alpha\rangle = |a_n\rangle \otimes |\phi\rangle$. Its evolution is given by the Schrödinger equation

$$i\hbar \frac{d}{dt}|\alpha(t)\rangle = H|\alpha(t)\rangle.$$

This can be solved by $|\alpha(t)\rangle = U(t)|\alpha(0)\rangle$, but one should now take into account the time dependence of the hamiltonian in order to compute the evolution operator $U(t)$. It obeys the Schrödinger equation

$$i\hbar \frac{d}{dt}U(t) = H(t)U(t).$$

As long as the interaction does not yet take place, i.e., before the time $-\varepsilon$, H is reduced to the free hamiltonian H_0 so that one has simply $U(t) = \exp(-iH_0 t/\hbar)$. Between the times $t = -\varepsilon$ and $t = +\varepsilon$, when $g(t)$ is nonzero, the intense signal $g(t)$ dominates all other effects and it modifies $U(t)$ according to

$$U(\varepsilon) = \exp\left(-\frac{i}{\hbar}\int_{-\varepsilon}^{+\varepsilon} H(t)\,dt\right) U(-\varepsilon) \approx \exp(i\lambda PA/\hbar) U(-\varepsilon), \quad (2.5)$$

where

$$\lambda = \int_{-\varepsilon}^{+\varepsilon} g(t)\,dt.$$

When acting upon an eigenstate of A with eigenvalue a_n, the exponential in the last term of equation (2.5) acts like $\exp(i\lambda P a_n/\hbar)$, which is the operator generating a translation of x over a distance λa_n. One has therefore

$$\exp(i\lambda PA/\hbar)\phi(x)|a_n\rangle = \phi(x + \lambda a_n)|a_n\rangle. \quad (2.6)$$

The wave function of the apparatus at the end of the measurement is therefore $\phi(x + \lambda a_n)$, i.e., a narrow wave packet centered around the position λa_n. Taking the amplification coefficient λ large enough, one can clearly separate (in principle) the final positions of the pointer associated with different eigenvalues of A. This means that the final position is a very good indication for the value of A that was realized in the input.

Reformulating the Results

Since this analysis shows that one can describe in principle a measuring process in the framework of quantum mechanics, it will now be convenient to eliminate all the inessential features of this description in order to see more clearly the questions it raises.

The simplest approach for so doing is to suppose that the pointer belongs to a voltmeter and to replace the old-fashioned dial voltmeter by a digitalized one. The apparatus is then able to show discrete numbers, which may be considered as the eigenvalues of some observable B belonging to M. Among the numbers that can be shown, we are only interested in those representing a digitalization of the data λa_n and we shall denote them by b_n, calling b_0 the initial number that was shown before the measurement.

The initial state of M is then $|b_0 r\rangle$, r representing uninteresting stuff about all the unessential variables in the wave function. The measurement process corresponds to the transition

$$|a_n\rangle \otimes |b_0 r\rangle \rightarrow \sum_{r'} c_{rr'}^{(n)} |a_n\rangle \otimes |b_n r'\rangle, \quad (2.7)$$

where the left-hand side represents an initial state of Q with the eigenvalue a_n for A and the neutral state of M. The right-hand side represents the situation after the measurement: Q is still in an eigenstate of A with the same eigenvalue and the record now shows the data b_n. Many things may have been modified inside the apparatus but this is included in principle in the coefficients $c_{rr'}$, which, by the way, are time-dependent. The important point is that the data should be clearly recorded, which assumes that B is a constant of motion for M when it is again isolated, i.e., B commutes with H_M.

Linearity of the Schrödinger equation implies that when the initial state of Q is not an eigenstate of A but an arbitrary superposition

$$|\psi\rangle = \sum_n \mu_n |a_n\rangle,$$

the final state is given by

$$\sum_n \sum_{r'} \mu_n c_{rr'}^{(n)} |a_n\rangle \otimes |b_n r'\rangle. \tag{2.8}$$

This troublesome expression contains the essential problems of measurement theory and we shall therefore have to analyze it in detail. However, before proceeding to the heart of the problem, it will be useful to notice a few consequences of what has already been obtained.

Taking the vectors $|b_n r>$ to be orthonormal, the conservation of norm resulting from the Schrödinger equation gives, when taking into account equation (2.7),

$$\sum_{r'} |c_{rr'}^{(n)}|^2 = 1. \tag{2.9}$$

Then one can ask what is the quantum probability to get the datum b_n in the final state whose wave function is given by the quantity (2.8). This probability is given by

$$\sum_{r'} |\mu_n|^2 |c_{rr'}|^2 = |\mu_n|^2,$$

which is the probability one assumed in Chapter 1 for the value of A to be a_n in the state $|\psi\rangle$. The measurement therefore has as a consequence that the abstract Born probability for A to have the value a_n is transferred to the probability of a phenomenon, namely that the record shows the data associated with this value.

The Crux of the Problem

One can try finally to simplify the formalism still further: The expression (2.8) for the final state will be useful when one considers the measurement as a preparation process. For the time being, it is still unduly complicated and a last impediment is still hiding the result.

It often happens that, when a measurement has been made on an atom, the atom gets lost in the walls of the apparatus. One might say that the apparatus has changed a little and its description needs a few more observables to be added to its already innumerable ones. So call M' this new apparatus to give some appearance of rigor and denote its eigenstates by $|b_n, r\rangle_{M'}$. We say that this is an experiment with a *lost measured system*.

In that case, the measuring process can be summarized by the transition

$$\sum_n \mu_n |a_n\rangle \otimes |b_0 r\rangle_M \to \sum_{nr'} \mu_n c_{rr'}^{(n)} |b_n r'\rangle_{M'}, \qquad (2.10)$$

where now the heart of the problem appears in full light: A *linear superposition of macroscopically distinct states of the measuring apparatus is obtained.*

5. The Problem of Macroscopic Interferences

What Goes Right

To realize a measurement, for instance with a Stern–Gerlach device, one successively sends a large number of identically prepared atoms towards the apparatus and registers the various data. One then finds two marked features: (i) the data occur in a random fashion so that their succession shows no regularity nor any correlation; (ii) the frequencies of the events are in agreement with the theoretical probabilities so that, in our notation, the frequency of a datum b_n is practically equal to the theoretical probability $|\mu_n|^2$.

It would then seem natural to assume that one is observing an ordinary random process where each individual event is due to unknown causes, which remain inaccessible but which are nonetheless described by ordinary probability calculus. Let us then examine whether this point of view is correct.

It will be convenient to simplify the density operator in which we are interested. We leave aside the degeneracy indices r', which play no role as long as one only observes the position of the pointer. It is then sufficient to keep the reduced density operator ρ_{red} that is obtained from the full density operator ρ of the apparatus in its final state by performing a partial trace upon the unobserved degrees of freedom, i.e.,

$$\langle b_n | \rho_{\text{red}} | b_{n'} \rangle = \sum_{r'} \langle b_n r' | \rho | b_{n'}, r' \rangle.$$

It is clear that the two operators ρ and ρ_{red} give the same results for the statistics of the data B.

If one assumes that ordinary probability calculus correctly describes the data, the information about their frequencies should be enough in principle to compute the reduced density operator directly. As a matter of fact, this is precisely the problem we solved in Chapter 1 when introducing the density operator. Since the probability of finding the datum b_n is $|\mu_n|^2$, one would expect to have

$$\langle b_n|\rho_{\text{red}}|b_{n'}\rangle = |\mu_n|^2 \delta_{nn'}. \tag{2.11}$$

This formula can be said to express the experimental results when one assumes that the apparatus is described by quantum mechanics and nevertheless the distribution of data is described by ordinary probability calculus, which assumes the different data to be clearly separated events.

Whether this common sense prediction is true or not can be checked by a direct computation of the reduced density operator. Since one knows explicitly the final state $|f\rangle$ of the apparatus, as given by equation (2.10), it is easy to obtain the full density operator $\rho = |f\rangle\langle f|$ and to compute the reduced density operator as its partial trace. A first feature of these results is quite satisfactory. It is concerned with the diagonal matrix elements for which equation (2.9) gives

$$\langle b_n|\rho_{\text{red}}|b_n\rangle = |\mu_n|^2, \tag{2.12}$$

in good agreement with the expected formula (2.11)

What Goes Wrong

Things are unfortunately much less satisfactory when the nondiagonal matrix elements are considered. Actually, the general formula, of which equation (2.12) is a special case, gives

$$\langle b_n|\rho_r|b_{n'}\rangle = \mu_n^* \mu_{n'} \sum_{r'} c_{rr'}^{(n)*} c_{rr'}^{(n')}. \tag{2.13}$$

It is clear that, contrary to what was expected from equation (2.11), the nondiagonal matrix elements (for $n \neq n'$) have no obvious reason to vanish. The existence of these nondiagonal elements shows that, at least at first sight, one cannot consider the final state of the apparatus to be correctly described by ordinary probability calculus or, in other words, as consisting of clearly distinct events.

To see the origin of this difficulty in a more physical way, let us consider the average value of an observable C belonging to the apparatus but not commuting with B. One might for instance look at the velocity of the pointer rather than its position. In that case, the "common sense" formula (2.11) gives

$$\langle C \rangle = \sum_n |\mu_n|^2 \langle b_n|C|b_n\rangle,$$

whereas the exact formula (2.13) gives

$$\langle C \rangle = \sum_{nm} \mu_n^* \mu_m \langle b_m | C | b_n \rangle.$$

This is exactly the kind of difference one meets in the interference theory, when one compares the points of arrival of a tennis ball going through two windows in a wall with the interference pattern of elementary particles. The interference pattern shows the existence of crossed terms in the probability, corresponding to nondiagonal terms in the density operator. The difficulty one finds now is therefore due to the existence of interference effects in the apparatus, which occur despite the fact that it is tremendously bigger than a photon or an electron. *The theory predicts possible interferences between macroscopically different states of the apparatus that are associated with different data.*

It is very difficult to have enough imagination to envision how this kind of situation would look. Maybe our ordinary vision of reality would be something like a photograph where several different pictures have been superimposed. As a matter of plain fact, such a situation has never been seen by any normal person. This should obviously carry much more weight than any kind of calculation; it is enough by itself to make us very cautious about the present results.

Some Ways Out

It was shown that the simple diagonal formula (2.11) rests upon three assumptions: the apparatus is well described by quantum mechanics, each individual event is well defined, and the totality of these events is correctly described by ordinary probability calculus. The last two assumptions cannot be wrong since the second one boils down to assuming the existence of experimental facts and the last one has been used as the basis of the multitude of experiments confirming the predictions of quantum mechanics.

Could it be that the first assumption is wrong, in which case the experimental data shown by an apparatus would not be described by quantum mechanics? This is the solution that was put forward by Bohr to become a rule for the Copenhagen school[7]—namely, *a macroscopic object is not subjected to the rules of quantum mechanics but only to classical physics.* It removes the difficulty, but it raises immediately a host of further questions: When is an object big enough to be classical rather than quantum? What happens near the transition between the quantum domain and the classical one; by what laws of physics is that transition described? Can one expect a formulation of physics standing upon two different categories of laws to be logically consistent?

This answer is not the only conceivable one. Another much simpler idea would be to assume that the complicated sum (2.13) over the degeneracy indices is always practically zero. This would be clearer if one replaces these abstract degeneracy indices by a complete set of observables commuting with B and among themselves. The sum would then be taken over their spectrum and one immediately realizes how rich and complex are the terms in this sum since they represent in full detail the quantum behavior of everything in the apparatus. Moreover, its vanishing would be just another of the frequent instances of destructive interference.

This answer is not easy to justify and it even looks as if there were a simple counterexample: consider the case when the apparatus M is at a zero absolute temperature so that its matter is necessarily in its ground state and the summation is reduced to a unique term, which cannot vanish.

Despite this apparently discouraging remark, this is nevertheless the correct answer as will be shown in Chapter 7. It will have to be investigated in much more detail before we will be able to show that the internal states of matter in an apparatus are orthogonal when they are coupled with two different macroscopic motions. More properly, they are not necessarily orthogonal, as the previous counterexample shows; they *become* orthogonal after a very short time as a consequence of Schrödinger's dynamics. This is the *decoherence* effect by virtue of which macroscopic interferences are destroyed. More than forty years were necessary for the effect to be well understood and its orders of magnitude estimated,[8] though this is one of the most efficient and rapid effects in physics. Curiously enough, its efficiency was responsible for its lack of observation: it had always already occurred before one could try to catch it. Hence one had to resort to theory rather than experiment in order to assert its existence, at least until very recently.

Two other categories of answers have also been proposed to this most basic problem of measurement theory:

The first one assumes the existence of *hidden variables*. These variables are supposed to provide the true description of reality and they either replace the description by the wave function or they add up to it. They are classically deterministic and their existence is the hidden cause of randomness in quantum measurements, in which probabilities then enter as usual as coming from an unknown or inaccessible mechanism. This implies that the description of the measured and measuring systems by their wave functions is only an incomplete representation of physics. Von Neumann had thought he could prove the impossibility of this assumption,[9] but he relied upon rather strong assumptions that were circumvented thirty years later by John Bell.[10] This approach will be considered in some more detail in Chapter 9.

Another "answer" was put forward by von Neumann[11] and by London and Bauer.[12] It was also supported by Wigner.[13] It assumes that a result becomes only actually "known" when the consciousness of an observer has registered it. This act of consciousness would not be described by physics and it would be able to break the chain of physical laws. This highly idealistic approach has fascinated many people in spite of its ad hoc character and there is even some irony in having von Neumann as one of its supporters. Von Neumann was later to become one of the founding fathers of computing science and, by now, most experimental data are recorded in the memory of a computer before anything else, with no intervention of a conscious observer. It is then hard to believe that, by looking at some listings, I suddenly elevate them from potentiality to actuality and the power of my brain brings forth the uniqueness of reality. This approach had additional difficulties in that it overemphasized the information content of the wave function and the role of the observer, both giving rise to some sort of anthropocentrism in physics.

THE COPENHAGEN INTERPRETATION

6. The Prescriptions of Measurement Theory

Bohr's Interpretation, or the Copenhagen Interpretation

What is called the Copenhagen interpretation is mainly the work of Niels Bohr. It came partly out of discussions with his many collaborators and visitors at the Copenhagen Institute. Bohr expressed himself on the subject at various meetings and later published several articles and comments,[14] but he never wrote a systematic and complete exposition of his views. These systematic explanations, so necessary for both guiding the work of physicists and the teaching of physics, were first given by some of his collaborators. They eventually became common knowledge with their appearance in various textbooks, where each author tried to explain them according to his own inclinations.

The general trend of these books is remarkable in that, even now and so long after the discovery of quantum theory, one still finds explicit references to the few writings by Bohr—commenting upon them, interpreting them, and retaining them as a necessary reference. Such an attitude had never been seen before or elsewhere in physics, or at least it had never lasted long. Ordinarily, a physical theory is presented in a sufficiently precise and consistent way so that one does not feel it necessary to quote the founders; rather one only strives to maintain their spirit and inspiration. Contrary to this sane custom, the most complete books devoted to the Copenhagen interpretation are all reprints of original articles or learned commentaries, becoming more and more commentaries upon commentaries as time went on. They devote much space to an endless discussion of the difficulties facing interpretation, often involving arguments where the philosophy of science becomes more important than physics itself.

Recent progress allows one to at last get rid of this kind of gloss, though it does not diminish our awareness of Bohr's genius. The discovery of new, more deductive methods together with the use of techniques that were developed in the interim, have shed a new light upon the problems he saw so acutely. Whatever changes were necessary, he had practically always made the right guess.

One may wonder whether Bohr, cocreator of quantum mechanics throughout his life, had not acquired an insight far deeper than what the linear trend of a discourse can render. Maybe he was the one who "saw" best what was the right interpretation, though explaining it from A to Z was too difficult and perhaps even impossible with the means at his disposal—impossible because some obstacles needed the work of many people over decades and the use of mathematical techniques that were unknown at his time. Perhaps the right interpretation was accessible to the intuition of a thinker who could throw a bridge over an abyss which was familiar to him as to nobody else. If this guess is true, it might help one to understand why Bohr was so firm in his statements and so confident in their validity, whereas he could almost never produce a satisfactory systematic proof to justify them.

The later writings by his many friends, followers, and interpreters are however far from showing complete unity and they disagree over some important points, sometimes to the point of contradiction. Hence there does not exist a unique Copenhagen interpretation but various more or less complete versions of it. We shall not attempt the task of pursuing their differences or discussing their details and, in order to find our way in their labyrinth, we shall rather follow a lucid analysis of the works by Bohr, Heisenberg, and Pauli, as given by Hans Primas.[15]

Measurement Prescriptions

What is most important in practice is of course to state the rules permitting a correct use of experimental results. This is the zero level of interpretation, where it appears as a necessary connection between the calculations coming from the theory and the experimental data. Because of both their lack of derivation from a deeper foundation and the problems they raise, we shall call them prescriptions.

These prescriptions are mainly concerned with so-called "ideal" experiments, similar to the ones that were described by von Neumann's theory in which each eigenvalue of an observable is associated with a specific datum. More precisely, one assumes the following repetition rule: *If one repeats immediately the same measurement after having performed an individual measurement for an observable A, one must get again the same result.*

According to the formalism of the previous section, this means that, if the state of the measured system Q is initially an eigenstate of A before the first measurement, it must still be in the same eigenstate at the end of the measurement. This is also what Pauli[16] called a measurement of the first kind or of *Type I*. Lüders[17] proposed to make this notion of ideal measurement more precise by giving it a mathematical form: if $|\psi\rangle$ is the wave function of the measured system Q before a measurement, then the Q component of the final wave function multiplying a wave function for the apparatus showing the datum b_n is given by

$$E(a_n)|\psi\rangle, \tag{2.14}$$

where $E(a_n)$ is the projector associated with the measured eigenvalue a_n.

One should stress that this condition can be checked in principle by solving the Schrödinger equation describing the interaction between the measured system and the apparatus. It is satisfied in the special case of the measurements described previously. Although this category of measurement is rather narrow and far from covering all the realistic cases, it is well suited to a discussion of the principles involved in measurement theory and it will be good enough for our present purpose.

One can then state the rule giving the probability distribution of the experimental data:

Prescription 1. (Probabilities) *The probability of finding the value of the measured observable A in some real domain J is given by the average value of the projector associated with that observable and that interval, i.e.,*

$$p(J) = \langle \psi | E(J) | \psi \rangle. \tag{2.15}$$

All that needs to be added from an empirical standpoint is a second prescription expressing what happens during a sequence of measurements, i.e., how to predict the result of the measurement of a second observable after a first measurement has been done:

Prescription 2. (Wave Packet Reduction) *Suppose one performs a series of measurements of an observable A over a sample of identically prepared systems in the state $|\psi\rangle$ and this is followed by a measurement of another observable B upon a subsample for which the first measurement gives a value of A in a domain J. Then the probability of finding the value of B in a domain K is given by*

$$p(K) = \langle \phi | F(K) | \phi \rangle, \tag{2.16}$$

where $F(K)$ is the projector associated with the domain K and the observable B. The so-called reduced wave function ϕ to be used in equation (2.14) is given by $|\phi\rangle = CE(J)|\psi\rangle$. The coefficient C insures that the state vector ϕ is normalized so that $|C|^{-2} = \langle \psi | E(J) | \psi \rangle$.

When the initial state is described by a density operator ρ rather than a wave function, equation (2.15) must be replaced by $p(J) = \text{Tr}(E(J)\rho)$, together with a similar form of equation (2.16), where the outgoing density operator is given by

$$\rho' = CE(J)\rho E(J) \quad \text{and} \quad C^{-1} = \text{Tr}\{E(J)\rho E(J)\}.$$

These two prescriptions are enough for the practice of physics. The first one is valid whether or not the measurement is of Type I, whereas the second one must assume explicitly that the first measurement is of Type I. A very large number of experiments have confirmed their validity and their practical value. There was only one trouble—nobody could understand why they held and what they meant.

What Is There to Understand?

The difficulties arising from the first rule have already been mentioned: It ignores the quantum superpositions of macroscopically distinct states of the apparatus and proceeds as if they did not exist.

The second rule is what is commonly called the reduction of the wave packet or, for short, *reduction*. It is also sometimes called the collapse of the wave function. It amounts to admitting that the superposition (2.8) of different states of the apparatus at the end of a measurement can be restricted to the unique one among them corresponding to the observed data, namely

$$|\phi\rangle = C|a_n\rangle\left(\sum_{r'} c^{(n)}_{rr'}|b_n r'\rangle\right).$$

It is clear that this prescription cannot follow directly from the Schrödinger equation, if only because the coefficient C depends nonlinearly on the initial wave function whereas the Schrödinger equation is linear. This transition therefore does not belong to elementary quantum dynamics. But it is meant to express a physical interaction between the measured object and the measuring apparatus, which one would expect to be a direct consequence of dynamics. Thus reduction appears as one of the major problems in the theory.

It remains that, as a practical rule, it is in very good accordance with experiments. As a matter of fact, when other measurements are performed on the object Q after a first one has been made, the reduction rule predicts perfectly well the frequencies of the results and this has been checked in innumerable experiments involving atoms, molecules, solids, radiation, and elementary particles. The objective is not therefore to replace the rule but to understand it.

7. The Copenhagen Epistemology

Stating the Theses

For a long time, the main trend of physics was not to consider the difficulties in the interpretation of quantum mechanics as accessible to the theory. Most people did not expect that theory would be able to solve them by its own means. The problem was rather believed to be an occasion for revising the nature of science itself, as well as the philosophic theory of knowledge underlying it. Not all the founding fathers, however, agreed with this point of view. Some of them, like Einstein and Schrödinger, believed that something was still missing in the theory, that it was incomplete. Whether it meant that new concepts should be added to it or that the ones already existing had to be improved, these were two possible options and each of them could be pursued without excluding the other. As for Dirac, he believed that these problems would finally find a theoretical solution but the times were not yet ripe for it.

We now report on the point of view of the majority by reviewing what the Copenhagen interpretation had to say. It will be given here, as already mentioned, by following the synthesis offered by Primas. Here are the theses of the Copenhagen interpretation as he reviewed them:

1. The theory is concerned with individual objects.
2. Probabilities are primary.
3. The frontier separating the observed object and the means of observation is left to the choice of the observer.
4. The observational means must be described in terms of classical physics.
5. The act of observation is irreversible and it creates a document.
6. The quantum jump taking place when a measurement is made is a transition from potentiality to actuality.
7. Complementary properties cannot be observed simultaneously.
8. Only the results of a measurement can be taken to be true.
9. Pure quantum states are objective but not real.

These theses need some comments to clarify their meaning and, eventually, to pinpoint their weaknesses. We will take the liberty to say briefly "Copenhagen" in order to avoid a fastidious repetition of locutions such as "the Copenhagen school" or "the Copenhagen interpretation." We even avoid quoting Bohr explicitly since this is not an occasion for scholarship but only for discussing some concepts.

A Theory of Individual Objects

Copenhagen considers that the theory applies to an individual physical object, which can be identified. The recent continuous observations of a unique atom support this point of view. An opposite viewpoint considers that, since the theory deals with probabilities, it must proceed as done in statistical mechanics where, when using Gibbs's formulation, one only refers to a collection of identically prepared systems. This approach is clearly at variance with classical mechanics for which (at least) big objects are clearly identified as unique. It would therefore exclude the classical realm from the quantum one, since it would ask, for instance, for a collection of identically prepared copies of the planet Earth before asking whether its existence and its behavior is consistent with quantum mechanics.

Probabilities Are Primary

This point was already mentioned in Chapter 1 when probabilities were said to be intrinsic, which is the same idea: The probabilities of quantum theory do not reflect some ignorance of the observer or of the theorist but a characteristic feature of Nature itself or, if one prefers, of the relation between reality and its mathematical description by the theory. It is impossible to predict anything except probabilities, as far as circumstantial experiments are concerned; one can say that the theory is complete when it predicts these probabilities for all conceivable measurements.

The Frontier between the Observer and the Observed

According to Heisenberg, the physical world can be split in two parts, the observed object and the observing system. A frontier separates the modes of description that one should use for each one of them—quantum or classical, respectively. This point can be made clearer by going back to the detection of a photon by the retina. There is no doubt that at each end of the process the description is well defined: the photon must be treated by quantum mechanics and the nervous signal is an electric current, which is a classical concept. But where is the frontier? One might describe a pigment molecule by quantum mechanics and this is what one does when trying to compute its absorption band from first principles. One might also consider it as working in an essentially deterministic, classical, manner. If hesitating, the cells or the lamella containing the pigments, or a whole stack of lamella could be treated classically. Heisenberg would say that this is left to our free choice.

This point of view clearly lacks a quantitative criterion. One has a feeling that every choice in the level of the frontier will correspond to a different error, which will be decreasing when one considers larger and larger observing systems.

Bohr's point of view on the subject was more radical and it is worth much consideration. According to him, quantum mechanics only refers to

THE PROBLEMS OF MEASUREMENT THEORY

the classical properties of a macroscopic system and it is expressed by statements relevant to that domain. It is impossible to reach by induction anything explicit concerning an atomic object from the knowledge of some classical data. There is no frontier as Heisenberg would have it, but an experiment should rather be considered as an indivisible whole: atomic objects have no specific properties of their own and only factual phenomena exist. They concern the totality of the atomic object and the macroscopic apparatus altogether. The concept of "phenomenon" should be strictly reserved to what is directly perceived and it should always be expressed by a statement expressed in the language of classical physics.

Observations Are Expressed by Classical Statements

The necessity of classically describing a datum has already been mentioned, but what are the real reasons for it? They go deep into the foundations of the theory of knowledge and its relation with ordinary language and logic. An experiment is an action, if only to set up an apparatus, and this can only be expressed by ordinary language (for instance in a user's manual), not by mathematical symbols. As for the data, they are facts and as such, they have absolute certainty, they are registered, and they have no trace of a probabilistic uncertainty. It is clear that only the language of classical physics can express this.

It should be stressed that classical physics, in this regard, is not so much Newton's mathematical framework, but something rather deeper. It was best described by David Hume, the philosopher, when he said that, all around us, there is a world of plain facts; we develop our intuition by looking at them, whereas our language, as well as our common sense, refers only to them.

It may be worth mentioning here an interesting idea by Fock.[18] According to him, it might be that classical physics is trustworthy, even if it is not absolutely certain, because the errors resulting from its approximate character are negligible. This prophetic point of view will be developed in detail later.

Irreversibility of a Measurement

For an experimental datum to be used some time after its occurrence, it must be recorded on some document. It may be written, shown on a dial, or kept in a computer memory or in the brain of the observer. Whatever is done, the production of such a document is essentially an irreversible process. This remark has often led to the idea that there exists a strong connection between the problems of measurement theory and those of irreversible thermodynamics.

Quantum Jumps

A measurement involves an action upon the measured object and it is often difficult to say how strong it is and how deeply it perturbs the object. It is assumed in any case to provoke a reduction of the wave packet, which is in no manner a consequence of quantum dynamics but a jump Bohr considered to be a new type of physical law. This jump is a random effect. Its probability follows from the theory but its mechanism or its actuality is supposed to exceed the reach of theory.

Complementarity

Bohr defined *complementarity* as a mode of thought admitting various concepts that are mutually exclusive while being nevertheless necessary for a full understanding of *all* experiments. The best example is provided by radiation, which can be considered as a wave or as a collection of particles according to the experiment. True to his conception of an experiment, Bohr considered the choice of one mode of thought rather than another to be completely dictated by the experimental environment. Radiation does not have a unique conceptual framework but one or the other must be used in each case: waves when using a receiving antenna, particles when using a photomultiplier. These concepts have only a cognitive value in definite conditions.

Truth

Complementarity and truth are intimately related notions. Heisenberg stressed the idea that, as long as one is restricted to talk only about phenomena, they become automatically the only elements of truth at our disposal. One cannot say for instance that a particle has a straight line trajectory if it is not detected and the track is not seen, as happens in a vacuum. There is no meaning in saying that a photon goes along a definite arm of an interferometer as long as this is not confirmed by observing a phenomenon signalling it.

Moreover, to say that something is "happening" or "occurring" has no meaning, except when this is a way of describing a phenomenon. One cannot say anything true about an atom between the time when it is prepared and the time when it is measured. This is not legitimate. An often quoted example concerns a book on a table in a closed room. After closing the door, one knows that the book is still in the same place, because this is a classical statement involving certainty and it is legitimate in the classical world (relying, by the way, upon determinism). Nothing of that sort can apply to an atom except at the few times when it is directly observed. Is it still there or not? Heisenberg considered the question as meaningless.

This point of view sometimes leads to rather curious consequences. There exists for instance a particle, the meson K^0, which is a superposition of two modes: a K_S^0 having a short lifetime and decaying sometimes into two pions and a K_L^0 having a longer lifetime and decaying into three pions. If one follows Heisenberg, one can truly talk of a K_S^0 when one sees two pions. But it does not mean that there was actually a K_S^0 before that, because it would be a sin. Since the K_S^0 disappears when it decays, nothing can of course be said about it at later times. More ephemeral than a mayfly, the K_S^0 has only known an instant and, similarly, many quantum objects can only be sadly recalled upon a tombstone where the two dates coincide.

The Last Thesis

This thesis, according to which the quantum states are objective, though not real, is very interesting but it would take us outside the proper domain of physics into philosophy. We shall therefore leave it aside, not to exclude it forever but to keep it waiting till the last chapter of this book.

A Critical Appraisal

Though the theses were not given here with the exact words of the founding fathers, they nevertheless are true to the original texts. When one ponders the strange and sometimes extreme character of some of them, one can readily appreciate why the Copenhagen interpretation has made many physicists uneasy for more than sixty years.

The criticisms, justifications, and philosophical comments written on the subject could fill a library. We prefer not to treat now the subject of philosophical meaning, since this might go as far as discussing the theory of knowledge, realism, or the existence of a quantum world. Some oceans are too full for another drop to be added and we shall try to remain on the grounds of physics.

To do so, we will primarily concentrate upon two practical prescriptions. The first one assumes ordinary probabilities to be reliable so that it throws out the horrid problem arising from quantum superpositions of macroscopically distinct states of a measuring apparatus. It rejects therefore as not even worth mentioning direct consequences of the theory. To say that quantum mechanics does not apply to the measuring device because it stands on the other side of Heisenberg's frontier does not add anything significant. The essential point is that one has a beautiful theory but one keeps to its consequences or leaves them out according to convenience. Such a procedure can only raise suspicions as to its logical consistency: If one is imposing classical physics when quantum mechanics finds a difficulty, is this not using double-talk? If I may be allowed a questionable analogy, it looks like a constitution where the elections are held by quantum rules though one adds that their results will be worthless when they disagree with the views of the classical government. In other words, the Copenhagen interpretation boils down to sometimes giving credit to the theory it wants to interpret and sometimes rejecting it.

The second prescription concerning reduction is at least as troublesome. Here is a beautiful theory agreeing remarkably well with everything that is known. It rests upon a dynamics, which is expressed by the Schrödinger equation. And suddenly one rejects this equation as being inappropriate when one is dealing with a few special interactions to be called measurements. One nevertheless continues to apply it confidently to a host of other interactions between a quantum object and a macroscopic one as long they do not generate a phenomenon. This lack of consistency can be shown by an example: The refraction index of a gas or a crystal can be computed using quantum mechanics and is due to an interaction between the photons and the collection of atoms comprising a macroscopic system. No interaction of that kind is amplified to give rise to a phenomenon so that no reduction is supposed to occur. This is fortunate because otherwise, if a reduction were to occur, there would not exist any transparent medium except the vacuum; they would all become opaque or at best translucent. On the other hand, when there is some smoke in the gas so that it can be seen, reduction is supposed to take place since radiation is actually detected. Why? How should one classify reduction when it is claimed to be a new kind of physical law, yet its only role is to contradict or limit the other laws while it is put by fiat outside the reach of experimental check?

These unsettling remarks are often expressed in a cautious way, with scholarly words blunting their sharpness. One should however be thankful to John Bell for having made them plain. If we have also chosen to expose them, it was not for the sake of provocation but to show that there exist obvious problems that many physicists prefer to avoid or to circumscribe. It is not however a lack of respect for the great elders to point a finger towards an unsatisfactory aspect of their heritage. Bohr, Pauli, Heisenberg, and Fock sought consistency without completely obtaining it. To follow them by adding more and more comments, being more and more philosophical, would not be another work of science. Such a process is well known to yield any initially desired conclusion if one keeps at it long enough.

One may prefer to think that Copenhagen left us with a vision and a program. Part of it is to be kept, another to be abandoned but, in priority, something more is to be built. Bohr left us with a problem belonging both to theory and the philosophy of science: Formulate the interpretation of quantum mechanics in a sufficiently clear and deductive way so that it appears in full light as a consistent construction. Then, and only then, philosophy may have its say because it can offer us nothing but new inconsistencies as long as the science upon which it relies is inconsistent.

8. Schrödinger's Cat and Wigner's Friend

Two striking illustrations of the difficulties of interpretation have been given by Schrödinger[19] and by Wigner.[20] They do not raise really new questions but they make the ones already encountered even more conspicuous. They are moreover very well known and so intriguing that they must be described in a book such as the present one, even if very briefly.

Schrödinger's Cat

A cat is imprisoned in a closed box containing a weak radioactive source. A devilish device can be triggered by a radioactive decay to send cyanide in the box. That's all.

All that was said about superposition can be repeated: The radioactive source is at every time in a state of superposition involving a nondecayed state and a decayed one. So, the cat, if described according to quantum mechanics, is also in a state of superposition involving the two cases where she is either alive or dead. When an observer opens the box and looks at its content, there is according to the Copenhagen interpretation a sudden reduction of the wave function and then, but only then (according at least to von Neumann), the cat becomes really dead or is still alive. Before that, she was neither alive nor dead, or rather she was partly both.

Heisenberg would also say that only when the box is open it becomes true that the cat is dead (or alive). Bohr would rather say that the cat, being macroscopic, is a classical cat and it must at all times be truly alive or truly dead. What is your opinion?

Wigner's Friend

This is an example where the situation is made even more striking because of an exchange of information between two conscious beings.[21]

One considers three systems: Q is a quantum system upon which the measurement of an observable A is performed. O is the observer. F is either the apparatus measuring A or a friend of the observer who has been asked to make the measurement in his place. It will be simpler to take the case when A has only two eigenvalues, say $n = 0$ and $n = 1$, this being a so-called yes or no measurement. When the measurement is made, F writes a message s_n on a placard for O to see it, to tell him what he has found.

If the initial state of Q is an eigenvector $|\psi_n\rangle$ of A, the final state of $S + F$ will be given, disregarding irrelevant observables describing all this in more detail, by a ket $|\psi_n\rangle \otimes |s_n\rangle$. The observer O considers the message as a given fact. If the initial state of Q is

$$\sum_n c_n |\psi_n\rangle,$$

the final state of $Q + F$ will be

$$\sum_n c_n |\psi_n\rangle \otimes |s_n\rangle. \qquad (2.17)$$

But the observer O is not so much interested in the messages sent by his friend F. He prefers to look thoroughly at his friend's features or, if one prefers, to measure him. This means that he can get a sufficient knowledge of the density matrix summarizing the various hidden inside feelings of F during the whole series of experiments. So, he can make sure that the state of $Q + F$ is indeed the pure state (2.17) and not a statistical mixture meaning only that F has registered 1 or else he has registered 0 in each case.

However, when F is really a person and not an instrument, O can ask him how he (F) felt when he interacted with Q. Since F is a person, he had definite conscious sensations at that time. Either he felt that he sent one signal or that he sent another. Then, the outcome of what F tells O is that he is sure that he (F) should be described by a mixture, whereas O has got the conviction that this is not true by his private investigations upon his friend.

So, the conclusion is that, if O firmly believes in quantum mechanics, he cannot believe what his friend tells him, because he has evidence to the contrary. How solitary he feels to be the only one to really know while nobody else is like him. This is what philosophers call solipsism, the state of the mind who feels he is completely alone and that the other minds do not exist by themselves but they are only what he himself thinks of them.

Here again, the answer will be left for the next chapters and the macroscopic character of one's best friends, together with the corresponding decoherence of their states, will be found essential. This was not however Wigner's conclusion who rather advocated when he wrote this a change in the laws of quantum mechanics.

WHAT AN INTERPRETATION SHOULD BE

The Copenhagen interpretation is obviously unsatisfactory, or at least something is still lacking in it. Before attempting to improve it, it will be convenient to ponder the nature of the problem so as to delineate a program. We will therefore consider two questions, both verging upon the frontier of physics and epistemology, though we shall not yet take sides on the questions discussed in the last section. We shall mainly try to understand why physics needs an interpretation, so as to make clear what an interpretation really is and what criteria it should obey.

9. Why Does Physics Need an Interpretation?

By an *interpretation*, we shall mean a translation of the empirical language describing the experiments and the phenomena in a way involving only the formal concepts of the theory. This definition will soon be justified and it is only given right from the beginning for the sake of clarity.

From Reason to Science

It may look surprising that physics, though a science dealing with the simplest objects, ends up demanding a further elaboration to become understandable. How can it be that one needs an interpretation to relate the formalism to experiment, to translate from one into the other the two languages describing on one hand the concepts and on the other hand the data and experiments?

A hint of an answer is in the very structure of physics. This is why we shall now come back to these foundations, even if it may look like recalling trivialities.

The aim of physics is to order empirical data so as to unveil the laws of the real world and to understand it more easily with its multitudinous objects, as it is perceived by our senses or investigated by sophisticated devices increasing the reach of the senses and allowing their impressions to become quantitative. The primary subject of this study, be it called reality or nature, is both obviously extant and out of the reach of a definition: it is here and that's all. One can however see that there exist regularities in it, some laws giving a meaning to the program of physics and enjoining us to explore and to coordinate these constancies of reality. At the origin of physics is reality, the existence of phenomena, and the regularities of facts.

The regularities of facts play no doubt a large part in the origin of our language, which allows us to express them and to make use of them: For instance, though no stone is identical to another, all stones have common qualities of hardness, compactness in their shape, a similar behavior under the action of heat and so on, allowing us to assemble them under that name of "stone" just as we say "air" or "water" to summarize an indefinitely repeated experience of all our senses. The regularities of facts are also to be assumed responsible when a language can model reality by making statements whose only meaning is their correspondence with observed facts, as when we say that stones can fall or wood can float. These regularities are so pervasive that they imprint in our brains images recalling many facts already seen, so that reasoning becomes possible. One can say, for instance, that if the stone gets free, it will roll along this slope and stop near that brook. In other words, our faculty of reasoning is constrained as well as generated by the regularities of phenomena. One can then say that science, in its most elementary form, begins by a mental representation of reality, which is possible because reality itself is so full of regularities as to allow language, logic, and reasoning to describe it faithfully. In that sense, science is a representation of reality that is expressed by a language whose main structure is logic.

This elementary and primary form of science was considerably enriched when it began to make systematic use of mathematics. Mathematics is invaluable when it comes to enclosing in a unique concept a vast field of reasoning, giving it a controlled rigor and finally including naturally numerical data. This was how physics got its universality and power. It became possible by such means to have better access to the notion of physical law and to arrive at some very powerful ones, the most striking historical example being the discovery of the laws of mechanics in Newton's times. The initial project of science, which was to understand reality, took then a new form: to get a representation of reality by means of mathematical analogs. One came to associate with a specific object, be it hypothetical or actual, a star or an atom, several purely mathematical concepts such as its mass, the coordinates defining its position, and so on. Purely mathematical operations were used for making precise other concepts such as velocity or acceleration as time derivatives of the position coordinates. Finally, all these concepts were related in the framework of a few laws or principles, such as the basic laws of dynamics.

If this is science, it is clear that one should be careful in distinguishing between reality and its representation. Their difference is much more marked than when men had only a "natural" visual representation and a rather simple language because the new representation is no longer a simple mental image consisting of accumulated remembrance but a collection of much more abstract concepts that are submitted to the exacting rules of mathematics, while they are able to mimic with an extraordinary precision and universality the behavior of reality itself.

The agreement between reality and its models was apparently not much a source of questioning for most scientists, except for a few thinkers, until a time came when the mathematical models broke free from the bounds of common sense. This occurred first with special relativity and then with general relativity. Ordinary intuition and common sense were no longer sufficient to give a familiar content to the concepts of the theory. It was no longer obvious what one should mean by time and distance. Moreover, the mathematical quantities occurring in the theory needed much more thinking before one could associate them with observable reality, with experiment. These developments made quite clear the difference between reality and its representation and, for the first time, one was faced with a problem of interpretation, which was to relate the empirical langauge and the theoretical concepts. The interpretation of relativity consisted then in giving a more intuitive, more concrete meaning to the mathematical concepts in terms of which the theory was written, so as to reexpress by their means what can be seen and measured. This was done without excessive trouble. However the necessity of interpretation became much more acute with the advent of quantum mechanics.

Common Sense

Before approaching the questions belonging properly to quantum mechanics, it will be useful to ask ourselves how it could happen that such a wide gap occur between the initial straightforward form of science, classical and so close to what is seen, and the present mature science where common sense itself becomes questionable. This is important because it amounts after all to asking the question: why give an interpretation?

One cannot completely avoid asking more closely what is common sense. It is now known to be not completely trustworthy when it says, for instance, that a photon or any particle must be in only one place at a given time. It remains nevertheless necessary since it agrees with everything we can see and do. Its questioning by modern science can only leave us very uneasy and this needs some clarification. When it comes to defining common sense, one might say that it is essentially the spontaneous initial representation of reality every one of us builds up as an infant, mankind itself having elaborated upon it during its own infancy and cast it into the mold of language. Psychologists, after Jean Piaget, have shown that some concepts are built up in a child's mind from a direct perception of his environment. Hence the concept of object seems to be not completely innate, even if the genetic structure of our brain is ready to bring it forth; the notion of weight is acquired at a later time, sooner than the notion of volume, not to say the notion of density if one arrives at it. Of course, as soon as a child is able to understand what is said, his own mental representations are strongly influenced by those of his community. Without trying to link the preconceptions of physics to any particular school in psychology, one can nevertheless assume that the ordinary notions man uses to understand what he observes are developed very early and have no reason to be conserved all through his life (or the lifetime of mankind) other than their agreement with what we usually see around us in ordinary circumstances. Nothing however tells us that this representation should remain valid if cars were to approach the speed of light or a tennis ball to become as small as an electron.

It is therefore almost obvious that the representation associated with common sense has no reason to be universally valid, even if many people have tried to turn this assumption into some philosophic principle or another. This is not the main question—the right one is to understand how science has been able to disentangle itself from the bounds of common sense, even daring to contradict them. As a matter of fact, the question is to understand what is the nature and the power of the scientific method.

The Scientific Method

There has never been complete agreement concerning the exact nature of the scientific method.[22] Some modern authors even go as far as questioning its existence but we shall not enter into this kind of polemic. The best description of the method was given by Einstein and it will be adopted here, although we formulate it in a slightly different manner.[23]

The scientific method works in four steps, namely: (i) Empirical observation, (ii) Conceptual formulation, (iii) Logical elaboration, (iv) Experimental verification. The empirical stage of science most often goes through a preliminary observation of the facts, a systematic acquisition of data. It also formulates some basic questions. Additionally it provides a summary of what has been observed in terms of empirical rules, sometimes already quantitative though not yet provided with a theoretical framework. As an example, such are the observations of planetary motion by Tycho Brahe and the three rules by which Kepler synthesized them. The conceptual step consists in inventing (creating?) some mathematical objects and some logical concepts aimed at representing the essentials of reality, together with some principles relating them. Considering again the case of mechanics, this is the long process culminating with Newton's laws. The elaboration step consists in developing the theoretical consequences of the assumed principles and in applying them to specific cases. It is mostly deductive. As an example, one may mention the calculation by Leverrier and Adams of the anomalies in the motion of Uranus, which led to the discovery of Neptune. Finally, crowning the whole construction, the last step consists in submitting all these duly worked out consequences to the verdict of experiments.

One of course does not pretend that the process of scientific discovery should pass through all these steps in a prescribed order: history has its own ways where circumstances, creativity, or inertia play their role. One should also not assign a kind of hierarchy among them since, according to the domain of science one is considering, one or another step may be almost trivial, whereas its accomplishment elsewhere may be a feat for mankind. The important thing is that no science has reached maturity before these four steps have been accomplished successfully.

Why Give an Interpretation?

This schematic description of the scientific method may help us better understand a few features of science and see the origin of interpretation. It should be noticed first of all that the contact with reality governs the game at the beginning and at the end since experiments are crucial during the initial empirical stage as well as the final stage of verification. There is little to be said about elaboration, although it is what is best explained in textbooks.

What is most interesting for our purpose is the conceptual step, which seems to obey no rule of conduct: Why will Newton give so much attention to acceleration, Einstein dream of a curved spacetime, Heisenberg play with noncommutative arrays of numbers, de Broglie assume an invisible wave, Feynman make electrons go backward in time to turn them into positrons, Gell-Mann use symmetries with no physical operations, Weinberg and Salam introduce gauge symmetries only to break them? What is important is not to answer these questions but to realize that at this step every possibility, even if crazy, is tried, combined with others, rejected, revived, and transformed. What is really extraordinary and worth pondering is that this game of riddles may succeed so systematically.

What is also worth remembering for our purpose is that at this essential step the concept finder may refuse to be bound by the constraints of common sense, as soon as they have shown their inability to take account of all the facts. This is where a basic conflict between creativity and conformism can take place. This is where something never seen before, something absurd at first sight, may appear, such as saying that a little piece of matter can follow two paths at the same time or that velocities do not add up. There is however an ultimate judge, which is experiment.

The construction of science is therefore in a subtle relation with common sense: In its construction of theory, science pays due respect to common sense as far as possible, but rejects it if necessary. Conversely, when it comes to relating theory and experiment, the ordinary conceptions and language of common sense regain their rights. Said otherwise, common sense is not an absolute rule for science. Indeed any philosophical principle which could take its place (whether it be deterministic causality or the distinguishability of objects) would not be either because experiment is the only unquestionable source of knowledge. But the result of experiment is after all only a fact or a collection of facts and it is impossible to express a fact by a mathematical formula: it can be seen; perceived; transmitted by a photograph, a picture or a record, or by words whose meaning is most ordinary. It is real, not conceptual, and above all not mathematical. So, in the end, the theoretical construction should be reexpressed in the usual common sense language of facts to be complete.

To summarize, the existence and the necessity of an interpretation are due to the very nature of physics. Physics proceeds by a theoretical synthesis based upon experimental data, but goes out of their realm when using mathematics. Each of these two aspects is expressed in a different language, each one being inescapable so that they have to be reconciled. Looking at it more closely, the problem of interpretation already existed in classical physics, where the electromagnetic field had for instance a few surprising features and Boltzmann's statistical mechanics was also very tricky. In the case of quantum mechanics, one is facing a much more difficult problem but, because of that, one may also learn from it much more about physics itself.

10. The Criteria of an Interpretation. Consistency and Completeness

What Is an Interpretation?

As long as science could be expressed easily within the language of common sense, no interpretation was really needed. This was more or less the case for most of classical physics. An interpretation becomes necessary when no pedagogical means can be found to express the basic concepts of the theory while staying within the bounds of common sense—in other words, when simplification becomes treason. It is then imperative to follow the opposite way, namely, to reformulate the categories of thought belonging to common sense that are used for action and experiment, within the conceptual language of the theory. An interpretation should then consist in recovering the language and the framework of common sense, so well suited to facts, experiments, and phenomena, from the principles of the theory. *Interpretation is therefore a translation of the language of facts into the formal language of a theory.*

Interpreting quantum mechanics is especially difficult. This is not only because its concepts are much more remote from intuition than anything else before. It also comes from the necessity of solving new problems that were not initially contained in the foundations of the theory: the theory is probabilistic in an essential way whereas phenomena and facts should be considered as having certainty. Determinism is an unavoidable feature of phenomena, since there would be no meaning in describing an experiment if someone repeating it didn't see his apparatus behaving in the same way under identical conditions. And what are the data if not something that has been recorded in a document so that this memory of the past can be used to revive it faithfully by what is after all a kind of determinism going backwards in time?

To summarize, the interpretation of quantum mechanics consists in reexpressing the phenomena and the data within the conceptual framework of the theory. This is made particularly difficult because along the way one should also reconcile the probabilistic character of the theory with the certainty of facts.

What Interpretation?

What criteria should be used to judge the quality of an interpretation? We reject a priori all kinds of philosophical criteria such as, for instance, demanding that the theory should be an exact description of reality, a complete explanation of it, or its intimate knowledge. To look for a realistic theory, rather than a positivistic one or an objective representation in terms of

THE PROBLEMS OF MEASUREMENT THEORY

models, may be an incentive for research. It cannot be an a priori condition for the ultimate form of the theory.

Interpretation is an intrinsic part of the theory. This is obvious if one agrees that it is based upon theory and it aims at reaching phenomenology. As a part of theory, it should share its two main criteria, which are agreement with experiment and consistency. Consistency is taken here with its usual logical meaning, a theory being consistent when it is free from any internal contradiction.

One should also expect the interpretation to be *complete*. This means that it gives explicit predictions, whatever the experiment being considered. This is obviously a necessary condition if the theory is to be fully exposed to the judgement of experiments, i.e., if it is what Karl Popper called a falsifiable theory.

Finally, the interpretation must be totally explicit in its description of phenomena. When it comes to a theory as universal as quantum mechanics, it is necessary that the interpretation offer a theory of phenomena, including a clear understanding of the status of determinism and a derivation of its empirical existence, whatever its limitations may be.

What about the Copenhagen Interpretation?

These criteria may look somewhat ambitious, considering that the Copenhagen interpretation does not fulfill any one of them. Its consistency is most questionable, since it was seen how the empirical rules upon which it rests are in conflict with the foundations of the theory.

Is it complete? One might have expected it until some key experiments, which were proposed by Anthony Leggett,[24] were realized. They will be considered in detail in later chapters so that a brief sketch will be enough here: Leggett suggested that some macroscopic devices involving superconductors, the so-called superconducting quantum interference devices or SQUID's, could in some conditions show typical quantum effects, such as having clearly separated energy levels or giving rise to a tunnel effect. This means that there is, at least in that case, no Heisenberg frontier between the microscopic and the macroscopic domains, nor is Bohr's point of view useful because one cannot tell here what is a phenomenon and what is not. These experiments have been now realized and they have fully endorsed Leggett's expectations.[25]

As a matter of fact, the Copenhagen interpretation does not offer a theory of phenomena. It just assumes that they exist and it puts them in parallel with the theory, without establishing a communication between them, except peremptorily.

It will be shown that one can build up an interpretation satisfying these criteria. Undoubtedly, it represents the outcome of more than sixty years of research and it would be unfair to slight the Copenhagen interpretation when compared to it. It confirms the Copenhagen empirical prescriptions, so well sustained by experiments, while removing the apparent contradictions they contained. It even recovers most of the Copenhagen preconceptions or theses, though they now follow from a strictly deductive process. Not all the Copenhagen theses however reappear. Some are rejected, others are completely reformulated, while still others are given a more limited domain of application or a new meaning. All this will be the subject of the rest of this book.

To summarize, the interpretation of quantum mechanics consists in expressing completely the properties of the experimental and natural phenomena in terms of the mathematical concepts entering the theory. The interpretation should be consistent (with no internal contradiction) and complete (giving an explicit prediction in every experimental setup). It should moreover provide a theory of phenomena and explain the determinism of classical facts. It will be shown that these goals can be reached and that their fulfillment is already seeded in the basic axioms of the theory. In other words, quantum mechanics is a theory providing its own interpretation.

PROBLEMS

Problems are marked with an asterisk when they border the frontier of research or require some nontrivial calculation.

1. *Imprecise Measurements**

The use of imprecise measurement devices can lead to results that are striking and apparently paradoxical.[26]

Consider a system of N 1/2-spins, N being a large number. A von Neumann measuring device is used to measure a component L of the total spin. This apparatus is imprecise in the following sense: the initial wave packet representing the momentum p of the pointer is strictly confined inside the domain $|p| < (N/2)^{-(1/2+\varepsilon)}$, where ε is small and one uses units where $\hbar = 1$. Show that the uncertainty for the position q of the pointer in the wave function $\phi(q)$ is then at least as large as $(N/2)^{(1/2+\varepsilon)}$. One takes the amplification coefficient λ in equation (2.6) equal to 1. The measurement of L is then bad enough that neighboring eigenvalues of L cannot be distinguished since they are separated by $\Delta L = 1$. The measurement however gives information about L since the relative error is of the order of L^{-1} when L is large.

One first measures L_x with a sharply precise device and it will be assumed that the result has the maximal value $N/2$. Show that this should happen in one case out of 2^N. The imprecise device is then used to measure the observable $A = 2^{-1/2}(L_x + L_y)$ representing the component of the total spin along the diagonal of the x and y axes, the coupling hamiltonian being $-\delta(t)pA$. Show that reduction of the wave packet implies that the system made up of the spins and the apparatus is left at the end in the state

$$|\psi\rangle = \exp(ipA)|L_x = N/2\rangle\phi'(q).$$

One finally measures the component L_y of the total spin with a sharp device and one selects once again the very rare cases where this value is equal to its maximum $N/2$.

(a) Show that

$$\phi'(q) = K\langle L_y = (N/2)|\exp(ipA)|L_x = N/2\rangle\phi(q) = KU\phi(q),$$

where K is a normalization factor.

(b) Let U be the operator acting upon the states of the measured spins. Show that U is equal to $\exp(2^{-1/2}ipN)$, up to corrections of order $N^{-\varepsilon}$. One may use as a first step the case of one spin where it can be shown that

$$|s_x = 1/2\rangle = \exp(i\sigma_y\pi/2)|s_z = 1/2\rangle,$$
$$|s_y = 1/2\rangle = \exp(-i\sigma_x\pi/2)|s_z = 1/2\rangle;$$

one may use the formula $\exp(i\mathbf{n}\cdot\boldsymbol{\sigma}\alpha) = \cos\alpha + i\boldsymbol{\sigma}\cdot\mathbf{n}\sin\alpha$, where $\mathbf{n}$ is a real unit three-dimensional vector and one may treat a component of the total spin as the sum of N individual components. Finally, show that one can use the approximations $\cos p = 1$ and $\sin p = p$ because of the assumptions first made about the support in momentum space of the initial wave function.

(c) Using $P = -i\partial/\partial q$, show that the pointer of the von Neumann device must be after the measurement in the position $q = 2^{-1/2}N$ with a probability $1 - \delta$, where δ is of the order of $N^{-2\varepsilon}$.

Noticing that the maximal value of A is associated with the pointer position $q = N/2$, what is the paradox?

Comment. This is not strictly speaking a paradox concerning a measurement but a statement about an experiment that only looks like a measurement. For instance, it is known that one can in principle add up waves down to a limiting wavelength in the form

$$\sum_{n=-N}^{N} c_n \exp[inx/L]$$

to get a function behaving almost exactly as a wave of a wavelength much smaller than L/N in a small interval in x. One could then be tempted to say that the momentum is much larger than expected if one tries to determine it by locating a (necessarily imprecise) measurement device in this small position interval. It would then be true that the result to be read on this apparatus would be much larger than the one associated with the constituting waves, but this would not be a momentum measurement.

2. *An Ideal Measurement of Energy*

The energy of a system Q with a hamiltonian H having a continuous spectrum is measured by an ideal von Neumann device M. The hamiltonian of M itself is 0. The $Q-M$ coupling hamiltonian is given by $H_{\text{int}} = -g(t)HP/p$, where $g(t)$ is a pure number equal to g between the times 0 and Δt and to 0 otherwise, P is the generator of translations for the pointer, and p is a fixed quantity having the dimensions of P. Let Δx be the width of the initial wave packet for the pointer. Show that the uncertainty ΔE about the measured value of the energy is given by $p\Delta x = g\Delta E\Delta t$ so that $\Delta E \Delta t$ can be made in principle as small as wanted.

3. *Another Form of the Bohr–Einstein Argument**

The spacetime metric in the case of a weak gravitational potential ϕ is known to be given by $ds^2 = (c^2+2\phi)dt^2 - dx^2$ and an energy E contributes to the potential as if it were a gravitational mass E/c^2.

The Einstein proposal for the measurement of the energy is reconsidered, the box being now at rest on the ground. Its mass is measured long before and long after the opening of the lid by measuring the small contribution to the gravitational potential it produces at some distance. Is it possible to recover Bohr's conclusion in that case?

Interlude

FIVE IDEAS FOR CONSTRUCTING A CONSISTENT INTERPRETATION. PROPERTIES, HISTORIES, LOGIC, CLASSICAL APPROXIMATIONS, AND DECOHERENCE

The aim of the present interlude is to provide a general orientation to the next five chapters where the foundations of interpretation will be given.

The previous chapter led us to look for an interpretation of quantum mechanics relying directly upon the basic principles of the theory, describing clearly every kind of empirical or experimental situation in terms of these concepts and, furthermore, being both consistent and complete. If such an interpretation exists, it cannot be exactly the Copenhagen interpretation. A systematic construction of the interpretation of quantum mechanics will now be attempted.

A peculiarity of this subject is worth noticing at the beginning. We are going to investigate, among other questions, the limits of validity of ordinary physical intuition and of common sense. We must therefore be careful not to introduce some uncritical or unconscious assumption that would come from uncontrolled common sense. Being physicists, we are also vulnerable to old habits imprinted by so many years of Copenhagen practice and teaching. This is why it will be safer to start from scratch. Accordingly, only the most basic physical concepts of the theory will be preserved and any hint of an interpretation in their formulation will be left aside. As a matter of fact, these principles amount to the adoption of the Hilbert space framework and Schrödinger's dynamics. This ascetic approach will be followed to make it clear that, even so, the principles contain in germ everything that is needed for their interpretation.

The program of constructing this interpretation will be therefore to draw consistently the consequences of the principles, the overall claim being that the basic theory is enough to provide its own interpretation without recourse to any outside empirical ingredient or even to common sense.

It was found that a full-fledged interpretation must rely upon at least five significant ideas, each one marking a decisive step in the construction:

Properties

A reliable definition of what is to be understood by a physical property of a quantum system must first be given. It may be noticed that, when doing so, one already departs from Bohr, who denied a priori the existence of such specific properties and reserved them for macroscopic systems.

This first step is necessary in view of the very abstract character of the theoretical foundations, which seem to have no direct relation with the ordinary experimental facts upon which physics is built. It would certainly be unwise to continue acting as was done long ago, when physics was going at a tremendous pace. Then one proceeded as if there were a direct and natural correspondence between wave functions, the Schrödinger equation, and all this abstraction with a voltmeter, an oscillograph, or an accelerator. Who can "see" clearly this correspondence? The gap is too wide. One must therefore begin very carefully so that even the first words to be pronounced should be duly chosen. These first words are intended to introduce the notion of property for an isolated system.[1]

A property asserts the value of some observable in some range of real numbers at a given time. Their most important feature is the existence of a correspondence between the properties and the projectors in Hilbert space, as was first noticed by von Neumann and also mentioned in Chapter 1. This correspondence is the first hint pointing to the fact that a convenient use of the mathematical tools coming with the basic theoretical concepts may provide the means of an interpretation.

A further inquiry into the nature of properties will provide a better understanding of what is the state of a system and what role should be given to probabilities in the theory. It should be stressed in this connection that the existence of probabilities is not included among the basic physical principles of the theory, but rather they are considered as belonging to the interpretation and, as such, as something to be derived from the principles.

Since the beginnings of quantum mechanics, it was assumed that the state of a system is the outcome of its preparation. This point of view is ultimately correct but it has the same inconvenience as the one we meet if we try to make use of a measuring instrument for asserting a property. We understand clearly what an instrument is as long as we are describing it by classical physics and by common sense, but not otherwise. When an instrument is considered as a quantum object, it is tremendously complicated and it would then be unwise to say that we still understand it. A careful approach to the notions of properties and of state must therefore start from ideas standing much nearer to the principles.

This is where probabilities enter. An important point should however be mentioned immediately about the nature of these probabilities: Since one does not understand yet what is a measuring device and what is a measurement, one cannot safely assume that the probability of a property is something that can be measured. Fortunately, probabilities are not necessarily empirical quantities. They can also be considered, from the standpoint of mathematics, as some kind of weight to be given to a property: a property can be impossible (probability zero), certain (probability one), or only possible. In that sense, a probability only measures the logical weight of a property and this point of view will become more and more important during the construction of an interpretation.

One can then make two more steps forward. The first one consists in defining formally the state of a system as data allowing one to attribute a probability to every property. The second one assumes that the words occurring in the statement of a property (the value of an observable A is in a domain D) make sense, particularly when one considers that the value of A might belong to two disjoint domains. It might be noticed by the way that asking for a legitimate meaning of the words amounts to introducing logic for the first time into the game of interpretation.

Then comes a remarkable result resting upon these assumptions which is due to Gleason. It says that the state of a system is necessarily defined through the knowledge of some density operator, in terms of which the probabilities of all properties are uniquely defined. Among other features, this result means that

- Probabilities provide a tool for constructing the language of physics long before they can be considered as empirically meaningful quantities.
- The expression for the probabilities first proposed by Max Born is an unavoidable part of an interpretation. If any probability should ever play a part in the theory, it can only be this one.
- The density operator is the basic notion one must associate with a quantum state and not simply a pure state represented by a wave function.

Histories

The second step will introduce another important notion: the histories of a quantum system. A history can be simply defined as a succession of different properties occurring at different times. If one were to compare a property with an instantaneous photograph of the system, a history would be something like a motion picture. As a matter of fact, it will be found ultimately that everything that can be said about the physical world (including its classical aspects) can be expressed in terms of such histories. Thus the whole language of empirical physics amounts to envisioning and discussing logically the possible occurrence of various histories. Examples will be given to show how this happens in practical situations.

One can sometimes assign a probability to a given history. It is quite significant however that this cannot be done for every kind of history: a history has a unique well-defined probability or none at all. Only a few histories allow this assignment of a probability and they may be said to be *consistent*. A typical inconsistent history (or rather an inconsistent set of histories) would for instance enumerate the various possible regions in the focal plane of a lens standing in front of an interferometer where a photon can arrive, while also trying to add other properties asserting that the photon went along only one arm of the interferometer. This kind of history has no sensible probability, and this has nothing to do with the presence or absence of a measurement device checking the properties to be mentioned. Their inconsistency probes much deeper into the structure of the theory. Furthermore, by straightforward calculations, one can make sure whether or not a history (or a set of them) can be given a probability and be consistent.

The histories, which should not be confused with Feynman's histories, were discovered again and again before being investigated systematically by Robert Griffiths, whose breakthrough contribution was the discovery of consistency.[2]

Logic

The significance of consistent histories and of their probabilities is revealed by the third idea: The use of a consistent family of histories is the only way to include the statements concerning a quantum physical system within the framework of logic.[3] They provide the only framework where one can state an assumption and derive its consequences with certainty.

Although this result is not too difficult to establish (fortunately, one needs very little knowledge of logic to master these questions), it provides a very useful guideline to the whole of interpretation. The reason why is easy to understand: One knows that common sense is not a reliable guide where quantum physics is concerned; macroscopic instruments, measurements, preparations, and so on are, as already said, too complicated to be understood right away. What can then be a reliable guide to avoid blunders and to get rid of inconsistencies? The answer is logic.

Logic is usually not of much use in physics, because its application is ordinarily more or less trivial. This is not true any longer in the case of quantum mechanics. Here the standards of logic become so constraining that they are creative: the possibilities are so constrained that they determine what interpretation should be.

This will be formulated explicitly by stating a unique and general rule for the interpretation of quantum mechanics. It asserts that any sensible description and any valid reasoning concerning a physical system should be set in a logically consistent framework. It turns out that this very simple rule contains in germ the usual prescriptions of measurement theory. It will also allow us to reconsider the various theses accompanying the Copenhagen interpretation. As a matter of fact, it can be used to develop the interpretation of the theory in a completely deductive way so that there is no longer any need for authoritative sayings by the founding fathers. One can check what they said, criticize it, correct it if necessary, or even reject it.

Classical Physics

The fourth step has to do with classical physics, which was often said to be the most difficult obstacle to an understanding of quantum mechanics. The Copenhagen school after all gave up including it into a consistent quantum framework, except for some rather vague statements about a dynamical correspondence relying upon Ehrenfest's theorem.

The next enterprise will then be to derive completely classical physics and its domain of validity from the basic principles of quantum mechanics. This will not only include dynamics but also the logical aspects of classical physics, how one can describe classical properties and relate them logically or, in other words, why and when common sense can be applied.[4] The necessary guidance will come from applying the previously mentioned logical rule of interpretation to macroscopic quantum systems. The derivation is not yet complete—for instance, one must rely upon the concept of object and the existence of collective observables, without disposing of a general justification of these notions, though their existence and their validity in any specific practical application are obvious enough.

Classical properties will then replace quantum properties. A classical property states that the collective coordinates and momenta of a macroscopic object are given within some fixed bounds, which are large when compared with the limits imposed by the uncertainty relations. Such a classical property can be proved to remain meaningful when considered as a quantum property because it can be associated with a definite quantum projector. The proof of this statement requires somewhat elaborate mathematical techniques in semiclassical physics but, fortunately, the essential ideas as well as the results can be grasped easily in an intuitive way.

Another significant result in the same direction is that the majority of macroscopic systems behave in a deterministic way. So, contrary to what was believed for a long time, there is no basic opposition between the probabilistic character of quantum physics and the deterministic character of classical physics. The price one must pay for this conciliation is only that determinism is no longer absolute, which is after all for the best. The conclusions one may draw from it are subject to errors arising from quantum fluctuations. Moreover, not all the macroscopic systems are found to be deterministic and this is found to be particularly true of chaotic systems, which do not obey determinism when considered from a quantum standpoint. This does not come as a surprise, though it is rather gratifying to obtain this result directly from quantum mechanics. It may also be added that Heisenberg's frontier between the quantum and classical descriptions of physics becomes now simply a quantitative matter, a change in the frontier implying only a change in the errors one can afford.

Another way of expressing these results is to consider a system consisting only of macroscopic objects with no chaos and no instability. It may be noticed that this excludes in particular the quantum measuring instruments, which always involve a metastable or unstable component. It is in fact the clearcut domain of classical physics, though considered now from a quantum standpoint. The logical rules of quantum mechanics together with a semiclassical analysis of the Schrödinger equation can then be used to show that the usual propositions of classical physics are valid from a quantum standpoint. They suffer of course from small probabilities of error because of the existence of quantum fluctuations, but this correction is so small that it is completely negligible in most cases. Another way of stating this result is to say that one can prove the validity of common sense in that case and also find its limits of validity through relying on only the basic principles of quantum mechanics together with its logical rules.

This is in fact they key to the success of the present approach: contrary to the previous attempts at understanding quantum mechanics, it does not assume common sense, it does not rely upon it. The ordinary custom was on the contrary to believe in common sense, perhaps sometimes only partly, perhaps by elevating a sublimated part of it to the level of a philosophical principle, the one most cherished by its believer. This shaky approach was used to criticize or to question quantum mechanics: see that it is not local, not separable, not causal, not so and so according to my definition; how could Nature tolerate this violation of a principle sanctified since Aristotle or Pythagoras? The present approach reverses completely the problem, which is often a good way for getting out of a mess: it lays down firm foundations, as close as possible to the basic acknowledged principles and it works from them by explicit construction and proofs. When finally it recovers common sense, this is not as a God-given law of reason, but something now controlled, limited, and justified when necessary by mathematical arguments; something moreover relying directly upon the basic principles of physics.

Decoherence

Last but not least, the fifth idea contributing to the construction of interpretation is decoherence, which was already alluded to in Chapter 2. It is a physical effect: the spontaneous dynamical diagonalization of the density operator for a macroscopic object, when only collective observables (i.e., classically meaningful ones) are taken into account.[5] It can take place for instance in a measuring apparatus and there it destroys extremely rapidly any possibility of observing a quantum superposition of distinct macroscopic properties. It comes from a destructive interference effect arising inside the matter of the system. It is so efficient that its observation was found to be extremely difficult until very recently, because one does not have enough time to catch it while it is at work before destroying every trace of the interfering situation existing before its action. It is basically responsible for the dissolution of Schrödinger's cat paradox: a cat is definitely either dead or alive, a pointer on a dial shows definitely some position and not another one simultaneously.

Synthesis

When taken together, these five ideas provide a complete theory of the classical phenomena. They also show that quantum mechanics contained from the beginning all that is necessary to understand what the instruments of physics or natural objects are, how they work or behave, how they can be described and understood by plain common sense. In other words, quantum mechanics can provide a theory of what is to be observed, of phenomena.

More work, though not really new ideas, is needed however before drawing from this background a consistent and complete interpretation and a few other questions will still have to be considered when we arrive at measurement theory. Here again, one will use a deductive approach. The ideas which have been introduced in the present interlude will be developed in Chapters 3 to 7, one idea per chapter. Once this is done, all the tools necessary for the applications of the theory will be available and measurement theory will become simply a deductive process.

3

Foundations and Properties

In order to start anew with the construction of a self-contained interpretation, we first restate in the present chapter the basic Hilbert space framework and the dynamics of quantum mechanics. The properties of a quantum system will then be defined and it will be shown how they can be expressed in mathematical terms. This will be used to define the state of a system and to assign well-defined probabilities to the properties.

As a matter of fact, the systematic construction of an interpretation of quantum mechanics is a rather long enterprise and one must therefore carefully lay down the foundations upon which it will be built. We assume that the Hilbert space framework and the Schrödinger dynamics are given. It was recalled in Chapter 1 how they were discovered through a combination of experimental results and inductive guesses. We will not go back to this historical perspective and the principles will be considered now as something that is already ascertained, the problem being to understand them fully. It may be noticed that the probability rules have not been included among the basic principles, because we intend to shed more light upon their meaning at the end of this chapter. In any case, in order to make things clear and to avoid any confusion between the basic principles themselves and some fragments of premature and uncontrolled interpretations which one might have in mind, it will be convenient to state the principles once again.

After doing this, we will give immediately the preliminary elements of a logical structure of quantum mechanics. They will be restricted in the present chapter to the properties of a system, a property stating that the value of an observable is in a set of real numbers at a given time. Care must be taken not to prematurely give an empirical meaning to a property and not to consider it as expressing the result of a measurement. This is because one cannot know yet what a measurement is, at least from a consistent quantum point of view, as long as one does not fully understand what a measurement device is, how it works, and why its working is consistent with the basic principles of quantum mechanics.

So we will try to rely only upon the basic principles, taking advantage as far as possible of their powerful hidden possibilities. As a matter of fact, one of the keys of this approach consists in recognizing that quantum mechanics

FOUNDATIONS AND PROPERTIES   111

contains the seeds necessary to grow its own interpretation and, as a first example of this, it will be seen how one can express formally a property in terms of a projector. Clearly, it means that something belonging already to the language of interpretation can be firmly anchored in the mathematical foundations.

We will then ask how it is possible to ascribe a probability to a property. This is where the theory begins to be probabilistic, not so much for empirical reasons but for the sake of getting a logical structure. The state of a system will be defined as something generating the probabilities of all possible properties and it will be shown that a few very simple logical requirements imply quite powerful conclusions, namely, that there should exist a unique density operator of the physical system and the probabilities can be expressed in terms of this operator in a unique and universal way. This shows again the remarkable power of a logical approach, since one finds that Born's probability rule is a logical necessity and not an axiom.

THE BASIC PRINCIPLES

1. The Mathematical Framework

We take for granted the notion of a physical system without trying to give it a more specific definition, since this would be tantamount to defining what physical reality is. No distinction will be made between systems that would obey quantum mechanics and others that would be governed by classical physics. *Every physical system, whether an atom or a star, is assumed to be described by a universal kind of mechanics, which is quantum mechanics.* This point of view implies that physics should deal with individual systems, as opposed to a statistical ensemble consisting of a large number of identical copies of the same system, because it would be too awkward to consider identical copies of the solar system as being the right thing to have in mind when discussing the planets. The same thing will be done for an atom or a particle, at least to start with, since nothing forbids introducing many identical copies later, if it may be useful.

We will be mainly concerned with isolated systems, suffering no influence from the external world. Every experimentalist knows how difficult it is to realize these conditions in practice and it will also be seen later on how infrequently perfect isolation occurs, if ever. It is not necessary however to be too demanding and it will be sufficient to consider systems for which the external influences are small enough to be neglected in a first approximation. Moreover, the systems being considered will not be assumed to have been isolated forever, since the beginning of time. They only happen to be isolated for the time being. Finally, it may be mentioned that this restriction to the case of isolated systems is only a momentary convenience when one starts to build up the theory of interpretation and it will be relaxed at a later stage.

One can then define the mathematical framework of the theory:

Rule 1. *The theory of a given individual isolated physical system can be entirely expressed in terms of a specific Hilbert space and the mathematical notions associated with it, particularly a specific algebra of operators.*

It was seen in Chapter 1 how this principle originated. One perhaps gets the feeling that, starting so abruptly, the theory appears almost aggressively abstract. It asserts in an extreme form how deep the problem of understanding reality is. We will not pause however to consider this aspect at present and only a few specifications will be added to the principle, mostly concerning its use: the word "entirely" occurring in it will be taken in its strongest sense, to mean that not only dynamics, but also the logical structure of the theory and the language one uses when applying it to observations and experiments will be cast into the mold of Hilbert spaces.

The specific algebra of operators mentioned in the rule depends upon the system to be described. Examples were given in Chapter 1 when individual particles were discussed and other ones will be treated later when we discuss macroscopic objects.

It could be said that Rule 1 defines in some sense a whole class of possible models for a given physical system, rather than assuming that there is an absolute Hilbert space representing it somewhere in Heaven. For instance, a hydrogen atom need not be described by the same Hilbert space framework in the cases when one takes the electron spin into account or not, when this electron is treated relativistically or not, when one takes into account the corrections coming from quantum electrodynamics or not. In the later case, one must use a very fat Hilbert space allowing for a possibly infinite number of photons, electrons, and positrons. When one goes even farther by looking at the quarks constituting the proton nucleus, the Hilbert space becomes even bigger. Rule 1 therefore defines a universal framework but it leaves open the version of it which is best suited for a specific purpose.

Although there are other ways of formulating the basic principles of quantum mechanics (using Feynman path formalism or abstract algebras of observables), they are all equivalent in the case of a nonrelativistic system with a finite number of degrees of freedom. As a matter of fact, the basic problems of interpretation are most conveniently set up in this framework and one will accordingly leave aside the case of systems having an infinite number of degrees of freedom or relativistic velocities, except for simple questions having to do with photons. This does not mean that there are no problems of interpretation outside this restricted framework, but they need most often more sophisticated techniques and, furthermore, they have not been much investigated.

2. Dynamics

Dynamics is governed by the Schrödinger equation:

Rule 2. *An isolated system is associated with a specific self-adjoint operator, which is its hamiltonian H. The vectors in the Hilbert space change with time according to the Schrödinger equation*

$$i\hbar \frac{d\alpha}{dt} = H\alpha.$$

Here again, some people prefer to start from the unitary group of time translations and get Schrödinger's equation from it, but this is unessential. What is much more important is the fact that time appears in quantum mechanics as a parameter that is not associated with an operator.

One will often use the formal solution of the Schrödinger equation in terms of the evolution operator $U(t)$ defined by $U(t) = \exp(-iHt/\hbar)$, from which one gets for the vectors in Hilbert space $\alpha(t) = U(t)\alpha(0)$. The notion of time-dependent Heisenberg operators, which was already considered in Chapter 1, will be taken for granted.

As for the hamiltonian, it may be assumed to be well known since one is dealing only with systems involving a finite number of nonrelativistic particles or photons. One may also sometimes depart slightly from Rule 2, when considering for instance a particle in a magnetic field. As a matter of principle, the field should be considered as generated by a coil or a magnet, which should be taken as a part of the isolated system to be taken into account. This procedure is however somewhat awkward and one does not gain much by using it rather than a much simpler treatment where the particle is considered as being isolated and submitted to a formal external field. This simplification, and similar ones, are always more or less trivial and don't need too much attention.

Finally, it will be convenient to state a third rule concerning noninteracting systems:

Rule 3. *Let two physical systems S' and S'' be associated with the Hilbert spaces $\mathbf{H}'$ and $\mathbf{H}''$ and the hamiltonians H' and H''. When they do not interact, they constitute both together a larger physical system S whose Hilbert space $\mathbf{H}$ is the tensor product $\mathbf{H}' \otimes \mathbf{H}''$, the hamiltonian H being the operator $H = H' \otimes I'' + I' \otimes H''$.*

One should also add to these three rules the bosonic or fermionic characters of the particles. It means, according to Rules 1 and 3, that one should only use a Hilbert space basis consisting of symmetric or antisymmetric tensor products of one-particle states and, according to Rule 2, that the hamiltonian commutes with a permutation operator exchanging identical particles. Together these rules are enough to determine the mathematical framework of physics to be used, as well as the dynamics.

THE PROPERTIES OF A SYSTEM

We now come immediately to the simplest logical aspects of quantum mechanics, with the first hints that the abstract formalism contains in germ a faithful description of all the real phenomena of physics.

3. Properties[1]

We shall first define an *observable* as being any self-adjoint operator A acting in the Hilbert space of the physical system. This name is slightly misleading, in so far as it seems to presume that the value of any observable could be known with the help of some measurement. This is what von Neumann assumed explicitly, namely that every observable is, in principle, measurable. The name itself bears testimony to this kind of assumption. Von Neumann did not make very precise what is meant by something being measurable *in principle*. However, his theory of measurement, as explained in Chapter 2, can shed some light upon the question: An observable is mathematically measurable in the sense that one can write a coupling hamiltonian with an imaginary system, to be called the measuring apparatus, and there is a quantum coordinate of this apparatus whose final value corresponds to the initial value of the "measured" observable. This has obviously nothing to do with a real measurement with a realistic apparatus.

It will be found later on that von Neumann's assumption is untenable as far as real measurements are concerned. In other words, not all observables can be observed. However, this does not preclude an observable being something one can talk about in a consistent way. One might accordingly think of the word "talkable" as perhaps more proper and more general than "observable", while some operators could have values really accessible by measurement and would be the true observables. This vocabulary would be however rather awkward and, furthermore, the present terminology has become universally accepted so that we shall continue to use it. As a matter of fact, most observables, whatever they may be, will only be used as mathematical toys for doing logic, at least in the next few chapters, and their exact status does not matter before one is able to really say what a measurement is.

The spectrum σ of an observable A generally consists of a discrete part σ_d and a continuous part σ_c. One shows in mathematics that the continuous spectrum can in general be split into two different parts: the completely continuous spectrum, which is Lebesgue measurable, and the singular spectrum, which is not (though it is measurable in the generalized Stieljes–Lebesgue sense).[2] These subtleties are sometimes significant and there are cases where the singular spectrum is not empty and it is physically meaningful. This seems to happen in the case of some observables occurring in the description of chaotic systems and also in the case of layers of atoms standing at the surface of a crystal, when the lattices of the layer and of the crystal are incommensurate. It will however be more convenient to leave aside these exotic questions and to consider only observables having at most a discrete and a completely continuous spectrum.

A *value* of an observable A will be defined as any real number belonging to its spectrum. Let then D be a subset of σ. It should be Lebesgue-measurable if it belongs to the continuous spectrum, but it will be better to ignore this kind of mathematical refinement, which does not add anything to physics, at least concerning the questions we are interested in. According to the spectral theorem given in Chapter 1, there is a well-defined projector E in Hilbert space that is associated with the set D. It projects upon the subspace **M** generated by the eigenvectors of A (in Dirac's sense) corresponding to eigenvalues belonging to D. More explicitly, one has

$$E = \sum_{a_n \in D';r} |a_n r\rangle\langle a_n r| + \int_{D''} \sum_r |ar\rangle\langle ar|\, da,$$

where D' is the discrete part of D, D'' is its continuous part, and r is a degeneracy index.

Properties

A property of a physical system refers to an isolated system S, an observable A and a real set D. It may be described by a sentence stating that "the value of the observable A is in the set D." This notion was introduced by von Neumann. It can clearly be associated with a well-defined mathematical object, i.e., the projector E already mentioned.

When one examines this notion more carefully, one can notice several interesting aspects of it; the first one is that the formalism offers in this way the possibility of a descriptive language. When one looks at the sentence or *elementary predicate* (as it was called by von Neumann) "the value of A is in D," one sees that all the words or symbols occurring in it have already a meaning within the conceptual framework of quantum mechanics; the observable A is formally a self-adjoint operator, its values are its eigenvalues, and D is a real set. So everything is defined except for the word "is."

One will not of course give a definition of the verb "to be," if only because it would necessarily use what is to be defined when it would begin by "to be is . . . ." Fortunately, the formalism is quite helpful by providing a projector as the mathematical equivalent for the full sentence expressing the property. One does not need therefore to define the verb "to be" but only to know how to use it when it occurs in the statement of a property. This is quite simple, if somewhat abstract, because the rules for using it will turn out to be the rules for using a projector as a mathematical object. One should not be afraid to see that a meaningful sentence is identified, at least to begin with, with a mathematical quantity. For instance, every sentence of the present book was typed upon a computer and through the process it became translated into a sequence of bits: a number. What is really significant is that, in the case of quantum mechanics, one will be able to follow all the subtleties and the rigor of a sensible language by keeping them in correspondence with specific mathematical forms and particularly with projectors.

In any case, the correspondence between properties and projectors is worth looking at in more detail, if only to show up to what point it is exact and free from ambiguity. It may happen that two different predicates are associated with the same projector: for instance, the two predicates saying respectively that "the value of A is in the interval $[-1, +1]$" and "the value of A^2 is in the interval $[0, 1]$" have the same projector. But everybody will agree that, though the predicates are differently worded, the properties are the same.

As one understands the meaning of a predicate, it is always possible to make a more or less trivial computation and to assert that it is the same thing to say the value of A is in D or to say that the value of the function $f(A)$ is in the domain $f(D)$, at least when the inverse function exists and $f^{-1}(f(D)) = D$. This is precisely the case where one can prove that the two properties have the same projector by using the spectral theorem, so that there is a much closer correspondence between a projector and the meaning of a predicate than with the predicate itself.

Conversely, every projector E in the Hilbert space can be associated with a property. This can be shown easily by taking E itself as the observable and the single eigenvalue $\{1\}$ for the corresponding set. The property can then be formulated by "the value of E is in $\{1\}$."

One may also remark that to associate in this way a rather intuitive physical notion with a well-defined mathematical object offers a typical example of what was meant when Rule 1 asserted that the Hilbert space framework should be enough to encompass all of physics, including its language.

Properties of Properties

We shall collect here a few more or less unrelated remarks:

There is a close correspondence between the logical aspects of the predicates and the algebraic relations among projectors. Consider for instance two disjoint subsets D' and D'' in the spectrum of A. According to the spectral theorem, the associated projectors satisfy the equations

$$E''E' = E'E'' = 0. \tag{3.1}$$

It will be said that, when equation (3.1) is satisfied, the corresponding properties are *mutually exclusive*. This is easily understood: What is more exclusive after all than two sentences where one states that the value of a quantity is somewhere while the other sentence says that the same quantity is somewhere else? The union D of D' and D'' is also associated with a projector E, which is the sum of E' and E''. This may be taken as a hint that to say that the value of A is in D means either that it is in D' or that it is in D''. So, logic is beginning to enter timidly.

Similarly, the set $\overline{D}$, complementary to D in σ, is associated with the projector

$$\overline{E} = I - E, \tag{3.2}$$

and one may consider it to be associated with the negation of the property associated with E. This is because saying that "the value of A is not in D" is tantamount to saying that "the value of A is in $\overline{D}$."

Let us also notice something that will be useful later on: Given two mutually exclusive projectors E' and E'', i.e., obeying equation (3.2), one can always find an observable A such that E' and E'' are the projectors associated with disjoint sets of values for A. The proof is easy by introducing three distinct real numbers a', a'', and b, taking $\overline{E}$ to be the complement of $E' + E''$, and putting $A = a'E' + a''E'' + b\overline{E}$.

It is also often convenient not to restrict the domain D to be a subset of the spectrum of A and to associate a projector with every real set D given by

$$E = \sum_{a \in D \cap \sigma} \left(\sum_r |a; r\rangle\langle a; r| \right).$$

Finally, one can extend the notion of properties to the case of two commuting observables A and A' or to any number of commuting observables. Rather than specifying a range of values for A and another one for A', one can just as well use a more general two-dimensional domain containing several values for A and A'. For instance, in the case of the three-dimensional position of a particle, the property stating that its position is in a domain V of space is associated with the projector

$$E = \int_V |x\rangle\langle x| \, d^3x. \tag{3.3}$$

4. Probabilities and States

To define the state of a quantum system is a touchy problem. The pioneers of the theory considered it to be defined by a wave function, or by a vector in Hilbert space. Later on, by taking into account the case when the system is prepared in a random way, von Neumann defined the state more generally by a density operator. According to him, the notion of state is basically probabilistic. It represents data allowing the prediction of the probabilities for all the possible results of all possible measurements. This could be called a predictive state.

The state depends however upon the actual conditions under which the system was prepared. This might be called a historical notion of state. To become effective, this kind of definition would have to take into account how the isolated system, for instance an atom, was extracted from a medium where it was interacting with many others and how it behaved when crossing a variety of experimental devices.

So there are two possible notions of states: a predictive one, directed towards the future, and a historical one, remembering the past. As a matter of fact, these notions coincide, but to prove this is quite an accomplishment of the theory of interpretation and it will be shown only later on. In order to follow a clearcut line of reasoning, one must adopt at least temporarily only one definition. The historical approach is too difficult to follow, as one does not yet know what a measurement or a preparation really is. Hence one must start from the predictive definition of a state.

The power of a logical approach will appear here for the first time, because it will be shown that a few sensible assumptions, obviously inspired by the aim of preserving a meaning for the language of properties, are enough to specify uniquely the description of the state and the specific form of quantum probabilities.

Probabilities and States

We will say that *the state of a system is well defined when one can assign a definite probability to every conceivable property.*

The probabilities one is looking for are by necessity rather formal, since one cannot yet assume that a measurement process can be realized so as to give them an experimental meaning. As a matter of fact, it will be shown in Chapter 7 that there exist many properties to which one can assign a probability but which cannot be tested by a measurement, even as a matter of principle. We shall therefore assume that, for some deep reasons, perhaps to be clarified later on, the construction of a language describing the facts of empirical physics must use probabilities. Later, when the interpretation is complete, some probabilities will become empirical, with the meaning of predicting the frequencies of random experimental events. For the time being, they are just mathematical tools, what a mathematician would rather call measures, which turn out to be a necessary step in constructing the language of physics. This is why the constraints to be imposed upon them by logic will be so important.

What Probabilities?

What should one expect from a purely formal notion of probability? Every property is associated with a projector so that one is looking for a probability $p(E)$ for every projector E. Let us guess what conditions it should satisfy.

One should first expect that $p(E)$ depends only upon the property to be considered. For instance, if **M** is the subspace onto which E is projecting,

FOUNDATIONS AND PROPERTIES

$p(E)$ depends only upon **M** and not upon a choice of basis in it, since this would mean that the property alone is not sufficient to define completely the probability.

Then we ask that $p(E)$ be a positive number smaller than 1, like any probability: $0 \leq p(E) \leq 1$. This can be made more precise in view of a few trivial remarks: If the property states that the value of an observable A is in the empty set, the corresponding probability should obviously be 0. But the corresponding projector is the null operator so that $p(0) = 0$. Similarly, when the property states only that the value of A is an arbitrary number, whatever it may be, the associated projector is the identity operator and the property is necessarily realized, so that one must have $p(I) = 1$.

The most important condition is additivity: Let D' and D'' be two disjoint real sets and D their union; the probability that an observable A has its value in D should be the sum of the probabilities for this value to be in D' or in D''. These properties correspond to two commuting projectors E' and E'' whose product is 0 (mutually exclusive projectors). Conversely, it was noticed previously that one can associate two mutually exclusive projectors with two different sets of values for some observable A, so that one can state additivity as follows: $p(E' + E'') = p(E') + p(E'')$, for every pair (E', E'') of mutually exclusive projectors.

5. Gleason's Theorem

The Theorem

It is remarkable that these mild assumptions are enough to entail the existence of a density operator and a specific form for the probabilities. This result, both beautiful and difficult, is due to Gleason and its precise statement is given in the Appendix. Its conclusion is that the probability is necessarily of the form $p(E) = \mathrm{Tr}(\rho E)$, where ρ is a positive definite operator with unit trace: $\mathrm{Tr}\,\rho = 1$, to be called of course the density operator.

The theorem does not say how to construct it nor does it give any hint how to relate it to a preparation process. It only says that ρ should exist or else it would be impossible to assume the existence of probabilities for the properties. This operator defines therefore completely the state of the system.

The simplest case occurs when the state is completely defined by a normed vector $|\alpha\rangle$ in Hilbert space, in which case one has

$$\rho = |\alpha\rangle\langle\alpha| \qquad (3.4)$$

and one says that this is a pure state. The name comes from historical reasons and there is no essential privilege attached to it. As usual, a nonpure state will be called a mixture. This is also only a traditional name and it does not mean that such a state should by necessity represent a statistical ensemble, as one can often read here and there.

The State as a Property

Equation (3.4) shows that the density operator for a pure state is a projector of rank one (i.e., projecting onto a one-dimensional subspace). It can be therefore associated with a property. It means that the fact of being associated with the specific state vector $|\alpha\rangle$ is a property of the system having probability 1. It also sometimes happens that one can summarize a whole preparation process by asserting simply a property of the system, to be called an initial property. If one denotes by E_0 the projector representing it, one can then generalize equation (3.4) by putting

$$\rho = \frac{E_0}{\text{Tr } E_0},$$

the trace in the denominator insuring that the trace of the density operator is unity. This formula is clearly restricted only to the case when the initial property has a finite-rank projector.

Although the probabilities one is dealing with are still purely formal, it is interesting to compare them with what is assumed in elementary quantum mechanics. In the case of a pure state, the property associated with a projector E has the probability

$$p(E) = \text{Tr}\{|\alpha\rangle\langle\alpha|E\} = \langle\alpha|E|\alpha\rangle.$$

When E itself projects upon a normalized vector $|a\rangle$, one gets $p(E) = |\langle\alpha|a\rangle|^2$, i.e., one gets the well-known elementary formula for probabilities, which was thought for a long time to be an independent axiom of the theory. For a particle having a state vector α or a wave function $\alpha(x)$, the property stating that the position observable $\mathbf{X}$ is in a given volume V has the probability

$$p(E) = \int_V \text{Tr}\{|\alpha\rangle\langle\alpha|x\rangle\langle x|\}d^3x = \int_V |\alpha(x)|^2 \, d^3x,$$

where one has used equation (3.3) for the associated projector. This is of course Born's formula.

6. Taking Time into Account

What was done up to now has not made any reference to time. We now introduce it. From an intuitive standpoint, the probability of a property depends upon the circumstances taking place at the time when it is supposed to occur, i.e., upon the state at that time. One should then use the Schrödinger equation in order to specify the state operator at a given time.

Let us consider to begin with a pure state that is defined at time 0 by a normed vector $\alpha(0)$ so that $\rho = |\alpha(0)\rangle\langle\alpha(0)|$. Schrödinger's equation gives the value of this vector at a time t as $|\alpha(t)\rangle = U(t)|\alpha(0)\rangle$. Hence, if one writes, $\rho(t) = |\alpha(t)\rangle\langle\alpha(t)|$, one gets $\rho(t) = U(t)\rho U^{-1}(t)$. This result is easily extended to an arbitrary density operator, as can be shown by writing the density operator at time 0 in terms of its eigenvalues and eigenvectors:

$$\rho = \sum_n w_n |\alpha_n(0)\rangle\langle\alpha_n(0)|.$$

One will then consider that the probability of the time-dependent property stating that "the value of the observable A is in the domain D at time t" is the probability of the time-independent predicate with projector E saying that "the value of A is in D," as obtained under circumstances where the density operator is $\rho(t)$. As a result, one has $p = \text{Tr}\{\rho(t)E\}$. If one introduces the Heisenberg time-dependent projector $E(t) = U^{-1}(t)EU(t)$, this probability can also be written as

$$p = \text{Tr}\{\rho(t)E\} = \text{Tr}\{\rho E(t)\},$$

the density operator ρ occurring in the last term being taken at time 0, which is of course only a convenient reference time to be chosen once and for all.

APPENDIX: GLEASON'S THEOREM

This theorem is due to Gleason.[3] A clearer proof was given by Jost.[4] Its precise formulation will be given here to complete what was mentioned in the text.

Let **H** be a Hilbert space with dimension N (finite or countably infinite) larger than 2. Let p be a probability measure defined upon the closed subspaces of **H** that is additive for a finite or infinite number of mutually orthogonal subspaces. Then there exists a trace-class positive operator ρ with unit trace such that $p(E) = \text{Tr}\{\rho E\}$, E being the projector upon a closed subspace.

The condition $N \geq 3$ is due to some peculiarities of the cases $N = 1$ and $N = 2$. When $N = 1$, there is only one nonzero projector and the theorem is void of content. When $N = 2$, every nonzero projector that is not equal to the identity can be written as $E = \frac{1}{2}(I + \boldsymbol{\sigma} \cdot \mathbf{n})$, where **n** is a unit three-dimensional vector. The only projector that is mutually exclusive with E is associated with the vector $-\mathbf{n}$. One can then take for $p(E)$ any function $p(\mathbf{n})$ such that $p(\mathbf{n}) + p(-\mathbf{n}) = 1$. One can however easily remove the condition $N > 2$ on physical grounds by considering a system S having a one- or two-dimensional Hilbert space together with another system with which it does not interact.

Under some technical restrictions, the theorem can be extended to non-separable Hilbert spaces.[5]

4

Histories

A history of a physical system is simply a series of properties occurring at different times. The name clearly expresses the idea, suggesting a sequence of significant events occurring in the course of time. The primary interest of this notion is its ability to describe in detail all that happens to a system during its evolution. It works equally well for a microscopic or a macroscopic system and the concept is so natural that it was invented and rediscovered many times by different people. This obviousness will become clear in the first part of the chapter when a realistic example introduces the concept of history. It is also important that one can assign a specific probability to a given history. This was also discovered many times, more or less as a consequence of the Copenhagen interpretation, each event mentioned in the history being considered as the result of some measurement. Though this point of view is not necessarily misleading, this way of introducing probabilities is much too narrow because histories turn out in fact to be more fundamental than measurements and they are prior to them in the order of concepts. They play a basic role in the systematic construction of an interpretation where, as a matter of fact, they constitute the basic tools of the language of physics. This is probably the reason why their probabilities are completely defined by a few logical requirements. Furthermore, as was first shown by Robert Griffiths, not all histories (or families of histories) make sense, because the necessary additivity of their probabilities is valid only under very specific consistency conditions.[1] This means that the language of quantum mechanics is very constrained, making clear what makes sense or not without having to call for measurements to do it.

We already met another kind of history in Chapter 1, where the Feynman histories, also called Feynman paths, were mentioned. The histories to be considered in the present chapter are different and they belong to the logical structure of the theory rather than its dynamical structure. Since Feynman histories play only a modest role in the framework of interpretation whereas Griffiths histories are essential, it will be convenient to reserve the name history for the latter in the rest of this book.

THE NOTION OF HISTORY

1. Experiments and Histories

When we give the name history to what happened in some country or to some people in the course of time, the idea is perfectly clear. Everything that has taken place is, in one way or another, a part of history or of some history. This notion also applies to what happens when an experiment in physics is performed, as we shall now see.

Tell Me a Story

Let us suppose for the sake of argument that we are interested in a simple nuclear reaction such as

$$n + p \to d + \gamma. \tag{4.1}$$

It is a neutron-proton collision producing a deuteron and a photon. In order to realize it, one can start with a beam of neutrons with well-controlled momentum and detect the products of the reaction while also measuring their momenta. Since it is known that energy and momentum are conserved in a collision, it is not necessary to detect both the photon and the deuteron and it is enough for instance to measure the photon momentum.

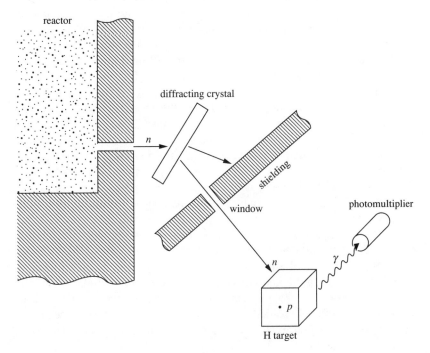

4.1. *An experiment in nuclear physics*

Among the many different devices that could be used for this kind of experiment, one can resort to the one shown in Figure 4.1: The neutrons are produced in a nuclear reactor R. Some of them cross the wall of the reactor through a channel C. It will be convenient to assume that a door P can be opened or closed at will in order to let the neutrons go through the channel at a well-defined time, though this is unnecessary for an actual experiment and it is assumed only for the sake of convenience. It will be assumed that this door was kept open during a short time interval around time 0.

The neutrons can then cross a velocity selector S. It consists for instance of a silicon monocrystal diffracting the neutrons. Since the wavelength of a neutron depends upon its velocity and it also determines the directions in which the wave is diffracted, the velocity can be selected by selecting the outgoing direction of the neutrons. This is accomplished by a window F in a shielding wall, so that the neutrons crossing the window have a rather well-defined velocity. The nuclear reaction can then take place in a target T containing liquid hydrogen, the protons taking part in the reaction being the nuclei in the hydrogen molecules. Finally, outgoing gamma rays can be detected and their energy is measured by a battery of counters surrounding the target.

The only observable phenomenon that can be directly ascertained is that one gamma detector D has reacted, yielding a value for the gamma ray energy. One knows consequently that a photon was emitted in the direction of this detector. It is then rather clear that one can tell the story that happened, which is a very plain story though still worth telling. Let us see how some Scheherazade of physics could narrate it, disregarding the risk of making her sultan sleepy:

"A neutron crossed the channel during the short time when the window was open. A moment later, its wave met the silicon crystal and it was diffracted. Then the neutron crossed the window, then it entered the target. It collided with a proton. The nuclear reaction (4.1) took place and a photon was produced. A very short time later, the photon entered in the detector D. The counter registered it and gave the value of its energy."

This kind of story is essentially what a physicist tells himself or what he says when he tries to communicate what happened during the experiment. From the supposed fact that the neutron has crossed the velocity selector, he concludes that he knows its initial velocity prior to the collision. Since the proton is essentially at rest, he can then assert the whole kinematics of the reaction from the knowledge of the photon momentum. He can also tell approximately at what time the various events occurred. A history is no more than that, namely something quite commonplace but irreplaceable, so obvious that our mind relates to it almost automatically. But does it have anything to do with science?

The Subtleties of Histories

However obvious it may look at first sight, the story we just told is full of ambushes and of subtleties when one takes quantum physics seriously into account. It treats rather carelessly the neutron sometimes as a particle (when it crosses the door or the window) and sometimes as a wave (when diffracting on the crystal). Moreover, the story cares very little about what the Copenhagen interpretation gave us the right to say and what it forbade. Except for one phenomenon, which is what is shown by the detector, everything in it is just imagining the world of particles and even assuming that we know things about it that are forbidden by the Copenhagen rules. Nevertheless, experimental physics is quite naturally described by such stories, giving a meaning to elementary events and allowing us to interpret them so that each piece of the story is "something", an event taking place at some time that is described by a simple property of the system.

2. Definition of Histories

A history has been defined in general as a sequence of properties of an isolated system occurring at different times. This is obviously a generalization of the concept of property where now several properties are considered together rather than just one. A property occurring at a specific time can then be compared with a photographic snapshot while a history looks like a succession of snapshots, i.e., a motion picture.

More precisely, let us assume that the state of the system is given at an initial time 0 by a density operator ρ. We shall consider a sequence of times $t_1, t_2, \ldots, t_n$, ordered in such a way that $0 \leq t_1 < t_2 < \cdot < t_n$ and a property of the system will be asserted at each time. It may be noticed that the times are discrete so that the comparison with a motion picture, rather than with a play in a theater or a scene in real life, is quite appropriate.

Consider some observables $A_1, A_2, \ldots, A_n$, one for each time. Several commuting observables can be introduced at the same time and one may think of them as being essentially one observable taking its values in a many-dimensional space. We will also introduce a family of real sets, for instance, some intervals $D_1, D_2, \ldots, D_n$ for the values of the observables. If all the observables are position operators, their values will be defined within some volumes. It will not be assumed that the various observables A_k commute with each other. One should also remember that with each given observable A_k and domain D_k, one can associate a projector E_k in Hilbert space and, at a given time, a time-dependent projector $E_k(t_k)$. It corresponds to the property according to which "the value of A_k is in the set D_k at time t_k." A history involves only a finite number of elementary predicates so that, as a mathematical object, it consists simply of a collection of time-dependent projectors $\{E_1(t_1), \ldots, E_k(t_k), \ldots, E_n(t_n)\}$.

Just as one can express a property by a predicate, a story can also be expressed by a sentence, which would here be: "the value of A_1 is in D_1 at time t_1, the value of A_2 is in D_2 at time $t_2,\ldots$, the value of A_n is in D_n at time t_n." In the next chapter, after learning how to introduce logic in these sentences, it will be possible to say it slightly differently, asserting that "the value of A_1 is in D_1 at time t_1 *and* that of A_2 is in D_2 at time $t_2 \ldots$ *and* that of A_n is in D_n at time t_n." It might look pedantic not to say this immediately but one should not forget that understanding quantum mechanics is a tricky business and everything that is set forth, every little step, must be carefully justified by being shown to be compatible with the basic axioms of the theory. No slip of the tongue due to carelessness, no statement considered as "obvious" because men always thought it to be that way, should be tolerated without running the risk of a blunder.

THE PROBABILITIES OF HISTORIES

3. Logical Criteria for the Probabilities

The next step will be to define a probability for histories. From an empirical standpoint, one can effectively consider that a story, such as the one we discussed when considering a nuclear reaction, is more or less probable. The neutron could have missed the window, it might not have collided with a proton or, even if it did, the collision could have been a simple scattering with no reaction. If the reaction took place after all, the photon could have gone in another direction. All that is obviously a matter of probability.

It should be stressed however that it is impossible to rely uniquely upon such empirical considerations when one starts building an interpretation. This is because what we really know is very little and much of what we say when telling a story is rather speculative. Moreover, we only know how to assert what we actually see, such as for instance what is shown by a counter, by using a simple empirical language far removed from the basic formalism of the theory. Accordingly, when considering an experiment, one can only guess that probabilities might be useful and significant but one cannot tell beforehand what these probabilities are explicitly.

Moreover, it should be stressed once again that we don't yet know how to interpret an experiment and, for instance, how a detector works, because we only know the basic principles of the theory presently. One of the main goals of interpretation is precisely to understand experiments in a way consistent with the formalism. Accordingly, we should be careful not to assume that we know what happens when an experiment is actually performed. When talking of probabilities, this means in particular that we cannot yet give them an empirical content such as would be the case if we were able to understand what a series of identical measurements with results giving empirical frequencies is. The probabilities we shall be able to conceive will therefore once again remain purely theoretical and they will remain at the level of a purely mathematical notion, mathematics and logic being still better guides than experiment at this very early stage of interpretation.

Graphical Representation of Histories

It will be convenient to represent histories by graphs. As an example, one may consider a history involving only two times ($n = 2$): the domains D_1 and D_2 for the values of the two observables are two intervals. One can then associate the corresponding history with a rectangle in a two-dimensional plane projecting onto the axes over these two intervals. This construction is easily generalized to any number n of reference times, the rectangle being replaced by an n-dimensional box (which is the direct product of the intervals from a mathematician's standpoint). When the domains D_1 and D_2 are more complicated than intervals, their direct product is a set consisting of rectangles, some of which may be reduced to a segment or a point when there are discrete eigenvalues. The corresponding regions might look complicated if one were actually to draw them, but this is of no consequence for our reasonings and we shall continue to call them boxes. Two histories are said to be *disjoint* when their boxes do not intersect. It may happen that, when adding two disjoint boxes, one generates another box. When the initial sets are intervals and the boxes are rectangles, this happens when the two rectangles stand side by side. The history associated with this larger box will be said to be the *union* of the two previous histories.

Tautology and Nonsense

Let us now consider an arbitrary number of reference times, say n. When it happens that the projector for a property $E_k(t_k)$ coincides with the one immediately following it, i.e., $E_{k+1}(t_{k+1})$, it looks as if nothing new is asserted: the history only repeats at time t_{k+1} what was already known from what took place at time t_k, except that time evolution according to the Schrödinger equation is taken into account. Nothing else took place in between and stating twice the same property is therefore pure redundancy. One will say that this is a *tautology*, which is the technical name in logic for a trivial repetition. Similarly, when two neighboring projectors $E_k(t_k)$ and $E_{k+1}(t_{k+1})$ are mutually exclusive so that they contradict each other, one will say that this is a situation of nonsense.

It should be stressed how important it is to consider two immediately successive times, while nothing happens between them, no intermediate property being introduced. Consider for instance the case of a spin-$1/2$ particle. To say that a spin component is $+1/2$ at some time and it is still equal to $+1/2$ a moment later when nothing happened in between is obviously a tautology. Its triviality comes from dynamics according to which a spin component is a constant of motion when there is no magnetic field; saying it again does not introduce anything new and just repeats what was already asserted. If on the contrary something happened or some statement was asserted in between, the repetition is no longer a tautology. For instance, if the particle has suffered a collision, it is now significant information to state that the spin component has not changed during the collision and it is no more a tautology.

Criteria for the Probabilities

Let us now state the conditions upon the probability $p(h)$ of a history h so that it is sensible from the standpoint of probability calculus and also of logic. Probability calculus as such imposes only three conditions upon probabilities, namely, positivity, normalization, and additivity. Positivity means that the probability $p(h)$ should be positive or 0. Normalization means that the trivial story asserting nothing worth mentioning is equal to 1. More precisely, this trivial history states that each observable has its value anywhere in its spectrum so that each set D_k is the whole real set (or the spectrum of A_k). The corresponding projectors are all reduced to the identity operator and their sequence is $(I, I, \ldots I)$. One can therefore write normalization as the condition $p(I, I, \ldots I) = 1$. As for additivity, which is by far the most important assumption here, it occurs when the union h of two disjoint histories h' and h'' is also a history and it says that

$$p(h') + p(h'') = p(h). \tag{4.2}$$

These three conditions would be all that is needed from a mathematical standpoint in order to define a measure upon histories.

Let us now consider the logical conditions to be expected from the probabilities:

1. When a property occurring in a n-times history h is trivial, i.e., when its associated projector is the identity operator, it will be assumed that $p(h)$ is equal to the probability of the $(n-1)$-times history one obtains by suppressing the property.
2. When two immediately successive properties give rise to a tautology, one can suppress one of them, for instance the second one, and the probability of the tautological history is the same as the probability of the $(n-1)$-times history so obtained.
3. In the special case where the initial state is itself described by a property (i.e., when the density operator is proportional to a projector) and when the property holding at time t_1 repeats it tautologically, one can also suppress the first property to get the probability. More generally, when the first property does not add anything new when compared to the initial data, i.e., when one has $E_1(t_1)\rho E_1(t_1) = \rho$, one can com-

pute the probability of the history by considering only the $(n-1)$-times history where the first property has been suppressed.
4. When two immediately successive properties lead to nonsense, the probability should be 0.

It may be noticed that these conditions can be used to relate together histories with different numbers of reference times. The assumptions can then be completed by considering the case when $n = 1$:

5. For a one-time history ($n = 1$), which necessarily coincides with a unique property with a projector $E(t)$, the probability is given as before by $p = \text{Tr}\{\rho E(t)\}$, according to Gleason's theorem and noticing that the assumptions of this theorem are also of a logical nature.

One will of course assume, as was the case for a unique property, that the probability of a history should only depend upon the properties it mentions and nothing else, i.e., only the projectors.

Uniqueness of the Probability

These simple conditions happen to determine in a unique way the probabilities of histories. The proof of this result is given in Appendix B and we only make the following remarks here:

The presently known proof only applies to the case $n = 2$. It does not seem however that this limitation could be due to the existence of other possible forms for the probability when n is larger than 2. As a matter of fact, when all the observables are restricted to be either a position or a momentum, another proof can be given for any value of n yielding the same result, as will be shown in next section. It seems therefore that the present limitation to $n = 2$ in the general case is due to an inadequacy of the mathematical techniques entering the proof.

Another important point should be stressed: The proof does not use all the additivity conditions (4.2) but only a few of them—those where there is a difference between two histories h' and h'' only as far as the last statement (for $t = t_n$) is concerned. It was found in the case of the nuclear reaction that the last statement is the result of a measurement and this is quite often the case in general. Anyway, the essential point is that many additivity conditions remain to be checked once the form of a history probability has been obtained and, as will be seen, many histories do not stand the test.

The Form of Probabilities

The unique probability one finds for a history is given by

$$p = \text{Tr}\{E_n(t_n) \cdots E_k(t_k) \cdots E_1(t_1) \rho E_1(t_1) \cdots E_k(t_k) \cdots E_n(t_n)\}. \quad (4.3)$$

One should pay attention to the time ordering among the projectors. This order is compelling and essentially comes from the logical condition (2) concerning the repetition of an initial property. It will be shown in the next chapter that this implies the non-invariance of the logical structure of quantum mechanics under time reversal.

One may also notice a few features of this formula:

When the density operator represents a pure state so that $\rho = |\alpha\rangle\langle\alpha|$, one can write

$$p = \langle\alpha|E_1(t_1)\cdots E_k(t_k)\cdots E_n(t_n)E_n(t_n)\cdots E_k(t_k)\cdots E_1(t_1)|\alpha\rangle,$$

or $p = \|E_n(t_n)\cdots E_k(t_k)\cdots E_1(t_1)|\alpha\rangle\|^2$. Everything happens as if every new property had the effect of projecting the previous state so that, at the end, the probability of the history is just the elementary probability of the surviving vector.

Equation (4.3) can be slightly simplified by using the property $E^2 = E$ of the last projector $E_n(t_n)$, together with cyclic invariance of the trace, so that

$$\begin{aligned} p = \text{Tr}\{E_n(t_n)E_{n-1}(t_{n-1})\cdots E_k(t_k)\cdots E_1(t_1)\rho E_1(t_1) \\ \cdots E_k(t_k)\cdots E_{n-1}(t_{n-1})\}. \end{aligned} \quad (4.4)$$

In the special case when the density operator is completely defined by a property with a projector E_0 so that $\rho = E_0/\text{Tr}E_0$, one gets

$$\begin{aligned} p = \text{Tr}\{E_n(t_n)E_{n-1}(t_{n-1})\cdots E_k(t_k)\cdots E_1(t_1)E_0 E_1(t_1) \\ \cdots E_k(t_k)\cdots E_{n-1}(t_{n-1})\}/\text{Tr}E_0, \end{aligned}$$

an expression containing only projectors.

The probability is always positive or 0. This can be shown by putting $\Omega = E_1(t_1)\cdots E_k(t_k)\cdots E_n(t_n)$, so that one gets $p = \text{Tr}\{\Omega^\dagger \rho \Omega\}$, where the positivity of p is an immediate consequence of the positivity of ρ. More explicitly, one can introduce the eigenvectors $|j\rangle$ and eigenvalues w_j of ρ to write

$$p = \sum_{jm} w_j |\langle j|\Omega|m\rangle|^2 \geq 0,$$

so that the positivity of p follows from that of the w_j's. It is also obvious that, when all the projectors are replaced by the identity operator, the probability is 1.

4. Connection with Feynman Histories*

Let us now consider more closely the meaning of the probability of a Griffiths history in the simple case when $n = 2$ and both observables A_1 and A_2 are either the position or the momentum of a particle in a one-dimensional space. Quantum dynamics will be described by Feynman sums, as explained in Chapter 1. This will lead us to a more direct derivation for the form of the probabilities as well as an understanding of the additivity conditions.

Let us begin by a few useful formulas. The probability amplitude for a particle initially at position x at time 0 to be in position x_2 at time t_2 is given by

$$\langle x_2, t_2 | x, 0 \rangle = \sum_F \exp[iS_F(x_2, t_2; x, 0)/\hbar],$$

where the sum is performed over all the Feynman paths F going from (x, t) to (x_2, t_2). The classical action $S_F(x_2, t_2; x, 0)$ is computed along the path F. The probability of finding the particle in a small interval Δx_2 in the neighborhood of x_2 is given by $|\langle x_2, t_2 | x, 0 \rangle|^2 \Delta x_2$. It was shown in Chapter 1 that the matrix elements of the projector $E_1'(t_1)$ expressing that the position is in the domain D_1' at time t_1 (this being a property) is given by

$$\langle x_2, t_2 | E_1'(t_1) | x, 0 \rangle = \sum_{F'} \exp[iS_{F'}(x_2, t_2; x, 0)/\hbar], \tag{4.5}$$

the sum being now taken over the paths F' crossing the region D_1' at time t_1. Similarly, one has

$$\langle x_2, t_2 | E_1''(t_1) | x, 0 \rangle = \sum_{F''} \exp[iS_{F''}(x_2, t_2; x, 0)/\hbar],$$

the paths F'' crossing now the domain D_1'' at time t_1.

If one identifies a Griffiths history with the family of Feynman paths satisfying its properties, one may assume that the probability for the Griffiths history stating that "the value of X is in D_1' at time t_1 and in the interval Δx_2 at time t_2" is given by

$$|\langle x_2, t_2 | x, 0 \rangle'|^2 \Delta x_2, \tag{4.6}$$

the amplitude $\langle x_2, t_2 | x, 0 \rangle'$ being given by a sum over the Feynman paths crossing the interval D_1' at time t_1 and ending in Δx_2 at time t_2. Using Equation (4.5) and introducing the projector

$$E_2(t_2) = \int_{\Delta x_2} |x_2, t_2\rangle\langle x_2, t_2|,$$

the corresponding probability becomes $\mathrm{Tr}\{E_2(t_2)E_1'(t_1)|x, 0\rangle\langle x, 0|E_1'(t_1)\}$, where one recognizes the probability already written for a Griffiths history. In order to complete the proof, one introduces a normalized wave function $\psi(x)$ for the initial state in place of the unnormalized vector $|x\rangle$. The corresponding amplitude is

$$\langle x_2, t_2 | \psi, 0 \rangle = \int \langle x_2, t_2 | x, 0 \rangle \psi(x)\, dx.$$

The probability (4.6) becomes then $|\langle x_z, t_2 | \psi, 0 \rangle'|^2 \Delta x_z$ which gives, after restricting the integration upon x_1 to the interval D_1',

$$\text{Tr}\{E_2(t_2)E_1'(t_1)|\psi\rangle\langle\psi|E_1'(t_1)\} = \text{Tr}\{E_2(t_2)E_1'(t_1)E_0 E_1'(t_1)\},$$

with $E_0 = |\psi\rangle\langle\psi|$.

One can therefore say that a Griffiths history is in that case a class of Feynman histories. The restrictions to $n = 2$ and to a one-dimensional underlying space are easily removed. As for the observables, one can use indifferently position or momentum coordinates at all intermediate times. However, the consideration of more general observables raises serious difficulties so that one cannot use this method confidently to give a proof for the general expression of the probability.

The use of Feynman paths also gives a better understanding for the conditions coming from additivity: consider for instance two nonintersecting intervals D_1 and D_1' for the position. Let F' denote the family of Feynman histories crossing D_1', at time t_1, F'' those crossing D_1'', and F those crossing the union D_1 of D_1' and D_1''. Additivity of probabilities gives in that case

$$\left| \sum_{F'} \exp[iS_{F'}(x_2, t_2; x, 0)/\hbar] + \sum_{F''} \exp[iS_{F''}(x_2, t_2; x, 0)/\hbar] \right|^2$$
$$= \left| \sum_{F'} \exp[iS_{F'}(x_2, t_2; x, 0)/\hbar] \right|^2 + \left| \sum_{F''} \exp[iS_{F''}(x_2, t_2; x, 0)/\hbar] \right|^2. \tag{4.7}$$

Hence, *requiring the additivity of probabilities selects the cases where probabilities and amplitudes add up separately.*

These conditions are obviously very restrictive. It will be seen however that they are not as drastic as one might fear at first sight, particularly when one requires only that additivity holds up to a small error rather than enforcing it absolutely.

CONSISTENCY CONDITIONS

5. The Consistency Conditions for Additive Probabilities

It has been shown that the probability of a history, as given by equation (4.3), satisfies the positivity and normalization conditions. It satisfies additivity in a very restricted way, namely for histories differing only in their last property at time t_n and otherwise identical at all previous times, as shown by the proof given in Appendix B. It remains to find what restrictions are brought by the other additivity conditions and to give a form for these conditions that will be more convenient than the condition in equation (4.7) on Feynman sums.

A First Encounter

Let us first see what the additivity condition (4.2) becomes in the case of two histories h' and h'', when $n = 2$. We shall assume that the two boxes associated with these two histories are disjoint and contiguous.

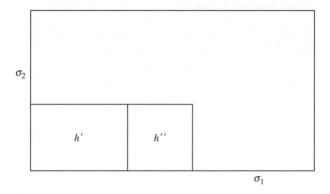

4.2. Two histories h' and h'' whose union is another history

Two cases are then possible: either the two boxes stand one upon the other or they are side by side. In the first case, the two histories have a common projector $E_1(t_1)$ and equation (4.4), where the last projector is written only once, shows immediately that additivity is satisfied. In the second case, the two families of projectors representing the two histories are respectively $\{E'_1(t_1), E_2(t_2)\}$ and $\{E''_1(t_1), E_2(t_2)\}$. Their union h corresponds to the history $\{E_1(t_1), E_2(t_2)\}$ where $E_1 = E'_1 + E''_1$. The corresponding probabilities are

$$p(h') = \text{Tr}\{E_2(t_2)E'_1(t_1)\rho\, E'_1(t_1)\}, \qquad p(h'') = \text{Tr}\{E_2(t_2)E''_1(t_1)\rho\, E''_1(t_1)\},$$

$$\text{and } p(h) = \text{Tr}\{E_2(t_2)E_1(t_1)\rho\, E_1(t_1)\}.$$

Replacing $E_1(t_1)$ by $E'_1(t_1) + E''_1(t_1)$ in $p(h)$, one gets

$$p(h) - p(h') - p(h'') = \text{Tr}\{E_2(t_2)E'_1(t_1)\rho\, E''_1(t_1)\} + \text{Tr}\{E_2(t_2)E''_1(t_1)\rho\, E'_1(t_1)\}.$$

The additivity condition is therefore

$$\text{Tr}\{E_2(t_2)E'_1(t_1)\rho\, E''_1(t_1)\} + \text{Tr}\{E_2(t_2)E''_1(t_1)\rho\, E'_1(t_1)\} = 0. \qquad (4.8)$$

These conditions were first discovered by Griffiths who called them *consistency conditions*. One knew in fact for a long time that some histories cannot have any meaning. This is the case for the the histories describing interferences when one tries to include in them the arm of an interferometer through which a photon went. One usually says that this kind of history is meaningless because there is no measuring device asserting which way the photon goes. The consistency conditions point towards another and probably deeper reason for rejecting such a history, which is the impossibility of assigning it a probability. This point will be analyzed in more detail in the next chapter where it will be found that it should also be rejected because it has no logical meaning.

Rewriting the Consistency Conditions

One can rewrite the consistency condition (4.8) in two different more convenient forms, each of which has its own advantages when it is generalized to n larger than 2. First use the self-adjointness of projectors to notice that

$$\begin{aligned}\mathrm{Tr}\{E_2(t_2)E_1''(t_1)\rho\,E_1'(t_1)\} &= \mathrm{Tr}\{E_2^\dagger(t_2)E_1''^\dagger(t_1)\rho E_1'^\dagger(t_1)\} \\ &= \mathrm{Tr}\{[E_2(t_2)E_1'(t_1)\rho\,E_1''(t_1)]^\dagger\} \\ &= [\mathrm{Tr}\{E_2(t_2)E_1'(t_1)\rho\,E_1''(t_1)\}]^*,\end{aligned}$$

so that the consistency condition (4.8) becomes

$$\mathrm{Re}[\mathrm{Tr}\{E_2(t_2)E_1'(t_1)\rho\,E_1''(t_1)\}] = 0. \tag{4.9}$$

One can also introduce commutators so that

$$\rho\,E_1'(t_1) = [\rho, E_1'(t_1)] + E_1'(t_1)\rho.$$

Using this expression in $\mathrm{Tr}\{E_2(t_2)E_1''(t_1)\rho\,E_1'(t_1)\}$, the second term in the right-hand side vanishes since the product $E_1''(t_1)E_1'(t_1)$ is 0 (as a product of mutually exclusive projectors). Proceeding similarly with $E_1'(t_1)\rho$ in the first term of condition (4.8), one can rewrite the condition by introducing a double commutator so that it becomes:

$$\mathrm{Tr}\{E_2(t_2)[E_1'(t_1),[\rho,E_1''(t_1)]]\} = 0. \tag{4.10}$$

These results are quite general and there exist various equivalent forms of the consistency conditions resembling either equation (4.9) or equation (4.10). They are given in Appendix A. Their most significant feature is that they are well-defined algebraic conditions that can be easily checked by a direct calculation. One may notice that condition (4.9) requires only a vanishing real part of a trace while equation (4.10) requires that a whole trace should vanish. This is because multiple commutators have selfadjointness properties so that the traces to be met are always real.

Families of Histories

Additivity obviously assumes that nonoverlapping histories represent so many different possibilities. Now, in probability calculus, one usually begins by considering a complete family of possibilities. For instance, one considers the various faces upon which dice can fall. This kind of approach will lead us to introduce a whole family of histories offering a complete sampling of histories. Geometrically, in the case $n = 2$, this can be obtained by covering completely the plane by a set of disjoint boxes.

All the histories in the family are associated with the same initial state of the system, the same sequence of times, and the same observables, the only difference being in the range of values for these observables. The simplest example of a complete family is due to Griffiths. It consists in slicing each spectrum, for instance the spectrum σ_k of A_k, into a set $\{D_k^r\}$ of disjoint domains covering the whole spectrum. These subsets of the spectrum may be conveniently a collection of intervals in the case of the position operator, each one giving the position within specific bounds. As a result, again in the case when $n = 2$, the two-dimensional space $\Xi = \sigma_1 \times \sigma_2$ is cut into a family of contiguous elementary boxes. The generalization to $n > 2$ is obvious.

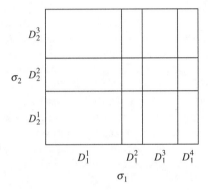

4.3. *The space of histories.* The spectra σ_1 and σ_2 of the two observables A_1 and A_2 associated with times t_1 and t_2 have been cut respectively into four intervals and three intervals. A Griffiths family consisting of twelve different histories is thereby generated.

Equation (4.3) gives a probability for each of these histories or elementary boxes. Any set in Ξ that is a union of several elementary boxes can be also assigned a probability, namely the sum of the probabilities for its constituting parts, at least when additivity is satisfied.

There are a priori as many consistency conditions as there are nonelementary boxes that can be made from elementary boxes. Alternatively, one can also say that there are as many consistency conditions as there are unions of histories in the family that are also histories, since histories and boxes are essentially equivalent. However all these consistency conditions might not be independent. They are given explicitly in Appendix A where the question of independence is also discussed. There are also other ways, sometimes more convenient ones, for cutting the space Ξ into a smaller number of boxes. This is however only a technical improvement and is left to the appendix.

Compound Histories

A geometric analogy has led us to consider the union of boxes or, in some sense, the union of histories. This kind of combination is also well known in probability calculus. For instance, when one plays poker, one is interested in combined possibilities that are more elaborate than just listing the five cards. More formally, probability calculus considers so-called compound events, as distinguished from elementary events. One may wonder what it means physically in the present case.

To answer this, we shall go back to our original experiment in nuclear physics. Consider the history already described, where the neutron crosses the velocity selector before hitting a proton and producing a photon to be detected by a detector D. Many other histories are conceivable. A complete family of them would take into account the possibility that: (i) the same first properties took place until the photon went in another direction, (ii) the neutron hit a proton but the photoproduction of a deuteron did not take place, (iii) the neutron did not interact with a proton, (iv) the neutron did not cross the window after going through the crystal. This is by the way an example of a complete set of histories of the type we alluded to a moment ago, more general and more flexible than Griffiths's construction. Anyway, it shows what a complete family can be. It will be found in the next chapter that taking a union of the associated boxes is essentially logical addition (using the logical operator "or"), as it occurs when one says that "the neutron hits a proton and it produces a deuteron *or* it scatters."

Consistency and Decoherence*

Gell-Mann and Hartle[2] have particularly stressed the case when consistency is due to a decoherence effect, as will be discussed in Chapter 7. This occurs only when at least one part of the system is macroscopic. One finds in this case that, when the real part of the trace in a consistency condition (4.10) vanishes, the whole trace also vanishes, or at least it is extremely small. Gell-Mann and Hartle therefore proposed to write down the consistency conditions as a vanishing of the full trace.

This point of view is convenient in the special case they considered but it will be better not to adopt it systematically for the time being. There are several reasons for that, one of them being not to mix the consistency conditions, which are universal matters of logic, with the decoherence effect, which is a dynamical effect restricted to macroscopic systems and, even so, not quite universal. Another reason is that extending the vanishing of the real part of the trace to the vanishing of the whole trace could lead to the omission of some cases where a simple system might be described consistently. Then one would be wrong in saying that something makes no sense without considering all the possibilities of consistency. These questions will be reconsidered in the next chapter.

APPENDIX A: GENERAL FORM OF THE CONSISTENCY CONDITIONS

1. Notations

We will write down various forms of the consistency conditions in the case where there are four reference times, which is sufficient to show what happens in the general case.

The observables are A_1, A_2, A_3 and A_4. The spectrum σ_1 is divided into disjoint subsets $\{D_1^r\}$, with similar notations for the other observables. In order to avoid heavy notation, the time dependence of the projectors will not be written explicitly since their lower index is enough to suggest it. Accordingly, a projector $E_1^r(t_1)$ associated with the set D_1^r will be simply written as F_1^r.

It will be convenient to use a specific notation for some multiple commutators occurring in the consistency conditions. Consider for instance a product of operators $ABCDE$. Keeping the first and last factors A and E, let us denote the intermediate product by $F = BCD$. The double commutator $[A, [F, E]]$ will then be written in the simplified form

$$[AFE] \stackrel{\text{def}}{=} [A, [F, E]].$$

It may happen that factors B, C, D are already themselves multiple commutators. If for instance $C = [G, [H, K]]$, we will write

$$[A, [F, E]] \stackrel{\text{def}}{=} [AB[GHK]DE],$$

where $F = B[G, [H, K]]D$.

2. The Main Results

The General Form of Consistency Conditions

Using the previous notation, all the consistency conditions can be written in a form involving traces of multiple commutators. They are:

$$\begin{aligned}
\text{Tr}(F_3^r F_2^s [F_1^q \rho_0 F_1^{q'}] F_2^s F_3^r F_4^u) &= 0 & q \neq q' \\
\text{Tr}(F_3^r [F_2^s F_1^q \rho_0 F_1^q F_2^{s'}] F_3^r F_4^u) &= 0 & s \neq s' \\
\text{Tr}([F_3^r F_2^s F_1^q \rho_0 F_1^q F_2^s F_3^{r'}] F_4^u) &= 0 & r \neq r' \\
\text{Tr}(F_3^r [F_2^s [F_1^q \rho_0 F_1^{q'}] F_2^{s'}] F_3^r F_4^u) &= 0 & q \neq q', s \neq s' \quad (4A.1)\\
\text{Tr}([F_3^r [F_2^s F_1^q \rho_0 F_1^q F_2^{s'}] F_3^{r'}] F_4^u) &= 0 & r \neq r', s \neq s' \\
\text{Tr}([F_3^r F_2^s [F_1^q \rho_0 F_1^{q'}] F_2^s F_3^{r'}] F_4^u) &= 0 & r \neq r', q \neq q' \\
\text{Tr}([F_3^r [F_2^s [F_1^q \rho_0 F_1^{q'}] F_2^{s'}] F_3^{r'}] F_4^u) &= 0 & r \neq r', s \neq s', q \neq q'.
\end{aligned}$$

If the spectrum of the last observable A_4 is divided into N subsets corresponding to N projectors F_4^u ($u = 1, 2, \ldots, N$), one needs only to write $N - 1$ conditions, the index u taking only the values $u = 1, \ldots, N - 1$.

These equations constitute a minimal system of necessary and sufficient conditions for the additivity of probabilities.

Griffiths Conditions

Griffiths has given another set of necessary and sufficient conditions for consistency. Their formal expression looks simpler than the previous ones involving multiple commutators. However, their number is in general much larger, which means that they are not in fact independent. They refer not only to the subsets $\{D_k^r\}$ in σ_k, but also introduce the sets $\{D_k^\rho\}$ obtained by taking any union of these basic subsets having the same fixed time-index k. This is indicated by a Greek subscript rather than a Latin one. Denoting by F_k^ρ the associated projector, Griffiths conditions are, in the case $n = 4$:

$$\mathrm{Re}\{\mathrm{Tr}(F_3^\rho F_2^\sigma F_1^q \rho_0 F_1^{q'} F_2^\sigma F_3^\rho F_4^u)\} = 0, \quad q \neq q'$$

$$\mathrm{Re}\{\mathrm{Tr}(F_3^\rho F_2^s F_1^\theta \rho_0 F_1^\theta F_2^{s'} F_3^\rho F_4^u)\} = 0, \quad s \neq s'$$

$$\mathrm{Re}\{\mathrm{Tr}(F_3^r F_2^\sigma F_1^\theta \rho_0 F_1^\theta F_2^\sigma F_3^{r'} F_4^u)\} = 0 \quad r \neq r'.$$

In general, they can be written as

$$\mathrm{Re}\{\mathrm{Tr}(F_{n-1}^\rho \cdots F_k^s \cdots F_1^\theta \rho_0 F_1^\theta \cdots F_k^{s'} \cdots F_{n-1}^\rho F_n^u)\} = 0, \quad s \neq s'.$$
(4A.2)

The two forms coincide when each spectrum is divided only into two subsets.

Other Bases

It is sometimes convenient to avoid dividing once and for all the direct product $\Xi = \sigma_1 \times \cdots \times \sigma_n$ into a regular array of boxes as in Figure 4.3. Figure 4.4 shows an example of a less systematic construction. The case most useful in practice is shown in Figure 4.5, in the case when $n = 2$. It may be called a *special family* of histories. It is particularly convenient when one wants to discuss only the logical aspects of a specific history. Its use is in fact necessary to discuss delayed-choice experiments (see Chapter 11). When one is interested, for instance, in a history involving the subsets (D_1, D_2, D_3, D_4), the space Ξ may be divided into the following five boxes:

$D_1 \times D_2 \times D_3 \times D_4$, $\quad \overline{D}_1 \times D_2 \times D_3 \times D_4$, $\quad \sigma_1 \times \overline{D}_2 \times D_3 \times D_4$,

$\sigma_1 \times \sigma_2 \times \overline{D}_3 \times D_4$, $\quad \sigma_1 \times \sigma_2 \times \sigma_3 \times \overline{D}_4$.

There are only $n - 1$ consistency conditions in that case, which is most economical when the calculations are to be done explicitly. In the case

$n = 4$, these conditions are:

$$\mathrm{Tr}(F_3 F_2 [F_1 \rho_0 \overline{F}_1] F_2 F_3 F_4) = 0, \qquad \mathrm{Tr}(F_3 [F_2 \rho_0 \overline{F}_2] F_3 F_4) = 0,$$
$$\mathrm{Tr}([F_3 \rho_0 \overline{F}_3] F_4) = 0. \tag{4A.3}$$

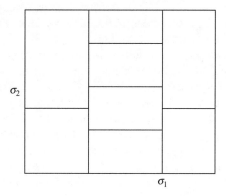

4.4. *A general basis of histories.* The splitting of the spectrum σ_2 depends upon the interval to be considered in σ_1. One can also consider the case when the choice of the second observable A_2 depends upon the first property.

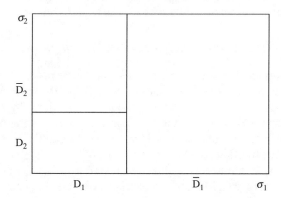

4.5. *A special basis of histories.* This is the most convenient family of histories when one wants to discuss the consequences of a specific history (which is defined here by the two intervals D_1 and D_2).

3. Sketch of the Proof

The consistency conditions are necessary and sufficient in order to insure the additivity of probabilities. Only the main features of the proof will be indicated here.

One considers explicitly all the boxes that can be obtained as unions of elementary boxes. Let them be called big boxes. To get the Griffiths conditions, each big box is divided into smaller boxes by planes perpendicular to the k axis, slicing σ_k along a few basic divisions. Each pair of elementary subsets $(D_k^s, D_k^{s'})$ belonging to the projection of the big box onto the k axis leads to a consistency condition belonging to the family of conditions (4A.2). The corresponding calculation is exactly the same as the one given in the text. These conditions are therefore necessary. They are also sufficient because they imply that the probability of a complex history is uniquely defined and it satisfies additivity.

To get the consistency conditions involving multiple commutators, one again considers all the big boxes. Select the ones projecting upon an elementary subset over each k axis, with $k = 2, 3, \ldots, n$. The corresponding consistency conditions, as obtained by a calculation similar the one given in the main text, are given by the first line in conditions (4A.1).

Now consider the big boxes projecting onto an elementary subset along the axes $k = 3, \ldots, n$. This gives the second line in the conditions (4A.1). One must also impose the conditions obtained during the first step when they are reconsidered for the boxes involved in the second step. This gives the fourth line in conditions (4A.1). The other conditions are obtained by the same process.

The proof of conditions (4A.3) follows exactly the same pattern, which is appreciably simplified because each spectrum is always divided into at most two subdomains. For more details, we refer the reader to the original papers.[3]

APPENDIX B: THE UNIQUENESS OF PROBABILITIES

Our goal in this appendix is to find the most general form for the probability of a history involving only two properties with respective projectors $E_1(t_1)$ and $E_2(t_2)$. These projectors will be denoted more briefly by F_1 and F_2. The Hilbert space **H** is assumed to be separable, its dimension being finite or infinite. The closed subspaces of **H** corresponding to the two projectors will be denoted by $\mathbf{M}_1$ and $\mathbf{M}_2$. The density operator ρ is given at time 0. The probability of the history can be considered as a function of operators $p(\rho, E_1, E_2)$ or $p(\rho, \mathbf{M}_1, \mathbf{M}_2)$ and one assumes the following conditions:

1. The probability has a universal form. This means that it has the same expression whatever the dimension of the Hilbert space and the choice of projectors. The main point of this assumption is to avoid having the formulas for the probabilities depend on some special geometric properties of the Hilbert space or on a specific pair of subspaces. This is because the ultimate goal is a logical construction having this kind of universality.

2. The probability depends only on its arguments (ρ, E_1, E_2) or, equivalently, on $(\rho, \mathbf{M}_1, \mathbf{M}_2)$. This means that it cannot depend on an arbitrary choice of bases for the subspaces or for the ambient Hilbert space.

3. The probability is nonnegative.

4. It is additive in ρ. This means that one can apply the principle of compound probabilities when considering two different initial states, thus

getting a statistical mixture of them. This can be written as the following condition:

$$p(q\rho + q'\rho', E_1, E_2) = qp(\rho, E_1, E_2) + q'p(\rho', E_1, E_2),$$

for any pair of states ρ and ρ' and two probabilities q and q' such that $q + q' = 1$ for their mixing. The construction of a density operator by the von Neumann method given in Chapter 1 immediately leads to this condition when it is applied to histories.

5. The probability reduces to $\mathrm{Tr}\{\rho E_1\}$ when $E_1 = E_2$. This is a tautology and the form $\mathrm{Tr}\{\rho E_1\}$ of a unique property comes from Gleason's theorem.

6. When the density operator has the special form

$$\rho = \frac{E_0}{\mathrm{Tr} E_0},$$

and is associated with an initial preparation predicate with projector E_0, one has

$$p(E_0, E_0, E_2) = \mathrm{Tr}\{\rho E_2\} \qquad (4\mathrm{B}.1)$$

$$p(E_0, \overline{E}_0, E_2) = 0. \qquad (4\mathrm{B}.2)$$

The first condition means that repeating the initial predicate at time t_1 is a tautology and negating it is nonsense.

7. One assumes furthermore that

$$p(\rho, E_1, E_2) + p(\rho, E_1, \overline{E}_2) = \mathrm{Tr}\{\rho E_1\}. \qquad (4\mathrm{B}.3)$$

This is additivity, but written for the last predicate only.

Since p is supposed to have a universal form, one can consider the case where the two subspaces $\mathbf{M}_1$ and $\mathbf{M}_2$ are in a so-called generic position, which means that the subspaces $\mathbf{M}_1 \cap \mathbf{M}_2, \mathbf{M}_1 \cap \mathbf{M}_2^\perp, \mathbf{M}_1^\perp \cap \mathbf{M}_2, \mathbf{M}_1^\perp \cap \mathbf{M}_2^\perp$ are all reduced to the vector 0. Halmos[4] has shown that in that case the subspaces $\mathbf{M}_1, \mathbf{M}_1^\perp, \mathbf{M}_2$ and $\mathbf{M}_2^\perp$ have the same dimension (finite or countable). There exists a Hilbert space $\mathbf{K}$ such that $\mathbf{H} = \mathbf{K} \oplus \mathbf{K}$ and a unitary transformation acting on $\mathbf{H}$, which brings the projectors E_1 and E_2 into the form

$$E_1 = \begin{bmatrix} 1 & 0 \\ 0 & 1 \end{bmatrix} \qquad E_2 = \begin{bmatrix} C^2 & CS \\ CS & S^2 \end{bmatrix}. \qquad (4\mathrm{B}.4)$$

The operator C is positive, it acts in $\mathbf{K}$, and its spectrum belongs to the open interval $]0, 1[$. The operator S commutes with C; it is positive and completely defined by $C^2 + S^2 = 1$, where 1 represents the unit operator in $\mathbf{K}$. One can then write the density operator ρ as

$$\rho = \begin{bmatrix} \rho_a & \rho_c \\ \rho_b & \rho_d \end{bmatrix},$$

where each operator ρ_j acts on $\mathbf{K}$.

Condition 2 implies that the probability depends only upon the five operators $\{\rho_j\}$ and C. When condition 4 is used, in the case where **H** has a finite dimension, one finds that the probability p is a linear function of the matrix elements of the four operators $\{\rho_j\}$ and one can always find four operators $\{A_j\}(j = a, b, c, d)$ acting on **K** such that

$$p = \sum_{j=a}^{d} \mathrm{Tr}\{\rho_j A_j\},$$

the traces being taken in **K**.

The self-adjoint character of ρ gives

$$\rho_a = \rho_a^\dagger, \qquad \rho_d = \rho_d^\dagger, \qquad \rho_b = \rho_c^\dagger.$$

Since p is real, we have

$$A_a = A_a^\dagger, \qquad A_d = A_d^\dagger, \qquad A_b = A_c^\dagger.$$

Condition 2 furthermore implies that the operators are functions of the operator C so that $A_j = f_j(C)$, from which one gets

$$p = \sum_{j=a}^{d} \mathrm{Tr}\{\rho_j f_j(C)\}.$$

Since one is looking for a universal form, one can let ρ vary and the positivity of p implies that the two functions f_a and f_d are nonnegative. Condition (4B.1) then gives

$$\mathrm{Tr}_K f_a(C) = \mathrm{Tr}_H\{E_1 E_2\} = \mathrm{Tr}_K(C^2).$$

Since the eigenvalues and the eigenvectors of C in **K** can also be made to vary continuously by choosing the properties and their projectors at will, this implies

$$f_a(C) = C^2. \tag{4B.5}$$

Similarly, condition (4B.2) gives $f_d(C) = 0$.

One can then use condition (4B.3) and write the projector $\overline{E}_2$ as

$$\overline{E}_2 = \begin{bmatrix} S^2 & -CS \\ -CS & C^2 \end{bmatrix},$$

which gives the following relation:

$$\mathrm{Tr}\{\rho_a[f_a(C) + f_a(S)]\} + 2\mathrm{Re}\,\mathrm{Tr}\{\rho_b f_b(C)\} = \mathrm{Tr}\rho_a.$$

From equation (4B.5) one has $f_a(C) + f_a(S) = 1$, and one gets $\mathrm{Re}\,\mathrm{Tr}\{\rho_b f_b(C)\} = 0$. Since the matrix elements of ρ_b can be made to vary continuously, we have $f_b(C) = 0$. So, finally, we obtain

$$p(\rho, E_1, E_2) = \mathrm{Tr}\{\rho C^2\}.$$

Using the Halmos form (4B.4) for E_1 and E_2, this can be recognized as

$$p(\rho, E_1, E_2) = \text{Tr}\{E_1 \rho E_1 E_2\},$$

which is the conclusion. This form is the only one that is consistent with the logical conditions introduced at the beginning.

This proof cannot be easily extended to more than two reference times. This is due to the difficulty of extending Halmos decomposition to that case, so that the trick of treating the probability as a function of only one operator C does not work. A larger number of projectors introduce several operators analogous to C and they do not commute, so that one cannot trivially define a function of them. One might be tempted to believe that this is only a technical difficulty and that the result is still valid for $n > 2$, but this would be wishful thinking. Fortunately, the case $n = 2$ is enough for most applications and the Feynman-path approach, though limited to the position and momentum observables, may be used to extend the result to any value of n.

Finally, although the proof as it is given here is rather elementary, it cannot be considered trivial because Halmos's theorem, upon which it rests, is a rather deep result.

PROBLEMS

1. The system is spin-$1/2$. Denoting by $\mathbf{n}$ a three-dimensional real vector with unit length, one considers three such vectors ($\mathbf{n}_0, \mathbf{n}_1, \mathbf{n}_2$). One also considers a family of histories for which the initial state is given by the property $\sigma \cdot \mathbf{n}_0 = +1$ and the properties at times t_1 and t_2 correspond to $\sigma \cdot \mathbf{n}_j = \pm 1 (j = 1$ or $2)$. Show that there is only one consistency condition, which is expressed by the geometric relation

$$(\mathbf{n}_0 \wedge \mathbf{n}_1) \cdot (\mathbf{n}_1 \wedge \mathbf{n}_2) = 0.$$

2. Consider the same problem for a neutral particle with gyromagnetic ratio g in a homogeneous magnetic field $\mathbf{B}$. Show that the consistency condition keeps the same form after replacing each vector $\mathbf{n}_j$ by a vector $\mathbf{n}'_j$, which is the solution at time t_j of the dynamical equation

$$\frac{d\mathbf{n}'}{dt} = g\mathbf{B} \wedge \mathbf{n}'$$

with initial value $\mathbf{n}_j$ at time 0.

3. Consider again Problem 1, this time with $n = 3$. Show that there are now two consistency conditions, which can be written as:

$$(\mathbf{n}_0 \wedge \mathbf{n}_2) \cdot (\mathbf{n}_2 \wedge \mathbf{n}_3) = 0$$
Either $(\mathbf{n}_1 \wedge \mathbf{n}_2) \cdot (\mathbf{n}_2 \wedge \mathbf{n}_3) = 0,$ or $\mathbf{n}_0 \cdot \mathbf{n}_1 = 0.$

5

The Logical Framework
of Quantum Mechanics

This is another key chapter for the construction of an interpretation. It is devoted to the logical content of quantum mechanics, which is perhaps less important than its dynamical content in most cases but which must come to the forefront when interpretation is concerned. The chapter is divided into four parts: The first one gives an elementary introduction to logic. The second one gives explicitly the logical constructions to be used in quantum mechanics, which provide a clearer meaning for Bohr's complementarity principle. Then, a universal logical rule providing a foundation for the process of reasoning in quantum mechanics will be stated. It is in fact the last basic axiom we will need to obtain a complete interpretation. The last part will be devoted to some examples and to further remarks for illustration and clarification.

ABOUT LOGIC

1. Why Does One Need Logic?

We saw in the previous chapter how a physicist describes what happens during an experiment. His intuitive approach to reality was better understood and also better controlled by inserting it in the framework of consistent histories. This success is not however sufficient to obtain a satisfactory interpretation, because science is not restricted to descriptions but it must also link together the properties of a physical system by reasoning. To describe is not enough; one must also know how to think and this is the aspect of physics to be investigated now.

This approach needs an explanation; otherwise it might look too far-fetched. When classical physics reigned alone over our understanding of reality, this kind of question was rather unimportant. How one is allowed to think about the physical properties of an object was more a matter of philosophy than of science and there remain today some traces of this attitude among physicists. The question cannot however be avoided any more in

THE LOGICAL FRAMEWORK OF QUANTUM MECHANICS

quantum physics, where it is clear that common sense has lost some of its ground and much of its grip. One should therefore be very careful about one's way of reasoning and the first caution to be taken is to make clear how to proceed.

It will be convenient, as a preliminary, to see how our faculty of reasoning is usually exerted in practice, if only to make clear what it is, how important it is, and also how nontrivial it becomes in a quantum framework. A simple way to do it will be to consider again the experiment in nuclear physics that was already described in the previous chapter. We saw how one can talk about it. The question is now how one can think about it.

An Example of Physical Thinking

When making an experiment, a physicist observes directly only a few phenomena, i.e., some conspicuous properties of the macroscopic instruments. These phenomena are quite insufficient to tell him everything he needs to know in order to understand what is going on. In the case of the nuclear physics experiment described in the last chapter, one can only see how the whole experimental setup is arranged and ascertain one fact, namely that some photon detector has reacted.

This is enough to give a physicist much food for thought and it will be interesting to see how he proceeds and to try decomposing his mode of thinking in some detail, even if it may look rather obvious at first sight. So, what does he think? Maybe something like this:

"Well, I see that this photomultiplier has registered an event. *Therefore,* it detected a photon. But *if* there was a photon, *then* it had to be produced in a neutron-proton collision. *Hence,* there has been such a collision. *Consequently,* a neutron entered the target. *If* a neutron entered the target, *then* it had to cross the window beforehand. *Hence,* it came out of the diffracting crystal with a velocity I can determine. *Since* I know the initial velocity of the neutron as well as the direction of the outgoing photon, I also know the whole kinematics of the reaction."

There is a significant difference between this and what was considered in the previous chapter. The history was then only a tentative description of what might have happened. Now, it is turned into an argument, something sure enough to carry the conviction that things not only might have happened that way but that they had to. Every time one says "therefore", "hence", "if ... then", "consequently", "since" or anything of that kind, one is reasoning about the system and trying to make sure that no doubt can be left.

Every time one of these little words was said, one was drawing what a logician would call an implication, or an inference or, in more common words, one was forging a link in an argument. Clearly, all of physics is full of similar reasonings, either deep or trivial, and there could be no science without them. When considering however that the neutrons and photons are quantum objects, one may wonder whether the reasoning is as convincing as it looks at first sight.

An Unbeliever's Approach

Any physicist who has participated in an experiment in quantum physics, or who had only to discuss it, will recognize in this example a significant part of the thinking usually devoted to an experiment. It is necessary in order to get the final experimental data and to justify the estimated errors accompanying them. When one looks however more carefully at the reasoning, one soon realizes that it is rather careless about the limitations coming from the quantum rules. As a matter of fact, it is mostly based upon some sort of common sense or craftmanship and it is far from being rigorous.

To get a better idea of the possible criticisms, one may call for the hero in a famous book by Voltaire, who was a Huron.[1] He was very remarkable for his common sense and, when he had become convinced of something, he knew how to keep to it firmly. So, let us call for a Huron coming straight out of a course in quantum mechanics to hear what our friend the experimentalist has just said. You can see him jump at almost every sentence:

"So," the Huron would say, "you have seen an electric signal in your photomultiplier. You say it was due to a photon. Why? Can you prove it? Why was it not a wave? Furthermore, what makes you think that the neutron followed a straight-line trajectory, like a small ball, when it crossed this window? You didn't see it. My teacher in quantum mechanics told me that one cannot know what happens to a neutron, except when there is an actual measurement to show it, and you say that it went along a straight line in a vacuum, like a little ball! You are really going too far! When it crossed the velocity detector, you say that the neutron was not a ball but a wave, and this is why you can tell its velocity. But you use it anyway as if it was the velocity of a ball. And you want me to buy that! Go back to school!"

If his adviser in quantum mechanics were then to enter, he would talk to him learnedly about a complementarity principle, according to which our way of thinking about a quantum object is determined by its experimental environment. The velocity selector compels the neutron to behave like a wave and the window constrains it to behave like a particle and so on.

"I have heard of that," said the Huron. "But I am a man from another age and I also know how this kind of thinking was called in my time: casuistry, jesuitism. It means that you choose the kind of axioms and the kind of reasoning you find better for your purpose, the ones allowing you to get the conclusions you wish. It means that you can look like a saint in public and enjoy yourself in private, according to what you might call the experimental environment. This was never believed to be a rigorous way of thinking nor even a correct one and you don't convince me. It would be better if you tell me whether or not this neutron is going along a straight line."

"Well," his adviser mumbles, "as a matter of fact, let's see... of course you cannot say it, at least when you have read Heisenberg carefully. But, you know, there is this Ehrenfest's theorem, just like classical stuff you know... and if one were to put a neutron detector along the way..."

"Do you mean something like in an interference experiment where this kind of detector kills the effect?"

"Quite so."

"And you tell me that putting a neutron detector there would justify your ideas that the neutron went along a straight line, when you know that inserting it into another setup would break the whole game. Why can you do it now and not in the case of interferences?"

"Because! Because one has always done it that way and it works, that's all. Good Heavens, if they were all like him, I would give up teaching quantum mechanics and go back to my lectures on celestial mechanics."

Is it because of the small number of Hurons, or because they are impressed by the last argument (it works!), that one is still always dealing with an experimental physics whose modes of reasoning superbly ignore the interdictions pronounced by the theory it is supposed to check. Everybody violates the law. Who cares? It works!

A true Huron, on the other hand, would not reject what the experimentalist says because this disagrees with what Bohr said, but he would also not believe all that Bohr said because he forbids too many things to the experimentalist for him to do his job. The theory more or less negates the means one must use to verify it. Surely, would say the Huron, there must be something else, not a set of compromises but a true principle allowing once and for all to make sure when a reasoning is right.

A Strategy

The main issue of the present chapter will be to put some order among these attempts at reasoning and, first of all, to look for a guiding principle. But how are we to proceed?

Since quantum mechanics leads us to question common sense, we shall have to reconsider it carefully under its noble and scientific guise, which is logic. Logic is precisely the art of reasoning. When it can be correctly stated, it is supposed to be rigorous whereas the reasoning of common sense, when applied to a quantum system, is too often groundless. So, one needs to find a firm ground upon which to erect the principles of logic most conveniently fitting the peculiar framework of quantum mechanics. This does not mean however that one should be ready to completely give up common sense. This is not because of conservatism but simply because one cannot do without common sense. It is intimately linked with our visual representation of the world; we use it to communicate what we have seen, what happened, what there is now. It is the language of facts and action, the only one allowing us to say that such and such an apparatus has shown a result, how this apparatus has been constructed, and what the directions are for using it. Common sense is essential because it is the framework of the only language dealing with empirical science. But interpretation is a direct link between theory and practice, so that it cannot ignore common sense, which is too useful, too flexible, and too rich in creative intuition, in one word, too necessary.

In any case, what else could be used to describe empirical reality? To discuss the experiment mentioned at the beginning, is one going to write a huge Schrödinger equation for the nuclear reactor, its walls, the crystal, the target, and the photomultiplier? Even if one were able to realize such a dubious feat, what would have been gained? How would one reconstruct the history of the reaction from the mess of a tremendously complicated wave function, which anyway treats all the possible events on the same footing? The right question to be asked is therefore whether common sense can be made rigorous and whether one can prove it starting from the axioms of quantum mechanics. If this can be done, one will probably find that common sense is not universally valid when it is conceived as a consequence of quantum mechanics but, after all, knowing the limits of its validity domain will also be probably quite instructive.

Accordingly, the basic strategy of the present approach is the following: *investigate the connection between quantum mechanics and logic well enough to get a clear and firm foundation for the corresponding logic and to prove the validity of common sense within the bounds of a clearly restricted domain.* This program is far from trivial and the present chapter will only set its foundations by making explicit the kind of logic one can confidently use in quantum mechanics. As far as the problem of "proving" common sense is concerned, it is intimately linked to the justification of its own framework, which is classical physics, and it will be considered in the next two chapters.

2. What Is Logic?

Before entering into the subject, one must make clear what one means when referring to logic. One must make sure that it is essential to physics and not only a game for some highbrow eggheads and, if only for that, one must show that logic is after all something rather simple and straightforward.

THE LOGICAL FRAMEWORK OF QUANTUM MECHANICS

Logic is defined most often as the art or science of reasoning. Like every science, it can become very abstract and technical when it is left to the refinements of specialists. Nevertheless and quite fortunately, what one needs to know of it in order to apply it to physics is rather simple.

Logic can be applied to many fields of knowledge or to other questions having to do with practical life: set theory, law, how a computer works, and so on. Every such field of application offers what is called by the specialists a Denkbereich (i.e., a field of thought) or a universe of discourse. It consists in all the possible sentences (or propositions) having to do with the corresponding subject. When the rules of abstract logic are applied to such a field, one obtains what is usually called an interpretation of logic. For instance, the propositions expressing the basis and the possible consequences, whether true or not, of the Constitution of a country, constitute a universe of discourse and, when one treats these propositions according to the rules of logic, this is an interpretation, i.e., the abstract rules of formal logic become explicit modes of reasoning in the domain of constitutional law. We shall however avoid using the word "interpretation" with this meaning, even if it is a conventional one. This is because it would be awkward and confusing if one were claiming to interpret quantum mechanics by using some specific interpretations of logic. We will therefore simply speak of *a* logic, rather than more pointedly of an interpretation of logic. One might as well have chosen to speak of a representation of logic, an application or a realization of logic; but the words do not matter so much. Anyway, although one will use the same word for the general science and for its specific applications, there will be practically no risk of confusion.

The Ingredients of a Logic

A logic involves three basic ingredients, namely: (i) A field of propositions, (ii) Five logical operations and relations among propositions, and (iii) A truth criterion. A proposition is a sentence whose construction obeys precise rules in its grammar and its vocabulary, these rules depending upon the domain of application. A predicate expressing a property in quantum mechanics is an example of a proposition. A sequence of ones and zeros is also a proposition in computer science. A theorem in arithmetic is another example.

From given propositions, one can obtain other propositions by using three basic logical operations, which are usually expressed by the words "not...", "...and...", "...or...". One can also use two relations, one of which is expressed in words by "if..., then..." and the other one by "...is logically the same as...". The operations "and, or, not" are easy to understand: The negative or contrary of a proposition is also another proposition. Given two propositions (a, b), one can consider that the combinations "*a* and *b*", "*a* or *b*" are also some propositions. In a computer message where a proposition is associated with a number, the proposition not-a is often associated with the opposite number $-a$, while "a and b" is associated with the product of these two numbers and "a or b" with their sum. Many other variants are possible, but these are the ones most commonly used and realized by "logical gates", which are specific electronic devices performing these operations.

The relations of equivalence and implication are somewhat different from these operations. Equivalence says that: "a is (logically) equivalent to b"; it is written as an equality: $a = b$. An interesting example occurs in the framework of quantum mechanics: Let a denote the predicate according to which "the value of the observable A is in the interval $[-1, +1]$" and let b denote the other predicate according to which "the value of the squared observable A^2 is in the interval $[0, 1]$." One may say that these two predicates are logically equivalent so that $a = b$. The other kind of relation, "if a, then b", is denoted by $a \Rightarrow b$; it is called an inference or implication. One cannot overstress that implication is the most important tool of logic, the one from which all of logic is built. When it allows one to draw a conclusion from given assumptions, facts, or information, through a full and rigorous reasoning, either simple or elaborate, it accomplishes logic's mission.

These five operations and relations are governed by about twenty rules or so, which were initially put forward by Aristotle and Chrysippus in Antiquity before being more conveniently codified during the last century by Boole, Frege, and others. They are given for completeness in Appendix A. These are the formal rules of conventional logic, which is also said to be aristotelian, or boolean, or with an excluded middle. It will be simpler to say that this is the logic everybody can understand and uses every day without questioning or wondering about it. In what follows, a logic satisfying the totality of these conventional axioms will be said to be *consistent*.

The existence of a truth criterion was also mentioned as being an important ingredient of a logic. In arithmetic, for instance, one assumes that some basic propositions (the axioms) are true by convention. One can then use the rules of implication to derive from them other true propositions. The true propositions are also called theorems and the explicit link showing their truth from the assumed truth of the axioms is their proof. One might also assume that there exists a truth function $T(a)$ having only the range of values $\{1, 0\}$ (1 for true and 0 for false), which is defined upon the field of propositions and such that $T(a) = 1$ for the axioms. If one can show, as Gödel did, that a truth function exists, one thus gets two logics of truth in arithmetic, with different truth criteria, where the first criterion relies upon a proof and the second one upon an overall abstract truth function. That these logics are not identical is the result of a theorem by Gödel according to which there exist true theorems in arithmetic that cannot be proved by a finite argument.

The example of a truth criterion we have just given should not lead us to expect that the question of truth in physics will plunge into similar abysses.

THE LOGICAL FRAMEWORK OF QUANTUM MECHANICS

Nevertheless, this question of truth is somewhat more delicate than the other logical aspects of physics and we shall have to delay its discussion until Chapter 8. For the time being, one can only give a hint, which may look rather trivial—namely, a property in physics is true when it asserts an observed fact or when it is logically equivalent to a fact in a universal way.

An Example

All this is probably too abstract and it will be useful to give an example much closer to the usual life of physics. We shall therefore consider as an example a classical physical system having only one degree of freedom. It is for instance a pendulum oscillating in a plane, its motion obeying Newton's laws of dynamics with no damping.

A proposition in classical physics is best defined by associating it with a domain C in classical phase space. If the coordinates are denoted by (q, p), the corresponding classical proposition asserts that the coordinates of the system at a time t belong to the domain C. The simplest case occurs when the domain C is a rectangular box, in which case both coordinates are given, up to sharply defined error bounds, the coordinate q being for instance contained in an interval $[q_0 - \Delta q, q_0 + \Delta q]$ with similar bounds for the momentum. The proposition itself will be denoted for short by the corresponding domain and time, namely $[C, t]$.

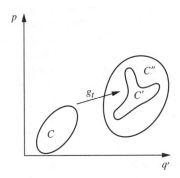

5.1. *The logic of classical mechanics.* A cell C in classical phase space is transformed into another cell C' by classical motion during the time t. The classical property stating that the coordinates (x, p) are in C at time 0 implies that (x, p) is in C'' at time t, if C' is contained in C''.

One can then define the elementary logical operations by borrowing some notions from set theory: The negation of the proposition $[C, t]$ is the proposition $[\bar{C}, t]$, where $\bar{C}$ is the complement of the domain C in phase space. The proposition "$[C, t]$ and $[C', t]$" may be defined in terms of the intersection of domains as $\{C \cap C', t\}$, whereas the proposition "$[C,t]$ or $[C',t]$" refers to their union and is given by $\{C \cup C', t\}$.

To define an implication $[C,t] \Rightarrow [C',t']$, one may use the knowledge of classical motion. Every point in C moves in a well-defined way during the time interval $t' - t$ and all the points belonging to C generate in that way another domain C''. One will then say that $[C,t] \Rightarrow [C',t']$ when C'' is included in C', the logical equivalence $[C,t] = [C',t']$ meaning simply that $C'' = C'$.

Though it would be easy to reformulate completely classical mechanics in this form, this reformulation would not bring anything new and its only interest is to provide an example of a much more general way of looking at science by resorting more explicitly than usual to logic. It shows by the way that *determinism is nothing but a logical equivalence between two propositions holding at different times*, namely $[C,t] = [C'',t']$ with the previous notation. This equivalence can be partly extended to the case when there is friction, the main difference being then that the cell C'' has a smaller volume than C. In the limit where the pendulum is finally at rest, one can still get an implication going from a given initial state towards the final one but it is no more a logical equivalence because one cannot turn the implication the other way round. It is of course easy to extend these remarks to an arbitrary number of degrees of freedom.

In classical physics, the truth criterion is completely external to the formalism: a property is said to be true when it is experimentally ascertained as a fact. This is what some logicians express in a flowery way by saying that the proposition "the rose is red" is true when the fact that the rose is red is observed as real. One can then derive many other true properties from the empirically asserted ones by using determinism and assuming explicitly the validity of classical physics. Two typical examples correspond respectively to prediction and retrodiction: for instance, when an apple falls, one can predict where and when it will hit the ground. The most important example of retrodiction occurs when one can know of a past fact by looking at a document recording it, whether it be a photograph, a note in a book, or a remnant of motion. This is how one can say that a book left on a table remains in the same place when nobody is looking at it. One may of course wonder how this extends to quantum mechanics but it will be necessary to hold one's breath till the next chapter before getting the answer.

Logic and Physics

The interest shown here in logic partakes of an important general trend in modern science. This point is certainly worth mentioning, if only to counteract a tendency among some people to think of physics as an isolated body of knowledge. It may also induce others to consider that an explicit account of logic in physics is neither new nor useless.

THE LOGICAL FRAMEWORK OF QUANTUM MECHANICS 153

Science had been using logic for a long time in the form initially given to it in Antiquity before the need for something more precise began to arise in mathematics. There were many occasions for this during the nineteenth century: the rigorous formulation of analysis as initiated by Weierstrass, the unintuitive character of noneuclidean geometries or of euclidean spaces with a dimension larger than three, the uncertain status of some arguments using recurrence, the intricacies of the infinite sets discussed by Cantor, and finally the meaning of the axiomatic method itself were among the main reasons to pay more attention to the basic tools of reasoning, i.e., to logic. The turning point was the advent of formal logic, its foremost contributor being Frege near the end of the nineteenth century and the beginning of the twentieth.[2]

Hilbert himself had worked upon the axiomatization of geometry and he launched an investigation of axiomatization that was to lead his former student Gödel to unexpected results. This was a part of one of Hilbert's grand programs. It aimed at setting all of mathematics upon a well-defined axiomatic basis, after having given a similar form to logic itself. This framework was even extended by Hilbert to theoretical physics, which was mainly at that time Boltzmann's approach to statistical physics and also a bit later the geometry of space-time (one should not forget his close friendship and partnership with Minkowski and his discovery of the equations of general relativity at the same time as Einstein). He wanted a physical theory to be a consistent mathematical construction as well as a good representation of physical reality and therefore involving the formal conditions arising from logic. The present formulation of quantum mechanics in terms of well-defined principles, whether it comes from Dirac in the Cambridge fashion (where Russell and Whitehead remained an inspiration) or from von Neumann, who was one of Hilbert's disciples, is an example of the influence of this point of view or at least of the spirit of these times.

Von Neumann was to give testimony of his regard for logic on at least four occasions. When he was still quite young, he made important contributions to axiomatic set theory. His treatment of predicates in quantum mechanics is another instance and Gleason's theorem is a direct outcome of the line he proposed. Later on, when he contributed most significantly to the advent of modern computing machines, he also proved his mastery of logic while showing the practical importance of that science. The fourth occasion is perhaps the least convincing one: Together with Birkhoff,[3] he proposed to make use in quantum mechanics of new unconventional logics lacking a few rules among the ones that are at the foundation of strict formal logic. Whereas a few convinced theoreticians continued to develop this kind of idea, a widespread feeling is now that this is much too complicated, more obscure than illuminating, and in fact not fruitful.

The place of logic in physics or, more properly, the physics of information, has been recently reconsidered from another standpoint.[4] Logic, information, and computation are strongly linked together and one may recall that electronic computers are essentially logical machines with their "and, or, not, nand, nor" gates. There is a lot of thinking presently devoted to this more pragmatic approach to logic, which goes in various directions and is not easily described in a few words. Although a few specialists of computers had long ago drawn some analogy between the action of the laws of physics in the real world and the action of a program in a computation, this remained extremely vague and probably useless. A more modest and also more fruitful approach was to allow an ideal universal computer (a Turing machine) to interact with an external physical world. This opened unexpected connections between information and physics such as the theoretical possibility of reversible computation, the existence of a lower limit to the generation of entropy in a computing machine, and a systematic treatment of the problem of Maxwell's demon in classical statistical mechanics, from which the notion of physical entropy can also be reconsidered. This is also a natural framework for investigating the informational aspects of life and, conversely, for allowing the consideration of observers in physics without having to introduce with them the problem of consciousness and its Pandora's box. This kind of research is also on the other hand linked with some approaches to cognitive science, particularly with artificial neuron networks.

As far as the interpretation of quantum mechanics is concerned, Gell-Mann and Hartle considered it from this point of view in different cases: One is the situation holding near the beginning of the universe in cosmology, when there is no conceivable measuring instrument and where Bohr's conceptual framework becomes completely inadequate. Another one consists in giving an objective meaning to an observer as an information gathering and utilizing system (IGUS), without calling for the intricacies of consciousness while obviously reducing an observer to be himself a part of the universe and nothing else.

There are still rather few unifying common features in this proliferating domain of research. One of them is obviously an increasing interest for the role of logic in physics, inspired by a pragmatic conception of logic relying upon the example of computers. Another one is an important mathematical concept, algorithmic complexity, which has found its way together with its applications in all these directions.

It would be probably premature to develop these new aspects of physics in the present book and we shall then leave them definitely aside, except for returning briefly to the concept of algorithmic complexity in Chapter 7 where it enters in a technical argument.

QUANTUM LOGIC

The next part of our program will be to define in a precise way what kind of logic is best suited to a deeper understanding of quantum mechanics and how it permits trustworthy reasoning concerning it. As a matter of fact, it will be seen that there are many possible logics for a given physical system, i.e., many ways of choosing a field of propositions to describe it and this is at the origin of complementarity. It will therefore be necessary to make sure that this multiplicity of logics cannot be responsible for paralogisms or paradoxes. Only when this is established and we have furthermore a complete theory of phenomena (or facts) will it be possible to define a truth criterion in quantum mechanics, and this will have to wait till Chapter 8. For the time being, it will be shown how one can define the relevant logics in terms of histories and also how one can express the logical operations and relations holding among them, the most important one being of course implication.

3. Defining a Quantum Logic

The Elements of a Logic

A specific logic purporting to describe and to discuss the properties of a given quantum physical system relies first of all upon a consistent family of histories for this system. This means that, according to the considerations of the previous chapter, one is interested specifically in a system in a well-defined initial state at time 0; only its properties at a well-defined sequence of times $t_1, t_2, \ldots, t_n (0 \leq t_1 < t_2 < \cdots < t_n)$ will be considered. These properties are concerned with specific observables $A_1, \ldots, A_n$ where, as usual, several commuting observables entering at the same time can often be considered as essentially a unique observable. There are also cases where the inverse point of view is convenient: in our example using a neutron-proton collision, one may be interested in two properties holding at the same time where, for one of them, photoproduction has occurred and one is interested in some property of the photon whereas, for the other one, there was just a neutron-proton scattering and one is interested in a property of the scattered neutron (the two projectors expressing these properties obviously commute). Finally, the properties to be mentioned in the various histories of the family are made precise by indicating in which ranges the values of these observables are supposed to be located. This is best obtained by decomposing the spectrum σ_k of each observable A_k into a complete set of disjoint domains $D_k^1, D_k^2, \ldots$ and we shall again make use of the corresponding decomposition of the set $\Xi = \sigma_1 \times \sigma_2 \times \cdots \times \sigma_n$ into disjoint boxes.

When two consistent families of histories differ in any detail, either because their respective initial states, the times they refer to, the observables, or the boxes are different, the corresponding logics will be considered as being a priori different. It might turn out that they are consistent with each other, but this will have to be proved and is not to be assumed.

The Field of Propositions

The propositions represent something slightly more general than histories and they have been already introduced in the previous chapter as compound histories. Each one of them can be associated with a set made from several elementary boxes. An elementary history associated with an elementary box is therefore the simplest example of a proposition. The propositions will be denoted by letters such as $a, b, c, \ldots$ and the associated sets will be denoted by $\alpha, \beta, \gamma, \ldots$.

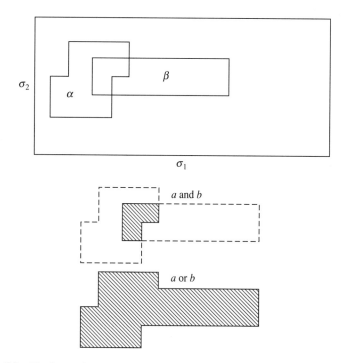

5.2. *The logical operations "and, or, not" for quantum propositions.*

The geometric interpretation and its direct connection with logical operators can be made clearer if one agrees to associate the complement of a set $\bar{\alpha}$ with the proposition not-a, or $\bar{a}$, negating the proposition a associated with the set α. Similarly, the proposition $c = $ "a or b", where (a, b) are two propositions associated with the sets (α, β) respectively, will be associated with the union of sets $\gamma = \alpha \cup \beta$. The proposition $d = $ "a and b" will be associated with the intersection $\delta = \alpha \cap \beta$. One may add a few remarks to this construction:

This use of "and" and "or" is quite similar to what is used in common sense. As an illustrating example, it shows how the following sentence can

THE LOGICAL FRAMEWORK OF QUANTUM MECHANICS

make sense in a quantum framework: A molecule of perfume came out of a phial in the bathroom at noon (i.e., one specifies the initial state), it was in the bedroom at 12:10 and in the sitting room at 12:15, or it was in the corridor at 12:10 and either in the sitting room or the kitchen at 12:15.

It is also well known that when one represents the elementary operations of logic "not, or, and" by the corresponding operations in set theory, the associated rules of formal logic are automatically satisfied. This is often used by the way to explain what these rules are. Accordingly, a compound history can be expressed in words by mentioning the possibility of several different elementary histories $h, h', h'', \ldots$. One can say this by the sentence: "h or h' or h'' or $\ldots$". Finally, an elementary property, stating for instance that "the value of the observable A_1 is in the domain D_1 at time t_1," is associated with a subset in Ξ given by $D_1 \times \sigma_2 \times \cdots \times \sigma_n$, i.e., by a slice in the cake Ξ. An elementary box can be considered as the intersection of several such slices, which shows that an elementary history can itself be expressed in words as the conjunction of various properties when saying that "the value of A_1 is in D_1 at time t_1 *and* the value of A_2 is in D_2 at time t_2 *and* $\ldots$."

Implication

It remains now to define implication and logical equivalence. An implication "if a, then b," to be written as $a \Rightarrow b$, can be defined using the probabilities of histories. It was shown in the previous chapter that a well-defined additive probability can be attributed to the various propositions (or compound histories) when the consistency conditions are satisfied. We always assume that these consistency conditions are satisfied and we shall leave aside the formal families of histories for which this essential restriction is not satisfied as strictly meaningless. This ostracism entails the rejection of many logics, some of which common sense might have been tempted to consider but which we will not aknowledge as logically meaningful.

Given two propositions a and b together with their associated domains α and β, one can define their joint probability $p(a \text{ and } b)$ as the probability measure of the set $\alpha \cap \beta$. As usual, the conditional probability $p(b \mid a)$ for b given a is defined by

$$p(b \mid a) = \frac{p(a \text{ and } b)}{p(a)},$$

assuming that $p(a)$ does not vanish.

It will be said that proposition a implies proposition b when p(a) is not 0 and the probability for b given a is equal to 1, i.e., $p(b \mid a) = 1$. The logical equivalence between two propositions (a, b) will then be identified with the two conditions

$$a = b : p(b \mid a) = 1, p(a \mid b) = 1,$$

or otherwise
$$a = b : a \Rightarrow b, b \Rightarrow a.$$
This new kind of logical equivalence has its roots in the bayesian theory of inference and it is not directly expressible by set theory because it does not mean that the two sets α and β are necessarily identical, but only that the set $(\alpha \cup \beta) - (\alpha \cap \beta)$ has zero probability.

Since these definitions may look so dry as to be almost unpalatable, it will be convenient to anticipate some results to be obtained later to give a hint of their flavor. Consider a Stern–Gerlach experiment where an atom crosses an inhomogeneous magnetic field and is registered by a detector. A number shown on a counter indicates that the detection has taken place. This is the experimental datum, i.e., a fact belonging essentially to the domain of classical physics. It can be compared with the result of the experiment, which is a genuine quantum property, namely that a spin component of the atom is for instance $+1/2$. It will be shown in Chapter 8 that data and results are logically equivalent. It means that each of them implies logically the other, though it is also clear that they are quite different. As a matter of fact, this logical equivalence will be one of the main keys to open measurement theory.

Consistency

One must make sure that a given quantum logic is consistent, by which one means that it satisfies the basic axioms of formal logic. A criterion for this is given by the following theorem:

Theorem. *A logic is consistent when its elementary histories satisfy the consistency conditions necessary for the existence of a probability.*

The consistency conditions, as they were given in detail in Chapter 4, Appendix A, are necessary and sufficient conditions for the probabilities to be additive. The present theorem tells us that they also insure the validity of the basic rules of logic, for which they are again necessary and sufficient conditions. The results found in Chapter 4 therefore acquire a new meaning since the consistency conditions now appear as the key to a consistent use of logic, or to the possibility of thinking rightly about quantum mechanics.

The proof of the theorem is easy. It amounts in practice to a verification of the basic rules of logic one by one, with no serious difficulty. This is done in Appendix B.

Approximate Implications

It is often necessary to relax some of the rigor in the strict logical considerations that were given up till now. For instance, one may be allowed a little freedom in an implication by letting it be valid only up to a very small probability of error ε: we say that a proposition a implies another proposition b up to an error ε if $p(b \mid a) \geq 1 - \varepsilon$.

The reason for this notion is mainly to encompass classical physics later on, i.e., to recover a good amount of common sense. An example will illustrate the problem: If one assumes that the motion of the Earth is basically governed by quantum mechanics, one must also consider as possible some aberrant effects with a very small probability, which look incredible from the standpoint of common sense. For instance, nothing forbids the Earth from leaving the neighborhood of the Sun by a quantum effect and going far away, ending its transition by revolving for instance around Sirius. This might be the consequence of a tunnel effect. When one computes the corresponding probability according to elementary quantum mechanics, it is found to be very small, of the order of ten to the power -10^{84}, and it becomes still much smaller when decoherence is taken into account. However small this probability may be, it is not strictly 0. One can therefore agree with common sense in saying that the sun will rise tomorrow, but one should allow for the possibility of such crazy events, at least as a matter of principle. Clearly, the simplest thing to do will be to predict that there will be a sunrise tomorrow up to that kind of error. Nobody should pay any attention to the corresponding risk, though a careful theoretical analysis cannot completely ignore it and must include it among the possible errors in its predictions.

Sometimes, the probabilities of errors will not be so ridiculously small and it will be a matter of judgement to decide whether or not they are worth being considered in more detail. As a matter of fact, they represent essentially what the older Copenhagen interpretation represented by Heisenberg's frontier between classical and quantum physics. One will see that the choice of this frontier is mainly a quantitative question, so that the larger a system is, or essentially the more classical it is, the smaller the errors in the logical consequences of the observed properties are. It will also be found that some probabilities are so small that they don't even have a physical meaning, so that they can be confidently neglected without resorting to an act of undue confidence. This is however a rather subtle point to be considered only at the end of Chapter 8.

4. The Complementarity "Principle"

Complementarity Rediscovered

With the help of the previous definitions and results, one is now able to avoid all kinds of logical mistakes when dealing with a quantum system. There is however something troublesome about the logics we have met, namely that there are too many of them at first sight. It turns out that quite different consistent logics corresponding to the same initial state, but with different observables, different domains for their values, or different times, can be used to describe one and the same physical system. For instance, it will be shown later on that, when a particle is emitted by the decay of an unstable nucleus and detected at some distance, one can choose to discuss either what was its position or its momentum at some given time before detection, but the two choices correspond to two different logics having not much in common.

This kind of multiplicity is not really new in the field of logic and one knows of many examples where a given subject can be considered differently according to the logic one chooses. For instance, the consequences of wedding are not the same according to the logic of a monogamic law as when it relies upon a polygamic or polyandric law. The ambiguities arising from some more subtle examples have even been at the origin of the idea of a field of propositions. Nonetheless, nothing of that kind was ever met in physics before the advent of quantum mechanics.

This is after all not so new in physics since Bohr realized long ago that a physical system can be described in many different ways and he called this the complementarity principle, as already mentioned in Chapter 2: the concepts one can use to describe a system depend upon the properties one can detect, which are ultimately defined by the measuring devices with which the system can interact.

The multiplicity of consistent logics is nothing but a rediscovery of the complementarity principle, though it has now lost its status of principle to become only a byproduct of logic. It will be seen more than once that this kind of occurrence is rather common when physics is approached from a logical standpoint and quite a few Copenhagen principles will be rediscovered so as to become a theorem, a side remark, or simply a convenience, not to mention the case when they can be disposed of.

In its present form, complementarity is not directly related to the presence of some specific experimental environment, as it was according to Bohr. Complementarity, as envisioned here, is much more intrinsic to the system itself and, in some sense, the various complementary logics are different possible ways of describing the system and discussing its properties. This means that one should be able after all to disentangle conceptually a quantum system from the experimental setup around it, contrary to Bohr's point of view. However, it should be mentioned that, when measurement theory becomes available in Chapter 8, it will become clear that the consideration of the experimental setup is necessary when one insists on obtaining a "true" knowledge of the quantum system. Bohr's assertions should therefore be reconsidered with some care and the point of view one may hold most safely about complementarity is the following:

There are many different possible ways to describe an isolated quantum system as such and they make sense, as far as logic is concerned, if they are consistent. They are however somewhat gratuitous because they only allow one to make assumptions and to draw consequences from these assumptions.

THE LOGICAL FRAMEWORK OF QUANTUM MECHANICS 161

These assumptions are justified when they originate from the occurrence of an observed fact, which can only happen when the system to be discussed is measured. In that case, the system that must be described logically is no longer the quantum system itself but the really isolated system consisting of the measured system together with the measuring apparatus. Moreover, when one wants to draw the consequences of an experimental fact, the logics to be considered are no longer completely arbitrary but they must obviously include the existing facts among the properties they consider. The corresponding "sensible" logics will be discussed in Chapter 8 after developing measurement theory and it will be seen that they considerably limit the properties of the measured quantum system one can assert to be true.

Operations upon Logics

One might be afraid that the occurrence of several different logics would give rise to many logical inconsistencies, paradoxes, or even contradictions. The worst event would be if two different ways of reasoning could lead to different conclusions when one is using two different consistent logics. In view of this danger, which would mean that the present approach is completely wrong, we shall initially discuss how two different logics can be related to each other.

Let a consistent logic be denoted by L. One may *extend* it by increasing its field of propositions. This can be done, with no change in the initial state, by increasing the number n of reference times, i.e., by keeping the times already occurring in L and adding others, together with some associated observables and related domains for their values. One can also use a finer splitting for the spectrum of an observable already considered, by cutting it into smaller pieces. These procedures work only of course when the new consistency conditions arising from the process are satisfied. The new logic L' will then be called an *extension* of L.

One can also go the other way by making L simpler. This can be done by suppressing a reference time or, what amounts to the same thing, by keeping only the identity operator among the family of projectors associated with that time. One can also use a coarser decomposition of a spectrum into bigger domains, which are obtained by combining some previous domains. The consistency conditions of the new logic are then among those of the initial one, so that consistency is automatic. Rather than slicing differently each spectrum one by one, one can also directly use bigger n-dimensional boxes. To prove that consistency is again automatic is less trivial but is still true. Whatever the method to be used, the new logic L' will be said to be a *simplification* of the initial one L. This procedure is very convenient when one only needs to discuss the logical content of a specific history belonging to some rather large logic, because then it is enough to take care of the properties occurring in that history together with their negations.

We will also say that a logic L *contains* a logic L' when they are both consistent and the field of propositions of L contains all the propositions belonging to L'. Two logics L' and L'' will be said to be *mutually consistent* when there is a third larger consistent logic L containing both of them. Finally, two logics L' and L'' will be said to be *complementary* when there is no larger consistent logic containing both of them.

These operations on logic are useful because they allow us to recognize when two logics are really different (more properly complementary) and when they essentially tell us the same things. In any case, all of this is pretty trivial. Considering a larger logic is rather similar to adding one of Euclid's books to the list of the previous ones one has already read and considering a simpler logic amounts to concentrating one's attention upon what is really important. It is only because thinking is an enterprise more open to failure in the case of quantum mechanics that one must show such a pedantic care.

The No-Contradiction Theorem

One can now remove the ghost of possible self-contradictions. This is afforded by the following no-contradiction theorem, the proof of which is given in Appendix C.

Theorem. *Let L and L' be two complementary logics, both containing the pair of propositions (a, b) in their fields. If the implication $a \Rightarrow b$ is valid in L, then it is also valid in L'.*

This result means that, when starting from some given assumption a, one will always obtain the same conclusion b, in spite of the multiplicity of logics. Of course, there may exist other logics containing a but not b, but this is not a self-contradiction; it is only complementarity and one must learn to live with it. In any case, when we shall obtain a criterion for truth later on, it will be seen that complementarity gives no hindrance to the existence of truth, i.e., to the agreement with reality.

The present theorem is a key result along the way that will show that quantum mechanics is completely consistent. It might look surprising that one dares to assert the impossibility of paradoxes in a theory that was so often reputed for them, but this is the way it is.

A FOUNDATION FOR INTERPRETATION

5. The Universal Rule of Interpretation

Stating the Rule

Having now at one's disposal the tools provided by consistent logics, it becomes possible to propose a new rule for quantum mechanics. It reads as follows:

THE LOGICAL FRAMEWORK OF QUANTUM MECHANICS

Rule 4. *Any description of the properties of an isolated physical system must consist of propositions belonging together to a common consistent logic. Any reasoning to be drawn from the consideration of these properties should be the result of a valid implication or of a chain of implications in this common logic.*

This rule calls for a few comments. It takes for granted the concept of a consistent logic together with all the notions underlying it. This includes the concept of histories, their probabilities, and their consistency conditions, as well as the notion of state as described by a density operator and coming from the logical requirements in Gleason's theorem. In any case, when one looks at these preliminary concepts, they are just a few definitions and a few properties following from these definitions and they do not represent new principles of physics by themselves, but only some convenient tools. This is also true of probabilities, which have been only up to now a mathematical convenience and need not therefore be postulated, since it was shown that their form is unique. When the present rule is stated, all these auxiliary notions are given a physical content, just as the Hilbert space and its auxiliary concepts were given a physical content in view of Rule 1. This includes in particular the notion of state, about which it is worth stressing once again that it need not be specifically assumed, since it is a necessary ingredient of probabilities. Whatever it may be, this discussion about what is an axiom and what is not is inessential, because it does not matter whether one states an axiom in just one sentence or whether it is split into several sentences.

What is really significant is that no other rule of quantum mechanics will be necessary from now on. Everything else will proceed from the four rules already stated, without any addition, through a purely deductive process. The last rule we have just stated contains in germ the Copenhagen rules for measurements, their proofs, their limitations, and their meaning. This is why it will be often called the universal rule of interpretation. It may look rather abstract at first sight but this is because it contains in a nutshell too many things that were once believed to be basically separate.

It may already be noticed that the rule reduces, at least in principle, every reasoning to a straightforward computation. This is because the consistency of a logic can be checked by computing its consistency conditions and the validity of an implication is asserted by computing the corresponding conditional probability. This means that one has at one's disposal a proposition calculus to which any reasoning and any description can be submitted. This may be very helpful when words become doubtful guides, when one does not feel sure about a statement or when two physicists disagree in their reasonings or their interpretations. One has only to sit down, to compute, and to conclude.

Of course, as happens for other instances of a proposition calculus, this one is not constructive but only normative. It does not say what an interesting physical problem or a nice experiment is. What one might wish to know about them remains part of our creative activity as physicists. This activity also suggests what would be worth checking or worth proving. How to proceed for a check or a proof can be left to the proposition calculus and, when this is done, there cannot remain any ambiguity in the final answer.

About the Structure of Quantum Mechanics

What was said about the noncreative character of the proposition calculus underlying the rule of interpretation does not mean that the rule itself has no proper fecundity. As a matter of fact, it will provide a very convenient organization of quantum mechanics, suggesting new problems and offering valuable hints about their solution. It is somewhat reminiscent of the thread Ariadne gave to Theseus to get out of the Labyrinth. The interpretation of quantum mechanics is so full of intermingled questions that an opportunity to organize them brings one already half-way to their answer. The rule of interpretation provides the necessary organization with a remarkable efficiency.

One can also take a look backwards so as to understand better the nature of each rule underlying quantum mechanics and to identify the related patterns of the theory: Rule 1, which associates a physical system with a Hilbert space and an algebra of operators, is a universal foundation setting up once and for all the mathematical framework of the theory. The first part of Rule 3, also concerning Hilbert spaces, makes more precise what this setup is as far as noninteracting systems are concerned. All the theory takes place in this universal mathematical structure. Rule 2, which is the Schrödinger equation, defines what dynamics is, as does the second part of Rule 3 concerning the hamiltonian of two noninteracting systems. Rule 4, the present one, is strictly logical. It introduces a logical pattern in the theory completing the dynamical pattern. It makes use of all the tools offered by dynamics, but it does not follow from dynamics. It brings a specific pattern to the theory; it contains the seed of interpretation. Notions such as the state of a system or the probabilities (which are only mathematical devices up to now) belong to the logical pattern of the theory, as well as other notions that are obviously logical such as properties, histories, and their relations.

One of the most essential differences between the dynamical structure and the logical one is their relation with time reversal. If one disregards the very small superweak interactions, dynamics is known to be time-reversal invariant. On the contrary, as could be suggested by the order of the projectors occurring in the probabilities of histories, *the logical structure is not time-reversal invariant*. This important point is discussed at length in Appendix D and it will be seen later on to have important consequences.

Finally, when given a set of principles or axioms, one often wonders whether or not they are independent. This question is relevant in the present case since several principles that were thought to be independent in the first formulations of quantum mechanics have already been found to follow directly from simpler assumptions. Furthermore, it has also been announced that the rules of measurement theory will be subsequently proved starting from the present ones. The independence of the axioms at the foundation of quantum mechanics is therefore a subject that is far from being settled at the present time and one cannot say much concerning this question as far as the present rules are concerned. There have been some interesting attempts at showing that Rule 1 is perhaps a direct consequence of some more fundamental considerations in formal logic.[5] They are however still inconclusive.

APPLICATIONS

6. Interferences

"How did you come in?"

"Most easily, through both the door and the window."

The whole difficulty of understanding how a particle can give rise to interferences, the necessity that it should go along both arms of an interferometer or that it should cross both slits in a Young interference device, is contained in this crazy dialogue. Perhaps nothing was more shocking to common sense or shows more clearly the drastic opposition between our habits of thinking and the ones authorized by quantum mechanics.

The theory of interferences and their logical treatment does not depend significantly upon the kind of particle one is considering so that we shall primarily choose to consider neutrons. These particles can be easily produced in a nuclear reactor. They cross the reactor wall through a narrow channel and their velocity is selected by diffraction, as already explained. One might also have used a simpler selecting device consisting of two rotating wheels, each one with a narrow window through which the neutron can pass. The conception of the preparing device is based purely in that case upon the idea of a neutron being a particle, i.e., a small chunk of matter. One could therefore expect that it should go through either one slit or the other when crossing a screen with two slits, as ordinary chunks of matter do.

Before considering the real problem, one must say a few words of caution concerning the language to be used or, if one prefers, the grammar. Ordinary language is much more careless than the one used by logicians. The story goes for instance that a logician was ill and, later on, a friend came to visit him and asked: "Hello, old fellow, are you still ill or have you now recovered?" The logician only answered "Yes." You and I would say that a stone can go through one window or another. The logician would be careful to distinguish that from saying either through that window or the other (the two windows can be facing each other so that it is possible that the stone goes through one and the other). It is however easy to avoid such ambiguous statements by being a bit careful about what one says once this little difficulty has been noticed.

After being prepared, the neutrons enter an interference device. It may be a very elaborate one. One may use for instance several crystalline silicon plates allowing different paths of diffraction as shown in Figure 5.3. The main practical difficulty is that the crystal lattices in the various plates must be absolutely parallel, but this can be obtained by sawing them out of a single big monocrystal of silicon. However, for simplicity, we shall mainly consider the idealized case of a two-slit device.

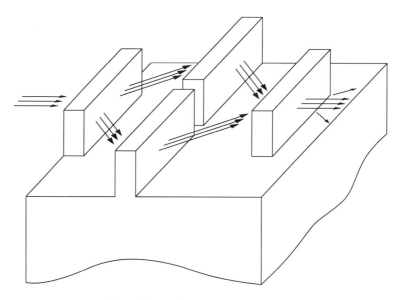

5.3. *A device for neutron interferences*

We can make the window in the reactor wall small enough and the selection mechanism at the entry of the device selective enough so that only a small number of neutrons will cross them per unit time. It is easy to realize conditions where only one neutron is inside the device while the next neutron to follow is still in the reactor. Then, the neutrons are really sent one by one.

The neutrons are detected by a battery of detectors located behind the slits. It will be convenient to assume them to be arranged along a plane parallel to the screen. We then observe the phenomena that were already discussed in Chapter 1: each detector reacts in a random way and there is no correlation among the events; their statistics clearly shows an interference pattern. If one slit is shut, the statistical distribution changes drastically and it has the shape of a single diffraction pattern.

Nobody can resist the feeling that a neutron should have passed through one slit only, whichever one, and not through both slits simultaneously. This feeling is perhaps even stronger in the case of the experiment shown in Figure 5.3, where the two paths along which the neutron can go are macroscopically separated along a long distance, because our mental representation is dominated by our visual experience. Our brain cannot help but produce this kind of image. There is no reason however to expect that this representation remains trustworthy in a case that is so far from the experience from which it originated.

The Copenhagen interpretation would say that only an experiment can give an answer to the question asking whether the particle passed through a definite slit. This would have to be realized by putting a detector along one path in order to register the neutrons crossing through it. When this is done, one finds that there is no interference pattern left but one only gets the addition of two diffraction peaks. The Copenhagen interpretation would then conclude that the question asking which way the neutron went is meaningless when there is no detector.

This does not however fully answer the question about the path followed by the neutrons. When there is a detector, the neutron interacts with it and both the neutron and the detector contribute to the isolated system one is dealing with. There are now some 10^{24} more particles or so in the system and it is an act of faith to assume that it still tells us something valuable concerning the case when there is only one neutron. One still does not know whether it is sensible or not to say that the neutron crossed only one slit, as long as no logical argument has yet shown that this statement is intrinsically meaningless.

This is where the universal rule of interpretation enters. It tells us that anything we can say about the properties of the system should belong to a consistent logic. In the present case, there are two kinds of properties one would like to state. The first ones concern detection. They assert the presence of the neutron at some time in one or the other of the detectors allowing one to see the interferences. The second ones state that the neutron crosses only one slit, or that it goes along only one path. It will now be shown that one cannot even formulate this second hypothetical type of property because there is no consistent logic involving them together with the empirical ones asserted by the detectors. The proof of this is quite simple in principle: One must construct a field of propositions corresponding to the smallest logic mentioning both types of properties. If it is not consistent, there will be no chance that any larger logic can be consistent. So one needs only to show that the consistency conditions are violated in the case of the simplest logic.

The calculation of the traces occurring in the consistency conditions and in the probabilities is sometimes tedious. It will therefore be convenient to consider the problem in its simplest conditions, while still keeping the main features of the experimental situation. We shall accordingly consider the case when, in place of two slits in a screen, there are two long channels along which the particle can go, a situation more akin to the experiment shown in Figure 5.4. This also fits the case of a photon that can travel along the two arms of an interferometer.

The calculations are practically the same as the ones met in the elementary theory of interferences. Let us first fix the notation: The incident neutron will be described by a pure state, its wave function being a wave packet. Taking a z axis along the direction of the initial beam and of the two channels, this wave packet has an average wave number k_0 along the z direction and we shall assume that its spatial extension in z is somewhat smaller than the length of the channels. Since quantum logics makes use of time explicitly, these assumptions make it clear that there is a time t_1 when the wave function of the neutron is entirely confined in the interior of the two channels.

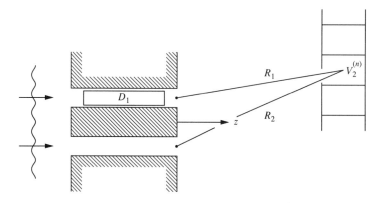

5.4. *Ideal description of an interference experiment*

When expressed mathematically, this means that the wave function at time t_1 is given by

$$e^{i(k_0 z - \omega_0 t_1)}[\Phi_1(x, y, z - v_0 t_1) + \Phi_2(x, y, z - v_0 t_1)], \qquad (5.1)$$

where the function Φ_1 is localized in the first channel and Φ_2 in the second one. The neutron velocity v_0 corresponds to a wave number k_0, ω_0 being the corresponding frequency. Except for their localization in the two different channels (as expressed by their dependency upon x and y) the two functions Φ_1 and Φ_2 are essentially similar and they are obtained from each other by a space translation.

THE LOGICAL FRAMEWORK OF QUANTUM MECHANICS   169

Let us now introduce a field of propositions together with the associated projectors. It will be better to avoid mentioning explicitly the neutron detectors as being physical objects, because one does not yet know how to describe such complicated systems. It will be enough however to state that the neutron is in the region of space occupied by a detector, without being more specific. So define several regions V_2^n, each one of them being a volume of space where there is a detector. It will be convenient to assume that their transverse dimensions (along the directions x and y) are small with respect to the distance separating two interference fringes. Their longitudinal size (along z) is taken to be large compared with the longitudinal extension of the wave packet. These assumptions imply that one can find a time t_2 when the wave function is completely located in the union of all these domains.

Let us assume that the wave function $\Phi_1(x, y, z - v_0 t_1)$ is centered at the midpoint of the first channel at time t_1, the origin of the z axis being taken at this midpoint. Let D_0 denote the distance from the plane of the counters to this origin. One can choose the times so that $t_2 - t_1 = D_0/v_0$, which is the classical time necessary to cross that distance with the average velocity. Finally, one makes the assumptions usually used in elementary interference theory, namely that the distance D from the slits to the plane containing the detectors is large when compared with the distance between the two slits.

In the elementary theory of interference, one considers an incident monochromatic plane wave. The wave at some point P behind the slits is a superposition of two waves coming from each slit. If R_1 and R_2 are the respective distances from P to the slits, the phases of these two waves are respectively equal to kR_1 and kR_2 up to an additive constant, k being the wave number. Assuming the amplitudes of the waves to be equal, the probability that the particle is at point P is proportional to

$$|e^{ikR_1} + e^{ikR_2}|^2 = 4\cos^2\left\{\frac{k(R_1 - R_2)}{2}\right\}. \tag{5.2}$$

Taking for the coordinates of the slits $x = \pm a, y = 0$, one gets

$$R_1 = [D^2 + (x-a)^2 + y^2]^{1/2} \approx D + \frac{x^2 + a^2 + y^2}{D} - \frac{2xa}{D},$$

$$R_2 = [D^2 + (x+a)^2 + y^2]^{1/2} \approx D + \frac{x^2 + a^2 + y^2}{D} + \frac{2xa}{D}. \tag{5.3}$$

The probability (5.2) is therefore proportional to $\cos^2(2kxa/D)$, which shows the existence of interference fringes.

Monochromatic plane waves can be added to build up a wave packet and the results then become time dependent. The central component with wave vector k_0 again gives the same result as before. It is now modulated in time and in the z direction so that, at time t_2, the wave function is

$$[e^{ik_0 R_1} + e^{ik_0 R_2}]e^{-i\omega_0 t_2}\Phi(x, y, z - v_0 t_2),$$

where the wave packet Φ is localized near the plane containing the detectors.

These elementary dynamical considerations are enough to describe the main characteristics of the experiment. One can compute the probability for the neutron to be in the region occupied by the nth detector, localized at time t_2 at a value x of the normal coordinate. It is proportional to the quantity (5.2) and this is the usual derivation of interference phenomena in the framework of quantum mechanics. One can also say that the probability of the properties stating that:

"the position X of the neutron at time t_2 is in the region $V_2^{(n)}$", (5.4)

is given by the quantity (5.2) up to a normalization constant.

Let us now introduce two space regions $D_1^{(1)}$ and $D_1^{(2)}$ coinciding with the interiors of the two channels 1 and 2. Let us also introduce the two properties stating that:

"the position of the neutron at time t_1 is in the first channel $D_1^{(1)}$";

"the position of the neutron at time t_1 is in the second channel $D_1^{(2)}$".

(5.5)

It will now be shown that there is no consistent logic containing both properties (5.4) and (5.5). The consistency conditions are given in that case by

$$\text{Re}[\text{Tr}\{E_1^{(1)}(t_1)\rho E_1^{(2)}(t_1)E_2^{(n)}(t_2)\}] = 0, \quad (5.6)$$

the projector $E_1^{(1)}(t_1)$ being associated with the first property (5.5), the projector $E_1^{(2)}(t_1)$ with the second property, and $E_2^{(n)}(t_2)$ with the property (5.4). These are all the consistency conditions and there is one of them for each detector. Since the initial wave packet is a pure state Ψ, the density operator is $|\Psi\rangle\langle\Psi|$. Using Equation (5.1) for the wave function at time t_1, one gets

$$E_1^{(1)}(t_1)|\Psi\rangle = e^{i(k_0z - \omega_0 t_1)}\Phi_1(x, y, z - v_0 t_1) \quad (5.7)$$

$$E_1^{(2)}(t_1)|\Psi\rangle = e^{i(k_0z - \omega_0 t_1)}\Phi_2(x, y, z - v_0 t_1). \quad (5.8)$$

When the wave function (5.7) evolves according to the Schrödinger equation, it becomes at time t_2 the component of the wave function originating from the first slit, whereas the function (5.8) generates the component coming from the second slit. They are respectively proportional to

$$|\Psi_1\rangle = e^{ik_0R_1}e^{-i\omega_0 t_2}\Phi(x, y, z - v_0 t_2)$$
$$|\Psi_2\rangle = e^{ik_0R_2}e^{-i\omega_0 t_2}\Phi(x, y, z - v_0 t_2). \quad (5.9)$$

The trace occurring in the consistency conditions (5.6) is then simply given by $\langle\Psi_2|E_2^{(n)}(t_2)|\Psi_1\rangle$, so that these conditions become $\text{Re}\langle\Psi_2|E_2^{(n)}(t_2)|\Psi_1\rangle = 0$. Using equations (5.9) and (5.3), this gives $\cos(4kax_n/D) = 0$, x_n

denoting the x coordinate of the center of the region $V_2^{(n)}$. It is clearly impossible to satisfy all these conditions.

We add a remark: the probability of detecting the neutron in the region $V_2^{(n)}$ is given by

$$p_n \approx \cos^2 \frac{2kx_n a}{D} = \frac{1}{2}\left\{1 + \cos\left(\frac{4kax_n}{D}\right)\right\}.$$

One sees that the consistency conditions are satisfied for a few values of x_n, precisely those where the sum of the probabilities agrees with the sum of amplitudes, i.e., the ones for which there is no interference effect.

7. The Straight-Line Motion of a Particle

Here is another problem that also gave rise to much discussion: A particle is produced near the origin of space, for instance by the decay of a radioactive nucleus. It is created in an isotropic S state (with orbital angular momentum 0) so that its wave function depends only on the distance r from the emitting nucleus (located at the origin of coordinates) and not upon the direction. A detector is located on the z axis at a large distance r_2 and it registers the particle at a time t_2. Now comes the question: can one assert that the particle was already undergoing a straight-line motion towards the detector before being detected?

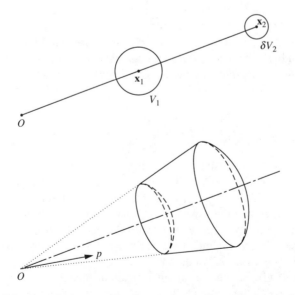

5.5. *Straight-line motion of an isotropically emitted particle*

This is a typical case of retrodiction, which means telling the past from the knowledge of datum. From the standpoint of the Copenhagen interpretation, retrodiction is impossible because it does not rely directly upon an actual measurement. Some physicists belonging to that school were however ready to consider the possibility of imaginary nondisturbing detectors, to be located along the way so as to give a meaning to retrodiction, though this does not stand upon a well-defined theory. Histories are on the contrary well suited to the problem, since it is a matter of logic having to do with properties occurring at different times.

When an actual experiment is made, one finds effectively that the particle goes essentially along a straight line. One can put for instance a bubble chamber in front of the detector. The track of the particle is then materialized by a straight array of bubbles. The theory of this effect was worked out for the first time by Mott who showed that the probability for the particle to ionize an atom far away from the straight line originating from the nucleus and ending at the detector is negligibly small[6] (one can easily understand this by using Feynman paths: the corresponding Feynman paths give a 0 sum in that case, as a result of their destructive interferences). The same argument could be used when an imaginary atom located somewhere in a vacuum is conceived, but it is far from obvious what status should be attributed to imaginary detectors in the framework of the Copenhagen interpretation. The main problem we have to consider is accordingly to find what one is allowed to assert when the particle moves in a vacuum between the nucleus and the detector.

This question becomes rather straightforward within the framework of the logical approach. Of course, some tedious calculations are necessary, but no more than the ones in Mott's theory. Since we are mainly interested here in understanding the basic notions associated with logic, these calculations will not be described in detail and only the ideas used and their results will be given. The relevant calculations are proposed as problems at the end of the chapter.

Straight-Line Trajectory

Let δV_2 be a volume of space containing the detector, which is taken for convenience to be very small. This volume is centered at a point x_2 with coordinates $x = 0, y = 0, z = r_2$, and the detection takes place at a time t_2. It will be assumed that the emission took place at time 0, the absolute value of the particle velocity v being rather well defined. The decay process will not be described in detail and one will assume without further justification that the time t_2 is essentially equal to r_2/v.

Let us now give a precise, albeit restricted, meaning to the statement according to which the motion follows a straight line. To do so, let us choose

THE LOGICAL FRAMEWORK OF QUANTUM MECHANICS 173

a point x_1 on the z axis, along the way to x_2. Let the time t_1 be defined as r_l/v. We shall also consider a sphere V_1 with its center at the point x_1 and a radius R_1. This radius will be taken much larger than the extension of the wave packet at time t_1, which includes both the effect of its initial extension and its quantum spreading between the times 0 and t_1.

Let us then define the two following predicates:

$$a = \text{"the position } X(t_2) \text{ is in } \delta V_2\text{"},$$
$$b = \text{"the position } X(t_1) \text{ is in } V_1\text{"}.$$

The question we are interested in is to decide whether the implication $a \Rightarrow b$ holds. To answer it, let us introduce a logic where the initial state is the S state of the particle emitted at time zero and two reference times t_1 and t_2 enter, the corresponding observables being the position in each case. This logic involves only two different properties at time t_1, namely b and its negation. Similarly, it refers to a and its negation at time t_2.

This simple type of logic was already considered in Chapter 4 and one knows that it requires only one consistency condition, namely

$$C = \text{Re Tr}\{E_1(t_1)\rho \bar{E}_1(t_1)E_2(t_2)\} = 0. \tag{5.10}$$

The probability of the property a is given by

$$p(a) = \text{Tr}\{\rho E_2(t_2)\} \tag{5.11}$$

and the probability of the proposition "a and b" is given by

$$p(a.b) = \text{Tr}\{E_1(t_1)\rho E_1(t_1)E_2(t_2)\}. \tag{5.12}$$

The calculation of the traces C and $p(a.b)$ is not trivial, though it is essentially the same as what must be anyway computed in Mott's theory. One can write for instance the second quantity explicitly as

$$p(a.b) = \text{Tr}\{U(t_2 - t_1)E_1 U(t_1)\rho U^\dagger(t_1)E_1 U^\dagger(t_2 - t_1)E_2\}.$$

Inserting formally an identity operator between each pair of factors and using its decomposition along the eigenstates of position, this can be written explicitly as

$$p(a.b) = \int_{\delta V_2} dx \int_{V_1} dy' \int_{V_1} dy'' \int_{R_3} dz' \int_{R_3} dz'' \langle x'|U(t_2 - t_1)|y'\rangle \cdot$$
$$\langle y'|U(t_1)|z'\rangle\langle z'|\rho|z''\rangle\langle z''|U^\dagger(t_1)|y''\rangle\langle y''|U^\dagger(t_2 - t_1)|x\rangle. \tag{5.13}$$

This is to be completed by the explicit form of the free particle evolution operator

$$\langle x'|U(t)|x''\rangle = (-2\pi i\, \hbar t/m)^{-3/2} \exp[im(\mathbf{x}' - \mathbf{x}'')^2/(2\hbar t)]$$

and by a specific assumption for the initial state, which is taken to be a pure gaussian state with wave function

$$\psi(x) = (\pi\delta^2)^{-1/4} \exp\{-r^2/4\delta^2 + ikr\}.$$

It corresponds to an uncertainty δ in the radial coordinate and a radial average velocity $k/m\hbar$.

One has then to integrate gaussian functions in several variables over somewhat complicated regions when computing the integrals in equation (5.13) and the similar expression for the trace C. This cannot be done exactly, but fortunately there exist reliable approximate methods (the stationary phase approximation and its variants) taking into account the smallness of Planck's constant, which makes the exponent in the gaussian very large (see the problems).

One can then show in the case of the consistency condition that the ratio $C/p(a)$ is a very small number ε of the order of $\exp(-R_1^2/\delta^2)$. It is furthermore a very rapidly oscillating quantity, the oscillation wavelength depending in a very sensitive manner upon the choice of the radius R_1. The reason for this kind of behavior will be explained in Section 10 where it will be shown to be representative of a very general situation. In the present case, one gets

$$C \approx p(a) \exp(-R_1^2/\delta^2) \cos\left(\frac{pR_1}{\hbar} + \alpha\right). \tag{5.14}$$

Here $p = mv$ and α is a phase, which could be made explicit but has no interest. If one chooses the radius R_1 of the sphere in such a way that the cosine in this formula vanishes, the consistency condition (5.10) is satisfied and the logic is consistent. It may be noticed that, because of the occurrence of Planck's constant in the argument of the cosine, there are many values of the exact radius R_1, all of them around a given macroscopic value, that are very near to each other and satisfy consistency, since a very slight change in the value of the radius is enough to make the quantity (5.14) vanish.

The logic being consistent, one can then consider the validity of the implication $a \Rightarrow b$ by asking whether the equation

$$p(b \mid a) = \frac{p(a.b)}{p(a)} = 1$$

is satisfied. The probabilities entering in this expression are given by equations (5.11) and (5.12). One first gets

$$p(a) = \int_{\delta V_2} |\Psi(x, t_2)|^2 \, dx.$$

Then compute approximately the gaussian integral $p(a.b)$ to get, in the leading approximation, $p(a.b) = p(a)$, from which one would conclude the

THE LOGICAL FRAMEWORK OF QUANTUM MECHANICS   175

validity of $a \Rightarrow b$. We can also find the small difference between these two probabilities in the following way: since the logic is consistent, probabilities are additive and therefore $p(a.b) + p(a.\bar{b}) = p(a)$. This gives

$$1 - p(b \mid a) = \frac{p(a.\bar{b})}{p(a)} = O(\exp\{-R_1^2/\delta^2\}), \qquad (5.15)$$

the last equality coming from an explicit evaluation of $p(a.\bar{b})$ using the same approximate methods.

One therefore finds that the implication $a \Rightarrow b$ is valid to order ε, the probability of error being given by equation (5.15). The conclusion is therefore that, at least in the present logic, the proposition stating that the particle crosses a well-defined space region along the way towards the detector is a logical consequence of the detection. This consequence involves a small but unavoidable probability of error but, because of the exponentially small value of this error, the risk of erring is in fact extremely small.

Variants

This conclusion calls for a few comments:

1. When the time t_1 is small, the sphere V_1 surrounds the origin of space and one can no more really say that this corresponds to a motion starting along a straight line. There is therefore a short interval of time Δt, just after emission, where one cannot decide in favor of a straight motion. This time interval increases when one wants the logical error ε to be smaller.
2. One might also consider several intermediate regions similar to V_1 along the way and corresponding times. This would show a motion along a straight line looking like a motion picture. One might even introduce a cylinder C_1 around the z axis and show that the particle is logically in this cylinder at any time t_1 such that $\Delta t < t_1 < t_2$. If the radius of the cylinder is R, one must take $\Delta t \gg m\delta^2/\hbar$. In that case, one does not specify the emission velocity.
3. One can also consider the case where the time of emission by the radioactive source is unknown. For a given emission velocity v, one then defines t_1 in terms of t_2 by $t_1 = t_2 - |\mathbf{x}_2 - \mathbf{x}_1|/v$.

    One can still obtain the same results as in the first example, i.e., the consistency of logic and the implication $a \Rightarrow b$. It is however necessary to use an explicit form for the wave function of a decay product[7] and the calculation becomes more cumbersome.
4. Another way of stating that the particle goes along a straight line would be to say that its momentum at a time t_1 prior to t_2 was directed along the z axis. To make this kind of retrodiction precise, one can define a domain B in momentum space expressing this property. It is convenient to take B as a thick spherical cap bounded by two spheres

centered at the origin of momentum and by a cone with a symmetry axis pointing toward the point x_2. Its apex angle θ must be much larger than $p\delta/\hbar$. The two spheres have radii $p - \Delta p$ and $p + \Delta p$, with $\Delta p \gg \hbar/\delta$. Introducing the property according to which

$$b' = \text{"the momentum } P(t_1) \text{ is in } B\text{"}$$

and choosing Δp and θ very exactly so as to satisfy the unique consistency condition, one can again show that $a \Rightarrow b'$.

5. It is also possible to introduce one or several intermediate properties occurring at different times and asserting that both position and momentum taken together agree with a straight-line motion at these times. This can be done by introducing "quasi-projectors" expressing the position of a particle in phase space, as will be explained in Chapter 6. The same kind of implications would again follow as before.

Despite the possibility of asserting in so many ways that the particle moves along a straight-line motion, one must be careful when appreciating the value of this conclusion. Notice that all these statements are in agreement with the overall classical idea of a straight-line motion with a velocity v, but they are in fact quite different from each other. Some of them mention position, others refer to momentum, some consider both position and momentum. They are complementary to each other. No specific proposition can state, without the ambiguity arising from complementarity, that the particle goes along a straight line. As a matter of fact, one might even introduce another consistent logic stating that the state of the particle is still isotropic at time t_1 (its angular momentum having the value 0) and this property would follow as a logical consequence of the preparation process. This means that all the properties that were mentioned up to now are, in the language to be used in Chapter 8, reliable but not true. This means that one can assert that a particle has a straight-line motion when it is not observed and this will never lead to any self-contradiction. One cannot however hold it as being absolutely true, as a feature belonging to physical reality. One might as well maintain on the contrary that the particle is still in an S state and this remains a matter of free choice among several complementary logics. It will be seen in Chapter 9 that the same kind of conclusions can be reached concerning the Einstein–Podolsky–Rosen experiment. In other words, there exist quantum properties that can be asserted reliably without running the risk of contradiction, but this does not necessarily mean that they are true or, in other words, real.

8. Decays

The usual treatment of a nucleus radioactive decay or of the decay of an excited atom is given in most textbooks, so that it does not seem necessary to repeat it here. It has however a few points of interest from the standpoint of logic and this is what will be considered. We shall recall as a preliminary a few key results of the usual approach.

When a nonperturbed state is prepared at time 0, a perturbation can provoke its decay towards other states in a continuous energy spectrum. In the case of beta decay, this perturbation is due to weak interactions and, for an excited atom, it comes from the coupling of the electrons with photons. The probability that the system has decayed after a short time Δt is given by

$$p(\Delta t) = \Delta t/\tau, \tag{5.16}$$

where the lifetime τ can be expressed in terms of the perturbation hamiltonian H_1 by the Fermi golden rule

$$\tau^{-1} = \frac{2\pi}{\hbar} \sum_k |\langle k|H_1|0\rangle|^2 \delta(W(k) - W_0).$$

In this expression, $|0\rangle$ represents the initial state with energy W_0 and $|k\rangle$ a decayed state with energy $W(k)$, k being a set of continuous indices representing, for instance, the momenta of the decay products.

This result is valid only when Δt is both small enough and large enough, i.e.,

$$\hbar/W \ll \Delta t \ll \tau. \tag{5.17}$$

In the first inequality, W is the total energy available to the decay products. The first inequality ensures the possibility of energy conservation during the decay and it is one of the many facets of the uncertainty relation for energy. It also ensures the outgoing character of the decay products, which forbids them regenerating to the initial state. The second inequality ensures the validity of first-order perturbation calculus. We shall restrict ourselves to the case when these two conditions can be simultaneously satisfied, leaving aside the decay of short-lived resonances by strong interactions, which is best treated by the techniques of collision theory.

It is interesting to notice that the probability $p(\Delta t)$ is smaller and even much smaller than the value given by equation (5.16) when Δt is of the order of $\hbar/W$ or less. This can be seen most easily by noticing that the decay amplitude is necessarily proportional to Δt when Δt is very small, so that the probability itself is proportional to Δt^2. It can be proved in general that[8] $p(\Delta t) \leq \Delta t/\tau$. This remark is at the origin of the so-called Zeno effect to be discussed in Chapter 11.

The exponential decay law is usually derived by considering the probability $p(t)$ for the particle to survive at time t and the conditional probability $p(t + \Delta t \mid t)$ for its decaying during the time interval $(t, t + \Delta t)$. Identifying $p(t + \Delta t \mid t)$ with the previous quantity $p(\Delta t)$, one gets

$$p(t) - p(t + \Delta t) = p(t).p(\Delta t), \tag{5.18}$$

which, in view of equation (5.16), immediately gives the well-known result

$$p(t) = \exp(-t/\tau). \tag{5.19}$$

This way of reasoning, simple as it looks, is somewhat puzzling from the standpoint of the Copenhagen interpretation. How is it possible to consider it ascertained that the particle has not decayed at time t and has decayed at time $t + \Delta t$ so as to give a meaning to $p(t + \Delta t \mid t)$? Furthermore, how do we assert its equality with $p(\Delta t)$? One might check experimentally that the particle decayed during the time interval $(t, t + \Delta t)$ but the physical system would not be the same as before; it would involve the apparatus that is checking the decayed/undecayed states and it is not obvious that the probabilities would be the same as in the absence of these devices. To say on the other hand that the particle was still intact at time t without a direct experimental check is after all a retrodiction, something that is in principle forbidden by Copenhagen. One might say that this is obvious, but why? Because a particle can only decay and that's all! But what about the time reversibility of quantum mechanics? Maybe the particle decayed some time before and then it was regenerated from its decay product, or even an almost-decay took place virtually. It looks like a questionable use of common sense to accept a straightforward meaning for $p(t + \Delta t \mid t)$ without more questioning.

To be sure, one might also consider measuring $p(t)$ by making many experiments, all done at the same time t after preparing the system and then repeating them systematically at the time $t + \Delta t$. But the first measurement could perturb the state and there is no reason to assume that the correlation for $p(t + \Delta t \mid t)$ and for $p(\Delta t)$ will be identical.

These various difficulties were of course noticed long ago and a satisfactory solution is known. It consists in solving explicitly the Schrödinger equation, though it is somewhat more complicated than the elementary calculation.[9] It gives directly the value of $p(t)$ for any time t satisfying the first condition (5.17) and it coincides of course with equation (5.19). It remains interesting nevertheless to justify the ordinary approach in a satisfactory manner and this is what will now be done with the help of quantum logic.

To say that the particle has not decayed at time t and that the decay products are present at time $t + \Delta t$ amounts to giving a history of the system. It is easy to express these properties by using the two projectors $E_1 = |0\rangle\langle 0|$ and

$$E_2 = \sum_k |k\rangle\langle k|,$$

together with the evolution operators $U(t)$ and $U(\Delta t)$.

There is only one consistency condition for the simplest logic containing these properties and it is given by

$$\text{Re Tr}\{E_1 U(t)\rho U^\dagger(t)\bar{E}_1 U(\Delta t)E_2 U^\dagger(\Delta t)\} = 0, \tag{5.20}$$

where in view of the nature of the initial state, one can identify the initial density operator ρ with the projector E_1.

This condition is satisfied when both t and Δt are much larger than $\hbar/W$, the essential point being that the decayed states communicating with the state $|0\rangle$ through $U(t)$ or $U(\Delta t)$ are outgoing states, which cannot regenerate the state $|0\rangle$. This point can be justified by a formal calculation but we shall admit it presently. Once it is known that the logic is consistent, one can compute without any difficulty of interpretation the probability $p(t, t + \Delta t)$, which is the probability of the history asserting the properties $E_1(t) = 1$ and $E_2(t + \Delta t) = 1$, as well as the conditional probability $p(t + \Delta t \mid t)$. Equation (5.18) then follows immediately.

We thus obtain a satisfactory logical framework for the formulation of the decay problem. It should be mentioned however that the consistency condition (5.20) is only satisfied approximately. This raises the question of what should be done about a logic when it is only approximately consistent. This point will be considered in the next section.

9. Approximations in Logic

A remarkable feature was met in the applications when some approximate implications occurred. Although one tried to be rather careful with the consistency conditions, there were also indications that they might be only approximate in some cases. These remarks suggest at least two lines of development, one being to understand more systematically the results we found and the other introducing the concept of approximation in logic.

As long as one is dealing with a finite-dimensional Hilbert space, the logical procedures are quite straightforward: either a consistency condition is satisfied or it is not and an implication is valid or not. Examples are given in the problems at the end of this chapter for a spin-$1/2$ system.

This simple situation is no longer the case when one considers a system with continuous degrees of freedom. Rather than finding exact implications $a \Rightarrow b$, i.e., a conditional probability $p(b \mid a)$ exactly equal to 1, one often gets only an almost exact implication, i.e., $p(b \mid a) \geq 1 - \varepsilon$, where ε is a small positive number. This means that the logical consequences of the assumptions one accepted or the measurements one made are not absolute. There is a very small but finite probability for the logical consequences to be incorrect. This kind of limitation seems to be unavoidable, particularly when classically meaningful situations are concerned. This is not however completely negative since this margin of freedom is precisely what is needed to reconcile quantum theory with the determinism of classical physics, as will be seen later on.

One may however be puzzled by the extreme and unrealistic restrictions that were found necessary to ensure a perfect validity of the consistency conditions. When discussing for instance the motion of a particle along a straight line, the consistency conditions were satisfied exactly when the particle was assumed to pass through a perfectly defined sphere on its way towards the detector. This was obtained however for extremely precise values of the sphere radius, or the position of its center, or its exact shape if it is not a perfect sphere. When for instance the radius is slightly modified, the trace occurring in the consistency condition has the form $\varepsilon \cos \phi$, where ε is an exponentially small number and ϕ is a phase which varies very rapidly under a slight change in the frontier of the spherical region. It was found possible to choose a perfect sphere, to fix exactly its center and then to choose its radius with perfect precision so as to verify exactly the consistency condition, but why such an unrealistic precision?

This situation raises an obvious question: is it sensible to give these details so sharply, only to get at the end an approximate implication, with the possibility of an error, however small, remaining? This extreme precision looks excessive, if only because no experimental device will ever be able to make it sure. Wouldn't it therefore be enough to accept an approximate consistency condition, since ε is very small, without having to refine artificially the hypotheses so that the cosine in $\varepsilon \cos \phi$ be strictly equal to 0?

The Origin of the Results

Before considering the basic aspects of this question, it will be useful to see why the quantity $\varepsilon \cos \phi$ occurs when one computes a consistency condition. Consider the case when the only observables of interest are position and/or momentum variables. One can then use Feynman path integrals to compute the probability of a history as well as the traces in the consistency conditions. Once the reference times $t_1, t_2, \ldots$ and the reference observables $A_1, A_2, \ldots$ (all of them being either a position or a momentum) have been chosen, it remains to choose the domains $D_1, D_2, \ldots$ for the values of these observables, i.e., some windows to be crossed by the Feynman paths. The property stating for instance that the x component of the position lies between 2 and 3 microns at time t_1 has a probability that can be obtained by restricting the Feynman paths to those crossing the window $2 \leq x \leq 3$ at time t_1. If there is a classical path among them, its phase will be stationary and the conditional probabilities for a sequence of semiclassical properties will be almost equal to unity.

As an example, one may go back to the properties of the motion of a particle along a straight line, as discussed in Section 7. The wave function at time t_2 is given in terms of the initial wave function by

$$\psi(\mathbf{x}_2, t_2) = \int G(\mathbf{x}_2, t_2 \mid \mathbf{x}, 0)\psi(\mathbf{x}, 0)d\mathbf{x},$$

where the propagator $G(\mathbf{x}_2, t_2 \mid \mathbf{x}, 0) = \langle \mathbf{x}_2 | U(t_2) | \mathbf{x} \rangle$ can be written as a sum over the Feynman paths going from the point $(\mathbf{x}, t)$ to the point $(\mathbf{x}_2, t_2)$:

THE LOGICAL FRAMEWORK OF QUANTUM MECHANICS

$$G(\mathbf{x}_2, t_2 \mid \mathbf{x}, 0) = \sum_\Gamma \exp(-i\Phi_\Gamma).$$

The phase associated with a Feynman path Γ is given by

$$\Phi_\Gamma = \frac{1}{\hbar} \int_\Gamma \frac{1}{2} mv^2 dt.$$

The probability of a Griffiths history expressing that the position is in a sphere V_1 at time t_1 and in the small volume δV_2 at time t_2 can be written as a trace or, equivalently, as a Feynman path integral

$$p = \int \Psi^*(\mathbf{x}')\Psi(\mathbf{x}) \left\{ \sum_\Gamma \exp(-i\Phi_\Gamma) \right\} d\mathbf{x}\, d\mathbf{x}'\delta V_2, \qquad (5.21)$$

where the paths now come and go: they start from the point $\mathbf{x}$ at time 0, then cross V_1 at time t_1, arrive at the point x_2 at time t_2, then they go backward in time to again cross V_1 at time t_1 and end up at the point $\mathbf{x}'$ at time 0. One might as well write this in the form:

$$p = \mathrm{Tr}\{E_1(t_1)\rho \bar{E}_1(t_1) E_2(t_2)\}. \qquad (5.22)$$

The consistency condition is similarly given by

$$C = \mathrm{Re}\, T = \mathrm{Re}\, \mathrm{Tr}\{E_1(t_1)\rho \bar{E}_1(t_1) E_2(t_2)\} = 0.$$

and one can also express the trace T occurring in it in a form analogous to the integral (5.22). In that case, the paths also come and go, still starting from $(\mathbf{x}, 0)$, crossing V_1 at time t_1, reaching $\mathbf{x}_2$ at time t_2, going back in time and then passing *outside* V_1 at time t_1 before ending again at $(\mathbf{x}', 0)$.

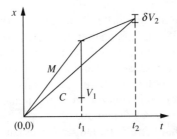

5.6. *Feynman paths and consistency conditions.* There exists a classical path going through the domain V_1 standing on the way from the point $(0,0)$ to (dV_2, t_2). The trace occurring in the consistency condition for the corresponding history is mainly controlled by the paths remaining close to this classical path on the direct way from $(0,0)$ to (dV_2, t_2) and avoiding V_1 on the way back.

It was shown that the two properties about the position of the particle:

$$a = \text{``}\mathbf{X}(t_2) \text{ is in } \delta V_2\text{''}, \qquad b = \text{``}\mathbf{X}(t_1) \text{ is in } V_1\text{''}$$

were such that $a \Rightarrow b$. The initial wave packet $\psi(\mathbf{x}, 0)$ was localized near the origin, so that all the relevant paths essentially started from the point $(0, 0)$ and went back to it at the end. The volume V_1 was chosen so that, in spite of the finite extension of the wave packet, there was always a classical path contributing to the integral (5.21). This "classicality" of the history "a and b" was essentially responsible for the implication $a \Rightarrow b$. The position, the shape, and the size of V_1 were chosen in such a way that the paths avoiding one of the windows interfere destructively to give an overall probability of order ε.

When considering the analogous path integrals for the trace T occurring in the consistency condition, one meets the opposite situation: no classical trajectory can come back from $(\mathbf{x}_2, t_2)$ to $(\mathbf{0}, 0)$ while staying out of (V_1, t_1). The paths contributing to the integral belong to the category that was left out in the calculation of the probability p. The destructive character of interferences is such that the paths coming nearest to (V_1, t_1) dominate in the evaluation of ε and the same interferences are responsible for the difference between p and unity and for the value of T. This is why only one parameter ε gives the same order of magnitude for $1 - p$ and T.

How to Deal with Approximations in Logic

Although these results were obtained in a special case, it is clear that they are valid generally. The leading terms in the consistency conditions will always be small and rapidly oscillating when one considers an implication suggested by classical physics. They will be the outcome of interferences in a Feynman sum over paths, systematically destructive when there is no classical path among them.

Following Gell-Mann and Hartle,[10] one might therefore adopt a pragmatic point of view. It does not matter after all whether a probability is 10^{-20} or 10^{-31}; no experiment will ever show the difference. In the same vein, it does not matter whether the logical operations are open to similar errors, because nobody will ever make sure empirically that the reasoning is incorrect to that order. Accordingly, an approximately consistent logic is just as good as a perfectly consistent one.

It may be noticed that there is no longer any essential difference in that case between saying that the real part of a trace in a consistency condition is small or that the whole trace itself is small. So, one can just as well write the consistency conditions as the approximate vanishing of the whole trace, without any special regard for its real part.

Some people might be tempted to say that rigor is more important than empirical success. They would accept an uncertainty in the ultimate conclusion of a reasoning but they would ask for logic to be absolutely above suspicion. Even they can be reconciled with a practical point of view in the following way: one knows that logic is perfectly rigorous as long as the probabilities are strictly additive. If one is given a finite field of propositions with imperfectly additive probabilities, one can always define other probabilities, very near to the given ones and perfectly additive. To do that, it will be enough to displace very slightly a domain or to very gently change its boundary. It is therefore possible, in all practical applications, to use a perfectly consistent logic if one asks for it, though with an excessive and pedantic precision in the choice of the properties to be considered.

Conclusion and Outlook

When complementarity was discussed, it was mentioned that, in spite of the great variety of consistent logics, only a few of them will turn out to be useful when we are able to take due care of measurements. This is very similar to the Copenhagen interpretation, which says that the choice of a specific description of a system is more or less dictated by the existence of the measuring devices surrounding it. Complementarity is therefore some sort of an initiation ordeal one must have gone through in order to become an adult in quantum mechanics.

There is something similar concerning the kind of rigor we have tried to maintain. Rigor is useful in the mathematical aspects of physics but it is of much less importance when it comes to practice. The Byzantine discussions about the exact region where a property takes place were only meant to make sure that rigor is possible when estimating the unavoidable probabilities of error in a logical argument. As a matter of fact, the author confesses that they were safeguards against anticipated criticisms. They had to be mentioned because more than sixty years of discussion about the interpretation of quantum mechanics have made some people very demanding as to the consistency of a new approach. Once it has become clear that the theory is completely consistent, it is easier to dispel this kind of uninspiring logomachy. One may even hope that the formal considerations occupying too large a part of the present chapter will soon be considered as what they really are, something more or less trivial and rather boring.

APPENDIX A: FORMAL LOGIC

A logic is associated with a specific field of propositions. These propositions are denoted by letters a, b, and so on; the field of propositions, i.e., their set, is denoted by P.

For any proposition a belonging to P, there exists another unique proposition, denoted by $\bar{a}$ or $-a$, which is its negation. Given two propositions a and b, there exists another proposition in P which may be expressed as "a and b" and is denoted by $a.b$. There is also another proposition which is expressed by "a or b" and denoted by $a + b$.

An equivalence relation exists among propositions. It is denoted by the equality sign ($=$). The simplest rules of logic concern equivalence. They are:

$$a = a; \tag{5A.1}$$

$$\text{when } a = b \text{ and } b = c, \text{ one has } a = c. \tag{5A.2}$$

The logical operators (and, or) have the following properties:

$$a.a = a, \tag{5A.3}$$
$$a.b = b.a \quad \text{(commutativity)} \tag{5A.4}$$
$$a.(b.c) = (a.b).c, \quad \text{(associativity)} \tag{5A.5}$$
$$a + a = a, \tag{5A.6}$$
$$a + b = b + a \quad \text{(commutativity)} \tag{5A.7}$$
$$a + (b + c) = (a + b) + c, \quad \text{(associativity)} \tag{5A.8}$$
$$a.(b + c) = a.b + a.c. \quad \text{(distributivity)} \tag{5A.9}$$

The properties of negation are the following:

$$\bar{\bar{a}} = a, \tag{5A.10}$$

$$\text{when } a = b, \text{ one has } \bar{a} = \bar{b}. \tag{5A.11}$$

De Morgan's rules relate negation with conjunction and addition; they are the following:

$$\overline{(a.b)} = \bar{a} + \bar{b}, \tag{5A.12}$$
$$\overline{(a + b)} = \bar{a}.\bar{b}. \tag{5A.13}$$

When a proposition a implies another proposition b, this is written as $a \Rightarrow b$ or expressed in words by "if a, then b." This relation obeys the following rules:

The couple of implications $(a \Rightarrow b, b \Rightarrow a)$ is equivalent to $a = b$,

$$\tag{5A.14}$$
$$a \Rightarrow (a + b), \tag{5A.15}$$
$$a.b \Rightarrow a, \tag{5A.16}$$
$$\text{when } a \Rightarrow b \text{ and } a \Rightarrow c, \text{ one has: } a \Rightarrow b.c, \tag{5A.17}$$
$$\text{when } a \Rightarrow c \text{ and } b \Rightarrow c, \text{ one has: } (a + b) \Rightarrow a, \tag{5A.18}$$
$$\text{when } a \Rightarrow b, \text{ one has: } \bar{b} \Rightarrow \bar{a}, \tag{5A.19}$$
$$\text{when } a \Rightarrow b \text{ and } b \Rightarrow c, \text{ one has: } a \Rightarrow c. \tag{5A.20}$$

These rules define the simplest version of formal logic. In spite of their simplicity and their intuitive character (they can be easily checked on a graphical representation where a proposition corresponds to a subset of a given set) they constitute the essential building blocks of logic. It may be noticed that we did not try to give here a minimal set of rules: for instance, the rules (5A.6), (5A.7) and (5A.8) for logical addition follow from the rules (5A.3), (5A.4) and (5A.5) for conjunction by using de Morgan's rules.

Truth Function

Although this cannot be done in the case of quantum mechanics, one often uses in logic what is called a truth function. It is a function defined upon the propositions and taking only the two values 1 or 0. A proposition a is said to be true when $T(a) = 1$ and false when $T(a) = 0$.

One can define the logical operations by the truth value of their arguments. For instance, the truth table of negation is the following:

$T(a)$	$T(\bar{a})$
1	0
0	1

The truth table of "or" is:

$T(a)$	$T(b)$	$T(a+b)$
1	1	1
1	0	1
0	1	1
0	0	0

The table for "and" is:

$T(a)$	$T(b)$	$T(a.b)$
1	1	1
1	0	0
0	1	0
0	0	0

The implication $a \Rightarrow b$ is then defined by $T(b) = 1$ whenever $T(a) = 1$. In the treatment of logical propositions on a computer, one often considers an implication $a \Rightarrow b$ as being also a proposition belonging to the field. It is defined as a function of the two arguments (a, b) such that

$$T(a \Rightarrow b) = T(\overline{(a.\bar{b})}).$$

We shall not use however truth functions because the propositions occurring in quantum mechanics do not generally take only the two values "true" and "false." Most often, they have only some probability of being true and another of being false. Under these conditions, implication does not stand upon the same level as ordinary propositions, though one can still use the axioms of logic as they have been given at the beginning.

APPENDIX B: FORMAL LOGIC AND CONSISTENCY

The purpose of this appendix is to prove that the consistency conditions ensure the validity of the rules of formal logic so that the associated quantum logics are thereby consistent.

The propositions $a, b, \ldots$ are associated in a one-to-one way with subsets $\alpha, \beta, \ldots$ belonging to the direct product Ξ of the various spectra. When the consistency conditions are satisfied, one can define a probability for each proposition and it obeys the basic rules of probability calculus. The logical notions for "not, and, or, if..., then," and logical equivalence are defined as in Section 3 of the present chapter.

We already know that the operations of set theory satisfy the corresponding formal rules for negation, conjunction, and logical addition, where the equivalence of two propositions a and b is understood as the identity of the corresponding sets α and β. The present theory is however probabilistic and two propositions are said to be logically equivalent when their representative sets differ only by a set of probability 0. We say that $a \Rightarrow b$ when

$$p(a + b) = p(a.b). \tag{5B.1}$$

This convention does not give rise to any special difficulty as far as the logical properties of negation, conjunction, and addition are concerned since, as a matter of fact, one can always add or subtract a set with zero probability to any domain α, without any effect upon the corresponding properties. The equalities among sets expressing the basic logical properties of negation, conjunction, and addition still coincide with their logical analogs since one can go from an equality between sets to an equation such as equation (5B.1) by suppressing all the sets (boxes) with zero probability from Ξ. A formal proof is easy but it will not be given.

This being understood, one can consider that the formal rules from (5A.1) to (5A.14) are satisfied with our conventions. It remains to prove the rules (5A.15) to (5A.21) where implication enters.

The following lemma will be useful:

Lemma. *When one has an implication $a \Rightarrow b$, the probability $p(a.\bar{b})$ is 0. The converse statement is true when $p(a)$ is is different from 0.*

Proof. The existence of the implication assumes that $p(a) \neq 0$ and that $p(a.b) = p(a)$. Define a proposition c as

$$c = (a.b).(a.\bar{b}) = a.(b.\bar{b}). \tag{5B.2}$$

The proposition $b.\bar{b}$ has probability 0, so that $p(c)$ is 0. On the other hand, one has

$$(a.b) + (a.\bar{b}) = a.(b + \bar{b}) = a.$$

Using the additivity of probabilities for the two disjoint propositions $a.b$ and $a.\bar{b}$, one has

$$p(a) = p(a.b) + p(a.\bar{b}) \qquad (5B.3)$$

and equation (5B.2) then gives $p(a.\bar{b}) = 0$, as stated in the lemma.

Conversely, if $p(a.\bar{b}) = 0$, equation (5B.3) shows that $p(a.b) = p(a)$, meaning that $a \Rightarrow b$ when $p(a)$ is not 0.

Using the lemma, one can verify the axiomatic properties of implications by taking them one by one:

Property (5A.15): $a \Rightarrow b$ and $b \Rightarrow a$ is equivalent to $a = b$.

By assumption, $p(a)$ and $p(b)$ are not 0. Therefore $p(a.b) = p(a) = p(b)$. But one has $p(a + b) = p(a) + p(b) - p(a.b)$, so that $p(a + b) = p(b) = p(a.b)$. This coincides with the definition (5B.1) for the logical equivalence of two propositions.

Property (5A.16): $a \Rightarrow (a+b)$. It follows from $(a+b).a = a$, which gives $p((a + b) | a) = 1$.

Property (5A.17): $a.b \Rightarrow a$. It is equivalent to the equation $p((a.b).a) = p(a.b)$, which is a direct consequence of $(a.b).a = a.b$.

Property (5A.18): When $a \Rightarrow b$ and $a \Rightarrow c$, one has $a \Rightarrow b.c$. By assumption,

$$p(a) \neq 0, p(a.b) = p(a.c) = p(a). \qquad (5B.4)$$

The lemma gives $p(a.\bar{b}) = 0$ and $p(a.\bar{c}) = 0$, so that $p(a.b.\bar{c}) = p((a.b).\bar{c}) = 0$. This gives

$$p(a.b.c) = p(a.b.c) + p(a.b.\bar{c}) = p(a.b) = p(a), \qquad (5B.5)$$

the last equality resulting from the assumptions (5B.4). Equation (5B.5) means that $a \Rightarrow b.c$.

Property (5A.19): When $a \Rightarrow c$ and $b \Rightarrow c$, one has: $(a + b) \Rightarrow c$. By assumption,

$$p(a) \neq 0, p(b) \neq 0, p(a.c) = p(a), p(b.c) = p(b).$$

From the lemma, we have $p(a.\bar{c}) = p(b.\bar{c}) = 0$, giving

$$p((a + b).\bar{c}) = 0. \qquad (5B.6)$$

One can write

$$a + b = ((a + b).c) + ((a + b).\bar{c}),$$

from which one gets, taking equation (5B.6) into account, $p(a + b) = p((a + b).c)$, which means that $(a + b) \Rightarrow c$.

Property (5A.20): When $a \Rightarrow b$, one has: $\bar{b} \Rightarrow \bar{a}$. By assumption, one has $p(a) \neq 0, p(a.b) = p(a)$. We will not consider a trivial proposition b having probability 1 because it does not give rise to any useful implication, so that one can take $p(\bar{b})$ to be different from 0. Using the lemma, one has $p(a.\bar{b}) = 0$. De Morgan's rule then gives

$$p(\bar{a}.\bar{b}) = 1 - p(a + b) \tag{5B.7}$$

One knows in general that $p(a + b) = p(a) + p(b) - p(a.b)$, which gives, in view of the assumption (5B.7),

$$p(\bar{a}.\bar{b}) = 1 - p(a + b) = 1 - p(a) - p(b) + p(a.b) = 1 - p(b) = p(\bar{b}),$$

i.e., the desired implication.

Property (5A.21): When $a \Rightarrow b$ and $b \Rightarrow c$, one has: $a \Rightarrow c$. By assumption, one has

$$p(a) \neq 0, p(b) \neq 0, p(a.b) = p(a), p(b.c) = p(b).$$

From the lemma, $p(a.\bar{b}) = p(b.\bar{c}) = 0$. Therefore, $p(a.\bar{b}.c) = p(a.b.\bar{c}) = 0$. One can then write

$$p(a.c) = p(a.b.c) + p(a.\bar{b}.c) = p(a.b.c),$$

but

$$p(a.b.c) = p(a.b.c) + p(a.b.\bar{c}) = p(a.b),$$

since $p(a.b.\bar{c})$ is 0. Therefore $p(a.c) = p(a.b) = p(a)$, establishing the implication $a \Rightarrow c$.

APPENDIX C: THE NO-CONTRADICTION THEOREM

Given two consistent logics L' and L'', both involving a couple of propositions (a, b), the no-contradiction theorem states that if the implication $a \Rightarrow b$ holds in L', then it also holds in L''.

The two logics are assumed to refer to the same initial state, at the same initial time. The times occurring in L' will be denoted by $\{t'_k\}$, $k = 1, 2, \ldots, n'$, the corresponding observables being $\{A'_k\}$. The direct product Ξ' of their spectra is cut into boxes according to a well-defined scheme. Similar notations will be used for L''.

The propositions a and b do not refer in general to all the values of the time occurring in the logics. They specifically mention some of them together with some associated observables. The direct product of the spectra for these selected observables will be denoted by Π. The other values of time and the other observables occurring in L' and L'' do not play any role. They do not enter in the propositions a or b, except in an insignificant way, where an identity operator takes the place of a nontrivial projector. One can therefore suppress them completely and this amounts to introducing two reduced (simplified) logics L'_r and L''_r, which are sublogics of L' and L'', respectively. The consistency conditions of a simplified logic belong to the family of consistency conditions for the initial logic so that they are automatically satisfied. The reference space for both logics L'_r and L''_r is the space Π. The difficulty of the problem comes from the fact that the decomposition of Π into boxes can be different in the two cases.

It will be useful to make a few general remarks at the beginning: It was shown in Chapter 4, and particularly in Appendix 4.A, that the decomposition of a space such as Ξ' into boxes can be made primarily in two different ways. The first one, which was introduced by Griffiths, consists of decomposing once and for all each spectrum into disjoint sets and to generate the boxes in Ξ' by slicing each coordinate axis along the frontiers of these subsets. It looks like cutting a piece of butter into cubelets with a knife. This is obviously the most direct way to proceed and we shall call this a Griffiths decomposition. It was also shown that this is not the simplest way to proceed and that fewer consistency conditions occur when one proceeds otherwise. The most efficient decomposition is a family of larger and larger boxes, giving what was called a special decomposition in Appendix 4.B. It is particularly convenient when one wants to singularize the role of a specific history.

Special decompositions are sometimes necessary and one cannot always resort to a Griffiths decomposition, though the proof of the no-contradiction theorem is much simpler in the latter case. An example of the necessity of non-Griffiths bases will be shown in Chapter 11 when we discuss delayed-choice measurements.

It may happen that a different choice of decomposition has been made initially in Ξ' and in Ξ''. As a consequence, the two reduced logics do not refer to the same decomposition of their common reference space Π. In the following proof, two cases will be covered: (i) The two decompositions of Π are obtained from two Griffiths decompositions of Ξ' and Ξ'', and (ii) There is a global reference history of the system mentioning the properties which (together with their negations) involve everything that is mentioned in the propositions a and b. A corresponding special decomposition of Ξ' has been used initially and the decomposition of Ξ'' belongs either to the special type or to Griffiths's type.

One might also consider other cases but there comes a point where more and more generality turns into more and more obscurity and this will be enough for all practical applications. The proof is rather cumbersome, mostly because of a necessarily heavy notation and also because of the geometric complexity of the situations one can meet. Except for these unfortunate intricacies, it is in fact rather straightforward.

a) First Case

Both decompositions of Π arise from an initial Griffiths decomposition. This is the simplest case. Let α and β be the sets associated with the propositions

a and b in Π. They are the same in both reduced logics. These sets are n-dimensional polyhedra with each of their faces normal to some coordinate axis, at least when the spectra are continuous. When a spectrum has a discrete part, one can consider it as a subset of the real set $\mathbf{R}$ and draw formal frontiers orthogonal to the coordinate axes separating the domains α and β from their complementary sets. The general case is thereby reduced to the case where all the spectra are continuous.

Let us consider the faces of the boundary of α that are normal to the j axis. Their projections on this axis constitute a discrete point set that may be used to slice the spectrum σ_j into a complete set of disjoint intervals. Proceeding in the same way for each coordinate axis and taking the direct products of all these intervals, one obtains a decomposition of Π into a Griffiths family of boxes, to be called the big boxes. In view of their construction, the big boxes are unions of smaller boxes already present in the initial decomposition of Π coming from the logic L'. Because of the consistency conditions satisfied by L', the probability $p'(a)$ of the proposition a, as computed in L', is uniquely defined as the sum of the L' probabilities for the big boxes occurring in α. One can proceed in a similar way and compute the probability $p''(a)$ for the proposition a in the logic L''. But such a big box corresponds to a history and its probability does not therefore depend upon the logic in which it is embedded so that $p'(a) = p''(a)$.

Replacing the set α by $\alpha \cap \beta$, one similarly gets $p'(a.b) = p''(a.b)$. The conditions $p'(a) \neq 0$ and $p'(a.b) = p'(a)$ are equivalent to the implication $a \Rightarrow b$ holding in L' by assumption. Since these probabilities are the same in L'', the implication $a \Rightarrow b$ also holds in L''.

b) Second Case

We consider first the case where the initial decompositions of Ξ' and Ξ'' into boxes are both special. Our main goal will be again to show that $p'(a) = p''(a)$. Let us as before first slice each spectrum σ_j by projecting the faces of the graph α normal to it. One can then take all the possible direct products of these intervals in all the spectra so as to construct a complete set of big boxes covering Π. The boxes of this category belonging to the set α will be denoted by G^k.

A box G^k is a direct product of several intervals I_j^k belonging to the various spectra $G^k = I_1^k \times I_2^k \times \cdots \times I_n^k$. Let $E_j^k(t_j)$ be the projector associated with the interval I_j^k. One can define the operator

$$F^k = E_n^k(t_n) E_{n-1}^k(t_{n-1}) \cdots E_1^k(t_1), \qquad (5C.1)$$

which occurs in the probability of the history associated with the graph G^k. Its probability is given by

$$p^k = \text{Tr}\{F^k \rho (F^k)^\dagger\}. \qquad (5C.2)$$

A difficulty arises at this point because the boxes G^k are not necessarily among the initial boxes in L' and L'', so that one cannot be sure that the consistency conditions hold for them. There is therefore no obvious guarantee that their probabilities are additive.

One can also start from the initial decomposition of Ξ' into boxes, as it is defined in the logic L'. When the observables that are not mentioned in the proposition a are suppressed, together with the corresponding dimensions in Ξ', one obtains a maximal decomposition of the set α as a union of boxes which will be denoted by $G'^{(\lambda)}$. In general, these boxes are not regularly piled upon each other, though the consistency conditions are automatically satisfied for them as a consequence of the consistency of L'. Each box $G'^{(\lambda)}$ is a union of boxes G^k and the inclusion of one box into another can be denoted as a relation of inclusion between their indices: $k \in \lambda$. A similar decomposition of α can be obtained by starting from Ξ'' and its decomposition into boxes, as defined in L''. The boxes G^k are still the same, since their definition is purely geometric, and we denote by $G''^{(\mu)}$ the boxes replacing $G'^{(\lambda)}$.

Let us now consider the special case when the initial state is defined by an initial projector E_0. The general case of an arbitrary initial density operator can be obtained from this simple case by adding probabilities, i.e., by the reduction of the density operator to a basis of its eigenvectors. Let $|\psi_m\rangle$ be an orthonormal complete family of eigenvectors of E_0 associated with the eigenvalue 1. Define the following vectors in Hilbert space:

$$|\phi_m^k\rangle = F^k|\psi_m\rangle, \qquad |\Phi_m^\lambda\rangle = \sum_{k \in \lambda} |\phi_m^k\rangle,$$

$$|\Phi_m''^\mu\rangle = \sum_{k \in \mu} |\phi_m^k\rangle. \tag{5C.3}$$

The probability associated with the box $G'^{(\lambda)}$, as computed in L', is given by

$$p'(G'^{(\lambda)}) = \sum_m \|\Phi_m^\lambda\|^2. \tag{5C.4}$$

We shall need the equation

$$\langle \Phi_m'^\lambda | \Phi_m'^{\lambda'} \rangle + \langle \Phi_m'^{\lambda'} | \Phi_m'^\lambda \rangle = 0, \tag{5C.5}$$

valid for $\lambda \neq \lambda''$. Its proof must distinguish the cases when the boxes $G'^{(\lambda)}$ and $G'^{(\lambda')}$, though different, have a common bottom and when they do not. The bottom is meant here as the last interval (in the order of increasing time) onto which the box projects. Let us first consider the case when there is a common bottom. The initial decomposition of Ξ' in boxes being special, i.e., ordered in increasing complexity with increasing time, means that the reduced space Π is also decomposed according to a special partition when one works in L'. We define the probability of $G'^{(\lambda)}$ by a formula analogous to equation equation (5C.2), using a convenient sequence of projectors as in equation (5C.1). These formulas only express that $G'^{(\lambda)}$ is a box associated with a history belonging to a consistent logic. Because of the special character of the initial decomposition, the families of projectors associated with the boxes $G'^{(\lambda)}$ and $G'^{(\lambda')}$ have the following properties: There is a time t_r such that these two families of projectors both reduce to the identity operator for the times $t_1, t_2, \ldots, t_r$, the two projectors associated with the time $t_r + 1$ being orthogonal (and in fact complementary) as well as the next ones. In that case, equation (5C.5) turns out to be a consistency condition for L' and it holds accordingly.

In the second case, the last projectors associated with the time t_n in $G'^{(\lambda)}$ and $G'^{(\lambda')}$ are orthogonal (and complementary). The scalar products in equation (5C.5) are 0. Equation (5C.5) therefore holds as announced.

Considering now the logic L'', one has:

$$p''(G''^{(\mu)}) = \sum_m \|\Phi_m^\mu\|^2 \text{ and}$$

$$\langle \Phi_m''^\mu | \Phi_m''^{\mu'} \rangle + \langle \Phi_m''^{\mu'} | \Phi_m''^\mu \rangle = 0,$$

when $\mu \neq \mu'$.

This gives immediately the equality between the probabilities $p'(a)$ and $p''(a)$:

$$p'(a) = \sum_\lambda p'(G'^{(\lambda)}) = \sum_{m\lambda} \|\Phi_m^\lambda\|^2,$$

as a result of equation (5C.4). Using the consistency conditions (5C.5), we get:

$$p'(a) = \sum_m \|\sum_\lambda \Phi_m^\lambda\|^2.$$

Equation (5C.3) then gives

$$p'(a) = \sum_m \|\sum_k \Phi_m^k\|^2.$$

Proceeding in the same way for $p''(a)$, one obtains the same expression, from which one gets $p'(a) = p''(a)$. As before, one can then show that $p'(a.b) = p''(a.b)$, which gives the implication $a \Rightarrow b$ in L'' as a direct consequence of the implication $a \Rightarrow b$ in L'. Finally, when L' uses a special basis of boxes and L'' a Griffiths one, one obtains again the same result since the consistency of the Griffiths basis implies the consistency of a special basis simplifying it.

APPENDIX D: LOGIC AND TIME REVERSAL

Neither the probability of a history nor therefore quantum logics are invariant under time reversal, which shows a strong difference between dynamics and logic. It has interesting consequences, to be considered in Chapter 7, and particularly the fact that the directions of time in thermodynamics and in logic must coincide.

The Operation of Time-Reversal

Time reversal was introduced in quantum mechanics by Eugene Wigner.[11] Consider for simplicity the Schrödinger equation for a spinless particle in an external potential,

$$i\hbar \frac{\partial \psi(x,t)}{\partial t} = -\frac{\hbar^2}{2m}\Delta\psi(x,t) + V(x)\psi(x,t). \tag{5D.1}$$

Replacing t by $-t$ does not change the operator in the right-hand side whereas the time derivative operation in the left-hand side changes sign. The new equation one obtains,

$$-i\hbar \frac{\partial \psi(x,-t)}{\partial t} = -\frac{\hbar^2}{2m}\Delta\psi(x,-t) + V(x)\psi(x,-t),$$

coincides formally with the one that is obtained by taking the complex conjugate of equation 5D.1:

$$-i\hbar \frac{\partial \psi^*(x,t)}{\partial t} = -\frac{\hbar^2}{2m}\Delta\psi^*(x,t) + V(x)\psi^*(x,t).$$

So if $\psi(x,t)$ is a solution of the Schrödinger equation, the function $\psi^*(x,t)$ is a solution of the time-reversed Schrödinger equation.

To go from a function $\psi(x)$ to its complex conjugate $\psi^*(x)$ is also an operation K acting in the Hilbert space L^2 of square-integrable wave functions. Let us write it as $\psi^* = K\psi$.

This transformation K is not an operator. It is antilinear, i.e., it conserves the sum of two functions but a multiplication by a complex number is replaced by a multiplication by the complex conjugate:

$$K(\psi_1 + \psi_2) = K\psi_1 + K\psi_2, \qquad K(c\psi) = c^*K\psi.$$

Doing this operation twice, one gets back the initial function, so that $K^2 = I$. Rather than changing the vectors in Hilbert space, one can equivalently change the operators according to the transformation $A \to KAK$. This change in an operator can be made clearer in a few examples. In the case of the momentum operator, one has

$$KPK\psi = \left[\frac{\hbar\partial}{i\partial x}(\psi^*(x))\right]^* = -\frac{\hbar}{i}\psi'(x) = -P\psi,$$

where $\psi'(x)$ is the derivative of $\psi(x)$ and the operation K is interpreted as taking the complex conjugate of anything to its right-hand side.

Accordingly, the momentum operator changes sign under time reversal. One can show in the same way that the position operator is invariant as well as the energy, at least in the present case. The components of the angular momentum change sign, so that this should also be true for the spin operators. When a constant electromagnetic field acts upon the system, the hamiltonian is no longer invariant. This can be seen in the case of a magnetic field $\mathbf{B}$ produced by a solenoid. The time-reversed situation corresponds to an inversion in the direction of the electric current and therefore to a field $-\mathbf{B}$. To get time-reversal invariance of the hamiltonian, one must replace $\mathbf{B}$ by $-\mathbf{B}$ in it, the electric field remaining the same.

These considerations are easily extended to a system containing any number of spinning particles. Spin transforms like an angular momentum and a state where $S_z = 1/2$ is replaced by another where $S_z = -1/2$. The general case has been considered by Wigner, who has shown that time reversal acts upon an arbitrary state vector by complex conjugation followed by a unitary transformation: $u \to VKu$. It is still antilinear. When acting upon operators, it gives

$$A \to A^t = VKAKV^{-1}. \tag{5D.2}$$

The unitary operator V is completely defined if one demands that the position operators are invariant, the operators of momentum and spin changing sign. When the system includes a quantized electromagnetic field, the fields $(\mathbf{E}, \mathbf{B})$ are replaced by $(\mathbf{E}, -\mathbf{B})$ and the potentials $(\Phi, \mathbf{A})$ by $(\Phi, -\mathbf{A})$. When this is done, the hamiltonian of an isolated system is invariant. More precisely, if one considers the Schrödinger equation, one has

$$i\hbar\frac{\partial}{\partial t}VK\psi = i\hbar VK\frac{\partial}{\partial t}\psi = VKH\psi = VKHKK\psi$$
$$= VKHKV^{-1}VK\psi = H^tVK\psi.$$

Experiments have shown that $H^t = H$, except for a superweak interaction that can be neglected for the applications we have in mind.

Time-Reversal of Properties

A time-independent projector E associated with an observable A and a subset D of its spectrum is transformed according to the general rule (5D.2):

$$E \to E^t = VKEKV.$$

THE LOGICAL FRAMEWORK OF QUANTUM MECHANICS   195

One can also replace A by A', but there can also be a change among the subsets where the values of A are given. For instance, the momentum P becomes $-P$ by time reversal and an eigenvector $|p\rangle$ becomes $|-p\rangle$. The projector $|p\rangle\langle p|$ becomes $|-p\rangle\langle -p|$ and $E = \int_B |p\rangle\langle p| dp$ becomes $E' = \int_B |-p\rangle\langle -p| dp$. This can also be written as $E' = \int_{B'} |p\rangle\langle p| dp$, where the set B' is obtained from the set B by inverting the direction of the momentum axes.

When an external field acts upon the system, the hamiltonian H is replaced by H' where, for instance, the magnetic field has changed sign. In this general case, the evolution operator is replaced by

$$U(t) \to U'(t) = VKU(t)KV^{-1} = VKe^{-iHt/\hbar}KV^{-1}$$
$$= Ve^{iHt/\hbar}V^{-1} = e^{iH't/\hbar}. \quad (5D.3)$$

This amounts to a simultaneous replacement of H by H' and of t by $-t$. The change in a time-dependent projector is therefore given by

$$E(t) \to E'(t) = U'^{-1}(t)E'U'(t) = E'(-t). \quad (5D.4)$$

Time-Reversed Histories

Consider for simplicity a history whose initial state is specified by an initial predicate with projector E_0, the initial time being now written as t_0. This history involves three reference times $t_1, t_2,$ and t_3 such that $t_0 \leq t_1 < t_2 < t_3$. Its probability is given by

$$p = \text{Tr}\{E_3(t_3)E_2(t_2)E_1(t_1)E_0(t_0)E_1(t_1)E_2(t_2)E_3(t_3)\}/\text{Tr}E_0. \quad (5D.5)$$

A time-reversed history would assume that the predicate with projector E'_3 defines the initial state and, from there on, the history runs backward. One must use a projector E'_3 rather than E_3 to reverse the velocities and the spins. The times are also now counted backward, which amounts to a replacement of t by $-t$. The time-dependent projectors describing the reversed history are therefore given by equation (5D.4): $E'_k(-t_k) = E'_k(t_k)$, the prime index in the second projector meaning that the range of values for the observable may have been inverted. The probability of the time-reversed history is therefore given by

$$p' = \text{Tr}\{E'_0(-t_0)E'_1(-t_1)E'_2(-t_2)$$
$$E'_3(-t_3)E'_2(-t_2)E'_1(-t_1)E'_0(-t_0)\}/\text{Tr}E'_3. \quad (5D.6)$$

Using equations (5D.3) and (5D.4) together with cyclic invariance of the trace and the unitarity of V, one finds that the trace occurring in the numerator of equation (5D.6) is equal to the one occurring in the numerator of equation (5D.5), so that

$$p' = p\frac{\text{Tr}E_0}{\text{Tr}E_3}.$$

The only difference between the two probabilities comes from their denominators and it only depends therefore upon the preparation predicates.

Is This Noninvariance?

Is such a minor modification really significant? This is the question we shall now consider under various aspects. One might come back to the definition of a probability, as given in the previous chapter. This is not exactly the form that was first stated by Griffiths and the whole discussion of the present question rests upon this difference and its meaning. Rather than writing down an expression similar to equation (5D.5) for the probability of a history, Griffiths used the expression

$$p_G = \text{Tr}\{E_3(t_3)E_2(t_2)E_1(t_1)E_0(t_0)E_1(t_1)E_2(t_2)E_3(t_3)\} / \text{Tr}\{E_0(t_0)E_3(t_3)\},$$

(5D.7)

which he called a conditional probability. He meant a probability where the initial state and the last property are given. It is easy to show that this expression is time-reversal invariant. A question therefore arises: Should one use the noninvariant formula (5D.5) for the probability p or the invariant formula (5D.7) giving p_G? Clearly, we must see where their difference lies. The expression giving p_G was constructed precisely so as to be invariant under time reversal. It looks therefore more attractive at first sight. One should however consider the consequences of each choice, because a history has no interest in itself and it becomes significant only when it is used in a logic and it is linked to the facts.

The Formal Approach

Let us first consider this question from the standpoint of pure formal logic. The history one started with was contained in a logic involving some properties along with their negations. Together with the final property, it therefore also contained its negation with projector $\bar{E}_3(t_3)$. One often draws an implication from the last property, which frequently represents a measurement. But then Griffiths's definition might give trouble with the rules of logic. This is because one of these rules states that $a \Rightarrow b$ should be equivalent to $\bar{b} \Rightarrow \bar{a}$. But Griffiths convention would give $p_G(E_3(t_3)) = 1$, and therefore $p_G(\bar{E}_3(t_3)) = 0$. Hence there cannot be any implication where the conclusion is $\bar{E}_3(t_3)$, when one starts from a proposition having a nonzero probability. Formally, Griffiths probability is therefore inconsistent with a rigorous use of formal logic.

From a more physical standpoint, one must resort to a result that will be established in Chapter 8: an experimental datum, for instance the reading of a counter, is logically equivalent to the really interesting physical result, which is a property of the measured system and not of the measuring apparatus. This equivalence is the outcome of implications going in both directions and requires in particular an implication going from the result towards the data. It would be impossible to prove it correctly in the framework of Griffiths probabilities for the formal reason just mentioned.

The Irrelevance of Griffiths Probability

In view of the importance of time-reversal invariance or noninvariance, one must still consider carefully the assumptions underlying Griffiths's formulation. Nothing forbids us to complete the history by an empty predicate (with the identity as its projector) at a time t_4 later than t_3. This is still the same history, as far as logic is concerned, but its probability is now given by p and no longer by p_G. One will have moreover to add the negation $\bar{E}_3(t_3)$ to the new logic. This shows that the relevant question is not really the form of the probability but the content of the logic and Griffiths's formulation is in that sense somewhat artificial.

The Artificial Character of an Initial Predicate

One can also learn more by considering the role and the meaning of the initial projector E_0 that was introduced. The associated property does not belong to the propositions of the logic under study. Effectively, one cannot introduce its negation by a projector $\bar{E}_0$, which would have to be available in order to draw correct implications from it. This negation projector does not belong in general to the trace class, so that it does not correspond to a density operator. It represents neither a proposition in the logic nor the initial state of another logic and therefore it lies outside the realm of logic. As a consequence, no implication can be drawn directly from the initial predicate, whereas the most interesting conclusions come most often from the last predicate. This clearly shows a marked difference between them.

Notice furthermore that the initial state operator is very seldom proportional to a projector, except in the case of a pure state. In general, it is obtained by a partial trace over the density operator of a larger system. The symmetry that was introduced by Griffiths between the first and the last predicate is therefore quite artificial since it would be a symmetry between a general positive operator (the density operator) and a projector. It was introduced by him in order to preserve formally time-reversal invariance but it became possible to appreciate it better when quantum consistent logic was more exactly formulated. As a matter of fact, the probability p (rather than p_G) can be obtained as the unique one satisfying a few simple logical conditions, as was shown in Chapter 4. The existence of this proof would have been sufficient to reject the probability introduced in the pioneering work of Griffiths, but the question of time-reversal invariance is so important that we felt it worthwhile to consider its various aspects.

Outlook

The noninvariance of logic under time reversal immediately raises the question of the relation between the direction of time occurring in logic and the one coming from thermodynamics. This will be considered in Chapter 7. It may also be added that the logical direction of time is universal, i.e., it must be the same when several different systems are considered together. This is obvious when they can interact but, even in the case of noninteracting systems, one must still use the same logical direction of time. This comes from Rule 3 according to which they can be treated as making a unique system, for which the logical direction of time must be chosen once and for all according to the proof of uniqueness for the probability that was given in Chapter 4.

PROBLEMS

Problems 1 to 4 discuss the logical status of some properties relative to spin, according to complementarity.

1. Consider the spin states of a spin-$1/2$ particle. Consider a logic L having for its initial state the pure state $S_x = +1/2$, two reference times t_1 and t_2 associated with the observables $A_1 = A_2 = S_z$, the different values of these observables being $\pm 1/2$. Show that L is consistent. Prove the implication: If $A_2(t_2) = +1/2$, then $A_1(t_1) = +1/2$.

2. The notations are the same as in Problem 1. The observable A_1 is now identified with S_x and the corresponding logic is denoted by L'. Prove its consistency and compute the conditional probability for the property $A_1(t_1) = +1/2$, given that $A_2(t_2) = +1/2$.

3. Show that the rule (5A.20) of formal logic, as given in Appendix A, ensures that no implication can take as its first predicate a predicate expressing an initial state of the system.

4. The initial state of a spin-$1/2$ particle is the pure state $S_x = +1/2$ and S_z is equal to $+1/2$ at a time t_2. Show that, within some logics involving only this particle, one can prove logically that $S_z = +1/2$ at a previous time t_1 but not that $S_x = +1/2$ at that same time. (It will be found in Chapter 9 that both properties can be proved in two different logics involving the measuring devices. Then they will be found to be both reliable but not true.)

THE LOGICAL FRAMEWORK OF QUANTUM MECHANICS   199

5. This problem considers the straight-line motion of a particle behind a diffracting device.

A neutron is in an initial state which is defined by a localized wave packet. It is diffracted by a crystal plate behind which its wave function becomes

$$2^{-1/2} \exp(i\mathbf{k}\cdot\mathbf{x})\phi(\mathbf{x}) + \exp(i\mathbf{k}'\cdot\mathbf{x})\phi'(\mathbf{x}),$$

$\mathbf{k}$ and $\mathbf{k}'$ being two different wavenumber vectors having the same length with their directions along two different Bragg directions. The wave packets $\phi(\mathbf{x})$ and $\phi'(\mathbf{x})$ are practically the same when leaving the slab and they are centered at the origin of coordinates. One assumes that the wave function behaves in a semi-classical way, the wave packet $\phi(\mathbf{x}, t)$ remaining centered at the point $\mathbf{v}t$ and $\phi'(\mathbf{x})$ at the point $\mathbf{v}'t$, the velocities $\mathbf{v}$ and $\mathbf{v}'$ being associated with the wave-numbers $\mathbf{k}$ and $\mathbf{k}'$, respectively. One considers two times $t_1 < t_2$ and four volumes in space: V_1 contains the support of the wave packet $\phi(\mathbf{x}, t_1)$ and V'_1 the support of $\phi'(\mathbf{x}, t_1)$, V_2 and V'_2 being similarly defined with respect to time t_2. A logic L is then defined in terms of the initial state, the times t_1 and t_2, the position observables at both times, and similar partitions of space, where, for instance for time t_1, space is divided into the sets V_1, V'_1, and the complementary set V''_1. Show that this logic is consistent. Prove the implication: If $\mathbf{X}(t_2)$ is in V_2, then $\mathbf{X}(t_1)$ is in V_1.

6.* Perform the calculations that are involved in the proof of the straight-line motion of a particle described in Section 7, considering the case of the particle position in space. One can use an isotropic initial wave function that is given by

$$\psi(\mathbf{x}) = K \int \exp(i\mathbf{p}\cdot\mathbf{x}/\hbar - x^2/4\delta^2)\, d^2\mathbf{p},$$

the integral being performed upon the directions of the vector $\mathbf{p}$ with a fixed length. One can also use the explicit expression for the matrix elements of the evolution operator

$$\langle \mathbf{x}'|U(t)|\mathbf{x}''\rangle = (-2\pi i\hbar t/m)^{-3/2} \exp[im(\mathbf{x}' - \mathbf{x}'')^2/(2\hbar t)].$$

The integrals encountered in the calculation of the consistency conditions and the probabilities can be evaluated by the saddle-point or Laplace method. It works as follows. For large values of λ, an integral

$$J = \int_\Omega \exp[-\lambda\phi(x)]\, d^q x$$

can be estimated under the following conditions: (i) $\phi(x)$ is an analytic function having a stationary value at a point x_0 whose real part is in the domain of integration Ω; (ii) the matrix A having for its coefficients the second derivatives of $\phi(x)$ at the point x_0 has a positive definite real part. Then one has approximately, for large positive values of λ,

$$J \approx (2^q \pi^q / \lambda^q \det A)^{1/2} \exp(-\lambda \phi(x_0)).$$

For the consistency conditions, one is led to integrals reducible to one-dimensional integrals such as

$$J' = \int_0^\infty \exp[-\lambda \phi(x)]\, dx,$$

where now the minimum of $\mathrm{Re}\phi(x)$ in $[0, +\infty]$ is reached at the boundary value $x = 0$. An approximate value for the integral J' for large positive values of λ is given in that case by

$$J' \approx [\lambda \phi'(0)]^{-1} \exp(-\lambda \phi(0)).$$

7.* Prove the properties relative to the momentum in a straight line motion, as they are described in Section 7, by using the methods of Problem 6.

6
Recovering Classical Physics

It is sometimes said that the most difficult problem in quantum mechanics is to understand classical physics. This may be true since at least two copious chapters will be needed to elucidate it. We will adopt here a point of view rather opposite to the one stressed by Bohr when he assumed that physics obeys two different kinds of laws, either the classical or the quantum ones according to the kind of physical object under consideration. It will be assumed here on the contrary that there is only one kind of laws given by the principles of quantum mechanics. The first hint that this point of view is correct was given by Bohr himself when he stated the correspondence principle, namely that the results of quantum mechanics must tend to those of classical physics when Planck's constant can be considered as negligible. The validity of this correspondence was later shown by Ehrenfest to be correct, as far as dynamics is concerned. One might then wonder why Bohr continued to keep apart a classical world and a quantum one after Ehrenfest's theorem had been obtained.

His reasons were somewhat compelling, as was already mentioned in Chapter 2: quantum mechanics can only deal with probabilities whereas the data resulting from experiment are plain facts leaving no possible doubt. In order to take care of the facts, one needs a framework where the properties one observes can be held as true with no trace of ambiguity, no probabilistic fuzziness. This is the main reason why one must use classical physics, not so much because it obeys Newton's laws of motion but because it deals with perfectly clearcut properties. In that sense, Ehrenfest's theorem is useless because it cannot help us to get rid of the probabilistic aspects of quantum mechanics.

Another aspect of experimental facts is that they can and must be registered. When for instance one performs a series of measurements in order to obtain the value of a quantum probability, each individual datum must be registered until at the end the whole collection of data is used to compute the frequencies. A record must be such that its reading faithfully recovers the fact that was registered and this is clearly a form of determinism. So, here again, what one needs basically is not really to have Newton's laws of motion as such, but their deterministic character. It can be recalled for comparison that determinism was shown to be a logical equivalence in the preceding chapter.

These remarks show that the main problem of the correspondence between classical and quantum physics is a matter of logic: the possibility of holding a classical property to be true and logically deducing a past fact from its present record. Of course, one should also recover classical dynamics, which is known to be phenomenologically valid, but logic should be given a priority over dynamics in a consistent justification of classical physics.

Naturally one cannot expect to recover classical physics as something absolute since, for instance, it is obvious that a record can happen to be spoiled by a quantum fluctuation. Determinism cannot therefore be absolutely reliable and there is always a tiny probability that it can lead to an error. Accordingly, classical logic, which is plain logic applied to the propositions of classical physics, can only be approximate and one of our tasks will be to get an estimate for the corresponding errors.

One cannot even expect a sweeping theorem stating once and for all that every macroscopic object obeys classical physics as soon as it is big enough, when, for instance, the number of its atoms is large enough. There are two reasons for this. The first one comes from chaotic systems: it turns out that their classical dynamical evolution ends up showing significant differences at the level of Planck's constant after a finite time. Another even more cogent reason is that one now knows examples of superconducting macroscopic systems behaving in a quantum way under special circumstances, as will be discussed in Chapters 10 and 11. The theorems predicting classical behavior of a macroscopic quantum system must therefore rely upon specific dynamical conditions, which will have to be made clear, though they hold very frequently.

It will be convenient to tackle these problems in two chapters and our purpose in the present one is only to investigate the classical limit of quantum mechanics in the situation where no spurious quantum effect can become manifest at a large scale, i.e., when essentially nothing similar to a quantum measurement can occur. This is the case one meets most often in practice since after all nobody ever suspected the existence of quantum effects before the discovery of radioactivity so that, in a nutshell, the purpose of the present chapter is to understand what goes on under ordinary circumstances. It should also be mentioned that this excludes more generally many cases where an unstable system (e.g., in a metastable state) is brought to a macroscopic change by a quantum fluctuation, though the question whether this is a measurement is not perfectly clear. The present investigation will be morever limited by a technical restriction, namely that friction and its dissipative effects are neglected. It will be easier to get rid of this unrealistic assumption in the next chapter, when the extension of the theory to a situation occurring after a quantum measurement will be considered. The reason for this delay in including friction is that decoherence, which will then be our main interest, has the same dynamical origin as dissipation.

This program will be executed in four steps. The first one has to do with objects, a pretty obvious notion for everybody but not so clear from the standpoint of quantum mechanics. An object can be a table, a voltmeter, a computer, a stone, and so on. It is the kind of thing one describes in classical terms by a finite number of coordinates. On the contrary, a physical system, as defined in Chapter 3, is a collection of particles that can assume many different characteristics. The theory to be given can probably be extended to "classical objects" having an infinite number of degrees of freedom, such as a fluid for instance, but the state of the art does not yet allow one to do it explicitly. It will be shown how to describe an object by collective observables, which are more or less the kind of variables one uses in classical physics, though one will have to clear up their correspondence with the associated quantum observables.

The second part deals with classical properties. As already said in the preceding chapter, they state typically the values of the position and momentum coordinates within well-defined bounds. Our main result will be to show that they are also quantum properties, namely that they can be associated with a quantum projector. This approach is therefore clearly in opposition to the one by Bohr, who accepted classical properties as being primary while forbidding a consideration of the microscopic world as such. Here one relies upon the universal interpretative rule to describe both the microscopic and the macroscopic worlds so as to include also classical properties. There are of course limitations to the simultaneous consideration of both position and momentum coordinates, if only because of the uncertainty relations, and some errors are unavoidably generated. They will have to be estimated.

The third part is concerned with dynamics. Since our aim is to recover the logical aspects of classical physics and, particularly, determinism when it is valid, we cannot be content with Ehrenfest's theorem and we shall have to compare classical cells in phase space, as they move under classical motion, with their associated projectors, as they change under a quantum evolution. When motion does not alter the correspondence between cells and projectors, classical physics is found to be valid under both its dynamical and logical aspects. We shall have of course to find under what conditions this is satisfied and what errors are involved. It turns out that the results one obtains are pretty natural and easy to understand, though their proof is by now still very technical. The best way to prove them relies upon the techniques of microlocal analysis (also called pseudo-differential calculus), a powerful tool that was devised in the last few decades in mathematics. Because of the resulting difference in the levels of the results and the proof, we shall try to understand the results with the help of simple physical arguments whereas the techniques and the proof will be left to the appendices for the mathematically inclined reader.

It will be shown finally that classical physics and its classical properties provide in most cases an excellent recipe for describing a macroscopic reality and reasoning about it, while still considering quantum mechanics and its universal rule of interpretation as the only firm basis. This is a somewhat startling result since it reverses a very old trend in physics, which was always to start from classical logic (i.e., common sense) as a primary knowledge directly inspired by observation (and the working of our brains) and to try understanding everything in terms of it. One considers here on the contrary that the logic of common sense is a secondary feature relying upon a somewhat deeper logical structure having its roots in physical reality. This point of view has obviously significant consequences for epistemology, but they will be left aside until the last chapter of the book.

OBJECTS

1. Orientation

Objects versus Physical Systems

Let us consider a very ordinary object—for instance, a bottle, and let us look at it from the standpoint of an atomist. This is not a new game since it was played by Leucippus, Democritus, Epicurus, and Descartes, to name but a few, but we shall play it in a modern way. From the standpoint of quantum mechanics, the bottle is certainly something different from what we defined previously as a physical system, namely a collection of elementary particles. The possible states of a large physical system have a tremendous diversity and the bottle is only one among many shapes its particles can take. They could as well appear as broken glass, fused glass, three smaller bottles, or two plates. At high temperature, it could be a gas or a plasma. If one were to redistribute chemically the atoms in a different way, one might get a heap of quartz, some potash, and who knows what. After redistributing the nucleons among the atomic nuclei, one might get a rose in a cup of gold.

There is every reason to believe that the concept of object refers primarily to the *state* of the system of particles. One of the simplest objects one can think of is a hydrogen atom. Although it is not macroscopic, one feels tempted to say that a proton and an electron constitute an object when they are in a bound state, whereas they are rather two different objects when they are not bound. An ammonia molecule NH_3 has also bound states and it can tell us something more: in the subspace of the Hilbert space corresponding to the ground state and the states nearest to it, some observables play an important role. They are the center-of-mass coordinates and the Euler angles describing the orientation of the molecule, together with their conjugate observables: the total momentum and the components of the angular momentum. Other observables can also respect the integrity of the molecule as an object and they describe it on a finer scale. This is the case for instance for the distances between the nuclei whose values can vary within a limited range when the molecule vibrates.

These observables are more or less specific for the object. Leaving aside the center of mass, the total momentum, and angular momentum (which are all linked to very general conservation rules) the other observables we mentioned are not particularly significant when the ammonia molecule becomes a system of errantly moving electrons and nuclei. The Euler angles, for instance, do not mean anything particular nor are they even well defined.

A ball of metal would show the same features. There would be again some overall observables allowing one to describe the dynamics of the ball, such as the position of the center of mass and some orientation angles, or an electric charge. There would also be some object-fixing observables such as the distances between the nuclei, each one of them having a limited range of variation. On the other hand, there would remain many other observables such as the positions of the electrons in the ball, the vibration amplitudes, and so on.

From a theoretical standpoint, an object is therefore presumably associated with a family of states and a privileged category of observables. The matter from which it is made, i.e., the particles inside it, must be described for its own sake by other observables.

Formalization

Let us try to give a more precise form to these vague ideas. It will be assumed that a supposedly well-defined object **O** is associated with a specific subspace $\mathbf{H_O}$ belonging to the Hilbert space of the corresponding physical system. The projector upon this subspace may be called the *object projector*. It commutes with a large number of observables allowing one to describe the object as such as well as the particles in it. Similar considerations can of course be made when several objects are considered together.

We shall be interested mostly in macroscopic objects made of a large number of particles that are bound together. It will be assumed that there exist some observables that are particularly well suited to the description of the object as such and of its motion; they will be called the *collective observables*. They are easily identified when one actually sees a real object: they are the coordinates of the center of mass, some orientation angles, some distances between outstanding reference points (as, for instance, the distance between the ends of a spring), the orientation of a wheel in a clockwork, the electric charge on a capacity, and so on. They are in fact familiar since they are exactly the kind of observables one uses in classical mechanics. One can also refine the description: for instance, once having defined the position and the orientation of a tuning fork as a solid object, one can also describe its vibrations by taking into account the distance between the two tips.

In many cases, one knows how to express the collective observables in terms of the particle observables, the most obvious example being the center-of-mass position. One can then perform a change of variables in the wave function to get two classes of position observables providing a complete system of commuting observables. The first class contains the collective observables and the other the so-called microscopic observables describing the particles inside the object, its matter. As was shown by the example of the tuning fork, there is some amount of freedom in the choice of the first class.

A deeper theoretical investigation will probably provide in the future a more systematic approach to these notions. There is every reason to believe that, the larger the number of collective observables one is considering, the larger their quantum fluctuations can be. In the Born–Oppenheimer approximation, the positions of the atoms themselves are treated as collective observables, while maintaining their quantum character. The collective character of observables is therefore presumably a quantitative matter and, when one chooses to include some set of them in the description while excluding finer ones, this is probably associated with the choice of an order of magnitude in the errors following from a classical treatment. This conclusion is at least strongly suggested by specific examples using the results of the present chapter.

There is no known method for constructing the collective observables starting from the first principles and using a universal algorithm that would allow us to order the coordinates in terms of a decreasing "collectivity". There has been some progress in that direction in nuclear physics, where the collective properties of nuclei and particularly their shape are of direct physical interest, but this is perhaps too special a case and it is not yet solved completely.[1] The most tantalizing results have been obtained by the mathematician Charles Fefferman.[2] He proved, for instance, that ordinary electrons and nuclei are organized in clusters that can be identified with atoms, molecules, or larger clumps including solids, when the total energy is near the ground state energy. He also found a method for approximately diagonalizing a general hamiltonian by using canonical transformations that are suggestive of a systematic introduction of collective coordinates. Unfortunately, his techniques are not explicit and one cannot yet be sure that they are really a step towards the solution of the problem of main interest in physics. A criterion for recognizing collective observables, though not by a constructive method, is however proposed in Appendix D.

Whatever it may be, there is no evidence anywhere that the simplifying assumptions we have made might produce serious difficulties and they will be applied without further ado, except for some elaboration.

2. Quantum Mechanics of Collective Observables

The objects we shall be interested in are the ones occurring usually in experimental physics: mirrors, accelerators, voltmeters, ammeters, detectors, diodes, etc., as well as recording devices: photographic films, magnetic memories, counters, and so on. These are the familiar and concrete objects we shall mainly have in mind, though a vast number of other natural or man-made objects could be described as well by the same means. Objects containing fluid parts and therefore requiring a continuous set of coordinates will be left aside because the presently available techniques are not yet able to cover this case.

In the subspace of the Hilbert space associated with the object (or collection of objects), one introduces collective position observables $Q_1, Q_2, \ldots, Q_n$ together with canonical conjugates $P_1, \ldots, P_n$. The number n is called the number of (collective) degrees of freedom. We denote by H_c the part of the total hamiltonian depending only upon the collective observables. The remaining part $H - H_c$ depends generally upon both the collective and the microscopic observables. It describes the internal motion of particles and dissipation amounts to an energy exchange between the collective motion and the internal motion. In the present chapter, we neglect the coupling between the collective and the microscopic observables. This means that the discussion is restricted to the academic case where $H - H_c$ does not depend upon the collective observables, so that the overall motion is decoupled from the internal motion of particles.

The collective coordinates commute with the microscopic coordinates and, in the absence of dissipation, the hamiltonian is the sum of a collective hamiltonian H_c and a hamiltonian H_e depending only upon the microscopic coordinates. It is convenient, under these conditions, to assume that we have two abstract physical systems that are described by the collective and the microscopic observables, respectively. They do not interact so that, according to Rule 3, the Hilbert space is a tensor product of a collective Hilbert space $\mathbf{H}_c$ and an internal Hilbert space $\mathbf{H}_e$ and one can restrict attention to the collective system only.

It can be described by wave functions $\psi(q)$ depending upon the variables $q_1, \ldots, q_n$, collectively denoted by q, which are the possible values of the coordinate observables $Q_1, \ldots, Q_n$. We denote by Γ a space parametrized by these coordinates, which is the configuration space of classical physics. The probability for the coordinates to be in an infinitesimal domain $d^n q = dq_1 \ldots dq_n$ around a point q is proportional to the square of the wave function and is given by

$$dp = |\psi(q)|^2 \mu(q) \, d^n q.$$

The choice of the weight function $\mu(q)$ is such that, under a Schrödinger time evolution,

$$i\hbar \frac{\partial \psi}{\partial t} = H_c\left(q_1, \ldots, q_n, -i\hbar\frac{\partial}{\partial q_1}, \ldots, -i\hbar\frac{\partial}{\partial q_n}\right)\psi,$$

the norm

$$\|\psi\|^2 = \int_\Gamma |\psi(q)|^2 \mu(q) \, d^n q$$

is invariant and the operator H_c is self-adjoint for the associated scalar product.

The set of functions defined over Γ and square integrable with the weight μ will be denoted by $L^2(\Gamma)$. It is the collective Hilbert space $\mathbf{H}_c$ of the object. The existence and unicity of a weight function satisfying the above conditions is nontrivial but they can be proved and the correct weight function will be given in the next section.

3. Classical Variables

What Is the Problem?

Classical physics is expressed in terms of some dynamical variables—the position and the momentum coordinates (q, p), which are pure numbers. It can also use more general dynamical variables that are more or less arbitrary real functions of q and p. Simple examples are provided by the energy and the components of angular momentum.

How is it possible to understand the efficiency of this classical description from the standpoint of quantum mechanics? This will be our first problem. It is more or less the inverse of another problem met in the early days of quantum mechanics when one wanted to find the quantum description of a system by starting from a classical model. This is how Schrödinger defined the hamiltonian as a function of the position and the gradient operator and it was called the quantization problem. At the present time, one knows that there exist a few fundamental interactions between particles and that quantum mechanics is the only basic form of physics. Accordingly, the whole of classical physics should now be viewed as a byproduct of quantum mechanics. This leads us to our first problem: how can one define the classical dynamical variable that may be associated with a given quantum observable? It will be noticed that the quantum observable, more fundamental, is taken first, while the classical variable is considered as a derived notion.

As already indicated, one associates the basic dynamical variables (q_j, p_j) with the canonical set of observables (Q_j, P_j). One also wants to associate a function $a(q, p)$ with an arbitrary observable A acting in the Hilbert space $L^2(\Gamma)$.

Correspondence Criteria

On what criteria should one establish the correspondence between dynamical variables and observables? One can be guided by a few simple considerations. In the case of the hamiltonian, for instance, one can change the reference units in which to express the energy. When this is done, the hamiltonian H becomes cH, where c is the ratio between the previous energy unit and the new one. The same is true for the classical energy $h(q, p)$, which becomes $ch(q, p)$. If the system consists of two noninteracting objects, the hamiltonian is a sum $H_{c1} + H_{c2}$ and the classical energy is similarly $h_1(q, p) + h_2(q', p')$. It seems therefore advisable to ask for a linear correspondence between the classical and quantum quantities. Furthermore, one obviously demands that self-adjoint operators be associated with real dynamical variables and conversely.

Another criterion, put forward by Dirac, asks for a correspondence between the transformations performed on both types of quantities. One can pass from one set of observables to another by a unitary transformation U so that A becomes $A' = UAU^\dagger$. Similarly, one can pass from a classical dynamical variable $a(q, p)$ to another by performing a canonical transformation on the basic variables (q, p). We will therefore ask that a unitary transformation generates a canonical one. This will assert that the correspondence is not restricted to a specific choice of coordinates.

However, this is asking too much because the correspondence between unitary and canonical transformations is not so universal as one might believe. It is only true in general to leading order in $\hbar$. There is however one case in which it can be made exact, which is when the unitary operator U has the form $\exp\{iF(Q, P)\}$, F being a second-order polynomial in Q and P where the noncommuting factors are combined in such a way that the whole quadratic form is self-adjoint. An explicit calculation shows that in that case the new observables (Q', P') are linear combinations of (Q, P) with constant coefficients. When furthermore the same linear transformation is considered as acting upon (q, p), it is a canonical transformation because it preserves the Poisson brackets. This is a case where unitary transformations and canonical transformations correspond exactly and their family is called the Heisenberg group.

One will therefore ask that the correspondence $A \to a(q, p)$ satisfies the following criteria:

1. It is linear.
2. The observables Q and P correspond to q and p, respectively.
3. It can be inverted, at least for a sufficiently wide category of observables (dequantization being the inverse of quantization).
4. A self-adjoint operator is associated with a real dynamical variable.

5. A unitary transformation belonging to the Heisenberg group generates the corresponding change of canonical variables in the arguments of a function $a(q, p)$.

It turns out that this defines the correspondence in a unique way under very mild conditions, namely that the configuration space is an infinitely differentiable manifold.[3]

The Answer

To write down the correspondence between observables and dynamical variables in the general case is not too difficult, but it requires a few specific notions in differential geometry and analysis. It is much easier to write it down when the configuration space is simply an n-dimensional euclidean space. This is why we shall stick to that case, noticing however that there is no essential difference with the general case, as shown in Appendix A. The observables of position will be accordingly denoted by X_j and the associated dynamical variables by x_j.

The correspondence is the following: Let Ω be an arbitrary operator in Hilbert space, either self-adjoint or not. It will be convenient to use the Dirac eigenstates of position $|x\rangle = |x_1, \ldots, x_n\rangle$ with the normalization convention $\langle x|x'\rangle = \delta(x - x')$. One can then associate a unique function $\omega(x, p)$ with the operator Ω, which is given by

$$\omega(x, p) = \int \langle x|\Omega|x'\rangle \delta\left(x - \frac{x + x'}{2}\right) e^{-ip\cdot(x-x')/\hbar} dx\, dx'.$$

One finds immediately that the operators X_j, P_j, and I, correspond to the functions of x and p given by x_j, p_j, and the function identically equal to 1, respectively. A more detailed study of this correspondence is given in Appendix A.

We thus obtain a universal linear correspondence between operators and functions, in which we shall systematically denote the operator and the function by the same letter, in capital or in lower case, respectively. In particular, the *Hamilton function* will be defined as being associated with the collective hamiltonian by

$$h(x, p) = \int \langle x|H_c|x'\rangle \delta\left(x - \frac{x + x'}{2}\right) e^{-ip\cdot(x-x')/\hbar} dx\, dx'.$$

This correspondence between operators and functions was first proposed by Wigner.[4] He introduced it with the purpose of associating a classical probability distribution with a given density operator and this specific application remains an important concept in statistical mechanics. The idea was generalized later on to almost every operator by Hermann Weyl,[5] who investigated how it extends to the product of operators.

RECOVERING CLASSICAL PHYSICS

This "Weyl calculus"[6] has now become a special case of a very powerful branch of analysis, which is particularly useful in the study of linear partial differential equations.[7] It is called microlocal analysis or pseudo-differential calculus and some of its key results, which are useful in semiclassical physics, are given in Appendix A. It systematically breaks down the study of a partial differential equation, such as the Schrödinger equation, to calculations dealing with functions rather than operators. It makes no essential distinction between a differential operator such as the hamiltonian of a nonrelativistic system and much more general so-called pseudo-differential operators (as for instance the square root of a laplacian or a Green's function), whence its name of pseudo-differential calculus. Its other name of microlocal analysis comes from a basic technique allowing one to perform the calculations in a localized region of phase space rather than in the whole space, as explained in Appendix A. Since the coordinates in a bounded region of the configuration space can always be associated with euclidean coordinates in a similarly bounded region, the techniques we shall use can be extended to an arbitrary system though we shall only develop them in the euclidean case.

Within the last few decades, mathematicians have found very general and very powerful theorems to deal with this kind of calculus and this has led to such beautiful results that it is sometimes called the "analysis of the seventies". One can learn a lot about the properties of an operator by looking at its associated function and one can also perform semiclassical calculations that were almost impossible beforehand, while getting estimates for the errors.

Symbols

One should however mention a slight drawback. These tools apply directly only to infinitely differentiable functions $a(x, p)$. This appears to be a severe restriction when one realizes that for instance a Coulomb potential does not satisfy it. One should not forget however that we only want to apply it to the collective observables, in which case the interesting dynamical variables are usually very smooth. Moreover, even when applying these techniques to an elementary system such as a particle, nothing physically significant is changed if one rounds off a potential in a very small region, though rigor would demand that a proof be given for the presumed insensitivity to the approximation.[8] In any case, we shall disregard these questions.

An operator will be called a *pseudo-differential operator* when it is associated with an infinitely differentiable function satisfying moreover some boundedness conditions, together with its derivatives, and also some further mild conditions concerning their asymptotic behavior. Such a function is called the *symbol* of the pseudo-differential operator. The explicit conditions are given more precisely in Appendix A.

Classical Motion

In the general case where the configuration space is not assumed to be euclidean, the position coordinates q and the momenta p parametrize the phase space. Knowing the Hamilton function, one can introduce the classical equations of motion through the usual Hamilton equations

$$\dot{q}_j = \frac{\partial h}{\partial p_j}, \qquad \dot{p}_j = -\frac{\partial h}{\partial q_j}.$$

These equations can be solved in principle to obtain the coordinates (q, p) at a time t for a point starting with the coordinates (q_0, p_0) at time 0. The corresponding mapping of phase space to itself will be denoted by g_t. By inverting the direction of time, one sees that the mapping is one to one. Another important property of this mapping is the conservation of an infinitesimal volume in phase space (Liouville's theorem), the volume being given by $dq_1 \, dp_1 \cdots dq_n \, dp_n$. The mapping also conserves energy so that $h(q_0, p_0) = h(q, p)$.

The Collective Hilbert Space*

Knowing the Hamilton function, one can use the kinetic energy to more precisely define the Hilbert space $L^2(\Gamma)$. We shall consider only the case where the Hamilton function can be written as a sum $h(q, p) = T(q, p) + V(q)$, where the kinetic energy $T(q, p)$ is a quadratic function of the momenta. The results are much more simply expressed when one uses the concepts and the notations of riemannian geometry, the coordinates $\{q^j\}$ being considered as contravariant and the momenta as the covariant coordinates of a vector $\{p_k\}$. The kinetic energy is written as

$$T(q, p) = g^{jk}(q) p_j p_k,$$

with Einstein's summation convention on repeated indices. The Hamilton equations give the velocities in terms of the momenta by

$$\dot{q}^j = \frac{\partial T}{\partial p_j} = g^{jk}(q) p_k.$$

This linear relation between momenta and velocities can be inverted by introducing the matrix with coefficients g_{jk}, the inverse of the matrix with coefficients g^{jk}, to give

$$p_k = g_{jk}(q) \dot{q}^j.$$

We can then write the kinetic energy as a function of the velocities in the form

$$T(q, \dot{q}) = g_{jk}(q) \dot{q}^j \dot{q}^k.$$

It is convenient to consider the configuration space as a riemannian space with the metric
$$ds^2 = g_{jk}(q)\, dq^j\, dq^k.$$
The differential operator associated with the kinetic energy, i.e., the quantum kinetic energy, is then given by $T_{op} = -\hbar^2 \Delta$, where Δ is the laplacian operator on the riemannian space, or more explicitly:
$$\Delta = g^{-1/2} \frac{\partial}{\partial q^j} \left[g^{1/2} g^{jk} \frac{\partial}{\partial q^k} \right].$$
The quantity g is the determinant of the matrix with coefficients g_{jk}. This gives the Schrödinger equation in the form
$$i\hbar \frac{\partial \psi}{\partial t} = -\hbar^2 \Delta \psi + V\psi.$$

Conversely, one can start from this Schrödinger equation and derive from it the riemannian metric and the classical Hamilton function. The scalar product that is conserved under time evolution is given by
$$<\psi, \phi> = \int_\Gamma \psi^*(q)\phi(q) g^{1/2}\, dq.$$

CLASSICAL PROPERTIES

It was shown in the previous chapter that a classical property may be expressed by asserting that the coordinates (q, p) are in a given domain in phase space. This is very different from a quantum property, which is associated with a projector in Hilbert space. It is very important to get a better idea about their possible correspondence. This is because physics is far from being only a formal manipulation of some mathematical formalisms among which one establishes some correspondence. It deals primarily with the physical properties of real objects and this is what interpretation is all about. One therefore feels an urgent need to explicitly translate a classical property, as it can be seen empirically, into something that makes sense from the standpoint of quantum mechanics. It will now be shown that a classical property can be expressed as a quantum property by associating it with a projector in the collective Hilbert space. One will also find what limitations should be imposed upon a classical property if it is to make sense in the quantum framework.

4. A Return to Classical Logic

Quantum mechanics and classical physics are so different that one must find a way to translate the language of the latter into that of the former. This cannot be only a mathematical correspondence but also a connection in the way they are understood. This can be done if one can find a correspondence between the logical frameworks of the two theories. We already sketched in the last chapter what the logic behind classical mechanics is and how it fits with common sense. It will be called *classical logic* in the following and its field of propositions extends to everything that may be stated according to classical physics. It will be useful to describe it again in greater detail.

Since we know that we shall have to refer to domains in phase space, let us make more precise a special category of them to be called the *cells*. This is a domain in one piece (connected) and without holes (simply connected). There is an extreme classical viewpoint according to which one can give with perfect precision the values of all the position and momentum coordinates. It will not be considered seriously, however, since, in addition to the impossibility of conceiving how the coordinates could be so well known, this is obviously inconsistent with the limitations coming from the quantum uncertainty relations. One can however resort to a more modest and more realistic description of a macroscopic object by giving each position coordinate q_j within an interval of width Δq_j and each momentum p_j in an interval Δp_j. This kind of assertion does not conflict with quantum mechanics, as long as each product $\Delta q_j \Delta p_j$ is large when compared to h. It defines a cell in phase space having the shape of a rectangular many-dimensional box and a classical proposition is associated with it. This proposition asserts that the classical coordinates (q, p) belong to a cell, namely the box C_0.

This kind of proposition, when extended to a cell with an arbitrary shape and referring to a specific value of the time, is enough to describe all the properties which are relevant to classical physics. For instance, after a time t the classical motion will transform the box C_0 into another domain. When there is a potential barrier, the initially connected domain can become broken into several disconnected domains, each of which is a cell. We shall leave this interesting case aside for separate consideration, if needed, and we shall consider only the case when the transformed domain is again a cell, to be denoted by C_t. If the initial cell is a rectangular box, the new one will have a less simple shape, which shows why one must consider more general types of cells. It will therefore be convenient to assume that the initial cell C_0 has a more or less arbitrary shape. The property according to which (q, p) is in a cell C at time t will be denoted by $[C, t]_c$, the index c recalling that it belongs to classical logic.

The cell C_0 cannot however be completely arbitrary. The uncertainty relations, for instance, forbid taking its dimensions arbitrarily small. This problem will have to be considered later in greater detail but we shall leave it aside momentarily to concentrate upon classical logic only.

It was already shown using the notions of set theory that the usual logical operations are quite simple when one deals with domains in phase space.

The relations of logical equivalence and implication are defined by

$$[C_1, t_1]_c = [C_2, t_2]_c, \text{ when } C_2 = g_t[C_1],$$

and

$$[C_1, t_1]_c \Rightarrow [C_2, t_2]_c, \text{ when } C_2 \supset g_t[C_1], \tag{6.1}$$

respectively, g_t being the mapping of phase space onto itself generated by classical motion during the time $t = t_2 - t_1$.

The fact that one uses discrete values of the time and not continuous values might seem to be a limitation, though it would be easy to introduce a continuously infinite set of properties indexed by a continuous time and to consider the basic space to have the coordinates (q, p, t). This is not however what quantum mechanics will give us, so that one may wonder whether one really needs to consider a continuous time when stating the logical consequences of classical mechanics. A little reflection shows that it is never so in practice. Every observation or measurement always takes a finite time and one cannot say very exactly when a property holds. On the other hand, the differential equations of dynamics are solved numerically by using discrete time values, even if these values are very close.

It is worth stressing again that determinism is so strongly imprinted in classical physics that it is given the status of a logical equivalence. Of course, it may be limited by chaos, but this does not change its nature. The implication (6.1) shows what determinism is: the property holding at the earlier value of time t_1, say $[C_1, t_1]_c$, implies that at a later time t_2, one has the property associated with the cell $g_{t_2-t_1}[C_1] = C_2$. Conversely, the property $[C_2, t_2]_c$ implies the previous property $[C_1, t_1]_c$, which is why they are logically equivalent. Of course, this is due to the one-to-one character of the mapping of phase space onto itself by classical motion, but the realistic meaning of the two implications is quite different. In the first case, this is determinism *stricto sensu* where one predicts with certainty the future from what is known about the present. In the second case, the given property plays the role of memory allowing one to recover the past from the present. True enough, memory, whether in our brain, a computer memory, or a document, is frozen by an irreversible process, but this aspect of the question will be taken up in the next chapter. What is important is to realize that determinism and memory are two aspects of the same physical laws.

Determinism is also essential for the meaning of truth as commonly used according to classical views. We ascertain a truth by observation but, when we take what has been seen as granted or when we think of it at a later time, this no longer relies upon what is observed but upon the memory of what was observed. Accordingly, *the notion of truth would be extremely deficient and useless, were it not for classical determinism, even if it is imperfect.*

5. The Quantum Form of a Classical Property

The Problem

The simplest properties of a classical system state that the coordinates (q, p) are in a cell C in phase space. Such a property does not involve time, though it mentions momentum and it belongs therefore to kinematics. It will be denoted by $[C]_c$ and our goal will be to associate it with a projector $E(C)$ in the collective Hilbert space $L^2(\Gamma)$.

Is this possible? We shall try to answer this question by using a few elementary considerations rather than giving the hard proofs. The main steps of the argument can be briefly described as follows:

- One must first dispose of a convenient semiclassical approximation for the time-independent Schrödinger equation and the associated semiclassical states.
- One must then find out what conditions are necessary for a cell in phase space to be associated with a projector. This can be done using semiclassical states.
- One must clear up the question of uniqueness: Is there only one projector answering the question? Do other operators, which are not projectors but very similar to them, also answer the question? Which ones?
- Given a projector, how can one say to which cell it corresponds?
- Finally, what errors are unavoidably occurring?

These considerations will of course not lead to a proof of the results, but only to a correct guess. The proof itself relies upon different methods and it will be given in Appendix B.

The Semiclassical Approximation

A time-honored approximate method for solving the Schrödinger equation is the BKW method (from Brillouin, Wentzel, and Kramers).[9] It can be used as a convenient guide in the present case. Consider for instance a simple one-dimensional time-independent Schrödinger equation

$$\frac{\hbar^2}{2m}\frac{d^2\psi}{dx^2} + (E - U)\psi = 0,$$

with a potential $U(x)$. We look for a solution in the form $\psi = \exp(i\sigma/\hbar)$ and expand the unknown function σ in powers of $\hbar$:

$$\sigma = \sigma_0 + (\hbar/i)\sigma_1 + \cdots.$$

Identifying the various powers of $\hbar$ in the Schrödinger equation for σ:

$$\sigma'^2/2m - i\hbar\sigma''/2m = E - U(x),$$

one gets to zero order

$$\sigma_0'^2/2m = E - U(x).$$

Hence $\sigma_0 = \pm \int p(x)\,dx$, with $p(x) = [2m\{E - U(x)\}]^{1/2}$.

At the next order, one gets a differential equation for σ_1 that is easily solved and the result is given by

$$\psi = Ap^{-1/2}\exp\{(i/\hbar)\int p\,dx\} + Bp^{-1/2}\exp\{-(i/\hbar)\int p(x)\,dx\},$$

A and B being constants. The approximation can be shown to be consistent if the force $-U'(x)$ varies sufficiently slowly in a de Broglie wavelength $h/p(x)$ or, using the value of $p(x)$, when:

$$m\hbar|U'|/p^2 \ll 1. \tag{6.2}$$

At the "turning points" where $p(x)$ vanishes so that $U(x) = E$ at a barrier, this condition cannot remain valid but a more detailed study near these points gives boundary conditions for the solution.

Bohr's Quantization Rule

Consider the special case where the particle moves in an attractive potential well, which is regular enough to satisfy the condition (6.2). Classically, a bound particle would oscillate between two values $x = a$ and $x = b$ at which one has $E = U(x)$. In that region, $p(x)$ is positive and it can be shown that the boundary condition at $x = a$ selects a solution

$$\psi = Ap^{-1/2}\sin\{(1/\hbar)\int_a^x p\,dx + (\pi/4)\},$$

whereas the boundary condition at $x = b$ gives

$$\psi = Bp^{-1/2}\sin\{(1/\hbar)\int_x^b p\,dx + (\pi/4)\}.$$

Since it must be the same function, the arguments of the two sines must be the same up to a multiple of π, which gives a quantization condition for the energy

$$\int_a^b p\,dx = \pi\hbar(N + 1/2),$$

N being an integer. This can also be written as an integral along a curve γ representing the trajectory of the particle in phase space:

$$\oint_\gamma p\,dx = h(N + 1/2). \tag{6.3}$$

One recognizes Bohr's old quantization rule, except that the value of the integral is no longer an integer N but a half-integer $N + 1/2$, which is of little importance when the motion is semiclassical, i.e., when N is large.

The Number of Semiclassical States

A trajectory γ associated with a large value of N can be considered as the boundary of a cell C in phase space, inside which all the trajectories associated with smaller values of N are contained. One can then say that N quantum states are essentially localized in C. If one notices that the integral in equation (6.3) is the phase space area of C, one finds that the number of quantum states that are essentially localized in C is given by the ratio $A(C)/h$, where $A(C)$ is the area of the cell and h is the ordinary Planck constant.

This is easily extended to an arbitrary number n of degrees of freedom and one finds that the number of quantum states in an arbitrary but sufficiently large cell C is given by

$$N(C) = \mu(C), \tag{6.4}$$

where $\mu(C)$ is the phase space volume of the cell in units of Planck's constant:

$$\mu(C) = \int_C (dq\, dp/h^n). \tag{6.5}$$

This assumes implicitly, however, that there exists some regular potential $U(x)$ associated with C, a condition which is somewhat difficult to express. The regularity conditions on the potential that are necessary for the validity of the semiclassical approximation suggest moreover that the boundary of C should be rather smooth.

It will be noticed that the formula (6.5) so obtained is additive: When a cell C is the union of two adjoining cells C' and C'', the number of states in C is the sum of the numbers of states in C' and in C''. This is trivially due to the additivity of their volumes.

What Cells?

The problem of finding a projector associated with a cell is simple when the cell can enclose all the quasi-classical trajectories of a regular potential up to a number $N(C)$ and no more. One can then introduce normalized semiclassical wave functions $|N\rangle$ and define the projector as

$$E(C) = \sum_{r=1}^{N(C)} |r\rangle\langle r|.$$

This procedure can of course only be used when the semiclassical approximation is valid for the majority of these states, which requires $N(C)$ to be large: $N(C) \gg 1$. In other words, the volume of the cell must be large enough.

This solution works when the cell is a rectangular box, since one can take a square well to obtain a good set of states. The general case is burdened by the condition of having a regular potential associated with C and this is too difficult. However, one can use additivity in the following way: The cell C is covered completely by a set of nonintersecting boxes, each with the same volume $N_0 h^n$, N_0 being a large number fixed once and for all. It is possible to get a coverage sufficiently close to C only when one has at least $N(C) \gg N_0$. In spite of the approximate character of this construction, it may give a good indication for the number of quantum states in C and the projector $E(C)$ can then be taken as the sum of the projectors associated with each box.

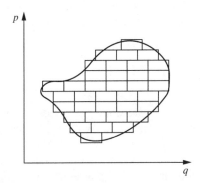

6.1. *A covering of a cell in phase space by rectangular blocks*

A different covering will give a different projector. Suppose two coverings differ in their internal boxes but they show the same boxes near the boundary. Then they will differ mainly by a different distribution of the basis vectors among the box projectors and the corresponding overall projectors will be essentially the same. This is at least what one can get from a direct calculation by splitting explicitly a big rectangular box into different rectangular pieces. Nevertheless, the overall projectors will certainly be different when the two coverings have different boundaries, because there are some semiclassical states belonging to one but not the other. This remark suggests that it might be better not to look for a unique projector, but rather for a whole family of projectors more or less equivalent to each other. It also raises the question of estimating the errors.

Estimating the Error Parameters

Let us first consider the error one may expect for the number of states in C. It depends upon the coverage and, if a coverage gives a value N for it, we expect N to be near the number $N(C)$ given by equation (6.4).

Consider as an example the case when there is only one degree of freedom and the boundary of C is an ellipse in phase space with its two axes along the q and p axes with respective lengths $2L$ and $2P$. We shall cover it by boxes with their sides parallel to the axes with respective length $l = \alpha L, p = \alpha P$. The larger the number of these boxes, the better will be the fit between the ellipse interior and the heap of boxes. It is necessary however that the semiclassical approximation applies to the individual boxes, which means that

$$N_0 = \frac{lp}{h} = \frac{LP}{h} \cdot \frac{1}{\alpha^2} \gg 1.$$

One knows by direct calculations that the value of N_0 need not be very large for the semiclassical approximation to be valid. Taking α somewhat larger than

$$\varepsilon = \left(\frac{h}{LP}\right)^{1/2}, \tag{6.6}$$

one may therefore hope for a satisfactory approximation.

In these conditions, the main error will come from the way the boxes are distributed around the boundary of C. The typical variation in N from one covering to another is essentially proportional to the number of boxes contributing to the boundary of the coverage, which is of the order of L/l, i.e.,

$$\frac{\delta N}{N} = O(\varepsilon). \tag{6.7}$$

This can be made a bit more precise by comparing two extreme coverings. The first one is an inside maximal covering, i.e., no box pops out of the cell but, if one were to add another box, it would necessarily cross the boundary. The second one is a minimal outside covering: C is completely contained inside it but no box can be removed without exposing a part of C. The difference in the number of their states is clearly of the order of the size of a box times the ratio of the length of the boundary to the volume of the cell.

This argument does not depend upon the shape of the cell but there is in any case a difficulty: although the volume of a cell is a well-defined quantity that may be obtained from Liouville's volume, the measure of its boundary is not so. In order to define it, one needs a metric in phase space and there is no natural one. Since the same difficulty will arise in a more careful analysis of the problem and the answer refers to its solution, it will be useful to give it. This will be done for an arbitrary number n of degrees of freedom.

Given a cell C, choose unit reference scales (L, P) for measuring the coordinates and momenta, assuming that all the coordinates have the same dimension (this can always be arranged). The scales are chosen so that

$$\left(\frac{LP}{h}\right)^n = N(C),$$

which is of course not sufficient to define them completely.

One then introduces a *classicality parameter* ε as given by equation (6.6). A metric is defined on phase space (in the case of a euclidean configuration space) by

$$ds^2 = \frac{dx^2}{L^2} + \frac{dp^2}{P^2}. \qquad (6.8)$$

One can then compute the $(2n - 1)$-dimensional volume of the frontier Σ of C in this metric. If one uses the dimensionless coordinates x_j/L and p_j/P, the volumes of Σ and C become dimensionless and one can define the geometric parameter

$$\theta = \frac{\text{Volume of } \Sigma}{\text{Volume of } C}. \qquad (6.9)$$

It should be clear that a value of θ of the order of one is the mark of a bulky cell and a large value of θ corresponds to an elongated or distorted cell. One can then completely determine L and P by asking that the ratio (6.9) be minimal. When this choice is made, one defines an *effective classicality parameter* as the product $\eta = \varepsilon\theta$. One could also, as a matter of principle, perform a canonical transformation upon (q, p) and retain a choice of coordinates giving a minimal value to η, but these refinements do not seem to be necessary.

The two classicality parameters turn out to control the various error estimates encountered. It is clear, for instance, from the above discussion that the estimate (6.7) can be refined to show that the relative fluctuations $\Delta N/N$ for different coverings are of the order of the effective classicality parameter θ. This may differ significantly from the previous estimate in terms of the classicality parameter ε when the shape of the cell is somewhat distorted.

In many cases, the parameters L and P that are found by this procedure are of the same order of magnitude as the leading geometric parameters of the cell, at least when its shape does not show very different features at different scales. It should be stressed that the two classicality parameters are proportional to the square root of Planck's constant. This does not come as a surprise since one knows from elsewhere[10] that semiclassical approximations are essentially expansions in powers of $\hbar^{1/2}$. The geometric parameter θ provides a good control also, since when it becomes very large because of the very distorted shape of a cell, the parameter η may become of the order of one in spite of the smallness of Planck's constant. There is obviously a limit where the covering of C by boxes becomes completely arbitrary and it will also be the limit where no sensible projector $E(C)$ can be defined.

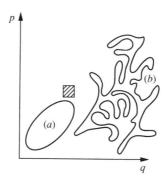

6.2. *A regular cell and a nonregular one.* The shaded area gives the value of Planck's constant h for comparison.

The Errors in the Projectors

The arbitrariness in the packing of boxes must correspond to a similar arbitrariness in the corresponding projectors. It was already indicated that the difference must occur from the boundary of the coverings but, when trying to express what it means that two projectors are near each other, we are immediately faced with the problem of finding what quantity is a good measure of this vicinity.

Let us therefore consider two projectors E and E' associated with two different coverings and let us begin with the case where the first covering properly contains the second one. It is then clear that the difference $E - E'$ is essentially a projector and one cannot use the Hilbert norm $\|E - E'\|$ as a measure of vicinity since it is equal to 1, just like the norms of E and E' themselves. The meaningful quantity in that case is the difference between the number of vectors or quantum states corresponding to each covering. It may be a large number but its ratio to $N(C)$ is small and of the order of η. Another way of stating this is to say that the ratio $\mathrm{Tr}(E - E')/\mathrm{Tr}E$ is of the order of η. This is particularly clear when one compares the projectors that are associated with a maximal inside covering and a minimal outside one, respectively.

RECOVERING CLASSICAL PHYSICS

Let us now consider the case when one of the coverages does not enclose the other. One cannot use the quantity $|\text{Tr}(E - E')|$ as a measure of the difference, because two projectors having no relation at all, except that they project onto two Hilbert subspaces with the same dimension, would give 0 for this quantity. This would be the case for two identical coverings for two identical cells far away from each other. The right answer must use the so-called *trace norm*, denoted $\text{Tr}|E - E'|$, which is the sum of the absolute values of the eigenvalues of $E - E'$. A vector belonging to the intersection of the two subspaces onto which E and E' project does not contribute to it and this eliminates most of the vectors associated with inside boxes, leaving only a difference coming from the boundary. So, one expects to find

$$\text{Tr}|E - E'| = N(C)O(\eta), \qquad (6.10)$$

to express that the projectors E and E' coming from two similar coverages are near each other.

The construction also implies that one cannot expect to end up with a unique projector but with a whole family of equivalent projectors, the equivalence meaning that every pair of them satisfies the estimate (6.10). This is a sensible definition of equivalence because the quantity $\text{Tr}|A|$ has the properties of a norm or, what amounts to the same thing, the quantity $\text{Tr}|A - B|$ is a good distance among operators satisfying the triangle inequality (see Appendix A for more details). Accordingly, keeping E fixed in equation (6.10) and letting E' vary, one finds that the projectors satisfying equation (6.10) form a neighborhood of E.

Quasi-Projectors

Since one is led to consider many equivalent projectors, why then stick necessarily to projectors? One might as well take some average of the projectors associated with different coverages and it would be just as good. This idea will lead us to the concept of a *quasi-projector*.

A quasi-projector is a self-adjoint operator having only discrete eigenvalues lying in the interval [0, 1]. Furthermore, in a gross way, it has many eigenvalues near 1, relatively few between 0 and 1 and the rest of them very near to 0. To state this property more precisely, one will say that a quasi-projector F is of rank N and order η when it satisfies the two conditions

$$\text{Tr} F = N, \qquad \text{Tr}(F - F^2) = NO(\eta),$$

N being a large number and η a small one.

This is exactly what one gets by taking averages of projectors upon various coverings, N being of the order of $N(C)$ and η the effective classicality parameter. The projectors themselves are quasi-projectors of a special type satisfying the condition $E = E^2$. The quasi-projectors are just as good as projectors when they belong to the equivalence class.

To state this last point more precisely, we will say that two quasi-projectors of order η are *equivalent* when

$$N - N' = NO(\eta), \qquad \text{Tr}|F - F'| = NO(\eta).$$

A Quasi-Projector Associated with a Cell

The last question to be clarified is the following: Assume that we are given an operator, for instance, by its matrix elements in the Hilbert space, and we are told that it is a quasi-projector associated with a cell C. The person who says that is known to be clever but to make frequent mistakes. How can we check that he or she is right? It is possible to check that the operator is a quasi-projector since this amounts to a straightforward calculation, but how can we make sure that it has something to do with the cell C? The interest of this question comes from the fact that, in a full-fledged approach to our basic problem, quasi-projectors will be constructed in a somewhat abstract way by means of their symbols and their exact connection with the cell will need elucidation.

It will be enough to consider the case of one degree of freedom. Given an arbitrary wave function $\psi(q)$, denote by B a rectangular box centered at the corresponding average values of Q and P, its sides being the uncertainties Δq and Δp. We shall say that ψ is well inside C when the box B is geometrically well inside C, meaning that the distance of its center to the boundary of C is large when compared with the uncertainties. One can define a wave function well outside C in a similar way and even restrict these considerations to gaussian wave functions. We say that a quasi-projector F is *associated* with a cell C when each wave function well inside (respectively outside) C is almost exactly an eigenfunction of F with the eigenvalue 1 (respectively 0).

The Result

The theorem answering the initial question should not come as a surprise after these considerations. Its statement is the following:

Theorem A.[11] *One can associate with a given cell C in phase space a class of equivalent quasi-projectors of rank $N(C)$ and order η.*

This theorem is of interest only when $N(C)$ is large and η is small. When this is the case, we say that the cell is *regular*. It means that the cell is big enough and its shape is not too distorted. One might also think that all its geometric characteristics, such as the curvature radii of its boundary and things like that, are large when measured with scales in which Planck's constant is of the order of one. Geometry is however a difficult subject and it is fortunate that one does not need to be more precise than giving the two numbers N and η. Theorem A applies to an arbitrary configuration space and not only to $\mathbf{R}^n$. Its proof relies upon microlocal analysis and it is given in Appendix B, in the case of an euclidean configuration space.

Two Distinct Cells

We will have to multiply the quasi-projectors associated with different cells when dealing with questions of logic. This is because the probabilities and the consistency conditions involve products of projectors that will be replaced by quasi-projectors. The logic of classical physics gives a hint about the necessary procedure. It was shown that, given two classical kinematic propositions $[C_1]_c$ and $[C_2]_c$, one can form another proposition "$[C_1]_c$ and $[C_2]_c$". The relation between the logical operation "and" and the intersection of sets tells us that this is the proposition "(q, p) is in $C_1 \cap C_2$". One may therefore expect that the product of two quasi-projectors F_1 and F_2, associated with C_1 and C_2, respectively, should be a quasi-projector associated with the intersection $C = C_1 \cap C_2$. This is what one finds when the two cells as well as their intersection are regular. We shall only state the result when the effective classicality parameters of C_1 and C_2 are of the same order of magnitude and the corresponding quasi-projectors will be denoted by F_1, F_2 and F. It will be convenient to assume $N(C_1) \geq N(C_2)$. One then gets

Theorem B. *Under the above assumptions, one has*

$$F = F_1 F_2 + \Delta F = F_2 F_1 + \Delta F',$$

with

$$\frac{\mathrm{Tr}|\Delta F|}{N(C_1)} = O(\eta), \qquad \frac{\mathrm{Tr}|\Delta F'|}{N(C_1)} = O(\eta).$$

When the intersection C is empty, one can take $F = 0$. For the union $D = C_1 \cup C_2$, one can write a quasi-projector G as:

$$G = F_1 + F_2 - F_1 F_2 + \Delta F'',$$

with similar bounds for $\Delta F''$. These estimates also give a bound for the commutator

$$\mathrm{Tr}|[F_1, F_2]| \leq K \eta N(C_1),$$

K being a constant of order one.

The case where the two cells are disjoint and sufficiently far away yields much better results for the estimates in Theorem B. This is particularly interesting, since it says with what confidence one can assert two clearly distinct classical properties as being distinct, when considered as quantum properties. It can be shown that the two products $F_1 F_2$ and $F_2 F_1$ are both extremely small. More precisely, let D be the dimensionless smallest distance between the two boundaries in the metric (6.8) and let the cells have the same volume N for simplicity. One then finds that $\text{Tr}|F_1 F_2|$ can be made as small as $N \exp\{-(D/\varepsilon)^2\}$, as shown in Problem 19.

6. The Construction of Quasi-Projectors

It will be useful to mention briefly how quasi-projectors are constructed in practice, if only because it sheds light upon some questions that will arise when we arrive at dynamics. There are essentially two methods for defining quasi-projectors, which will be described briefly in the case of one degree of freedom. The first method uses gaussian wave functions (also called coherent states) and the second one relies upon microlocal analysis so that a quasi-projector is defined through its symbol.

The Method of Coherent States

A simple gaussian wave function with average values q and p for position and momentum can be written as

$$g_{qp}(x) = \frac{1}{(\pi\sigma^2)^{1/4}} \exp\{-(x-q)^2/2\sigma^2 + ip \cdot x/\hbar\} \qquad (6.11)$$

where σ is the uncertainty Δx (divided by $\sqrt{2}$), which is fixed once and for all.

A quasi-projector associated with a cell C as given by

$$F = \int_C |g_{qp}\rangle\langle g_{qp}| dq\, dp/h^n,$$

as shown in the problems at the end of the chapter.

The Pseudodifferential Method

In this method, one tries to define a quasi-projector F associated with a cell C by its symbol $f(x, p)$. The simplest idea would be to take for $f(x, p)$ the characteristic function of the cell C, i.e., the function equal to 1 inside C and to 0 outside. However, it does not define a good operator because of its discontinuous character. One remedies to that defect by regularizing (smoothing) the characteristic function. This means that one still takes a function equal to 1 inside C and to 0 outside, except in a small region along the boundary of C where it goes smoothly from 1 to 0. This transition region is called the *margin* of the cell. Its width must be chosen conveniently so as to give the smallest possible errors and it turns out that the best value is the classicality parameter ε, in dimensionless units using the metric (6.8).

DYNAMICS

Since a classical property at a fixed time can be given a quantum meaning in terms of a class of projectors or quasi-projectors, these operators can be used in the framework of quantum logic. An obvious question is then whether or not the classical motion of a cell corresponds to quantum dynamics, i.e., whether it corresponds with the time evolution of its associated projector when considered as a time-dependent Heisenberg operator. Such a result would mean that the correspondence between the classical and quantum properties is preserved by time evolution. It will be shown that this is true, under precise conditions that are fortunately often met in practice. The corresponding errors or, if one prefers, the quantum corrections to determinism, will also be evaluated.

7. The Dynamical Correspondence

What Is the Problem?

The best guide to a logical investigation of dynamics is probably to start with classical determinism. It was shown to be a logical equivalence between two classical properties. It says that, if C_0 is a cell in phase space and its transform C_t by classical motion during a time t is also a cell, then the two properties expressing that "(q, p) is in C_0 at time 0" and "(q, p) is in C_t at time t" are logically equivalent. We know how to give a quantum meaning to these two classical properties and the main question is to find whether this equivalence is maintained when considered within the framework of quantum logic.

It will be assumed that both cells C_0 and C_t are regular, i.e., they satisfy two conditions: their volume is large (noticing that these volumes are the same because of Liouville's theorem) and both effective classicality parameters are small. This regularity assumption is required in order to associate a family of quasi-projectors with each cell, so as to give a quantum meaning to the corresponding classical property. We shall select arbitrarily a representative quasi-projector F_0 and F_t for each cell.

If determinism were perfect and one could use a unique projector in place of a class of quasi-projectors, then it would be easy to express a perfect logical equivalence between the two properties: The quantum time evolution of the operator F_0 is given at time t by

$$F_0(t) = U^\dagger(t) F_0 U(t),$$

where $U(t) = \exp\{-(i/\hbar)H_c t\}$. Perfect determinism would say that this is nothing but the property that is expressed by F_t, i.e.,

$$F_t = F_0(t). \tag{6.12}$$

This relation cannot of course be absolutely correct, if only because each quasi-projector is not uniquely defined. The interesting question is therefore to find how close the two sides of equation (6.12) are. One can rely upon the expression of equivalence in terms of trace norms and upon the fact that Liouville's theorem tells us that the traces of F_0 and F_t are the same. This naturally suggests a quantum version of determinism that would read

$$\mathrm{Tr}|F_t - U^\dagger(t)F_0 U(t)| = N(C_0) \cdot O(\zeta), \qquad (6.13)$$

ζ being a quantity depending upon the initial cell, as well as upon the hamiltonian and the elapsed time. Determinism would then be true, up to quantum correction of order ζ, which would of course be useful only when ζ is small. It will turn out that this guess is the right one.

The Difficulties

We indicated how one can define quasi-projectors, either in terms of gaussian wave functions or by starting from a microlocal symbol. The second method will turn out to be the most powerful one when a proof is needed. The first one, however, better shows the difficulties as well as giving a better understanding of the behavior of wave functions. This is why we shall rely upon it for an orientation.

An obvious reason why equation (6.13) might not hold well enough could be due to classical motion itself. The motion could break the cell or distort it so much that it does not remain regular. The cell is broken for instance when there is a potential barrier and some points in C_0 cross the barrier while others do not. This is however a case that can be treated for its own sake and offers in principle no essential difficulty. The cell can also become utterly distorted when one is dealing with chaotic classical dynamics and this is an important case to be considered further in a moment. Another difficulty might come from the spreading of wave packets. It is clear that, in equation (6.11) giving a projector as a sum over gaussian states, everything will be spoiled if the wave packets spread over distances comparable with the dimensions of the cell.

Similarly, equation (6.11) defines a quasi-projector as long as the wave functions are gaussian. But this is not a characteristic that is preserved by a Schrödinger equation, except in very special cases.[12] In general, dynamics has two effects upon an initial gaussian state, as can be seen when the system can be simplified to a representative particle moving in an n-dimensional space with a given potential: (i) a spreading of the gaussian wave packet, which can be completely obtained using only a knowledge of classical motion and, (ii) nongaussian corrections to the wave function; they are of a higher order in Planck's constant[13] and depend upon the third derivative of the potential.

It is rather difficult to take care of all these effects together and no intuitive approach, as was possible for Theorem A, seems to be available here. More powerful mathematics is necessary.

Chaotic versus Regular Systems

We now consider the limitations to correspondence resulting from a strong distortion of the cells during classical motion. Mathematicians have been interested for a long time in the problem of cell distortions under the effect of a hamiltonian flow. They have classified these flows into several categories, going from the most regular to the most irregular.[14] Everything is clearer when the phase space is bounded, which is the case to be considered. A phase space is bounded when the configuration space is bounded (this is the case of many objects after separating out the center of mass) and energy is also bounded.

The most regular dynamical systems are the completely integrable ones, i.e., they correspond to the case when there are as many constants of motion as there are degrees of freedom. An example is provided by the Kepler system for two bodies interacting through a gravitational force. When the equations of motion are not completely integrable, the flow tends to become more and more erratic as the number of constants of motion decreases. When there is no constant of motion except energy, one can have some very irregular systems.

There is a whole hierarchy of systems going towards increasing disorder: First come the ergodic flows, where almost all the trajectories $(q(t), p(t))$, at one time or another, come arbitrarily close to a given arbitrary point (q, p) corresponding to the same energy. Then come the mixing systems, where this covering of phase space becomes uniform, the trajectory of a point covering in a uniformly dense way the whole allowed phase space. Finally there are the so-called K-systems (named after the Russian mathematician Kolmogorov) or chaotic systems; their instability and disorder are so high that the distance between two points of phase space initially very near each other can increase exponentially with time.

The simplest example of a chaotic system is provided by the so-called Sinai billiards (from I. Ya. Sinai, a Russian mathematician). A point, which may be the center of a billiard ball, moves freely on a billiard table that is bounded outside by a rectangle and inside by a circle. When two balls are initially very near each other and have almost equal velocities (so that the two corresponding points in phase space are also very near each other), they rapidly diverge and their distance increases exponentially. This is due to the fact that the angle between the two velocities doubles at every collision with the circular boundary. Another formal example is a point moving along geodesics in a space with negative constant curvature.

230 CHAPTER 6

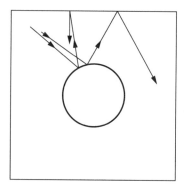

6.3. *Sinai's billiards*

It is obvious that the dynamical flow of a chaotic system cannot maintain the regularity of a given initial cell C_0 for a long time. Because of Liouville's theorem, the volume of the cell remains constant. In order to compensate for the exponential growth of the distance between neighboring points while keeping the same volume, the cell must become distorted into thin strips with an exponentially decreasing thickness and therefore an exponentially increasing length. Because phase space is finite, this strip must become more and more distorted in order to find room for its outrageous length. It is therefore clear that the transformed cell C_t, after some finite amount of time, will have characteristic local scales below Planck's constant.

From the standpoint of a mathematician, there are many more chaotic systems than regular ones. This means that, if one were to generate the Hamilton function at random, the chances would be very high that one would get a chaotic system. However, nature does not play that kind of game and the majority of ordinary objects around us are not chaotic, except maybe at a very small scale. Clockworks, computers, oscillographs, and planets have a predictable behavior, at least for times not too outrageously long. The letters on the pages of a book take centuries or millenia before waning. There are therefore a great many regular objects. It does not mean that chaos never appears at the macroscopic level in a finite time. The turbulent motion of water in a torrent, or of the atmosphere, are chaotic (the chaotic behavior of atmospheric motion is even often deemed to be responsible for the impossibility to predict the weather for more than a few days). Both categories of extreme physical objects, the regular and the chaotic ones, therefore exist in reality.

It is most probable furthermore that chaos plays an important role in dissipation. When one tries to describe in detail the small vibrations of a ball rolling on a table, taking into account the nonlinear effects through which

RECOVERING CLASSICAL PHYSICS

the various vibrations interact, one is still in a domain that is described by classical collective variables but also most probably already in the domain of chaos. This means that one should stick to the simplest and most obvious collective variables when approaching classical physics, leaving the almost infinite variety of special cases that can occur at small scales for a specific investigation that will be sketched in Chapter 10. In any case, one should be prepared to find regular systems for which correspondence is valid and other systems for which it is not.

Regular Dynamics

These remarks suggest that one should try to more properly define regular dynamics and restrict one's attention to it. There is unfortunately no convenient definition of regularity in the theory of classical dynamical systems. To restrict attention to the completely integrable systems would be far too narrow. Fortunately, the nature of our problem suggests a simple and efficient criterion.

Given a dynamical system, we say that its dynamics is *regular* for a given initial cell C_0 and a time t if all the domains $C_{t'}$ that are generated from C_0 by the hamiltonian flow during a time t' with $0 \leq t' \leq t$ are regular cells. This important definition calls for a few comments. Firstly, it was shown that the Hamilton function generating the flow is the symbol of the quantum hamiltonian H_c for the collective observables. The definition of regularity refers therefore indirectly to the hamiltonian operator. Secondly, it should be stressed that the definition also refers to an initial cell and to a finite time. This last restriction is important. It does not exclude chaotic systems from consideration, but it limits the time during which they can be expected to satisfy both quantum and classical dynamics. The most interesting systems are however once again the clockworks, planets, voltmeters, and so on, most of which are dynamically regular for very long times. On the other hand, meteorology has to do with a limiting time t of the order of one or two weeks at most.

So, it is not enough to state that the cells $C_{t'}$ are regular. We should also know how much they are so, because this will ultimately control the errors that are involved in classical mechanics.

The Limitations on ζ

The parameter ζ occurring in equation (6.13) will be called the *dynamical classicality parameter*. It has been mentioned that the microlocal definition of a quasi-projector for a cell C_0 used a margin in which the symbol $f(q, p)$ goes smoothly from 1 to 0, when (q, p) goes from the inside to the outside of the cell. It was also said that the best value for the width of the margin is given in dimensionless units by

$$\varepsilon = \left(\frac{\hbar}{LP}\right)^{1/2}. \tag{6.14}$$

When the hamiltonian flow is acting, the margin is deformed and its width is locally controlled by the first derivatives of $q(t)$ and $p(t)$ with respect to their initial values (q_0, p_0). One may therefore expect to find these derivatives occurring in the value of ζ. On the other hand, one cannot expect the validity of dynamics to be better than that for kinematics at each time so that ζ can in no case be smaller than the parameters $\eta_{t'} = \varepsilon\theta_{t'}$ controlling the projectors of the intermediate cells at times $t' (0 \le t' \le t)$.

The precise expression of ζ cannot be found intuitively from these simple remarks, as was done in the case of kinematics, and only a complete calculation can give it. This is done in Appendix B and, before stating the result, we shall again give a few necessary definitions: The scales (L, P) for the coordinates and momenta will be fixed by the previous choice with reference to the initial cell and a parameter ε will be defined in terms of them by equation (6.6). We will use dimensionless coordinates q/L and p/P, collectively denoted by X' so that X' represents $2n$ dimensionless coordinates in phase space. We again denote by $\theta_{t'}$ the ratio between the dimensionless volume of the boundary of an intermediate cell $C_{t'}$ and the volume of the cell.

Let us also introduce the $2n \times 2n$ matrix $M_{t'}$ whose elements are the first derivatives of $q(t')$ and $p(t')$ with respect to (q_0, p_0), all of them being expressed in dimensionless units. One will need the operator norm $G_{t'}$ of this matrix, i.e., the absolute value of its largest eigenvalue. Finally, one will define a quantity $<G_{t'}>$ as its average value over the boundary of the cell C_0 with a weight given by the elementary area of the boundary. Then we define the two quantities

$$\zeta_1(t) = \sup_{0 \le t' \le t} \left\{ \varepsilon \theta_{t'} \langle G_{t'} \rangle \right\},$$

$$\zeta_2(t) = \sup_{0 \le t' \le t} \left\{ \frac{\hbar^2}{[LP]^3} \varepsilon^{-2} \theta_{t'} \left(\sup_{(q,p) \in C_{t'}} \left| \frac{\partial^3 h(q,p)}{\partial X'^3} \right| \right) \langle G_{t'} \rangle t' \right\}. \tag{6.15}$$

The third-order derivatives of the Hamilton function are taken with respect to the above dimensionless variables. Finally, we define the dynamical classicality parameter by taking $\zeta = \zeta_1(t)$ and restricting the time t to values such that

$$\zeta_2(t) < \zeta_1(t), \tag{6.16}$$

which may be a very long time in view of the higher order dependence of $\zeta_2(t)$ on Planck's constant. It may be noticed that these quantities can be computed in principle by using only the results of classical calculations.

The Theorem

We can now formulate the theorem controlling the classicality of motion. Its proof is given in Appendix B and it might be used for obtaining greater precision if necessary for some applications, as well as numerical estimates.

We say that the dynamics is regular to order ζ for a cell C_0 during a time t if the condition (6.16) is satisfied, ζ being given as before and being a small quantity. Then one has:

Theorem C.[15] *Consider a macroscopic system having a regular dynamics to order ζ for an initial regular cell C_0 during a time t. Let $F(C_0)$ be a quasi-projector for this cell and $F(C_t)$ a quasi-projector for a cell C_t obtained by transforming C_0 by the classical motion during the time t. Then one has the inequality*

$$\frac{\mathrm{Tr}|F(C_t) - U(t)F(C_0)U^{-1}(t)|}{\mathrm{Tr}|F(C_0)|} \leq K\zeta,$$

K being a constant of the order of one.

This theorem applies to negative values of t if one replaces the factor t' by $|t'|$ in the expression of $\zeta_2(t)$. When its technical apparel is removed, the theorem merely means that the correspondence between classical and quantum mechanics is realized when the conditions of its application are valid.

Physical Significance of the Errors

It is interesting to interpret the results from a physical standpoint. In the dominant term $\zeta_1(t)$, a leading factor is the upper bound of the effective classicality parameters $\{\varepsilon\theta_{t'}\}$, which was to be expected. It is multiplied by an average of $G_{t'}$ over the cell boundary, which may be interpreted from two different standpoints. According to microlocal analysis, this is the average of the inverse dimensionless width of the margin in the transformed cell. As such, one might get a smaller bound for it than the operator norm of the matrix $M_{t'}$. If instead one refers to a quasi-projector in terms of gaussian wave functions, one recognizes this factor as being associated with the spreading of the wave functions initially centered near the boundary.

Although the term $\zeta_2(t)$ is generally negligible, it can also be interpreted in terms of gaussian wave packets. Note that, in view of the third derivatives in it, it vanishes when the kinetic energy does not depend upon the position and the potential is a polynomial of order 2. This is precisely a case where a gaussian wave function remains gaussian. It is therefore tempting to associate it with the nongaussian corrections that have been studied by Hagedorn.[16] Its proportionality to the time and to the third derivatives of the potential confirm this idea, though it has not been completely proved.

It should also be mentioned that the expressions (6.14) and (6.15) are obtained from the analysis given in Appendix B, after making explicit some of the calculations. Furthermore, a more careful investigation of the results of this proof might allow a reduction of some factors when the rough bounds that have been imposed become excessive. This has not been done here because it would be more appropriate to a specific computation where the proper improvements would be better defined and also because the relevant formulas tend to become too lengthy.

This theorem establishes the correspondence between classical physics and quantum mechanics. It contains a refined form of Ehrenfest's theorem as well as its logical interpretation. It shows that nonregular systems do not allow a good correspondence and this is usually the case for chaotic systems after a finite time. This raises the question of whether chaotic quantum systems are reliably described by classical physics when full chaos takes place. This will be considered in Chapter 10.

JUSTIFYING COMMON SENSE

Common sense, as meant in this book, is our usual way of describing ordinary objects and logically inferring the consequences of what is observed. It is also conventional logic as applied to the classical properties of rather big objects. From what has been shown up to now, one knows how to describe an object in terms of properties that are equivalently expressed in classical or quantum language. One also knows when and how dynamics preserves their equivalence. It is therefore possible to investigate whether common sense is an outcome of the universal rule of interpretation. In other words, the question is to discover why what is obvious (the world around us) is in agreement with what is universal (quantum mechanics).

8. Recovering Common Sense

The problem is the following. Consider several macroscopic objects interacting together so that they can be considered as a unique overall macroscopic system. They are assumed to be initially in a state that may be described in classical terms and one wants to know what classical properties holding at another time can be stated consistently and logically related to each other.

To say that the initial state may be described in classical terms can be expressed explicitly in the following form: Let ρ be the initial density operator of the system of objects. There exists a classical property, expressed by a quasi-projector F involving the collective variables, so that

$$F \rho F = \rho. \quad (6.17)$$

This condition only says that the classical property described by F holds at the initial time. It does not say that one knows everything about the system once this property has been stated. To give an example, the property could state that the initial situation of a pendulum is well described classically by its position and its velocity within specific error bounds and this is what is expressed by the quasi-projector F. If the pendulum is made of copper and it contains a cadmium atom as an impurity, it is clear that the spin state of the cadmium nucleus might be specified by the complete density operator ρ whereas F does not say anything about it.

The condition (6.17) generally does not hold exactly and the statement allowing one to use it in quantum logic is $\text{Tr}|F \rho F - \rho| < \eta$, the number η being smaller than the dynamical classicality parameter ζ corresponding to the initial cell associated with F and the time t during which the objects will be described. When this condition is satisfied, it can be said that one starts from a *state of fact* for the system of macroscopic objects.

In order to justify the reasoning of classical physics about these objects, proceed as follows: Explicitly introduce the various classical properties of interest; they are then reexpressed in a quantum framework by associating a specific quasi-projector with each one of them. The time at which they hold is taken into account by using the time-dependent Heisenberg form for the quasi-projectors and one thereby obtains a well-defined history. Other histories are also introduced in order to obtain a complete family, which is easily done by introducing the negation of every property in the basic history. One thus obtains the field of propositions of a specific quantum logic.

As an example, think of two people discussing billiards. The physical system under study consists of a few balls and a cue. The initial state of facts describes the initial position of the balls and it states that their velocities are very small according to classical standards. It also gives the initial position of the cue and its velocity, within some error bounds that are again very small according to classical standards. One of the two people is a master and he can tell his partner what will happen, which ball will hit which, how it will be rotating at that time, what will happen next, and so on (we disregard the fact that he takes friction into account and we have not yet included it). What he is describing is simply a history of the system. Perhaps he is not so sure of what is going to happen so that he says something like this: *if* ball number 6 hits ball number 8, *then* ball number 8 will hit ball number 5. He is accordingly considering several mutually exclusive histories in his reasoning. The game we are playing presently is to show that such a description can be given a meaning according to quantum mechanics, which is supposed to provide the only basic laws of science, and that the reasoning of the master (who is in fact relying upon classical physics) is also perfectly sane from the standpoint of quantum mechanics.

The procedure to be followed is rather clear. Having a field of propositions, we must make sure that the corresponding logic is consistent by checking its consistency conditions, up to some estimated error. Then, to substantiate the master's reasoning, we must compute the corresponding conditional probabilities allowing logical implications between various properties, once again up to some error. In particular, when two properties imply each other in both directions, they are logically equivalent and the most important case of that sort is classical determinism.

This program is of course sensible only when all the properties to be considered are described by regular cells. The essential condition allowing one to check consistency and to recover the implications of classical physics is the regularity of dynamics starting from the initial state of fact for the time during which the system is described. There is an upper bound to the errors involved in the various consistency conditions and for the validity of the implications (i.e., the difference between the conditional probabilities and 1). These errors are given by the dynamical classicality parameter ζ for the system under consideration, the initial state of facts, and the duration of the description. This program offers no special difficulty given our previous results, so that we only state the main results, their proof being given in Appendix C. These results are: (i) Classical logic, as it applies to macroscopic objects, is consistent; and (ii) The classical logical equivalence establishing classical determinism also holds as a quantum logical equivalence or, in other words, classical determinism is a consequence of quantum mechanics.

Classical logic and determinism are therefore direct outcomes of the universal rule of interpretation. This is in fact one of the reasons why this rule was said to be universal, since its consequences go beyond the probabilistic framework of quantum mechanics to reach almost certainty in classical situations though, of course, probabilities continue to rule over the possible errors.

The Limits of Classical Physics

One of the essential aspects of these results is that classical physics is not absolute, neither in its dynamical nor in its logical contents. Given an initial state of fact and a fixed time, there exists a finite, even if very small, dynamical classicality parameter ζ measuring the amount of logical inconsistency that might occur in a classical reasoning, i.e., the probability of making an error when asserting a conclusion according to classical physics. The errors are of two kinds. They can be due to a slight violation of the consistency

RECOVERING CLASSICAL PHYSICS

conditions or to a slight deviation from 1 of a conditional probability justifying a link in reasoning. The first kind of error gives rise to a break in logic so that some aristotelian axioms are at risk of being wrong. This is the kind of nonsense one gets when, for instance, it is not exactly the same thing to say that a ball is outside some region R and to say that it is not in R. It comes from the fact that a lot of equivalent quasi-projectors are supposed to state the same thing though they are slightly different. It all boils down to the fact that, when one must say in mathematical terms what the frontier of the region R is, one has to define it with an absolute precision though this is obviously absurd from a physical point of view. It is not surprising nor particularly troublesome that a rigorous logical construction brings this back to our attention by mentioning the corresponding inconsistency. It is also clear that it is of no practical importance.

The other source of error occurs when a conditional probability leading to an implication is not strictly equal to 1. It can happen, for instance, that a conclusion derived from determinism turns out to be wrong. The existence of this remote risk is unavoidable because of the possibility of quantum fluctuations and, here again, there is nothing surprising in the result. When a conditional probability $p(b|a)$ has a value $1 - \varepsilon$, the small number ε is the probability for the conclusion b not to happen because of some quantum fluctuation occurring after the realization of the initial condition a. These effects are completely negligible (and also irreproducible) and nobody in his right mind will ever attribute a failure of his car engine to a quantum fluctuation.

It should be stressed that the results indicated here assume that the system under consideration has regular dynamics. They do not hold when, for instance, there is no determinism at the classical level. Consider the case of a perfectly clean spark chamber without any external incoming charged particle, the electric potential being increased continuously. It is impossible to predict where the first spark will occur because it must necessarily arise from a fluctuation. Whether it has a quantum or a thermal origin is a matter to be left for a specific investigation if necessary.

This last remark calls attention to the fact that the probabilities of error given previously are due to pure quantum fluctuations, since the coupling between the overall behavior of the objects and their internal properties was neglected together with dissipation and friction. If one were to use the results for an exact estimate of errors, it is clear that one should also take into account the effects of thermal fluctuations. In any case, the conclusions are pretty obvious since they boil down to reobtaining determinism when it obviously holds from classical considerations, the errors being due to quantum and thermal fluctuations. The conclusions of this analysis are therefore very simple. The essential point is that one can now be sure that classical physics is in accordance with quantum mechanics, from a logical standpoint as well as a dynamical one, everything being perfectly consistent.

APPENDIX A: ELEMENTS OF MICROLOCAL ANALYSIS

Microlocal analysis, also called pseudo-differential calculus, is a vast and comprehensive collection of mathematical methods. It offers, among other applications, a very powerful technique for a systematic study of linear partial differential equations, including the Schrödinger equation. It is also remarkably well suited to a study of the correspondence between classical and quantum mechanics. This is the only aspect to be considered in the present appendix. The theory is quite elaborate and it makes use of many other fields of analysis. The proofs are correspondingly difficult and this is why we will only introduce the theory as providing a set of tools, stating the results that are most convenient for our purpose without entering into the details necessary for their proof.

1. Operators and Symbols

Consider an operator Ω in the Hilbert space $L^2(\mathbf{R}^n)$ of square-integrable complex-valued functions of n real variables $x = (x_1, \ldots, x_n)$. A function $\omega(x, p)$, where p is an n-dimensional vector $p = (p_1, \ldots, p_n)$, will be defined as the following partial Fourier transform of the matrix elements of Ω:

$$\begin{aligned} \omega(x, p) &= \int \langle x'|\Omega|x''\rangle \delta\left(x - \frac{x' + x''}{2}\right) e^{-ip\cdot(x'-x'')/\hbar} \, dx' \, dx'', \\ &= \int \langle x + \frac{y}{2}|\Omega|x - \frac{y}{2}\rangle e^{-ip\cdot y/\hbar} \, dy, \qquad (6\text{A}.1) \\ &= \int \langle p'|\Omega|p''\rangle \delta\left(p - \frac{p' + p''}{2}\right) e^{ix\cdot(p'-p'')/\hbar} \, dx' \, dx''. \end{aligned}$$

The "eigenvectors" of position and momentum are normalized according to the convention

$$\langle x'|x''\rangle = \delta(x' - x''), \qquad \langle p'|p''\rangle = (2\pi\hbar)^n \delta(p' - p''),$$
$$\langle x|p\rangle = \exp(ip \cdot x/\hbar),$$

where

$$p \cdot x = \sum_{k=1}^{n} p_k x_k.$$

The action of the operator Ω on a vector u in the Hilbert space, with wave function $u(x)$, gives a vector $v = \Omega u$. This can be written in terms of the wave functions by

$$v(x) = \int \omega\left(\frac{x+y}{2}, p\right) e^{ip \cdot (x-y)/\hbar} u(y) \frac{dy\,dp}{h^n}, \qquad (6A.2)$$

where $h = 2\pi\hbar$, as usual. The equivalence between the various forms of equation (6A.1) as well as equation (6A.2) can be derived from the well-known decompositions of the identity

$$I = \int |x\rangle\langle x|\,dx = \int |p\rangle\langle p|\frac{dp}{h^n}.$$

One easily finds that the operator X_k, which is the observable corresponding to the kth component of position, is associated with a function of (x, p) identical to x_k and the operator P_k is similarly associated with p_k. The identity operator is associated with the function identically equal to 1 and the operator zero with the function everywhere equal to 0. The operator $\Omega^\dagger$, the adjoint of Ω, is associated with the function $\omega^*(x, p)$, which is the complex conjugate of $\omega(x, p)$. A self-adjoint operator is therefore associated with a real function.

When the trace of Ω exists, it can be written as

$$\text{Tr}\,\Omega = \int \langle x|\Omega|x\rangle dx = \int \omega(x, p)\frac{dx\,dp}{h^n}. \qquad (6A.3)$$

Although it was said that $\omega(x, p)$ is a function, it is often only a distribution when the operator Ω is arbitrary. The most useful applications occur however when $\omega(x, p)$ is infinitely differentiable with respect to all variables and its derivatives have well-defined upper bounds. From here on, we shall assume that these conditions are satisfied and Ω will then be called a *pseudo-differential operator*, $\omega(x, p)$ being called its Weyl symbol or, more briefly, its *symbol*.

The bounds one can impose upon the derivatives of a symbol are very flexible and there are therefore many different classes of pseudo-differential operators. This flexibility is used with virtuosity by mathematicians, but it has a drawback because this high level of generality makes the concepts and the notations somewhat obscure. We shall therefore try to obviate this source of misunderstanding by sticking to a special class of symbols.

It will be convenient to introduce a scale L with respect to which the position variables x will be measured and another scale P for the momentum variables. This will allow us to use everywhere dimensionless variables, so that the formulas will become simpler. When this is done, Planck's constant $\hbar$ is replaced by the dimensionless number $(\hbar/LP)$. It will be assumed that the scales L and P are characteristic of a macroscopic situation so that $LP \gg h$, or, in dimensionless variables, $\hbar \ll 1$. From here on, x, p, and $\hbar$ will be treated as dimensionless.

We can now state the conditions to be imposed upon the symbols and their derivatives. A function $\omega(x, p)$ will be called a symbol of order m, m being a real number, when its derivatives of all orders have the following uniform bounds in phase space

$$\left|\partial_x^\alpha \partial_p^\beta \omega(x, p)\right| \leq C_{\alpha\beta}(1 + x^2 + p^2)^{(m-|\alpha|-|\beta|)/2}. \tag{6A.4}$$

We use the conventional notation where an index α denotes a collection of nonnegative integers $(\alpha_1, \ldots, \alpha_r)$ and $|\alpha|$ denotes the sum $\alpha_1 + \cdots + \alpha_n$. The derivatives are then written symbolically by taking

$$\partial_x^\alpha = \left(\frac{\partial}{\partial x_1}\right)^{\alpha_1} \cdots \left(\frac{\partial}{\partial x_n}\right)^{\alpha_n}.$$

In the bounds (6A.4), the quantities $C_{\alpha\beta}$ are constants depending only upon the indices α and β. They are called the *seminorms* of $\omega(x, p)$. The family of all symbols of order m is denoted by S^m. For instance, x and p are symbols of order 1 and p^2 is a symbol of order 2.

A special metric

$$ds^2 = dx^2 + dp^2. \tag{6A.5}$$

will be introduced more or less arbitrarily on phase space. In physics, the kinetic energy can be used to choose this metric in a better way, but we shall presently stick to the metric (6A.5) because the theorems to be given become somewhat more obscure, though also often more powerful, in the case of a more general riemannian metric on phase space.

The first theorem to be given relates some properties of an operator to its symbol:

Theorem A.1. *Consider a symbol belonging to the class S^m. When $m \leq 0$, the associated pseudo-differential operator is bounded. When m is strictly negative, the operator is compact.*

One may recall that a compact operator has a purely discrete spectrum, the eigenvalues accumulating only towards 0. The theorem shows that a pseudo-differential operator whose symbol is different from 0 only in a finite region of phase space automatically has a discrete spectrum. This is valuable information that can be obtained easily when the symbol is known.

2. Operator Products

Consider two pseudo-differential operators A and B with symbols $a(x, p)$ and $b(x, p)$ belonging to the classes S^m and $S^{m'}$, respectively. We are interested in the properties of the operator product $C = AB$.

RECOVERING CLASSICAL PHYSICS

We first introduce notation:

1. Given two points in phase space (x, p) and (x', p'), define an antisymmetric bilinear form, also called a symplectic form, by

$$\sigma(x, p; x', p') = p \cdot x' - p' \cdot x.$$

2. Consider the differential operator called the Poisson bracket and denoted by $\{\ \}$. It is defined on a product of two functions of (x, p) and it acts as a differential operator of the first order upon both terms in the product, one standing on its left and the other on its right, according to the convention

$$\{\ \} = \sum_{k=1}^{n} \left(\frac{\overleftarrow{\partial}}{\partial p_k} \frac{\overrightarrow{\partial}}{\partial x_k} - \frac{\overleftarrow{\partial}}{\partial x_k} \frac{\overrightarrow{\partial}}{\partial p_k} \right).$$

The direction of the arrows indicates which function is to be differentiated. For instance, one has $f\{\ \}g = \{f, g\}$, where $\{f, g\}$ is the ordinary Poisson bracket

$$\{f, g\} = \sum_{k=1}^{n} \left(\frac{\partial f}{\partial p_k} \frac{\partial g}{\partial x_k} - \frac{\partial f}{\partial x_k} \frac{\partial g}{\partial p_k} \right).$$

Similarly, one has

$$f\{\ \}^2 g = \sum_{jk} \left[\frac{\partial^2 f}{\partial p_j \partial p_k} \frac{\partial^2 g}{\partial x_j \partial x_k} + \frac{\partial^2 f}{\partial x_j \partial x_k} \frac{\partial^2 g}{\partial p_j \partial p_k} - 2 \frac{\partial^2 f}{\partial p_j \partial x_k} \frac{\partial^2 g}{\partial x_j \partial p_k} \right].$$

The symbol $c(x, p)$ for the operator $C = AB$ can then be formally written as

$$c(x, p) = a(x, p) e^{-i(\hbar\{\ \}/2)} b(x, p). \tag{6A.6}$$

This can be proved by having the operator A act first upon a wave function according to equation (6A.2), then letting B act on the wave function obtained. The symbol $c(x, p)$ is then computed by a partial Fourier transform, as given by equation (6A.1). Equation (6A.6) is not a strict equality, except when one of the symbols $a(x, p)$ or $b(x, p)$ is a polynomial in either x or p. In general, the difference between the two sides of equation (6A.6) is a function belonging to the class $S^{-\infty}$ so that it decreases faster than any power of x and p. Another form of equation (6A.6) can be written in terms of the Fourier transform $\tilde{c}(q, y)$ of $c(x, p)$ with respect to its two arguments. It reads

$$\tilde{c}(q, y) = \int \tilde{a}(q', y') \exp[i\sigma(y', q'; y'', q'')/2] \tilde{b}(q'', y'')$$

$$\times \delta(y - (y' + y'')) \delta(q - (q' + q'')) \frac{dy'\, dq'\, dy''\, dq''}{\hbar^n}. \tag{6A.7}$$

A most important feature of an expression such as equation (6A.6) should also be stressed: When the exponential is expanded as a series, one obtains an expansion of $c(x, p)$ in powers of $\hbar$, which is called a semiclassical expansion. The series however does not converge in general but it is only an asymptotic series. This means that it gives a good approximation for $c(x, p)$ when its first few terms are kept, but one cannot use it to an arbitrarily large order. This would make the series formal and useless were it not that one can give explicit error bounds when the series is truncated at a given order. So, let us write the series by exhibiting its N first terms together with a remainder:

$$c = ab - i\frac{\hbar}{2}\{a, b\} - \frac{\hbar^2}{2^2 \cdot 2!} a\{\ \}^2 b + \cdots + \frac{(-i\hbar/2)^{N-1}}{(N-1)!} a\{\ \}^{N-1} b + r_N. \tag{6A.8}$$

Then one has

Theorem A.2. *The product of two pseudo-differential operators A and B, respectively belonging to the classes $S^{(m)}$ and $S^{(m')}$, is a pseudo-differential operator belonging to the class $S^{m+m'}$. Its symbol can be written in the form given by (6A.8), the remainder $r_N(x, p)$ being bounded according to*

$$r_N(x, p) \leq K\hbar^N (1 + x^2 + p^2)^{(n+m-N)/2}. \tag{6A.9}$$

The constant K does not depend on (x, p) but it depends on $a(x, p)$, $b(x, p)$, and the order N. This is given explicitly by

$$K = \frac{c}{N!} \sup |\partial_x^\alpha \partial_{x'}^{\alpha'} \partial_p^\beta \partial_{p'}^{\beta'} \sigma^N(\partial_x, \partial_p; \partial_{x'}, \partial_{p'}) a(x, p) b(x', p')|; \tag{6A.10}$$

the supremum being taken upon all values of (x, p) and (x', p') and all the derivation indices $\alpha, \alpha', \beta, \beta'$ satisfying the condition

$$0 \leq |\alpha| + |\alpha'| + |\beta| + |\beta'| \leq n + 1,$$

n being the number of degrees of freedom. The constant c is of the order of one and it depends only on n and N. Its explicit value could be known in principle but it is not explicitly given in the mathematical literature, as far as we know. One may hope that a greater interest in the practical applications of these methods will lead to greater attention to these numerical considerations.

The term with the lowest power of $\hbar$ occurring in equation (A.16) is usually the dominant one and, in that case, one can use

$$K = \frac{c}{N!} \sup\{|\sigma^N(\partial_x, \partial_p; \partial_{x'}, \partial_{p'}) a(x, p) b(x', p'|\}.$$

RECOVERING CLASSICAL PHYSICS

Notice that the bound (6A.9) for the remainder is uniform, whatever the values of (x, p), whereas one expects that it should be smaller outside the intersection of the supports of $a(x, p)$ and $b(x, p)$. This question is relevant when one needs to integrate $r_N(x, p)$, as happens for instance when estimating a trace norm. Equation (6A.9) does not insure that the integral is finite because its right-hand side does not vanish at infinity. However, there exist other theorems showing that the quantity $|c(x, p)|$ and the corresponding derivatives decrease rapidly outside the intersection of the supports where the function $a(x, p)\{\ \}^{N-1}b(x, p)$ itself is 0. It in fact decreases faster than any inverse power of the distance between the supports, in most cases exponentially. These theorems are given explicitly in Hörmander's book (see the notes at the end of this chapter). One can say with certainty that $c(x, p)$ is bounded in the outside region by quantities of the form $K[d^2(x, p)/\hbar]^{-r}$, where the notation is the following: $d(x, p)$ is the distance from the point (x, p) to the intersections of the supports, in dimensionless coordinates; Planck's constant is also a dimensionless number; r is any integer or half-integer number; and K is a constant controlled by the seminorms of $a(x, p)$ and $b(x, p)$. This knowledge is sufficient for all applications, since it allows one to neglect troublesome terms without needing their explicit form. For an example, see Problem 19.

3. Norm Estimates

Consider now a symbol $a(x, p)$ that is bounded from above and below so that $c_1 \leq a(x, p) \leq c_2$. When one thinks of $a(x, p)$ as a dynamical variable representing the observable A, one is tempted to assume that similar bounds are valid for the operator. So one asks whether it is true that

$$c_1 I \leq A \leq c_2 I? \tag{6A.11}$$

It can be shown that this property is essentially correct, except that some higher corrections in powers of $\hbar$ must be taken into account.

The reason why this happens can be understood by considering the case when the lowest bound c_1 is 0, so that $a(x, p)$ is a nonnegative symbol. The first inequality in the hypothesis (6A.11) would then mean that A is a positive operator. To check that, introduce a nonnegative function $b(x, p)$, which is the square root of $a(x, p)$. If it is a symbol, i.e., if it is infinitely differentiable with seminorms bounds, it can be associated with a pseudo-differential operator B. Using Theorem A.2, one can then find an expression for the operator B^2. Its symbol is given by

$$b^2(x, p) - \frac{\hbar^2}{8} b\{\ \}^2 b + \cdots = a(x, p) - g(x, p),$$

where $g(x, p)$ represents the sum of the series starting from the second term on. This can be written in terms of operators as $A = B^2 - G$, showing that A is the sum of the nonnegative operator B^2 and of a correction having a symbol $g(x, p)$ of order $\hbar^2$. This very simple-minded approach clearly shows the kind of result to be expected and the order of magnitude of the corrections. It is of course not a proof, because usually $b(x, p)$ is not a symbol. This is shown by the example $a(x, p) = x^2$, which gives $b(x, p) = |x|$. This function is clearly not differentiable to an arbitrary order, so that it cannot be a symbol, and B is not a pseudo-differential operator, so that Theorem A.2 cannot be used. The exact result, which is known as Gårding's sharp inequality, is given by the following theorem, not given here in its full generality but in the form most convenient for our purpose:

Theorem A.3. *Assume that the symbol $a(x, p)$ is nonnegative and the associated pseudo-differential operator A is bounded. Then, whatever the normed function u in $L^2(\mathbf{R}^n)$, one has $< u|A|u > \geq -C\hbar^2$, where C is a constant controlled by the seminorms of $a(x, p)$.*

We shall not try to give an explicit expression for the constant, because all the applications we shall use will only be concerned with orders of magnitude. Furthermore, the estimates one can find in the mathematical literature are obscure, although there exists in principle an algorithm to obtain them.

The inequalities (6A.11) are therefore corrected by terms of order $\hbar^2$, as the previous simple argument suggested. As a special case, one gets an estimate for the Hilbert norm of operator A:

$$\|A\| \leq \sup(|c_1|, |c_2|) + C\hbar^2,$$

where the constant C has the behavior mentioned in Theorem A.3.

Note: In statistical mechanics, one often associates a density operator ρ with its symbol $f(x, p)$, which is called Wigner's distribution. It is known that the positivity of the density operator does not imply an exact positivity of the distribution. Gårding's sharp inequality shows essentially a reverse situation where a positive distribution does not exactly generate a nonnegative density operator. The corrections are however usually innocuous since they are of order $\hbar^2$.

4. Estimates of Trace Norms

Equation (6A.3) gives the trace of a pseudo-differential operator as the integral of its symbol over phase space. More is needed for logical applications, because one also wants to evaluate some trace norms. Let us first define this notion more precisely. Given a self-adjoint operator, its absolute value

$|A|$ is defined as a positive operator $|A|$ such that $|A|^2 = A^2$. In a basis where A is diagonal with eigenvalues $\{a_j\}$, $|A|$ is also diagonal with eigenvalues $\{|a_j|\}$. The trace norm of A is defined by

$$\|A\|_{\mathrm{tr}} = \mathrm{Tr}|A|. \tag{6A.12}$$

The operator A is said to be of trace class when the sum (6A.12) converges. It is then the sum of the absolute values of the eigenvalues. The main interest of this norm is due to the obvious master inequality

$$|\mathrm{Tr} AB| \leq \|A\| \mathrm{Tr}|B|. \tag{6A.13}$$

In the case of a non-self-adjoint operator A, one cannot use eigenvectors of A, though one still defines $|A|$ as $(AA^\dagger)^{1/2}$ with the same definition (6A.12) for the trace norm of A. The usual techniques of operatorial analysis can then be used to obtain an expression for the trace norm

$$\|A\|_{\mathrm{tr}} = \sup_{\{\phi_j\}} \sum |<\phi_j|A|\phi_j>|,$$

the supremum being taken over all possible orthonormal bases $\{\phi_j\}$.

The necessity of considering all possible bases, or of explicitly knowing the eigenvalues in the case of a self-adjoint operator, makes the trace norm inadequate for the use of ordinary operator techniques. This is probably why this notion is much less used than the Hilbert norm $\|A\|$, though it frequently provides a more physical expression for the proximity of two operators. One of the interests of microlocal analysis is to make this much more accessible.

Let us give more definitions and notation. The L_1 norm of a symbol $a(x, p)$ is defined as

$$\|a\|_{L_1} = \int |a(x,p)| \frac{dx\,dp}{(2\pi\hbar)^n}.$$

Denoting by B the basic ball in phase space defined by $x^2 + p^2 \leq \hbar^2$, we introduce the function

$$\delta a(x, p) = \sup |\partial_x^\alpha \partial_p^\beta a(x + x', p + p')|,$$

the supremum being taken over all the pairs of indices (α, β) such that $|\alpha| + |\beta| = n + 1$ and over all the points (x', p') in B.

One can then give an estimate for the trace norm of a pseudo-differential operator by using the following theorem, which is due to Hörmander:

Theorem A.4. *The trace norm of an operator with symbol $a(x, p)$ is bounded by*

$$\mathrm{Tr}|A| \leq C_1 \|a\|_{L_1} + C_2 \hbar^{(n+1)/2} \|\delta a\|_{L_1}, \tag{6A.14}$$

where the constants C_1, C_2 are of the order of one and do not depend on $a(x, p)$.

The first term in the estimate (6A.14) looks like the integral (6.3) for the trace. The second one, depending on the derivatives of order $n + 1$, is somewhat more surprising at first sight. This is the kind of error estimate one always finds when the proof rests upon Bernstein's theorem controlling the convergence of a Fourier series.[17]

5. More General Phase Spaces

The modern theory of pseudo-differential operators is also called microlocal analysis because it reduces the study of an operator to the consideration of the values of its symbol in various localized regions of phase space. The underlying technique is based upon the use of *partitions of unity:*

Consider a family of functions $f_k(x, p)$ having the following properties: Each function is a symbol vanishing outside an open set U_k of phase space where it is nonnegative. The union of all these open sets completely covers phase space and the functions add up everywhere to 1, namely:

$$\sum_k f_k(x, p) = 1. \tag{6A.15}$$

Such a family of functions is called a partition of unity and one can prove that it exists under a few simple conditions upon the open sets.

Given a pseudo-differential operator A with symbol $a(x, p)$, equation (6A.15) gives

$$a(x, p) = \sum_k a_k(x, p),$$

where $a_k(x, p) = a(x, p) f_k(x, p)$. If A_k denotes the pseudo-differential operator with symbol $a_k(x, p)$, one has

$$A = \sum_k A_k.$$

This equation reduces the study of A to the study of operators with localized symbols.

This method can be used to extend the theory to an arbitrary phase space. In fact, up to now, the configuration space Γ we considered was simply an n-dimensional euclidean space, so that the associated phase space was a $2n$-dimensional linear space. In classical mechanics, these are often more general configuration spaces: for instance, the configuration space of a spinning top is a three-dimensional space isomorphic to the rotation group $O(3)$ describing all the positions of the top in terms of Euler angles. Although the method of pseudo-differential operators might at first sight appear to be restricted to a euclidean configuration space $\mathbf{R}^n$ in view of its reliance upon Fourier transforms, this is fortunately not so because one can use partitions of unity.

In general, the phase space associated with a general configuration space Γ is parametrized locally by the coordinates of position and momentum. This is well known for instance in the case where Γ is a sphere, since choosing a set of coordinates is the same thing as drawing a map. In general, one proceeds as in geography by covering Γ by local maps, each one of them being defined in some open set V_k. The maps represent the whole configuration space. In each open set, introduce local coordinates (x_k, p_k), which belong to an open set U_k in $\mathbf{R}^{2n}$. A local symbol in V_k, i.e., a C^∞ function vanishing outside V_k, can be expressed as a function of the local coordinates, i.e., as a symbol in $\mathbf{R}^{2n}$ vanishing outside U_k. In order to have functions belonging to the class C^∞ (infinitely differentiable), the configuration space must also be a C^∞ manifold, which is rarely an inconvenient restriction. The study of the operator A is obtained by getting the properties of all its components A_k and then putting the pieces back together.

It is not particularly difficult to use these methods, either in principle or in practice. They rely however heavily upon differential geometry and this is too large an outside body of knowledge to be introduced in the present book. Furthermore, the results one obtains in that way are usually trivial extensions of what may be obtained more easily in the case where the configuration space is euclidean. This is why we shall stick to that case, while extending the results without much caution to the general case. Finally, this approach can be used to define a wave function depending on local collective coordinates in the configuration space. The corresponding Hilbert space is easily constructed together with the theory of pseudo-differential operators by a proper localization.

6. Summary

The content of Appendix A can be considered as a kit of tools with the following directions for use in physics, in view of the applications we need. Symbols are used to define operators and particularly quasi-projectors. Theorem A.1 can be used when one wants to show that some operator is bounded or even compact, for instance, to make sure in the later case that its spectrum is discrete. Theorem A.3 is used to get estimates for the Hilbert norms, although this will be made most often to show that they are negligible. A product of operators can be computed according to Theorem A.2 and this is extremely important, particularly for estimating errors. Finally, Theorem A.4 gives control over trace norms.

APPENDIX B: SEMICLASSICAL THEOREMS

This Appendix is devoted to the proof of Theorems A, B, and C occurring in the main text of the chapter. We do not insist on complete mathematical rigor for at least two reasons, the most cogent one being the inability of the author in this regard and the second one being the undue length and technicality of such a proof. We thought of referring the reader to original articles, but this was not convenient because the present proofs are somewhat better than the published ones, though very far from perfect. Every physicist knows furthermore that practical applications seldom fall right in the middle of the domain where a known mathematical theorem holds and one often needs to adjust it. This is why an explicit proof is useful and why it is given here.

1. Theorem A

Geometric Conditions

Let C be a cell in phase space, connected and simply connected, and let Σ be its boundary. Phase space will be taken to be $\mathbf{R}^{2n}$, the generalization to more complex cases being assumed to be more or less straightforward. One introduces two scales of length and momentum (L, P). They represent most often typical scales for the cell. With these units, the position and momentum variables (x, p) are dimensionless, as well as Planck's constant, and we shall assume that the value of Planck's constant is very small. It will also be assumed that the boundary Σ has only a small number of edges, like the case of a box though the faces might be curved, but the discussion will be made in the case when there is no such edge, the generalization being easy.

A metric in phase space is given by

$$ds^2 = dx^2 + dp^2 \tag{6B.1}$$

In the case of a general configuration space, one uses other, locally defined, metrics. The discussion will however be made only in the simple case of the metric (6B.1).

In view of defining conveniently a quasi-projector associated with the cell C, let us introduce two $(n-1)$-dimensional surfaces Σ_1 and Σ_2 enclosing the boundary Σ; Σ_1 lies inside the cell C and Σ_2 outside. They are respectively described by the equations

$$v_1(x, p) = c_1, \quad v_2(x, p) = c_2,$$

where v_1 and v_2 belong to the class C^∞, and c_1 and c_2 are two constants. The distance between Σ_1 and Σ_2 is practically constant and equal to a quantity ε to be fixed later on. The region M lying between the two surfaces is called the margin of the cell C.

Quasi-Projectors

The symbol $f(x, p)$ of a quasi-projector is defined as a C^∞ function equal to 1 in the region inside C having Σ_1 for its outside boundary and equal to 0 in the region outside C with inside boundary Σ_2. It is nonnegative and goes smoothly from 1 to 0 when (x, p) goes from Σ_1 to Σ_2 in the margin.

Using Theorem A.1, one finds that the operator F with symbol $f(x, p)$ is bounded (because $f(x, p)$ belongs to the class of symbols S^o); it is also compact (because $f(x, p)$ belongs to the class S^m for all negative numbers m). Its spectrum is therefore discrete. Theorem A.3 shows that its eigenvalues are contained in the interval $[0, 1]$, up to corrections of order $\hbar^2$ that will be neglected.

The trace of F is given by Theorem (A.3) so that

$$\mathrm{Tr} F = \mu(C), \tag{6B.2}$$

$\mu(C)$ being the volume of the cell C in units h^n. Since $f(x, p)$ lies between 0 and 1 inside the margin, equation (6B.2) is true up to corrections of relative order $\mu(M)/\mu(C)$.

Using equation (6A.2) giving the action of a pseudo-differential operator on a wave function, one shows easily that a wave function that is well inside (or outside) C is an eigenfunction of F with eigenvalue 1 (respectively, 0), up to corrections that are small in Hilbert norm.

The quasi-projector F is not exactly a projector and one must evaluate the operator $G = F - F^2$, which measures the nonprojective character of F. The most important point will be to evaluate its trace norm.

Estimating Trace Norms

Theorem A.4 gives estimates for trace norms. It will be convenient to add a few comments before using it. It reads

$$\mathrm{Tr}|A| \leq C_1 \int |a(x,p)| \frac{dx\,dp}{h^n} + C_2 \hbar^{(n+1)/2} \int \delta a(x,p) \frac{dx\,dp}{h^n}, \tag{6B.3}$$

where C_1 and C_2 are constants of the order of 1. The function $\delta a(x, p)$ is given by

$$\delta a(x, p) = \sup |\partial_x^\alpha \partial_p^\beta a(x + x', p + p')|, \tag{6B.4}$$

the supremum being taken over the points (x', p') inside the ball with center $(0, 0)$ and radius $\hbar$ as well as over all the derivation indices (α, β) such that

$$|\alpha| + |\beta| = n + 1. \tag{6B.5}$$

The first integral in the bound (6B.4) is the L_1 norm of $a(x, p)$. It will be convenient to use a specific notation for it, namely

$$\int |a(x,p)| \frac{dx\,dp}{h^n} = \|A\|_{L_1}.$$

The second integral in the bound (6B.4) comes from an estimate using Bernstein's theorem for the rate of convergence of a Fourier series. We shall accordingly call it the Bernstein norm of the operator A, though this name is somewhat inadequate because it is not really a norm but rather what a mathematician would call a seminorm. Nevertheless, we shall use the shorter name and we shall write

$$\int \delta a(x,p) \frac{dx\,dp}{h^n} = \|A\|_B.$$

The evaluation of L_1 norms and Bernstein norms is rather different so that they will have to be considered separately.

Structure of the Symbol of $F - F^2$

Using Theorem A.2, we find that the symbol $f^{(2)}(x, p)$ of the operator F^2 is given by

$$f^{(2)}(x,p) = f^2(x,p) - \frac{\hbar^2}{8} f(x,p)\{\ \}^2 f(x,p) + \cdots, \qquad (6B.6)$$

where there is no term of order $\hbar$ since the Poisson bracket $\{f, f\}$ vanishes.

First consider the contribution $g_1(x, p) = f(x, p) - f^2(x, p)$ to the symbol of $G = F - F^2$, keeping only the leading term in equation (6B.6). The function $g_1(x, p)$ is 0 everywhere, except in the margin M, since this is the only place where $f(x, p)$ is not equal either to 1 or 0. Inside M, $f(x, p)$ has its values in the range $[0, 1/2]$. Using equation (A.3) for the trace of an operator, we get

$$0 \leq \mathrm{Tr}\, G_1 \leq \int_M \left[\frac{1}{2}\right] \frac{dx\,dp}{h^n} \leq \frac{1}{2}\mu(M).$$

Let us now get a first estimate for the difference between $\mathrm{Tr}\, G_1$ and $\mathrm{Tr}\, |G_1|$. Note that $g_1(x, p)$ is nonnegative inside M, vanishes together with all its derivatives on the boundary of M, and is 0 outside M. Its square root is therefore also a nonnegative symbol vanishing outside M. Let $G_{1/2}$ denote the pseudo-differential operator associated with this symbol. Theorem A.2 gives

$$G_1 = G_{1/2}^2 + \Delta G_1, \qquad (6B.7)$$

RECOVERING CLASSICAL PHYSICS

where the symbol of the operator ΔG_1 is given by

$$\Delta g_1(x, p) = \frac{\hbar^2}{8} g_1(x,p)^{1/2} \{\ \}^2 g_1(x,p)^{1/2} + \cdots$$

Equation (6A.3) gives

$$\mathrm{Tr} G_{1/2}^2 = \int \left[g_1(x,p)^{1/2} \right]^2 \frac{dx\, dp}{h^n} = \mathrm{Tr} G_1$$

and equation (6B.7) shows that G_1 is a positive operator to which the operator ΔG_1 is added. This gives

$$\mathrm{Tr}|G_1| - \mathrm{Tr} G_1 \leq \mathrm{Tr}|\Delta G_1|.$$

One must then estimate the trace norm of ΔG_1. Its L_1 norm will be considered first.

The Structure of Some L_1 Norms

By definition

$$\|\Delta G_1\|_{L_1} = \int \frac{\hbar^2}{8} |g_1(x,p)^{1/2} \{\ \}^2 g_1(x,p)^{1/2}| \frac{dx\, dp}{h^n}. \tag{6B.8}$$

Integrals such as

$$\int |\phi(x,p) \{\ \}^2 \phi(x,p)| \frac{dx\, dp}{h^n}, \tag{6B.9}$$

will be met several times and they are worth discussing in some detail. The Poisson bracket operator is local. If one performs a change of coordinates (x, p) that is both linear and canonical (a Heisenberg transformation), one can bring one of the axes of the new coordinates along the normal to Σ_1 at a given point (x, p). The formulation of microlocal analysis is unchanged under the change of variables. This shows therefore that the Poisson bracket differential operator only involves one derivative in each direction, although it is of second order. No second derivative will ever be taken twice in the same direction.

Applying this simple remark to the integral (6B.8), one sees that each differentiation of the function $g_1(x,p)^{1/2}$ gives rise to a number of the order of 1 when the derivative is taken along a direction parallel to the boundary Σ_1 of the margin and of the order of ε^{-1} along the normal direction. When the Poisson bracket differential operator is iterated, as in the integral (6B.8), one gets a dominant term in the integrand of the order of ε^{-2}. This gives

$$\|\Delta G_1\|_{L_1} = O\left[\left(\frac{\hbar}{\varepsilon}\right)^2 \mu(M)\right]. \tag{6B.10}$$

The Second Term G_2 and Its L_1 Norm

Let us now consider the term G_2 in G coming from the second term in the semiclassical expansion (6B.3) of $F - F^2$. Higher order terms will be neglected because they give corrections of a higher power in $\hbar$. Its L_1 norm is given by:

$$\|G_2\|_{L_1} = \frac{\hbar^2}{8}\int\left|\sum_{jk}\left(2\frac{\partial^2 f}{\partial x_j \partial x_k}\frac{\partial^2 f}{\partial p_j \partial p_k} - 2\frac{\partial^2 f}{\partial x_j \partial p_k}\frac{\partial^2 f}{\partial p_j \partial x_k}\right)\right|\frac{dx\,dp}{h^n}.$$

It has the form (6B.9) that has just been discussed and the same approach as for equation (6B.10) now gives

$$\|G_2\|_{L_1} = O\left[\left(\frac{\hbar}{\varepsilon}\right)\mu(M)\right]. \tag{6B.11}$$

Bernstein Norms

We must now find estimates for the Bernstein norms of ΔG_1 and G_2. The derivatives occurring in equation (6B.5) are no longer restricted to act along canonically conjugate directions and they can be taken over all possible directions. All of them can in particular be taken along the normal direction, which gives for each of them a factor ε^{-1}. Except for that, the calculations are practically the same as before and one gets

$$\|\Delta G_1\|_B = \|\Delta G_1\|_{L_1} \cdot O\left[\left(\frac{\hbar}{\varepsilon^2}\right)^{n+1}\right] \tag{6B.12}$$

$$\|G_2\|_B = \|G_2\|_{L_1} \cdot O\left[\left(\frac{\hbar}{\varepsilon^2}\right)^{n+1}\right] \tag{6B.13}$$

Fixing the Classicality Parameter

We can now find the best choice for the classicality parameter ε. The L_1 norms get smaller with smaller values of ε. If however ε is too small, the Bernstein norms will become larger than the L_1 norm. A good compromise consists in taking their ratio to be of the order of 1, which can be obtained in view of the estimates (6B.12), (6B.13) by taking

$$\varepsilon = \left(\frac{\hbar}{LP}\right)^{1/2} = \hbar^{1/2}.$$

The last expression in this equation is a reminder of the use of dimensionless coordinates where Planck's constant is a pure number.

End of the Proof

The cell must have a large volume in terms of Planck's constant, which implies that the dimensionless value of $\hbar$ is a small number. One should also be able to insert the surface Σ_2 inside C, so that the minimum distance between two opposite points on Σ must be at least of the order of ε. In order to study the properties of the quasi-projector, let $A(\Sigma)$ denote the area of the boundary Σ (as measured in the given metric and therefore dimensionless) and let $V(C)$ be the dimensionless value of the volume. Let θ denote the ratio $A(\Sigma)/V(C)$. The number θ is of the order of 1 when the shape of the cell C is very regular (the curvature radii of its boundary being of the order of 1 when S has no edges) but it can also become very large when the shape of Σ is very distorted. The previous estimates now give

$$\frac{\mathrm{Tr}|F - F^2|}{\mathrm{Tr}\, F} = \frac{\mu(M)}{\mu(C)} = \varepsilon\theta,$$

which is the main result in Theorem A. The theorem has been given in the text by choosing the best values for the scales (L, P) so as to make the errors as small as possible.

Finally, it should be noticed that the definition of the symbol $f(x, p)$ is not unique since there is at least an arbitrariness in the precise choice of the function $f(x, p)$ and of the margin where it decreases from 1 to 0. It is easy to compare two different quasi-projectors F and F' and to give a bound for their difference, which gives

$$\frac{\mathrm{Tr}|F - F'|}{\mathrm{Tr}\, F} \leq (\varepsilon + \varepsilon')\theta.$$

The proof of Theorem B contains nothing new compared with the proof of Theorem A, except for the case of two disjoint cells that is considered in Problem 19.

2. Dynamics: Theorem C

We start with a regular cell C_0 and a quasi-projector F_0 associated with it as before. Consider the time-dependent operator

$$F(t) = U^{-1}(t)F_0 U(t), \qquad (6\mathrm{B}.14)$$

which evolves in time like a density operator. A time evolution similar to that of a Heisenberg operator would give similar results. The problem consists in showing that the operator given by equation (6B.14) is essentially a quasi-projector for the cell C_t that is obtained from C_0 by classical motion. This result assumes of course that classical motion is itself regular, with criteria to be made precise.

The Time Evolution of Symbols

The expression (6B.14) for $F(t)$ can be replaced by the differential equation

$$\frac{dF}{dt} = \frac{-i}{\hbar}[H,F]. \tag{6B.15}$$

The (collective) hamiltonian will be assumed to be a pseudo-differential operator with symbol $h(x, p)$. It will be assumed that this symbol belongs to the class S^2, because the kinetic energy is generally a quadratic function of momentum. This also implies that the potential energy does not increase faster than a constant times x^2 at infinity.

Equation (6B.15) can be replaced by an equation for the symbol $f(x, p, t)$ associated with $F(t)$. Using Theorem A.2, one gets a differential equation in time involving an asymptotic expansion in powers of $\hbar$:

$$\frac{\partial f}{\partial t} = -\{h, f\} - \frac{\hbar^2}{24} h(x, p)\{\ \}^3 f(x, p, t) + \cdots \tag{6B.16}$$

Liouville Approximation

Keeping only the leading term in Planck's constant in equation (6B.16), we get the Liouville equation

$$\frac{\partial f}{\partial t} = -\{h, f\}, \tag{6B.17}$$

the solution of which is easy: Let g_t denote the internal mapping of phase space generated by classical motion. The solution of equation (6B.17) is then given by

$$f(x, p, t) = f_0(g_t^{-1}(x, p)), \tag{6B.18}$$

where $f_0(x, p)$ is the symbol of F_0. It is easy to prove the following. Denote by $(x(t), p(t))$ the coordinates of a point undergoing classical motion. They must satisfy the Hamilton equations

$$\frac{dx_k}{dt} = \frac{\partial h}{\partial p_k}, \quad \frac{dp_k}{dt} = -\frac{\partial h}{\partial x_k},$$

and equation (6B.18) is equivalent to $f(x_0, p_0, t) = f_0(x(t), p(t))$. Taking the derivative of the right-hand side of this equation with respect to time and replacing the derivatives of $x(t)$ and $p(t)$ with the help of Hamilton's equations, one finds immediately that this function satisfies the Liouville equation (6B.17).

Geometric Properties

The function $f(x, p, t)$ has some of the properties of the symbol of a quasi-projector associated with the cell transformed from the initial cell by classical motion. Recall that, by construction, $f_0(x, p)$ is infinitely differentiable and differs from the constant values 1 or 0 only in the margin M_0 of C_0 having a width ε. Classical motion transforms this margin into another margin M_t for the transformed cell C_t.

One can ask whether the main topological and metric properties of the initial cell are conserved. This will be true if the jacobian matrix for the transformation from (x_0, p_0) to $(x(t'), p(t'))$ has uniformly bounded matrix elements when (x_0, p_0) is in C_0 and $0 \leq t' \leq t$. These matrix elements are the partial derivatives of the $2n$ functions $(x(t), p(t))$ with respect to the $2n$ variables (x_0, p_0). These conditions are always satisfied under the conditions stated in the main text of the chapter. In this case, the cell C_t is connected and simply connected. The transform of the inner boundary Σ_1 of the margin M_0 is a C^∞ manifold lying inside C_t, the transform of the exterior boundary lying outside C_t; none of them cuts itself nor do they cut each other. The margin M_t is bounded by them and it contains the boundary of C_t. Its width $\varepsilon(x_S, p_S)$ is however not a constant, i.e., it depends upon the reference point (x_S, p_S) on the boundary of C_t. It remains however always of the order of the initial parameter ε_o as long as the partial derivatives of $(x(t), p(t))$ with respect to (x_0, p_0) remain comparable with 1.

It is convenient to introduce the notation $\langle \varepsilon^\alpha \rangle$ for the average of a power $\varepsilon(x_S, p_S)^\alpha$ over the boundary Σ_t of C_t:

$$\langle \varepsilon^\alpha \rangle = \frac{\int_{\Sigma_t} \varepsilon(x_S, p_S)^\alpha dS}{\int_{\Sigma_t} dS}.$$

Note that this quantity can be computed by using only data coming from classical calculations. Finally, we shall assume without a proof that the function $f(x, p, t)$ is a symbol. The proof of this statement appears to be difficult and this is probably the place where the present analysis differs most from a rigorous mathematical treatment.

A Perturbation Method

The symbol $f(x, p, t)$ of the operator $F(t)$ will now be expanded in powers of Planck's constant according to

$$f(x, p, t) = f_1(x, p, t) + f_2(x, p, t) + \cdots$$

The various terms are assumed to satisfy the equations one obtains from the evolution equation (6B.16) by identifying the coefficients of the same power in Planck's constant. The first two terms are the following:

$$\frac{\partial}{\partial t} f_1(x,p,t) = -\{h(x,p), f_1(x,p,t)\},$$

$$\frac{\partial}{\partial t} f_2(x,p,t) = -\{h(x,p), f_2(x,p,t)\} - \frac{\hbar^2}{24} h(x,p)\{\ \}^3 f_1(x,p,t).$$

The corresponding initial conditions are given by

$$f_1(x,p,0) = f_0(x,p), \qquad f_2(x,p,0) = 0. \tag{6B.19}$$

The function f_1 has already been found to be:

$$f_1(x,p,t) = f_0(g_t^{-1}(x,p)) \tag{6B.20}$$

and, assuming that this is a symbol, the associated pseudo-differential operator will be denoted by $F_1(t)$.

Second-Order Approximation

Now consider the function $f_2(x,p,t)$. It is assumed to be a symbol and the associated pseudo-differential operator will be denoted by F_2. We will now try to find the properties of this operator. Let us first change variables (x,p) to new variables (x',p') such that $(x,p) = g_t(x',p')$. We write accordingly $f_2'(x',p',t) = f_2(x,p,t)$. Because of the relation between classical motion and Liouville's equation, the second equation (6B.19) giving the time evolution of f_2 becomes, with these new variables

$$\frac{\partial}{\partial t} f_2'(x',p',t) = -\frac{\hbar^2}{24} h(x',p')\{\ \}^3 f_0(x',p'). \tag{6B.21}$$

When writing this equation, we took into account the explicit form (6B.20) of $f_1(x,p,t)$ in terms of $f_0(x',p')$, as well as the conservation of energy insuring that $h(x,p)$ is equal to $h(x',p')$. However, we did not change the expression for the Poisson bracket operator, which remains given by its usual form in terms of the variables (x,p). To get a completely self-contained form of equation (6B.21), this operator must be reexpressed in terms of the variables x' and p'.

The change of variables in this differential operator is rather cumbersome and we shall only describe it in general terms. Let D_1 denote any matrix element of the jacobian for the transformation of (x,p) into (x',p'), i.e., any partial derivative of some function among the set (x',p') with respect to one of the $2n$ variables (x,p). The notation D_2 similarly represents an arbitrary second order derivative and D_3 a third order derivative. Conforming to this

RECOVERING CLASSICAL PHYSICS

convention, any first order derivative of the function f_0 will be denoted by f_0' and so on for higher derivatives.

With these conventions, the triple Poisson bracket occurring in equation (6B.21), when expressed completely in terms of the variables (x', p'), can be written as:

$$h\{\}^3 f_0 + D_2 D_1 D_1 h'' f_0''' + D_3 D_1 D_1 D_1 h' f_0''' + D_2 D_1 D_1 h''' f_0''$$
$$+ D_2 D_2 h'' f_0'' + D_2 D_1 D_1 h'' f_0'' + D_3 D_1 D_1 D_1 h''' f_0' \quad (6B.22)$$

The fact that the Poisson bracket reappears in the terms containing third derivatives of the hamiltonian and of f_0 is due to the invariance of this operator under a linear canonical transformation. The following terms have a complicated tensor character and this somewhat rough expression will be enough to discuss the main features of the equation giving f_2.

Let us write equation (6B.21) for f_2 in the abbreviated form

$$\frac{\partial f_2}{\partial t} = -\frac{\hbar^2}{24} A_3(h, f_0, t),$$

where $A_3(h, f_0)$ denotes the full quantity (6B.21) and the dependence upon time, as it enters through the value of the partial derivatives of (x', p') with respect to (x, p), has been made explicit. This gives immediately

$$f_2(t) = -\frac{\hbar^2}{24} \int_0^t A_3(h, f_0, t') dt'. \quad (6B.23)$$

Operator Estimates

One can use equation (6B.23) to obtain useful information about the operator F_2. One can estimate its L_1 norm and its Bernstein norm, as was done when proving Theorem A. The quantities to be integrated at a time t' are nonzero only inside the margin $M_{t'}$ because they depend only upon the derivatives of f_0 and not upon f_0 itself. Because of the rapid variation of f_0 near the boundary of C_0, its derivatives of highest order will dominate, namely the first three terms in the sum (6B.22) involving derivatives of third order. These leading terms are of order

$$\hbar^2 \mu(M) \langle \varepsilon^{-3} \rangle \Delta_3 ht,$$

where $\Delta_3 h$ represents a derivative of order 2 or 3 of the Hamilton function, which may be multiplied by partial derivatives of (x', p') with respect to (x, p). The estimate of Bernstein norms gives analogous results that are similar in form to the last expression, except for the occurrence of higher order derivatives. When this is compared with Tr F, the corrections are found to be of the order of $\hbar t$ and they remain small as long as the time is not too long.

It becomes more and more difficult to compute the next terms in the expansion of $f(x, p, t)$ in powers of $\hbar$. Having found that second order corrections can be neglected in trace norm under some conditions which remain to be given more precisely, we will simply assume that the next terms are less important.

The Properties of $F(t)$

The properties of the operator $F(t)$ are obtained by methods analogous to the ones already used in the proof of Theorem A, so that their description can be made shorter. The trace of $F(t)$ is computed by using equation (6A.3), which gives

$$\text{Tr } F(t) = \mu(C_t)h^{-n} = \mu(C_0)h^{-n},$$

to leading order in $\hbar$, the equality between the last two terms in this equation coming from Liouville's theorem.

One obtains similarly the leading term of

$$\text{Tr}|F(t) - F^2(t)| \approx \mu(M_t) = \mu(M_0).$$

The leading terms for the corrections are given by quantities similar to the ones met in equations (6B.10) and (6B.11), the only significant difference being the nonconstant width of the margin, which amounts to the replacement of a factor ε^{-2} by its average. One thus obtains

$$||\text{Tr}|F(t) - F^2(t)| - \mu(M_0)|| = O[\hbar^2 \langle \varepsilon^{-2} \rangle \mu(M_0)].$$

One can also compare the quasi-projector $F(t)$ with a quasi-projector F_t associated optimally with the cell C_t, using the construction given in the proof of Theorem A. One easily finds that the L_1 norm of their difference is of the order of $O(\mu(M_0)h^{-n})$, but that the Bernstein norm is of the order of $O(\hbar^{n+3} \langle \varepsilon^{-(n+3)} \rangle) \mu(M_0)$.

Conclusions

We see that, whatever the Hamilton function $h(x, p)$ and the regular cell C_0, there is always a time interval $0 \leq t \leq T$ during which the operator $F(t)$ generated by quantum evolution from a quasi-projector associated with C_0 behaves like a quasi-projector for the cell C_t generated by classical motion. Explicit bounds following from the above analysis are given in the text. They follow from the present estimates by a simple rewriting.

APPENDIX C: CONSISTENCY OF CLASSICAL LOGIC

The field of propositions to be used in an application of classical logic will refer to a sequence of times

$$0 \leq t_1 \leq t_2 \leq \cdots \leq t_r = T. \tag{6C.1}$$

We also consider a collection of cells in phase space $C_1, C_2, \ldots, C_r$, all of them being deduced from each other by classical motion at the corresponding times. In other words, one has $C_j = g_{t_j - t_k}(C_k)$, for any pair of times belonging to the sequence (6C.1). One can also construct a cell C_0 at time 0 by $C_0 = g_{-t_k}(C_k)$. Some optimal quasi-projectors associated with these cells will be denoted by $F_0, F_1, \ldots$. The dynamical classicality parameter of this sequence is the parameter ζ occurring in Theorem C, corresponding to a dynamics that is regular to order ζ for the cell C_0 during the time T.

We assume that the classical property $[C_0, 0]_c$ is valid to order ζ for the initial state, i.e.,

$$p_0 = \text{Tr}(\rho F_0) \geq 1 - c\zeta.$$

Equivalently,

$$p_k = \text{Tr}(\rho U^{-1}(t_k) F_k U(t_k)) \geq 1 - c\zeta,$$

for any time in the sequence.

In order to build up a field of propositions for a special application of classical logic, we must also introduce the negation of these classical properties. They can be used to construct many different histories and one can also produce more general propositions using the logical operation "or". These propositions are very similar to common sense reasoning, when many different possibilities are considered and compared.

It is easy to compute the probability for a history involving an arbitrary sequence of properties $[C_j, t_j]_c$ at a few times included in the sequence (6C.1). It will be shown that these probabilities are equal to 1 up to corrections of order ζ. On the contrary, when one or several negations $[\overline{C}_j, t_j]_c$ are introduced, the probability will turn out to be at most of the order of ζ. One can then compute the traces occurring in the consistency conditions to find that they are of the order of ζ, which means that classical logic is consistent to order ζ. Finally, one can compute some conditional probabilities in order to establish implications. For any couple of times (t_j, t_k), it will be found that

$$[C_j, t_j]_c \Rightarrow [C_k, t_k]_c,$$

these implications being valid to order ζ. This is the basis for determinism. Let us see now how this program can be fulfilled.

Explicit Calculation of Some Probabilities

Denote by **h** a Griffiths history according to which the system has the classical properties expressed by the cell C_1 at time t_1 and by the cell C_2 at time t_2. Its probability is given by

$$p(\mathbf{h}) = \text{Tr}(F_1(t_1)\rho\, F_1(t_1)F_2(t_2)).$$

Making time dependence explicit, this can be written as

$$p(\mathbf{h}) = \text{Tr}\{U^{-1}(t_1)F_1 U(t_1)\rho\, U^{-1}(t_1)F_1 U(t_1)U^{-1}(t_2)F_2 U(t_2)\}. \quad (6\text{C}.2)$$

Theorem C gives

$$U^{-1}(t_1)F_1 U(t_1) = F_0 + \Delta E, \quad (6\text{C}.3)$$

where

$$\text{Tr}|\Delta E|/N(C) \le c\zeta \quad (6\text{C}.4)$$

and $N(C)$ is the number of quasi-classical states corresponding to the volume of the cell.

Inserting the first term in the right-hand side of equation (6C.3) in equation (6C.2) giving $p(\mathbf{h})$, one gets

$$p(\mathbf{h}) = \text{Tr}\{F_0 \rho F_0 U^{-1}(t_2)F_2 U(t_2)\}. \quad (6\text{C}.5)$$

Using the initial property

$$F_0 \rho F_0 = \rho + \Delta\rho,$$

where

$$\text{Tr}|\Delta\rho| \le c\zeta, \quad (6\text{C}.6)$$

the leading term in the probability is

$$p(\mathbf{h}) = \text{Tr}\{\rho U^{-1}(t_2)F_2 U(t_2)\} = p_2, \quad (6\text{C}.7)$$

where p_2 is the probability for the property $[C_2, t_2]$. Note that equation (6C.7), which will be shown to be valid up to order ζ, is enough to prove the implication (to order ζ)

$$[C_2, t_2] \Rightarrow [C_1, t_1]$$

if the consistency conditions are satisfied.

We now evaluate the errors. The error in $p(\mathbf{h})$ is given by

$$\Delta p(h) = \text{Tr}\{\Delta F \rho F_0 F_2(t_2)\} + \text{Tr}\{F_0 \rho \Delta F F_2(t_2)\}$$
$$+ \text{Tr}\{F_0 \Delta\rho F_0 F_2(t_2)\} + \cdots,$$

where only first-order corrections have been written down explicitly. Using the master inequality $\text{Tr}(AB) \le \|A\|\text{Tr}|B|$, and the bounds on the trace norms that are given by equations (6C.4) and (6C.6) together with the bounds

$$\|\rho\| \le 1, \quad \|F_0\| \le 1, \quad \text{and} \quad \|F_2(t_2)\| \le 1,$$

one gets

$$\frac{\Delta p(h)}{p(h)} = \Delta p(h) = O(\zeta).$$

Consistency Conditions

We shall only consider explicitly the case of two times when there is only one consistency condition:

$$\gamma = \text{Tr}\{[F_1(t_1),[\rho,\overline{F}_1(t_1)]]F_2(t_2)\} = \text{Re } \kappa = 0, \qquad (6C.8)$$

with the notation $\kappa = \text{Tr}\{F_1(t_1)\rho\overline{F}_1(t_1)F_2(t_2)\}$, where the quasi-projector $\overline{F}_1(t_1)$ is associated with the domain complementary to C_1. Theorem B shows that it can be identified with $I - F_1(t_1)$ up to a correction of order ζ in trace norm.

The calculation of the trace κ is very similar to what has been done for the probabilities. First obtain a variant of equation (6C.5) given by

$$\kappa = \text{Tr}\{F_0\rho\overline{F}_0 U^{-1}(t_2)F_2 U(t_2)\} = O(\zeta), \qquad (6C.9)$$

the last equality resulting from $\rho\overline{F}_0 = F_0\rho F_0\overline{F}_0$. This is also true up to corrections of order ζ, to be treated as the previous correction $\Delta\rho$. Furthermore $\text{Tr}|F_0\overline{F}_0| = O(\zeta)$ from Theorem B. The result stated in equation (6C.9) is then a consequence of the master inequality (6A.13). When going from the form (6C.8) to the form (6C.9) of κ, one neglects a term in ΔF, the contribution of which is bounded again by the master inequality. Finally, one gets $\kappa = O(\zeta)$. This is the result we were looking for, showing the consistency of classical logic.

APPENDIX D: A CRITERION FOR THE EXISTENCE OF COLLECTIVE OBSERVABLES

The existence of collective observables is so important for the validity of classical physics that we shall indicate here as a conjecture a criterion allowing splitting a complete set of commuting observables into collective coordinates, microscopic (necessarily quantum) coordinates, and perhaps intermediate ones. The approach is unfortunately not constructive but at least it shows that collective coordinates are state-dependent (or more precisely, dependent upon the position in the full microlocal phase space for all particles) and that one can recognize them if one is lucky or clever enough to guess what they are. This criterion has been successfully checked in a few special cases but its proof looks nontrivial.

Start from the symbol of the full hamiltonian $h(q, p)$ for all the articles in the system. This is therefore a microlocal analysis formulation. Arbitrary canonical transformations (or at least "smooth" ones[18]) can be made (this is why the method is not constructive) and one wants to know whether the observable Q_1 associated with the first position variable q_1 is a collective observable. This is supposed to be local so that one asks the question for every fixed value of the other coordinates $\{q_j, p_j\}$, $j \neq 1$, and in some neighborhood of the (q_1, p_1) phase space. It may be noticed that a permutation of variables is a special and trivial canonical transformation so that all the observables can be checked for their collective character separately.

Consider the curve γ in the (q_1, p_1) two-dimensional phase space defined by the equation $h(q, p) = \text{constant} = K$, the variables $\{q_j, p_j\}$, $j \neq 1$, being fixed, and the constant being chosen so that γ passes through the point to be tested. Consider the region defined by $h(q, p) \leq K$ and, if it is multiply connected, retain only the connected part containing the point (q_1, p_1) to be tested on its boundary. Assume that it is a cell C (pathological or exceptional geometrical cases may arise but they are not considered). The conjectured criterion assumes that Q_1 is a collective observable when C is a regular cell. This holds only for quantum states for which the Wigner function (symbol of the density operator) is localized in a region of the full phase space where the collective character of Q_1 is valid. If the cell C has a volume of the order of Planck's constant, Q_1 is definitely a quantum observable; this is what occurs, for instance, for the coordinates of an electron in an atomic inner shell. Other possibilities might occur and they certainly occur in complicated cases (like in the case of glass?); they indicate again how delicate and subtle the problem of collective observables is, but also that there is no fundamental objection to their frequent occurrence.

This conjecture also suggests that the hierarchy of collective observables might be controlled by the (average) effective classicality parameter of the cell C.

PROBLEMS

1. Let N be a large positive number and ε a small positive number. Consider a self-adjoint operator F having a discrete spectrum, its eigenvalues being contained in the interval $[-c\varepsilon^2, 1 + c\varepsilon^2]$, the constant c being is of the order of 1. Assume that the trace of F is equal to N, the trace of $|F|$ is equal to $N + O(\varepsilon)$ and the trace of $|F - F^2|$ is of the order of $N\varepsilon$. Let N_0 be the integer nearest to N and let E be the projector having for its eigenvectors with eigen-

value 1 the N_0 first eigenvectors of F when they are ordered according to decreasing eigenvalues. Show that the trace of $|E - F|$ is of the order of $N\varepsilon$.

2. When a pseudo-differential operator Ω associated with the symbol $\omega(x, p)$ acts upon a wave function $\psi(x)$ to give another wave function $\phi(x)$, show that

$$\phi(x) = \int \omega\left(\frac{x+y}{2}, p\right) e^{ip\cdot(x-y)/\hbar} \psi(y) \frac{dy\,dp}{h^n}.$$

3. Let Ω be a pseudo-differential operator belonging to the trace class in the Hilbert space $L^2(\mathbf{R})$ and let $\omega(x, p)$ be its symbol. Using a basis $\{\phi_n(x)\}$ of eigenfunctions for a harmonic oscillator, show that

$$\sum_{n=0}^{\infty} <\phi_n|\Omega|\phi_n> = \int \omega(x, p) \frac{dx\,dp}{h}.$$

4. Show the following correspondence between an operator and its symbol: X_j is associated with x_j, P_j with p_j, the identity operator with the function identically equal to 1, and the operator P^2 with p^2 (for the last question, one can use Theorem A.2).

5. Let Ω be a pseudo-differential operator with symbol $\omega(x, p)$. Show that the adjoint of Ω has for its symbol the complex conjugate of $\omega(x, p)$.

6. Make the following conventions about Fourier transforms:

$$\hat{u}(p) = \int e^{-ip\cdot x/\hbar} u(x)\,dx, \qquad u(p) = \int e^{ip\cdot x/\hbar} \hat{u}(x) \frac{dp}{h^n}.$$

Let A and B be two pseudo-differential operators with symbols $a(x, p)$ and $b(x, p)$. Their product $C = AB$ has the symbol $c(x, p)$.

i. Prove the formula

$$c(x, p) = \int a(x + x', p + p') b(x + x'', p + p'')$$
$$\exp[2i\sigma(x'', p''; x', p')] \frac{dx'dp'dx''dp''}{(h/2)^{2n}},$$

where $\sigma(x'', p''; x', p') = p'' \cdot x' - p' \cdot x''$.

ii. Define

$$a_1(p', x') = \int a(x, p) \exp[i(p\cdot x' - p'\cdot x)/\hbar] \frac{dx\,dp}{h^n}.$$

Show that

$$c(x, p) = \int \frac{dx'dp'}{h^n} \frac{dx''dp''}{h^n} dx'''dp''' \delta(p'' + p''' - p')$$
$$\times \delta(x'' + x''' - x') \cdot e^{i(p'\cdot x - p\cdot x')/\hbar} a_1(p'', x'') b_1(p''', x''')$$

iii. Expand the last exponential in this integral and derive equation (6A.6) from it. Compute the first three terms in the series.

iv. Show that, to first order in Planck's constant, the symbol of the commutator $C = [A, B]$ is the Poisson bracket $(\hbar/i)\{a, b\}$.

7. Using Theorem (A.2) for the symbol of an operator product, prove the well-known property $\text{Tr}([A, B]) = 0$ by using an integration by parts.

8. Obtain the canonical commutation rules by using Theorem (A.2) for the product of two operators.

9. Consider a physical object having two degrees of freedom, its energy being purely kinetic and given by
$$h(x, p) = g_{11}(x)p_1^2 + 2g_{12}(x)p_1 p_2 + g_{22}(x)p_2^2.$$
Give an explicit expression for the action of its hamiltonian upon a wave function.

10. Consider two symbols $f_1(x, p)$ and $f_2(x, p)$ adding exactly to 1, together with the associated operators F_1 and F_2. Given a pseudo-differential operator A, define the operators $A_1 = F_1 A$ and $A_2 = F_2 A$. Show that A is equal to their sum.

11. Prove that the trace of the square of a pseudo-differential operator is given by
$$\text{Tr} A^2 = \int a^2(x, p) \frac{dx\,dp}{h^n}.$$
(Use Theorem (A.2) and integrate by parts.)

12. The "eigenstates" of momentum are normalized by $\langle p|p'\rangle = h^n \delta(p - p')$. Show that
$$\omega(x, p) = \int \langle p''|\Omega|p'\rangle e^{i(p''-p')\cdot x/\hbar} \delta\left(p - \frac{p' + p''}{2}\right) \frac{dp'\,dp''}{h^n}.$$

13. Let S be an n-dimensional surface in an $(n+1)$-dimensional euclidean space. The coordinates $(x_1, \ldots, x_{n+1})$ of a point in S are given as functions of n parameters $q^1, \ldots, q^n$ by $x_j = f_j(q^1, \ldots, q^n)$.

i. Show that the square of the distance between two infinitesimally nearby points in S is given by[19]
$$ds^2 = \sum_{jk} g_{jk}(q)\,dq^j\,dq^k$$
and express the coefficients g_{jk} in terms of the partial derivatives of the functions f_j. Show that the surface element in S is $g^{1/2}\,dq^1 \ldots dq^n$, where g is the determinant of the matrix with coefficients g_{jk}.

ii. The Hilbert space $L^2(S)$ is defined by the scalar product
$$\langle \Psi_1|\Psi_2\rangle = \int \Psi_1^*(q)\Psi_2(q) g^{1/2}\,dq.$$

RECOVERING CLASSICAL PHYSICS

Position "eigenstates" on S are defined with the convention $\langle q|q'\rangle = g^{-1/2}\delta(q-q')$. Show that

$$I = \int |q\rangle\langle q| g^{1/2}\, dq.$$

iii. Show that the symbol of the laplacian on S is

$$\sum_{jk} g^{jk}(q) p_j p_k,$$

where the matrix with elements g^{jk} is the inverse of the matrix with coefficients g_{jk}.

14. This problem and the next introduce another construction of quasi-projectors, which is due to I. Daubechies. Let $|g_{qp}\rangle$ denote the gaussian wave function

$$g_{qp}(x) = K^{-1}\exp(-(x-q)^2/4\Delta x^2 + ip\cdot x),$$

where one uses a system of units where $\hbar = 1$. Show that $\langle X\rangle = q$ and $\langle P\rangle = p$.

i. Compute the scalar product $\langle g_{qp}|g_{q'p'}\rangle$. Show that it tends towards 0 like a gaussian in $q-q'$ when $q-q'$ is large compared with the uncertainty Δx and $p-p'$ is large compared with Δp.

ii. Show that

$$\int_{\mathbf{R}^{2n}} \langle g_{qp}|g_{q'p'}\rangle (2\pi)^{-n}\, dq\, dp = 1.$$

Define a quasi-projector associated with a regular cell C by

$$F = \int_C |g_{qp}\rangle\langle g_{qp}| (2\pi)^{-n}\, dq\, dp.$$

iii. Show that F is self-adjoint and nonnegative.
iv. Compute the trace of F and show that it is equal to $N(C)$.
v. Compute the Hilbert-Schmidt norm of F. Show that the spectrum of F is discrete and its eigenvalues are nonnegative.
vi. Show that F becomes the identity operator when C is the whole phase space. Deduce from it an integral form for $I - F$ and show that all the eigenvalues of F are less than or equal to 1.
vii. Show that a gaussian state well inside (or outside) C is an approximate eigenstate of F with the eigenvalue 1 (or 0).

15. Consider the case $n = 1$ in the previous problem. The cell C is given by $x^2 + p^2 \leq R^2$, R being a large number. Show that F commutes with the operator $X^2 + P^2$. Deduce from it that the eigenfunctions of F are Hermite functions. Compute the nth eigenvalue of F. Show that it is given by the

incomplete gamma function

$$\lambda_n = (n!)^{-1} \int_0^{R^2/2} t^n \exp(-t)\, dt.$$

Deduce from it the behavior of λ_n as a function of n.

16. This problem shows that a gaussian state remains gaussian when the potential is a second-order polynomial.

Consider a quantum system with n degrees of freedom. Its hamiltonian is $P^2/2m + V(X)$, where $V(X)$ is a polynomial at most of second degree in the components of X. One denotes by V' the gradient of V and by V'' the matrix of its second-order derivatives. Let $(x(t), p(t))$ be classical solutions of the classical equations of motion and $S(t)$ the corresponding classical action between the times 0 and t. Consider matrices $A(t)$ and $B(t)$ having the initial values A_0 and B_0 satisfying the differential equations

$$m\, dA/dt = (i/2)B, \quad dB/dt = 2iV''A.$$

An initial gaussian state is given by

$$\Psi(x) = (2\pi)^{-n/4}(\det A_0)^{-1/2} \times \exp\left\{-(x - q_0|A_0^{-1}|x - q_0)/4 + ip_0 \cdot (x - q_0)\right\},$$

where $(x|A|x)$ is the quadratic form with argument x and matrix A.

Show that the state satisfying the Schrödinger equation and having this initial value is given by

$$\Psi(x, t) = (2\pi)^{-n/4}(\det A(t))^{-1/2} \exp\left\{-(x - q(t))|B(t)A(t)^{-1}|x - q(t))/4 + ip(t) \cdot (x - q(t)) + iS(t)\right\},$$

where $(q(t), p(t))$ is obtained from (q_0, p_0) by classical motion and $(A(t), B(t))$ from (A_0, I) by the above differential equations. Some indications concerning the solution of this problem can be found in the work by G. Hagedorn, Comm. Math. Phys. **77**, 1 (1980).

It is worth mentioning that similar results can be obtained for a particle in a Coulomb potential.[20]

17. Using the results of Problems 14 and 16, discuss the dynamical evolution of a quasi-projector for a particle in the case where it is free, where it is submitted to a constant force, or where it is harmonically bounded.

18. Consider a rectangular box in phase space and define a quasi-projector F as a product of two smooth functions in x and in p (if the x side of the box is the interval $[0, L]$, take a function of x that is equal to 1 within the interval and to 0 outside, except near the extremities where it goes smoothly from 1 to 0 inside an interval of length εL and do the same for p).

Show that the L_1 norm for the leading term of the operator $\Delta F = F - F^2$, if it were to be taken alone, would allow an error ε of the order of h/LP. What would it give in n dimensions?

RECOVERING CLASSICAL PHYSICS

By taking into account the Bernstein norm for the terms of all orders in ΔF, show that one is still led to take ε of the order of $(h/LP)^{1/2}$.

19. A symbol for a quasi-projector F of a rectangular one-dimensional cell C is defined as follows. Let L and P be the lengths of its two sides and define a parameter ε by equation (6.30) as usual. Define the two parameters $\sigma = \varepsilon L$ and $\tau = \varepsilon P$. Consider the gaussian weight

$$G(x, p) = A \exp\{-(x/\sigma)^2 - (p/\tau)^2\},$$

the coefficient A being chosen so that the integral of G with an element of volume $dx\,dp/h$ is 1. Denoting by χ the characteristic function of C, one defines the symbol of F as the convolution product $\chi * G$.

Consider another cell C' that is obtained from C by a translation in x, the two cells having an empty intersection and their shorter distance being denoted by D. Show that the symbol of the product FF' is everywhere bounded by a quantity proportional to $\exp(-D^2 P/L\hbar)$, decreasing very rapidly with the distance between the two cells.

Hint: When using the explicit form of a product given in the first question of Problem 6, all the integrations can be performed explicitly.

7

Decoherence

It was shown in the preceding chapter how to derive classical physics from quantum theory. The results covered the interactions between several classical objects but did not consider macroscopic properties that arise directly from a quantum interaction, as occurs after a quantum measurement. It was found in Chapter 2 that then there are in principle linear superpositions of macroscopically different states, which is a great difficulty for the interpretation of quantum mechanics and something obviously foreign to classical physics. It should also be stressed that the proof of the validity of classical physics given in Chapter 6 did not take friction into account. These two questions—how to understand the origin of friction in classical physics and how to disentangle a superposition of macroscopically different states—will be the main subject of the present chapter. They are taken together because it will be found that they are intimately related.

We must however go back beforehand to the notion of collective variables as it was previously introduced. We found it convenient to consider a macroscopic object (or a system made up of several such objects) as consisting formally of two dynamical systems, one of them being described by collective coordinates and the other one by microscopic ones. We shall call these two systems from now on the *collective system* and the *environment*, respectively. The second name might look misleading in that it suggests the idea of something external to the object whereas it is clear that the environment is more properly the internal part of the object, i.e., its matter. The name has however become conventional by now and we shall keep it as such. When the object to be considered actually has an external environment consisting, for instance, of the air or the light around it (one then speaks of an *open* system), this will also be included formally in the environment as meant before.

It was assumed up to now that the collective system and the environment are uncoupled. The total hamiltonian consisted only of two terms $H = H_c + H_e$, where the collective hamiltonian H_c depends only on the collective observables and the environment hamiltonian H_e only on the microscopic ones. The average value of H_c is the mechanical energy and the average value of H_e is the internal energy as they are used in thermodynamics. The absence of coupling between the two formal systems implies that there is no exchange between the two kinds of energy.

Thermal dissipation is due to the existence of a coupling between the two systems so that the real hamiltonian should be written as $H = H_c + H_e + H_{int}$, where the coupling hamiltonian H_{int} depends on both kinds of observables and can thereby produce an energy transfer between the two systems, thus allowing dissipation. In a few cases one knows explicitly what this coupling is though, in general, one can only guess what it looks like. It is pretty obvious that this coupling must be taken into account in order to describe the existence of friction effects in classical physics. Far less trivial is the fact that it has also a dramatic effect upon a superposition of two macroscopically different states, which is called *decoherence*. It destroys quantum interferences at a macroscopic level and one can roughly describe it as follows: even if it happens that the environment wave functions for two macroscopically different states are coherent at some time (i.e., even if their relative phase is well defined so as to allow for possible interferences), they will become very rapidly orthogonal because of their coupling with different values of the collective macroscopic observables.

Decoherence is a dynamical effect taking place in the bulk of matter. It is extraordinarily efficient, so much so that it was found extremely difficult to observe experimentally. It happens so quickly that one cannot catch it while it is acting and one can only accede most of the time to a situation where it has already occurred. The existence of the effect is quite consistent with observation since it is practically impossible to find any interference in macroscopic objects, except for some specially devised superconducting systems that were recently devised for that purpose. As a matter of fact, these exceptions confirm the rule since superconductors are precisely characterized by the absence of dissipation, i.e., by a very weak coupling H_{int}. Radiation offers another example of a macroscopic system showing interference (it is macroscopic when it consists of many photons), but this is again a case where there is no internal dissipation. Microscopic objects such as electrons, neutrons, and every kind of elementary particle, as well as atoms, can also show interference effects but they have no environment at all or they interact very weakly with it when interferences can be seen. Interference effects are also probably possible in principle for rather light molecules but not for complex ones because their own vibration degrees of freedom act practically as an internal environment would in a heavier object.

The study of the decoherence effect is far from trivial and it will be taken up in several steps. We will begin by trying to understand its origin, as far as can be done with only intuitive considerations. We then proceed to the study of more and more elaborate models in order to substantiate these ideas and to obtain quantitative estimates. The reason models play such a central role is that no general theory is yet available. This lack of a general setup is not due to a lack of ingenuity in the people who have investigated the problem but rather is a genuine difficulty. One is after all asking for a knowledge of the detailed phases in different parts of a wave function for a system involving a large number of particles and one wants to know how they change with time. This is a very difficult problem and one should marvel at the fact that it has been solved for rather realistic models instead of feeling frustrated for not having a complete theory.

Once the effect is essentially understood, we shall consider its relevance for the basic problem of measurement theory. A serious difficulty will then stand in the way: although macroscopic interference effects disappear when a collective observable is measured (so that Schrödinger's cat is necessarily dead *or* alive as ordinary probability calculus would assert as opposed to elementary quantum mechanics), a pure quantum state remains a pure state under time evolution so that, in principle, interference might still be found if one were to measure a more clever or finer observable. It will be shown however that the experimental devices that would have to be used to perform such a measurement would have to be tremendously big. They would have to be in most instances much larger than the whole universe. They would be too big for light to cross them in a short enough time while they are acting so that special relativity precludes their existence, or else so bulky and heavy that they would immediately collapse into a black hole under their own gravity. As a by-product of this analysis, it will be found that many conceivable observables cannot be measured, even as a matter of principle. The overall conclusion will be that decoherence provides a satisfactory answer for the oldest problem in quantum mechanics: macroscopic superpositions.

It might be noticed finally that not much will be said about the consequences of including friction in the foundations of classical logic as they were treated in the previous chapter. The existence of friction will be considered more or less as obvious and not developed extensively. This is because it does not add anything significant to what has been already understood. For instance, determinism cannot be used to reconstruct past motion once a system has come to rest under friction and, more generally, the uncertainties in reconstructing the past tend to increase as one tries to go farther back in time from a given situation. The author felt that substantiating such trivialities by numerical estimates was not worth the boredom it would generate. The important point is that records of past facts can exist in the framework of quantum mechanics and what is already known on the foundations of classical physics together with decoherence is enough to make it obvious.

ORIENTATION

We shall begin by noticing that a macroscopic object has, in some sense, a particularly large Hilbert space and a very crowded energy spectrum. These two closely associated peculiarities play an important role in the existence of decoherence. When N elementary physical systems are put together, each one having a Hilbert space with dimension n (which is momentarily assumed to be finite), the total Hilbert space has a dimension n^N. When N becomes very large so that the overall system is macroscopic, one can say, in a very rough way, that the Hilbert space becomes enormous. The energy levels become very close together, as may be seen from many different points of view. Take for instance as an elementary system an atom having only two states with respective energy 0 and W. The total energy must then lie in the interval $[0, NW]$ and there are 2^N different states so that the energy spectrum becomes extremely crowded, the average difference between neighboring energy levels decreasing exponentially with N. Consider now another example where each system is a quasi-classical particle having only a kinetic energy $p^2/2m$, all the particles being located in the same spatial volume V; the total momentum space is $3N$-dimensional and there are $(V^N d^{3N} p/h^3)$ quantum states per momentum volume $d^{3N} p$. The total number of these states having an energy smaller than a given value E is essentially the volume of a $3N$-dimensional sphere in momentum space and it increases like the $(3N/2)^{\text{th}}$ power of the energy E. In that case, the crowding of the energy spectrum becomes really tremendous. These very dense energy spectra are always expected in the case of a macroscopic system.

The closeness of energy levels has for a consequence an extreme sensitivity to perturbations resulting in very complicated wave functions. This is because the change in a wave function under any small perturbation depends, according to perturbation theory, upon the difference of the unperturbed energies occurring in a denominator. The effect of any sizable perturbation is therefore tremendously increased. As a result, two slightly different perturbations give rise to very different perturbed wave functions, which will usually be orthogonal in view of their large number of variables. This is what actually occurs in decoherence where the perturbations of the environment are due to its coupling to different values of the collective observables.

It will be convenient, for a first encounter with the effect, to consider the collective degrees of freedom of a macroscopic object as being absolutely classical. This means, for instance, that we consider a watch to be a collection of many atoms but we nevertheless agree to describe the clockwork by classical physics. If it is an old-fashioned mechanical watch, there must be a coupling between the motion of the wheels and the quantum state of the internal atoms so that energy is exchanged between the two of them. It results in friction and dissipation. When seen from the standpoint of the atoms or the matter constituting the watch, the motion of the clockwork is a perturbation and, because of the high density of the energy eigenstates, its overall wave function will be extremely sensitive to any slight change in the clockwork motion. When the watch starts from two states of the spring that are differently wound, even if only slightly, the wave functions of the bulk of atoms in the watch will become very different in the two cases. This does not mean of course that some drastic overall difference will become manifest at a large scale but, if one were able to compare the two wave functions, they would take quite different values as functions of the coordinates of the nuclei and electrons. One may wonder in these conditions if the wave functions for the two states of the watch with different initial winds of the spring will be able to interfere. The point is that interference demands a well-defined phase relation between the interfering wave functions and this kind of coherence, even if initially realized, will certainly soon be lost. One may therefore expect that the coupling between the collective motion and the atomic bowels of the object will tend to destroy the possibility of interference between two macroscopically different states.

This is the essence of the decoherence effect. It is a tendency of the environment wave functions to lose all remnants of an initial phase correlation when they are coupled with two different collective states of motion, even when the difference is quite small. This loss of phase correlation (in the high-dimensional internal configuration space) should have as a result that the environment wave functions will become rapidly orthogonal. In a very rough way, one might also say that the Hilbert space is so huge that two state vectors in it have a very small chance of not being orthogonal. Enormous Hilbert spaces, exponentially growing energy level densities, extreme sensitivity to perturbations, spontaneous tendency for two perturbed state vectors to become orthogonal under time evolution—all these features are strongly linked together and one cannot attribute the physical result to one of them rather than another. They are the common ground in which decoherence is rooted.

These considerations are strongly linked to the central problem of measurement theory. As was shown in Chapter 2, this has to do with the quantum superposition of macroscopically different states. Consider for instance a two-component state

$$|\Phi\rangle = a|\Phi_1\rangle + b|\Phi_2\rangle,$$

occurring after the measurement of a two-valued observable. In this expression, Φ_1 (respectively Φ_2) represents a state of a counter showing a datum 1 (respectively a datum 2). The quantities $|a|^2$ and $|b|^2$ are the corresponding probabilities. The density operator $\rho = |\Phi\rangle\langle\Phi|$ has a diagonal part

$$\rho_d = |a|^2 |\Phi_1\rangle\langle\Phi_1| + |b|^2 |\Phi_2\rangle\langle\Phi_2|.$$

It was already noted in Chapter 2 that this would be the exact expression of the density operator if the two states of the apparatus were described by conventional probability calculus as two distinct events. On the other hand, the nondiagonal part

$$\rho_{nd} = a^*b|\Phi_1\rangle\langle\Phi_2| + ab^*|\Phi_2\rangle\langle\Phi_1|$$

DECOHERENCE

represents the possibility of quantum interference between the two macroscopically different states.

One might expect that the decoherence effect will make the nondiagonal part vanish, suppressing the possibility of interference. This cannot however be rigorously true, since a quantum evolution under a Schrödinger equation cannot transform a pure state into a mixture. One can therefore only expect that the interference effects will vanish when one observes a collective observable without probing in detail the microscopic details of the wave function. This is after all what always happens in practice. The problem to be considered is therefore if, when, and how the expected vanishing of the nondiagonal *collective* terms will take place. This will be first investigated in a very rough manner so as to understand what is going on without delving into too many intricacies, which will however have to be introduced later on.

1. An Intuitive Approach

Let us consider an example. The object is a pendulum or, more precisely, a ball made of metal hanging on a metal wire. It can oscillate under gravity and it will be simpler to consider that the oscillations take place in a well-defined vertical plane so that the pendulum position is completely defined by the angle θ between the wire and the vertical direction. This angle is the only collective variable to be taken into account together with its canonically conjugate momentum, which is the component of the angular momentum normal to the plane of oscillations. All the other observables will be treated as being microscopic.

We first consider an initial state that is obtained by letting the pendulum go, with zero initial velocity, from an initial position θ_0. This is most simply described in the quantum formalism by using a gaussian wave function in the variable θ, with average value θ_0 and zero average momentum. As for the initial state of the environment, the simplest case occurs when the matter of the pendulum is initially at zero temperature so that the environment is in its ground state. The initial state of the whole object is then given by

$$|\Psi(0)\rangle = |\phi_{\theta_0}\rangle_c \otimes |0\rangle_E, \quad (7.1)$$

where the first state vector represents the gaussian collective wave function and the second one represents the ground state of matter.

Starting from these initial conditions, the pendulum oscillates. The oscillations deform the wire, particularly near its regions in contact with the ball and with the support. From a classical point of view, these are elastic deformations and, from a quantum point of view, they are excitations of the environment, i.e., phonons. They are generated by the geometric constraints at the extremities of the wire but one can as well describe them by a coupling hamiltonian between the collective pendulum and the internal environment.

When the elastic waves in the wire are damped, they communicate their energy to the atoms so that the whole matter of the pendulum is heated, even if very little. From a classical standpoint, this dissipation is due to a friction force, which is proportional to the velocity and opposite to it. The classical equation of motion for the pendulum is then given by

$$m\left(\frac{d^2\theta}{dt^2} + \lambda \frac{d\theta}{dt} + \omega^2\theta\right) = 0. \tag{7.2}$$

One has assumed in this equation that the free classical motion can be correctly described by a harmonic oscillator, though one knows that this is only approximate in the case of a real pendulum. This point is however of no particular importance for our purpose.

Let us consider a time t at which dissipation has begun to have some effect and a small part of the pendulum energy has been transferred to the environment. The latter is therefore no longer in its ground state but in another state we don't want to describe in detail right now, except for noticing that it has a classically well-defined average energy $W(t)$. As for the value of the angle θ, one expects it to be concentrated around an average value $\theta(t)$ given by the solution of equation (7.2) with the prescribed initial conditions. If one expands the solution of the full Schrödinger equation for the pendulum in the energy eigenstates of the environment, one expects to find something looking more or less like

$$|\Psi(t)\rangle = \sum_j c_j(t)|\phi_j\rangle_c \otimes |w_j\rangle_E, \tag{7.3}$$

where $|\phi_j\rangle_c$ represents a normalized wave function depending upon the variable θ, with average values $\langle\theta\rangle$ and $\langle\dot\theta\rangle$ near the classical values $\theta(t)$ and $\dot\theta(t)$; $|w_j\rangle_E$ denotes an eigenstate of the environment hamiltonian with energy w_j; and c_j is a probability amplitude. This amplitude is significantly different from 0 only for values of the internal energy w_j near the energy $W(t)$ classically transferred by friction during the time interval t, which is given by

$$W(t) = \int_0^t m\lambda\dot\theta^2\, dt = m\lambda\theta_0^2\omega^2 \int_0^t \sin^2 \omega t'\, dt'. \tag{7.4}$$

The last equality in equation (7.4) is valid only near the beginning of motion, when one can still replace the velocity by its value in the absence of friction in order to compute the first integral. It shows that the dissipated energy depends quadratically upon the initial angle. The initial density operator is given by

$$\rho(0) = |\Psi(0)\rangle\langle\Psi(0)|. \tag{7.5}$$

DECOHERENCE

Now comes an important though simple idea: when computing the statistical properties of the measurement of a collective observable A_c or when computing the probability of a property involving it, one is led to compute a quantity of the form

$$\text{Tr}(\rho A_c), \qquad (7.6)$$

where the observable A_c commutes with the environment observables. It may be a projector when one computes the probability of a collective property, or another observable such as the position or velocity of the pendulum when a measurement is to be made. It might also be a quasi-projector asserting simultaneously the values of both position and momentum within some error bounds in the case of a classical measurement. In any case, one can always compute the trace (7.6) in two steps:

1. One computes a so-called *collective density operator* (or *reduced density operator*) by performing first a partial trace of the complete density operator ρ upon the environment degrees of freedom, i.e., $\rho_c = \text{Tr}_E \rho$.
2. One can then get the desired average value by performing a final trace upon the collective Hilbert space to get $\text{Tr}_c \rho_c A_c = \text{Tr}(\rho A_c)$. The reduced density operator will play a central role in the theory of decoherence because it is the one losing its nondiagonal part by decoherence, as will now be shown.

By taking into account equation (7.5) for ρ and the explicit form (7.1) for $|\Psi(0)\rangle$, one finds that at the initial time the reduced density operator is given by

$$\rho_c(0) = |\phi_{\theta_0}\rangle\langle\phi_{\theta_0}|$$

showing that, even at the collective level, this is still a pure state. At a later time t however, equation (7.3) gives

$$\rho_c = \sum_j |c_j|^2 |\phi_j\rangle_c \langle\phi_j|_c,$$

which is a mixture.

Let us now consider a more interesting case when the initial state is a linear combination of two states of the previous kind, both with zero initial velocity and having initial positions θ_1 and θ_2, respectively. The environment will be assumed in both cases to be initially in its ground state so that the complete initial state is given by

$$|\Psi(0)\rangle = a|\Psi_1(0)\rangle + b|\Psi_2(0)\rangle,$$

where

$$|\Psi_1(0)\rangle = |\phi_{\theta_1}\rangle_c |0\rangle_E, \qquad |\Psi_2(0)\rangle = |\phi_{\theta_2}\rangle_c |0\rangle_E.$$

At a later time t, one has

$$|\Psi(t)\rangle = a|\Psi_1(t)\rangle + b|\Psi_2(t)\rangle,$$

where, for instance,

$$|\Psi_1(t)\rangle = \sum_j c_j^{(1)}(t)|\phi_j^{(1)}(t)\rangle_c \otimes |w_j\rangle_E, \qquad (7.7)$$

The collective states $\phi_j^{(1)}(t)$ give average values for θ and $\dot\theta$ that are practically equal to what is predicted by a classical motion starting from the initial position θ_1. As for the amplitudes $c_j^{(1)}(t)$ and $c_j^{(2)}(t)$, it should be noticed that they are only significantly different from 0 for values of w_j very near the respective average classically dissipated energies $W_1(t)$ and $W_2(t)$. They depend upon the motion to be considered and are generally different.

We consider again the full density operator $\rho(t) = |\Psi(t)\rangle\langle\Psi(t)|$ and compute the corresponding reduced density operator. At time 0 (since the state of the environment is everywhere the ground state), the partial trace over the environment is trivial and the initial reduced density operator is:

$$\rho_c(0) = |a|^2 |\phi_{\theta_1}\rangle\langle\phi_{\theta_1}| + |b|^2 |\phi_{\theta_2}\rangle\langle\phi_{\theta_2}| + ab^*|\phi_{\theta_1}\rangle\langle\phi_{\theta_2}| + a^*b|\phi_{\theta_2}\rangle\langle\phi_{\theta_1}|.$$

Note that it still represents a pure state and the nondiagonal terms are superbly present.

When the reduced density operator is computed at a later time, nothing much happens as far as the diagonal part is concerned since it is given by

$$\rho_{cd} = |a|^2 \sum_j \left|c_j^{(2)}\right|^2 |\phi_j^{(1)}\rangle\langle\phi_j^{(1)}| + |b|^2 \sum_j \left|c_j^{(2)}\right|^2 |\phi_j^{(2)}\rangle\langle\phi^{(2)}|.$$

It represents a probability distribution for the position that is concentrated around the two possible values one would expect from two classical motions starting with different angles, with respective probabilities $|a|^2$ and $|b|^2$, as can be checked by noticing that the normalization conditions for Ψ_1 and Ψ_2 give

$$\sum_j \left|c_j^{(1)}\right|^2 = \sum_j \left|c_j^{(2)}\right|^2 = 1.$$

The nondiagonal part of the reduced density operator is much more instructive. Equation (7.17) giving $\Psi_1(t)$ involves only the energy eigenstates of the environment with eigenvalues in the immediate vicinity of $W_1(t)$, whereas the analogous expension for $\Psi_2(t)$ involves energies that are near to $W_2(t)$. When t is large enough so that the difference $W_1(t) - W_2(t)$ is larger than the quantum fluctuations in the energy of the environment, the partial trace over the environment suppresses the nondiagonal terms. This can be seen by choosing the energy eigenstates of the environment as a basis

DECOHERENCE

for computing the partial trace. Considering, for instance, the nondiagonal part of ρ_c coming from $|\Psi_1(t)\rangle\langle\Psi_2(t)|$, one finds a contribution to the reduced density operator given by

$$ab^* \sum_j c_j^{(1)} c_j^{*(2)} |\phi_j^{(1)}\rangle\langle\phi_j^{(2)}|.$$

But we have just seen that the coefficients $c_j^{(1)}$ are different from 0 only when w_j is near to $W_1(t)$, in which case the coefficients $c_j^{(2)}$ multiplying them are 0. The nondiagonal elements must therefore vanish.

This *spontaneous dynamical diagonalization of the reduced density operator* is always the main manifestation of decoherence and it appears clearly here as a by-product of dissipation. It is also sometimes called a spontaneous generation of a superselection rule, but this calls for a notion[1] that is not really necessary in the present case as well as an applicability which is open to controversy, so that it will better be left aside. The existence of this effect does not require the dissipation to be large since all that is needed is that the difference $W_1(t) - W_2(t)$ is large when compared with the dispersion of the internal energy. This does not require that $W_1(t)$ and $W_2(t)$ be sizable when compared with the initial energy of the pendulum, which would be the case if one had to wait until a significant amount of dissipation has taken place. Diagonalization nevertheless requires a finite amount of time since it does not hold at time 0. This time must be long enough for the two dissipated energies to differ by more than their own fluctuations.

It is clear that, from a mathematical standpoint, the key of the argument is the high density of the energy spectrum allowing the energy fluctuations to be small compared with the average internal energy. We have been assuming that there is no external environment and, in particular, no thermostat with which the pendulum can interact. The presence of a thermostat would not spoil the results but the above argument would not work, showing that a more general argument should rely more explicitly upon the tendency of the different states in the environment to become orthogonal.

In spite of the rough character of this analysis, there is little doubt that its results are essentially correct and the calculations to be given later on will only confirm its conclusions and make them more quantitative. One can summarize this by saying that the nondiagonal terms in the reduced density operator vanish rapidly after a short time, under the condition that there is some dissipation.

The most obvious macroscopic physical systems for which dissipation is negligible are superfluids, superconductors, and also ordinary electromagnetic radiation, the list being not exhaustive. They are also the ones in which one might expect decoherence not to take place. This is, by the way, obvious as far as radiation is concerned since it can be used to produce macroscopic interferences. Except for them, it is extremely difficult to realize an experiment with something macroscopic under conditions that would be immune to decoherence and would show the existence of quantum interference. So, contrary to what was universally believed at the beginnings of quantum mechanics, linear superpositions of states are not present everywhere but they are on the contrary very rare and hard to preserve except in very small systems. Of course, this is true only as long as one does not probe the bowels of the system by looking more precisely at the state of the environment itself.

Remark. The understanding of decoherence is sometimes made more difficult when one has a previous knowledge of another physical effect: the existence of spin echoes. The aim of the present note is to avoid this kind of confusion. A typical spin echo experiment looks like the following. A solid sample containing a few atoms possessing a magnetic nucleus at the impurity level is used. The spins of these nuclei have only a very weak interaction with their surroundings, except under the influence of an external magnetic field. As a matter of fact, they do not experience the external magnetic field itself but each nucleus "sees" a local magnetic field resulting from the externally applied field and from the local change in the average electron polarization. The local fields are stationary and they have slightly different values at the various locations of the impurities. The experiment starts from an initial state where all the nuclear spins of the impurities are aligned along the same direction x by a nuclear magnetic resonance device (NMR). The external magnetic field is then rapidly modified so as to point in the z direction and, afterwards, each spin precesses in the xy plane with a proper frequency depending on its own local field. Since all these frequencies are slightly different from each other, the spins do not remain aligned as initially, their relative directions change and the total magnetization resulting from their addition decreases and finally vanishes. The specific idea of a spin echo experiment consists in reversing the external field at some time T after the vanishing of the total magnetization, thus provoking a reversal of all the local fields. Each individual spin precession is then exactly reversed in time and, after another time T, all of them are again parallel and the initial magnetization is recovered. This regeneration of a magnetization that was apparently irretrievably lost has been called a spin echo because magnetization comes back like an echo.

It is tempting to say that spin precession constitutes an interaction with the environment and this is correct as far as the conventional meaning of the word "environment" is concerned. It might also be said that the vanishing of the total magnetization is due to a loss of phase coherence between the various spins. One should however be cautious about the precise meaning of the words in each case. The very large number of degrees of freedom is an essential ingredient of decoherence theory as it is discussed in this

DECOHERENCE

chapter. In the case of spin echoes, it turns out that the local fields, in spite of their diversity, depend altogether only on three degrees of freedom, namely the components of the external field. As a matter of fact, true decoherence also occurs in the magnetic sample because of the direct magnetic coupling between the nuclear spins and the spins of electrons in their vicinity. It takes however a rather long time because of the weakness of these couplings, leaving the possibility of spin echoes taking place beforehand.

SOLVABLE MODELS

A more consistent approach to decoherence will be now developed by considering that macroscopic systems are also governed by quantum mechanics. It was already noticed that, in spite of the presumably very wide domain of validity of the decoherence effect, there is still no general proof covering all the cases where it seems to occur and there are only a few more or less specific models having a wide domain of application. The corresponding calculations are always delicate and often difficult. The reason for such an unfortunate situation is rather obvious since one needs to control the behavior of a system with a very large number of degrees of freedom. In spite of this complexity, one wants to get at fine details of the wave functions to show their tendency to become orthogonal.

In view of this, we shall not try to give all the calculations in detail, except for the simplest models, and our only aim will be to make the methods and the results understandable. Two simplified models will now be shown. The first one is rather formal but it shows clearly the role of the phase relations between the vector states. The second one is again the example of the pendulum, in an ideal case where the environment consists only of elastic waves not interacting.

2. A Simple Model

The Model

The simplest model[2] consists in replacing the collective subsystem by a two-state quantum system. It will be called the "spin" by analogy with the states of a spin-$1/2$ particle. A basis in its two-dimensional Hilbert space will be denoted by $\{|+\rangle, |-\rangle\}$. The environment is modeled by a bath of many similar two-state systems, to be called the atoms. There are N atoms, each denoted by an index k and associated with a two-dimensional Hilbert space with a basis $\{|+\rangle_k, |-\rangle_k\}$. The corresponding dynamics is also very simple. When there is no coupling with the environment, the two spin states have the same energy, which is taken to be 0. Similarly, all the atoms have zero energy in the absence of coupling. The whole dynamics is therefore contained in the coupling, which is supposed given by the following interaction hamiltonian:

$$H_{\text{int}} = \hbar \sum_k g_k \sigma_z \sigma_z^k.$$

In this notation, σ_z is analogous to a Pauli spin matrix. It has the eigenvalue $+1$ for the spin eigenvector $|+\rangle$ and -1 for $|-\rangle$; it acts as the identity operator on all the atoms. The operators σ_z^k are similar; each acts like a Pauli matrix on the states of a specific atom k and as the identity upon all the other atoms as well as upon the spin. A more explicit form of the coupling hamiltonian using tensor notation is

$$H_{\text{int}} = \hbar \sum_k \left(g_k \sigma_z \otimes \sigma_z^k \otimes \prod_{j \neq k} I_j \right),$$

the matrices σ_z and σ_z^k being now ordinary 2×2 matrices, as well as the identity matrices I_j.

Notwithstanding its abstract character, the model still retains some physical meaning. If, for instance, the two states $|\pm\rangle$ are associated with the two states of circular polarization for a photon and if the two-state "atoms" represent a collection of left-oriented or right-oriented molecules, the model might provide a very sketchy representation of a polarization analyzer.

Decoherence

We start from a normalized initial state

$$|\Psi(0)\rangle = (a|+\rangle + b|-\rangle) \prod_{k=1}^N \otimes [\alpha_k |+\rangle_k + \beta_k |-\rangle_k].$$

It is easy to solve the Schrödinger equation and one gets, for the state at time t:

$$|\Psi(t)\rangle = a|+\rangle \prod_{k=1}^N \otimes [\alpha_k \exp(ig_k t)|+\rangle_k + \beta_k \exp(-ig_k t)|-\rangle_k]$$

$$+ b|-\rangle \prod_{k=1}^N \otimes [\alpha_k \exp(-ig_k t)|+\rangle_k + \beta_k \exp(ig_k t)|-\rangle_k].$$

Writing the complete density operator $\rho(t) = |\Psi(t)\rangle\langle\Psi(t)|$, one can take its trace over the atomic degrees of freedom to get the reduced density operator

$$\rho_c(t) = |a|^2 |+\rangle\langle+| + |b|^2 |-\rangle\langle-| + z(t)ab^*|+\rangle\langle-| + z^*(t)a^*b|-\rangle\langle+|,$$

DECOHERENCE

where

$$z(t) = \prod_{k=1}^{N} [\cos 2g_k t + i(|\alpha_k|^2 - |\beta_k|^2) \sin 2g_k t]. \tag{7.8}$$

The complex number $z(t)$ controls the value of the nondiagonal terms. It depends only upon the probabilities $\{|\alpha_k|^2, |\beta_k|^2\}$ in the initial state of the environment and not on the phase of the probability amplitudes. Now assume that, in physical examples representing a practical case, the initial state of the environment will be statistically random so that the values of $(|\alpha_k|^2, |\beta_k|^2)$ are generally different from the trivial pair of values $(1, 0)$ or $(0, 1)$. This means that each factor in the product (7.8) has an absolute value smaller than 1 except when t is not 0. When the phase $2g_k t$ passes through a multiple of π, the factor containing it takes a value ± 1 but most of the other factors have a modulus smaller than 1, at least if the coupling constants g_k are distributed irregularly. The product of many uncorrelated complex factors smaller than 1 in modulus is a very small number. This can be made more explicit by assuming a random distribution of the coupling constants. One can then show that

$$z(0) = 1; \quad |z(t)|^2 \leq 1; \quad \langle z(t) \rangle = \lim_{T \to \infty} \int_0^T z(t)\, dt = 0;$$

$$\langle |z(t)|^2 \rangle = 2^{-N} \prod_{k=1}^{N} \left[1 + (|\alpha_k|^2 - |\beta_k|^2)^2\right].$$

Theoretical and numerical evaluations of $z(t)$ show that, if the initial state is a statistical mixing where the initial probabilities $\{|\alpha_k|^2, |\beta_k|^2\}$ are distributed uniformly in the interval $[0, 1]$, $z(t)$ will start from the value 1 at time 0, its first derivative being 0; it then decreases rapidly to end up with fluctuating values of the order of $2^{-N/2}$. There is therefore decoherence when N is large enough.

What Can Be Learned from this Example?

This simple model also shows the existence of a spurious phenomenon: The function $z(t)$ is a so-called multiperiodic function. This means that $z(t)$ has a finite (though very large) number of definite frequencies. It is known that such a function will always come back arbitrarily near to its initial value at one time or another. This means that the fluctuations of $z(t)$, when they seem to have reached the level of a thermal motion, are in fact not so arbitrary. At least in the present model, they may for a short time again become as large as the initial value $z(0)$. This would occur for a very large value of the time, but it would nevertheless be a time at which the initial reduced density operator would reappear with its full-fledged nondiagonal elements.

This phenomenon is very similar to a Poincaré recurrence as it is known in classical dynamics: a dynamical system obeying the Hamilton equations of motion and not going to infinity always comes back arbitrarily close to its starting point in phase space, if one waits long enough.

Decoherence is mainly due to the existence of a very large number of degrees of freedom in the environment. It also requires that the coupling between the collective subsystem (i.e., the "spin" here) and the environment should be somewhat disordered (the coupling constants g_k have been assumed to be distributed irregularly). These are ingredients to be met again and again. It may also be noticed that, though the model is very simple from a physical standpoint, the mathematical study of the decoherence factor $z(t)$ is already not quite trivial. The model is therefore instructive, but it is oversimplified since, for instance, the energy spectrum of the environment is trivial with only one eigenvalue, 2^N times degenerate. It does not furthermore give quantitative information that one could apply to a realistic case. This is why, though the model might be investigated in more detail, we shall leave it aside to consider other models more akin to physical reality.

3. Another Example: The Pendulum

We now reconsider the example of the pendulum, which was already used to introduce the decoherence effect. It was found convenient to treat the pendulum as an exact harmonic oscillator, in the absence of coupling with the environment. This will still be done, the mass being denoted by m and the frequency by ω. The problem can be solved completely if one assumes that the environment is a collection of noninteracting harmonic oscillators, which might represent the phonons in the wire and the ball, in the approximation of linear elasticity theory. It has been solved in great mathematical detail by Hepp and Lieb[3] but the considerations to be given here are not intended to reach the same level and they only aim at a rough understanding of the results.

Coherent States

Before stating the assumptions of the model, it will be convenient to recall a few points concerning the theory of an elementary one-dimensional harmonic oscillator. The position operator of the oscillator will be denoted by X and its conjugate momentum by P so that $[X, P] = i\hbar I$. The hamiltonian is given by

$$H = \frac{P^2}{2m} + \frac{1}{2}m\omega^2 X^2.$$

DECOHERENCE

In the elementary theory of the harmonic oscillator, one introduces annihilation and creation operators a and $a^\dagger$, which are adjoint to each other, the annihilation operator a being defined by

$$a = (2m\hbar\omega)^{-1/2}\{m\omega X + iP\}.$$

They obey the commutation relation

$$[a, a^\dagger] = I \tag{7.9}$$

and the spectrum of the "number" operator $N = a^\dagger a$ is discrete and nondegenerate, consisting of the nonnegative integers $n = 0, 1, 2, \ldots$. The hamiltonian can be written in terms of this operator as

$$H = \left(N + \frac{1}{2}\right)\hbar\omega,$$

so that the normalized eigenvectors $|n\rangle$ of N are also the eigenvectors of the hamiltonian. The state $|0\rangle$, corresponding to $n = 0$, is the ground state. All these normalized eigenvectors can be expressed by the simple formula

$$|n\rangle = (n!)^{-1/2}(a^\dagger)^n|0\rangle, \tag{7.10}$$

as one easily verifies by using the following commutation relations resulting from equation (7.9):

$$[N, a] = -a, \qquad [N, a^\dagger] = a^\dagger.$$

We now introduce the notion of a *coherent state*. This is a normalized state to be denoted by $|z\rangle$, which is characterized by a complex amplitude z. It is an eigenvector of the (non-self-adjoint) annihilation operator a associated with the (complex) eigenvalue z. This definition can be expressed by the eigenvalue equation

$$a|z\rangle = z|z\rangle, \tag{7.11}$$

with the condition

$$\langle z|z\rangle = 1. \tag{7.12}$$

To show the existence of the state vector $|z\rangle$, look at its expansion in the vector basis $\{|n\rangle\}$, say

$$|z\rangle = \sum_{n=0}^{\infty} c_n |n\rangle.$$

Since it is known that $a|n\rangle = n^{1/2}|n-1\rangle$, equation (7.11) defining the coherent state gives $n^{1/2} c_n = z c_{n-1}$, so that

$$c_n = z^n (n!)^{-1/2} c_0.$$

The first coefficient c_0 can then be obtained from the normalization condition (7.12), which gives up to an irrelevant phase factor

$$1 = \sum_{n=0}^{\infty} |c_n|^2 = c_0^2 \left\{ \sum \frac{|z|^2}{n!} \right\} = c_0^2 \exp(|z|^2).$$

So taking c_0 to be real and positive, $c_0 = \exp(-|z|^2/2)$ and

$$|z\rangle = \exp(-|z|^2/2) \sum_{n=0}^{\infty} \frac{z^n}{(n!)^{1/2}} |n\rangle. \qquad (7.13)$$

This can also be written, in view of the explicit form (7.10) of the vector $|n\rangle$, as

$$|z\rangle = \exp(za^\dagger - |z|^2/2)|0\rangle. \qquad (7.14)$$

Writing the eigenvalue equation (7.11) defining the coherent state $|z\rangle$ as a differential equation using $P = (\hbar/i)\partial/\partial x$, one finds that the wave function of the coherent state satisfies

$$\left[x + (\hbar/\omega m) \frac{d}{dx} \right] \psi_z(x) = z \psi_z(x).$$

From this we get, up to a normalization coefficient,

$$\psi_z(x) = A \exp\{i p_0 x/\hbar - (x - x_0)^2 (2\hbar/\omega m)^{-1}\}.$$

The quantities x_0 and p_0 are the mean values of position and momentum for this gaussian state and they are given in terms of the complex amplitude z by

$$z = (2m\hbar\omega)^{-1/2} \{m\omega x_0 + i p_0\}. \qquad (7.15)$$

We will also need the scalar product of two coherent states, which is easily obtained from their explicit gaussian wave functions or from their expansion (7.13). It is given by

$$\langle z|z'\rangle = \exp\{-|z - z'|^2/2 + i\phi/2\}, \quad \text{where} \quad \phi = \text{Im}(zz'^* - z^*z'). \qquad (7.16)$$

Note that the coherent states are not orthogonal. However, their scalar product becomes exponentially small when the difference in their amplitudes $|z - z'|$ is large. Finally, note that a coherent state remains coherent under time evolution: Let $|z_0\rangle$ be an initial state and look for a solution of the Schrödinger equation in the form $|z(t)\rangle$. The Schrödinger equation gives

$$i\hbar \frac{d}{dt} |\psi\rangle = \left(a^\dagger a + \frac{1}{2} \right) \hbar\omega |\psi\rangle.$$

If one assumes that $|\psi\rangle = |z(t)\rangle$, equation (7.11) defining a coherent state gives

$$\left(a^\dagger a + \frac{1}{2}\right)\hbar\omega|z(t)\rangle = \left(a^\dagger z(t) + \frac{1}{2}\right)\hbar\omega|z(t)\rangle,$$

whereas the explicit form (7.14) for this state gives

$$i\hbar\frac{d}{dt}|z(t)\rangle = i\hbar\dot{z}(t)a^\dagger|z(t)\rangle.$$

This last equation holds if one assumes that the modulus of $z(t)$ remains a constant. The Schrödinger equation is then satisfied if one takes

$$z(t) = z_0 e^{-i\omega t}; \qquad |\psi\rangle = e^{-i\omega t/2}|z(t)\rangle. \qquad (7.17)$$

The constancy of $|z(t)|$, which was taken as an assumption, is then found to be satisfied. The first equation (7.17) shows that $z(t)$ is the complex amplitude one usually associates with a classical oscillator when describing its position and momentum as the real and the imaginary parts of a rotating complex number. It follows from this that, at least as far as coherent states are concerned, there is a great similarity between the classical and the quantum evolutions of a harmonic oscillator.

A Model for the Pendulum

Let us return to the pendulum having its own harmonic collective motion, its environment consisting of noninteracting phonons. This is an overall system of harmonic oscillators where the only coupling takes place between the collective motion and the various phonons. We shall describe the collective oscillator by creation and annihilation operators a and $a^\dagger$ and a frequency ω. The environment will be treated as a collection of N oscillators having their frequencies ω_k more or less densely distributed. Creation and annihilation operators $b_k^\dagger$ and b_k ($k = 1, 2, \ldots, N$) will also be used to describe the oscillators in the environment. The total hamiltonian will be taken as

$$H = \hbar a^\dagger a + \sum_k \hbar\omega_k b_k^\dagger b_k + \hbar\sum_k (\lambda_k a^\dagger b_k + \lambda_k^* b_k^\dagger a),$$

where the first term is the collective hamiltonian (the energy of the center of mass in the present case), the second one is the hamiltonian for the environment as made up of N oscillators, and the last term is the coupling from which we expect to get both dissipation and decoherence. The coupling constants λ_k may be complex and they are assumed to be small, which means that the dissipative coupling influences the collective motion but does not dominate.

One can define a coherent state of the whole system as a normalized vector $|\alpha, \{\beta_k\}\rangle$ satisfying the eigenvalue equations

$$a|\alpha, \{\beta_k\}\rangle = \alpha|\alpha, \{\beta_k\}\rangle, \qquad b_k|\alpha, \{\beta_k\}\rangle = \beta_k|\alpha, \{\beta_k\}\rangle,$$

where α and the various quantities $\{\beta_k\}$ are complex. A calculation very similar to the one already made in the case of a single oscillator shows that such a coherent state is a solution of the Schrödinger equation if its parameters evolve in time according to the following system of differential equations

$$i\frac{d\alpha}{dt} = \omega\alpha + \sum_k \lambda_k \beta_k, \qquad i\frac{d\beta_k}{dt} = \omega_k \beta_k + \lambda_k^* \alpha. \qquad (7.18)$$

Once again, the quantum problem has been reduced to a classical calculation, which is why this model is simple. In the example that was considered in Section 1, the initial collective state of the pendulum had a gaussian wave function and the environment was in its ground state. This corresponds precisely to a coherent state where all the parameters β_k are initially 0.

One may notice a useful conservation property of the differential system (7.18): Given two solutions $(\alpha, \{\beta_k\})$ and $(\alpha', \{\beta_k'\})$, the quantity

$$\alpha^* \alpha' + \sum_k \beta_k^* \beta_k' \qquad (7.19)$$

is found to be a constant under time evolution. The differential system (7.18) was first introduced by Weisskopf and Wigner in a study of the effect of relaxation upon the width of an atomic emission line, relaxation being the damping of a physical quantity originating from an interaction with the environment.[4] A well-known relaxation effect is the broadening of spectral lines by atoms undergoing collisions; other ones look more directly like friction, as will be seen. One might think of solving the system (7.18) by diagonalizing the matrix associated with its right-hand side, but this would not easily yield the most useful results. Instead notice that there exists a solution exhibiting the collective harmonic motion as well as the effect of friction, which can be written as

$$\alpha(t) = \alpha(0) \exp(-i(\omega + \delta\omega)t - \gamma t) + \text{fluctuations}.$$

The first term corresponds to a classical motion of the pendulum showing a small shift in frequency as well as damping. The fluctuating term remains small in general and it can vary in a very complicated way, practically at random. This solution is only valid when the coupling constants λ_k are small compared with ω. When t becomes large enough and the damping is more or less complete, the fluctuations become of course dominant. The pendulum has then come to rest except for thermal motion, which is a fluctuation.

Justification of the Solution*

Let us show how this solution can be obtained. This exercise is not very important in itself but it will provide a few quantitative results helpful for a better knowledge of the efficiency of decoherence. The details of the present calculation can therefore be skipped since only the results are important. We change the functions to be studied by putting

$$\alpha(t) = e^{-i\omega t}\alpha'(t), \qquad \beta_k(t) = e^{-i\omega_k t}\beta'_k(t).$$

The Weisskopf–Wigner system becomes

$$i\frac{d\alpha'}{dt} = \sum_k \lambda_k e^{i(\omega-\omega_k)t}\beta'_k(t), \qquad i\frac{d\beta'_k}{dt} = \lambda_k^* e^{-i(\omega-\omega_k)t}\alpha'(t). \quad (7.20)$$

Integrate the second equation to get

$$\beta'_k(t) = -i\lambda_k^* \int_0^t e^{-i(\omega-\omega_k)t'}\alpha'(t')\,dt'.$$

Using this expression in the first equation (7.20), one gets:

$$\frac{d\alpha'}{dt} = -\sum_k |\lambda_k|^2 \int_0^t e^{i(\omega-\omega_k)(t-t')}\alpha'(t')\,dt'.$$

It will be convenient at this point to take into account the largeness of the number of oscillators in the environment and to assume that their frequencies have practically a continuous distribution. Denoting the average number of oscillators per unit frequency interval by $n(\omega')$ and representing the coupling constants λ_k as functions of $\omega' = \omega_k$, the summation upon the index k becomes an integral and one gets:

$$\frac{d\alpha'}{dt} = -\int_0^\infty |\lambda(\omega')|^2 n(\omega')\,d\omega' \int_0^t e^{i(\omega-\omega')(t-t')}\alpha'(t')\,dt'. \quad (7.21)$$

As a matter of fact, the integral over the frequencies does not extend to infinity but is cut off at a maximum value (the Debye frequency), which is very large as compared with the pendulum frequency ω. One can represent the same regularizing effect by taking the integration over ω' to extend formally to infinity while replacing ω' by $\omega' - i\varepsilon$, where ε is a very small positive number. This can be done only when the time t is positive, so that this procedure breaks time reversal. We will therefore have to come back later to this point since the results do not follow from quantum dynamics alone.

Integrating the integral (7.21) by parts over t', gives

$$\frac{d\alpha'}{dt} = -i\int_0^\infty |\lambda(\omega')|^2 n(\omega')\, d\omega' \{(\omega + i\varepsilon - \omega')^{-1}\alpha'(t)$$
$$- e^{i(\omega-\omega')t}(\omega + i\varepsilon - \omega')^{-1}\alpha'(0)$$
$$- \int_0^t e^{i(\omega-\omega')(t-t')}(\omega + i\varepsilon - \omega')^{-1}\frac{d\alpha'(t')}{dt'}\, dt'\}.$$

The second term in the bracket can be considered as a fluctuation, since it gives rise to a sum over all frequencies interfering destructively as soon as t is large enough compared with the inverse of the Debye frequency (which is in reality a very short time). The last term is small because only values of t' very near to t will fail to produce destructive interferences and, furthermore, the time derivative of $\alpha'(t)$ will be small under the conditions of a weak damping. One therefore neglects these terms to keep only the dominant one so that, approximately,

$$\frac{d\alpha'}{dt} = -i\int_0^\infty |\lambda(\omega')|^2 n(\omega')\, d\omega' (\omega + i\varepsilon - \omega')^{-1}\alpha'(t).$$

One can then use a well-known formula from distribution theory according to which

$$\frac{1}{(\omega + i\varepsilon - \omega')} = P\frac{1}{(\omega - \omega')} - i\pi\delta(\omega - \omega'),$$

where P denotes a finite integral kernel to be taken in the sense of Cauchy:

$$P\int_a^b (\omega - \omega')^{-1}\cdots d\omega' = \lim_{\varepsilon \to 0}\left\{\int_a^{\omega-\varepsilon}(\omega - \omega')^{-1}\cdots d\omega'\right.$$
$$\left. + \int_{\omega+\varepsilon}^b (\omega - \omega')^{-1}\cdots d\omega'\right\}.$$

So finally

$$\frac{d\alpha'}{dt} = (-i\delta\omega - \gamma)\alpha'(t),$$

where

$$\gamma = \pi\int_0^\infty |\lambda(\omega')|^2 \delta(\omega - \omega')n(\omega')\, d\omega', \quad \text{and}$$

$$\delta\omega = P\int_0^\infty |\lambda(\omega')|^2 \frac{1}{(\omega - \omega')}n(\omega')\, d\omega'.$$

It is possible to justify these results in a more rigorous way as Hepp and Lieb did. Also note that, when the sum over the oscillators is replaced by an integral, one implicitly assumes that the total number of oscillators is infinite. As a matter of fact, the overall solution is not affected by this approximation, though the fluctuations are more sensitive to it. Hepp and Lieb have shown that the overall corrections are of the order of $N^{-1/2}$ and therefore negligible when the number of degrees of freedom is large enough.

It should also be mentioned that the Weisskopf–Wigner differential system satisfies Poincaré's theorem, which means that the overall coherent state comes back after a long time arbitrarily close to its initial value. This is a kind of behavior one already met with the coupled spin-atoms model in Section 2, i.e., one might expect the occurrence of some giant fluctuations restoring the initial situation at some much later time, at least when N is finite. This difficulty is however due to the somewhat academic features of the model and we shall not pay further attention to it.

The Reduced Density Operator

It is now easy to get the reduced density operator for the pendulum starting in a collective gaussian state, its matter being initially at zero temperature. The state of the system at time t is given by

$$|\Psi(t)\rangle = |\alpha(t), \{\beta_k(t)\}\rangle.$$

The reduced density operator is obtained immediately, since the environment is in a pure state so that

$$\rho_c = \mathrm{Tr}_E(|\Psi(t)\rangle\langle\Psi(t)|) = |\alpha(t)\rangle\langle\alpha(t)|.$$

Hence the collective state of the pendulum is a coherent (gaussian) state, with average values for the position and momentum as given by classical motion with damping.

We now consider the much more interesting case where the initial state of the pendulum is a linear superposition of two macroscopically different states initially starting from different positions with zero average velocity. One then assumes that

$$|\Psi(0)\rangle = a|\alpha_1(0), \{\beta_k = 0\}\rangle + b|\alpha_2(0), \{\beta_k = 0\}\rangle,$$

a and b being probability amplitudes. The initial reduced density operator is given in that case by

$$\rho_c(0) = |a|^2 |\alpha_1(0)\rangle\langle\alpha_1(0)| + |b|^2 |\alpha_2(0)\rangle\langle\alpha_2(0)|$$
$$+ ab^*|\alpha_1(0)\rangle\langle\alpha_2(0)| + ba^*|\alpha_2(0)\rangle\langle\alpha_1(0)|,$$

plainly showing the existence of a quantum superposition. One can also compute the reduced density operator at a later time t. The calculation of its diagonal part is straightforward and gives

$$\rho_{cd} = |a|^2 |\alpha_1(t)\rangle\langle\alpha_1(t)| + |b|^2 |\alpha_2(t)\rangle\langle\alpha_2(t)|.$$

The nondiagonal part involves scalar products of different coherent states of the environment when one takes the partial trace. These scalar products are given by

$$\langle\{\beta_{k1}(t)\}|\{\beta_{k2}(t)\}\rangle = \prod_{k=1}^{N} \langle\beta_{k1}(t)|\beta_{k2}(t)\rangle.$$

Using equation (7.16) for the scalar product of 2 one-dimensional coherent states, one gets

$$\prod_k \langle\{\beta_{k1}(t)\}|\{\beta_{k2}(t)\}\rangle = \exp\left(-\sum_k \tfrac{1}{2}|\beta_{k1}(t) - \beta_{k2}(t)|^2 + i\phi\right),$$

where ϕ is an uninteresting phase. This can be simplified using the conservation equation (7.19), which gives

$$\prod_k \langle\{\beta_{k1}(t)\}|\{\beta_{k2}(t)\}\rangle$$

$$= \exp\left(-\tfrac{1}{2}\{|\alpha_1(0) - \alpha_2(0)|^2 - |\alpha_1(t) - \alpha_2(t)|^2\} + i\phi\right).$$

By the known time evolution of the pendulum amplitude, this last equation can also be written as

$$\prod_k \langle\{\beta_{k1}(t)\}|\{\beta_{k2}(t)\}\rangle = \exp[-\tfrac{1}{2}\{|\alpha_1(0) - \alpha_2(0)|^2 (1 - e^{-2\gamma t})\} + i\phi],$$

so that finally the nondiagonal part of the reduced density operator is given by

$$\rho_{cnd} = \{ab^*|\alpha_1(0)\rangle\langle\alpha_2(0)| + ba^*|\alpha_2(0)\rangle\langle\alpha_1(0)|\}$$
$$\times \exp[-\tfrac{1}{2}|\alpha_1(0) - \alpha_2(0)|^2 (1 - e^{-2\gamma t})].$$

The phases have been included in the probability amplitudes a and b for convenience. The initial amplitudes of the pendulum $\alpha_1(0)$ and $\alpha_2(0)$ can be reexpressed in terms of the average initial positions, using Eq. (7.15). This gives for the decoherence factor:

$$\exp[-\tfrac{1}{2}|\alpha_1(0) - \alpha_2(0)|^2 (1 - e^{-2\gamma t})]$$
$$= \exp\left[-\tfrac{1}{4}\frac{m\omega^2}{\hbar}(x_1(0) - x_2(0))^2(1 - e^{-2\gamma t})\right].$$

The extraordinary efficiency of decoherence following from this expression is best seen in a numerical example: Take a pendulum with a mass of one gram and a period of one second. Its damping time γ^{-1} is taken to be very large, say one hour. Look at the decoherence effect one nanosecond after letting the pendulum go in a superposed state with initial positions differing only by one micron. One then finds that, even in these extreme conditions, the nondiagonal matrix elements of the reduced density operator have become smaller than $\exp(-10^3)$. As a matter of fact, the nondiagonal elements vanish exponentially within a characteristic decoherence time of the order of

$$4\hbar\tau[m\omega^2(x_1(0) - x_2(0))^2]^{-1},$$

where τ is the damping time of the pendulum. The decoherence time is in the present case 10^{-20} times smaller! It is even larger when the pendulum starts with a high enough temperature.[5] No effect so spontaneous, so efficient, and of so frequent occurrence is known in the whole field of physics.

Finally, it should be mentioned that decoherence can be even more effective when the system does not start from zero temperature. When the thermal energy kT is much larger than the quantities $\hbar\omega_j$, the decoherence factor is given by

$$\exp\left[-\frac{mkT(x_1(0) - x_2(0))^2}{\hbar^2} \cdot \frac{t}{\tau}\right].$$

MORE GENERAL MODELS

The "general" theory of decoherence is rather involved though it remains based upon special models, for reasons already explained in the introduction. The best approach rests presently on a method that was introduced by Feynman and Vernon[6] and extensively developed later on by Caldeira and Leggett.[7] It makes use of all the power of path integrals and the detailed computations are consequently rather cumbersome. This is why we shall only describe the main features of the results without considering the details.

4. The General Theory

The Framework of the Model

The most general model allowing a study of decoherence is restricted by several conditions:

1. The collective system to be considered is rather general, except for having a strictly finite number of degrees of freedom. This means that one can apply the results to any system consisting of solid objects if one does need to include the elastic effects, though a finite number of elastic modes could be included. The case of fluids, with their

formally infinite number of degrees of freedom, has not even been touched upon.

2. The environment is assumed to be a collection of noninteracting harmonic oscillators. They might be thought to represent phonons, though their lack of mutual interactions would then be a very poor approximation since it is known that phonon-phonon interactions are essential for getting a thermal distribution of energy.

    This restriction of the environment to a collection of harmonic oscillators has nothing to do with physics but it is prescribed by the mathematical techniques at our disposal. The point is that harmonic oscillators give rise to gaussian integrals in the Feynman path formalism and these are the only ones allowing explicit computations. Caldeira and Leggett have tried to extend the validity of this model by an argument showing its formal equivalence with a realistic environment, at least in the case of a superconducting system.[8] They propose to associate a formal harmonic oscillator having a frequency ω_k with every eigenstate of energy E_k of the environment, the ground state energy being taken to be 0 and ω_k being equal to $E_k/\hbar$. Their argument is however rather subtle and also of limited scope so that it will not be reproduced here.

3. The initial total density operator is a tensor product of a density operator for the collective system and a density operator for the environment. The initial collective density operator can be arbitrary. The initial state of the environment is supposed to be in thermal equilibrium at a given temperature T. The system remains nevertheless isolated and does not interact with a thermostat after time 0.

    This condition is again a technical restriction originating from mathematical reasons: One has to perform a partial trace over the environment in order to get the reduced density operator and this can only be done explicitly when the initial state has a Gibbs energy distribution. This is again due to the possibility of integrating explicitly gaussian functions, the Gibbs distribution for a system of harmonic oscillators being gaussian.

4. The coupling between the collective system and the environment has a special form. It consists of functions of the collective position coordinates multiplied by the position coordinates of the environment oscillators (or their momenta, though one case is exclusive of the other). Caldeira and Leggett have also given arguments in favor of this kind of coupling, both from a basic point of view by considering the physical meaning of the environment oscillators and from a phenomenological one by looking at the kind of phenomenological friction forces one can get. This specific kind of coupling has an important consequence, namely that the reduced density operator tends to become diagonal preferentially in the coordinate representation and only some time later also in the momentum representation.

The mathematical formulation of the model is therefore as follows. The collective system is described by its position coordinates $(q_1, \ldots, q_r)$ and their conjugate momenta $(p_1, \ldots, p_r)$. The collective hamiltonian is given by

$$H_c = \sum \frac{p^2}{2m} + V(q_1, \ldots, q_r).$$

This form of the kinetic energy may look somewhat restrictive but it is in principle easy to extend the results to a more general quadratic form in the momenta. The environment is a collection of N noninteracting harmonic oscillators, with formal masses m_k, frequencies ω_k, position x_k, and momenta p_k. The summations over k are replaced formally by integrals over the frequencies so that the number N of oscillators is formally considered as infinite. Contrary to what was done by Hepp and Lieb, no estimate of the corrections coming from the finiteness of N has been done in the general case. The coupling between the collective subsystem and the environment has the form

$$H_{\text{int}} = \sum_k G_k(q_1, \ldots, q_r) x_k.$$

There are further restrictions upon the magnitudes of the couplings G_k and one must also add compensating terms in the coupling hamiltonian to take care of renormalization.

*A Sketch of the Calculations**

The complete density operator is written in the coordinate representation as $\rho(Q', Q; t)$, where Q represents all the position variables, both collective and microscopic, i.e., the q_j's and x_k's. The matrix elements of the initial density operator $\rho(Q', Q; 0)$ and its value at time t, $\rho(Q''', Q''; t)$, are linearly related because of the linear character of the Schrödinger equation. This can be expressed in terms of a kernel J connecting them through the relation

$$\rho(Q''', Q''; t) = \int J(Q''', Q''; Q', Q; t)\, \rho(Q', Q; 0)\, dQ'\, dQ.$$

The evolution kernel J can be written as a Feynman path integral where two different families of Feynman paths enter, one of them for the evolution of the ket $|Q, t\rangle$ on the right-hand side of ρ and the other one for the evolution of the bra on the *left-hand* side. This gives:

$$J(Q''', Q''; Q', Q; t)\rho(Q', Q; 0) = \int_{Q'(0)=Q'}^{Q'(t)=Q'''} \Pi_t\, dQ'(t')\, \exp\{-iS[Q'(t')]/\hbar\}$$
$$\times \int_{Q(0)=Q}^{Q(t)=Q''} \Pi_t\, dQ(t')\, \exp\{+iS[Q(t')]/\hbar\}.$$

The classical action $S[Q(t)]$ is taken along the path $Q(t)$ and the normalization coefficient entering in the element of integration is implicit in the definition of $dQ(t)$.

The reduced density operator is obtained by integrating over the environment coordinates so that

$$\rho_c(q',q;t) = \int \rho(q',x,q,x;t)\,dx'\,dx.$$

It turns out that, because of the simplifying assumptions entering in the formulation of the model, this integration can be made explicitly and the reduced density operator at an arbitrary time t can be expressed in terms of the initial collective density operator by

$$\rho_c(q''',q'';t) = \int J_c(q''',q'';q',q;t)\rho_c(q',q;0)\,dq'\,dq.$$

The collective evolution kernel J_c can finally be written in a form analogous to a Feynman path integral over two different families of Feynman paths,

$$J_c(q''',q'';q',q;t) = \int \prod_{t'=0}^{t} dq(t')\,dq'(t')\,I[q(t'),q'(t')], \quad (7.22)$$

one path family starting from q at time 0 and ending at q'' at time t, the other one starting from q' and ending at q'''. The kernel $I[q(t),q'(t)]$ is called the influence functional because it represents the effect of the proper dynamics of the collective system when influenced by the environment. This influence is best seen by using new variables

$$X(t') = \tfrac{1}{2}[q(t') + q'(t')], \qquad y(t') = q(t') - q'(t').$$

The larger $y(t)$ is, the farther away are the paths entering in the ket and the bra in

$$\langle q'|\rho_c|q\rangle = \rho_c(q',q),$$

so that this variable provides a good measure for the nondiagonality of the reduced density operator in the coordinate representation. The influence functional has a form similar to a Feynman path integrand, namely the exponential of an action, except that its exponent contains both a real and an imaginary part:

$$I[q(t'),q'(t')] = \exp\{+iS_1[X(t'),y(t')]/\hbar\}\exp\{-S_2[X(t'),y(t')]/\hbar\}. \quad (7.23)$$

DECOHERENCE

The first term can be compared to an action and its explicit form is given by

$$S_1[X(t'), y(t')] = \int_0^t dt' \left\{ y(t')[m\ddot{X}(t') + \lambda \dot{X}(t')] \right.$$
$$\left. + V\left[X(t') + \frac{1}{2}y(t')\right] - V\left[X(t') - \frac{1}{2}y(t')\right] \right\}, \quad (7.24)$$

at least in the special case of a one-dimensional collective system. The coefficients $G_k(q)$ are simply taken to be proportional to q. The second factor in equation (7.23) is responsible for decoherence and its exponent S_2 is given under the same conditions by

$$S_2[X(t'), y(t')] = \int_0^t dt' \int_0^t dt'' K(t'' - t') y(t'') y(t'), \quad (7.25)$$

where

$$K(t) = \int_{-\infty}^{+\infty} \frac{d\omega}{2\pi} e^{-i\omega t} \gamma(\omega) \cdot \omega \coth\left(\frac{\hbar\omega}{2kT}\right). \quad (7.26)$$

The weight $\gamma(\omega)$ depends upon the couplings and on the level density of the environment in a way similar to what was found in the case of the pendulum. If, for instance, the interaction hamiltonian is given by

$$H_{\text{int}} = \sum_k g_k q x_k,$$

then one has

$$\gamma(\omega) = \frac{\pi}{2\omega} \sum_k \frac{g_k^2}{m_k \omega_k} \delta(\omega - \omega_k).$$

Furthermore, the friction force $-\lambda \dot{X}(t')$ occurring in equation (7.24) is shorthand notation for a kind of retarded force whose occurrence is well known in dissipation theory, namely, a force

$$-\int_{-\infty}^t \lambda(t - t') \dot{X}(t') dt',$$

where the Fourier transform of $\lambda(t)$ is $\gamma(\omega)$. These results are representative of what one obtains in the general case and they can be used to show how dissipation and decoherence take place.

Dissipation and Decoherence

One may notice that the function $K(t)$ given by equation (7.26) is the Fourier transform of a positive function. This implies that the quantity S_2, given by equation (7.25), is positive definite regardless of the function $y(t)$. The exponential of S_2 occurring in equation (7.26) will therefore strongly damp the contributions to J_c coming from two Feynman paths where the collective variables q and q' are too far away from each other. This means that the most important contributions will come from small values of $y(t')$. The given initial conditions fix unambiguously the dependence of the reduced density operator on $y(0)$ but, as soon as the time t increases, the exponential damping coming from the exponent S_2 will select contributing paths corresponding to smaller and smaller values of $y(t)$ or, in other words, nearer and nearer values of $q(t)$ and $q'(t)$. This means that the density operator becomes rapidly diagonal in the coordinate representation.

It is also interesting to find values of the position coordinates around which the matrix elements of the density operator $\rho_c(q''', q''; t)$ tend to concentrate. We shall assume for simplicity that the initial state is already essentially localized in position. Since one knows that only small values of $y(t')$ will contribute to the final result, one can expand the integrand in equation (7.24) for S_1 to first order in $y(t')$. This gives

$$S_1[X(t'), y(t')] = \int_0^t dt' y(t')\{m\ddot{X}(t') + \lambda \dot{X}(t') + \nabla \cdot V[X(t')]\}. \quad (7.27)$$

When this is reinserted into equation (7.22), using equation (7.23), it is clear that, as usual with this kind of oscillating Feynman path integral, the dominating terms will come from the paths that are able to interfere constructively. They correspond in the present case to functions $X(t)$ giving a practically zero value to the integral (7.27). Since $y(t)$ is also small, it means that $\rho_c(X(t), y(t); t)$ will be strongly concentrated after a short time around the value 0 for $y(t)$ (this is decoherence, i.e., a spontaneous dynamical diagonalization) and around values of $X(t)$ satisfying the equation of motion

$$m\ddot{X} + \lambda \dot{X} = -\nabla \cdot V(X)$$

describing classical motion with friction.

Other Approaches

The "general" theory of decoherence is severely limited because of its restrictions on the model of the environment and on thermal equilibrium for its initial state. Gell–Mann and Hartle have recently extended it to allow the treatment of other cases but they need to assume that decoherence takes place. They have difficulty proving this, though they can draw precise consequences of this assumption.[9]

It is rather clear that one limitation of the Feynman path approach is the necessity of explicitly performing the calculations. One therefore wonders if other mathematical techniques might do better. Microlocal analysis (as used in Chapter 6) is an obvious candidate because of its great versatility. It has been recently applied to the model of Section 3 and the calculations are found to be somewhat simpler than the ones by Hepp and Lieb or using Feynman paths.[10] What looks even more promising is the possibility of giving explicit criteria for the existence of collective observables and for sorting them out. It would however be premature to try to assess the generality of the method and the scope of its applications so that these developments must be left to the future.

5. Decoherence by the External Environment

We have considered up to now only decoherence effects originating from the internal environment of an object (except if one is ready to accept that an external environment can also be modeled by a collection of oscillators). We are now going to show how an external environment can also generate decoherence.

As a beginning, let us see why the existence of such an effect can be expected. If one considers, for instance, a solid object moving freely in a vacuum, one knows that the total momentum is exactly conserved. The hamiltonian does not depend on the position X of the center of mass and its dependence upon the total momentum P is completely contained in the kinetic energy $P^2/2M$. Accordingly, there is no term in the hamiltonian through which the collective center-of-mass observables could be coupled to the internal environment degrees of freedom. The center of mass of a freely moving object would not therefore show any decoherence effect entirely due to internal coupling. One might therefore expect that interference effects could be shown by freely moving macroscopic objects and one might even imagine seeing an isolated object in a quantum superposition state or, to report a half-jesting question by Fermi, why is the moon not fuzzy? Similar considerations could be made about the orientation of a solid object (why are the craters of the moon not fuzzy?). They are however less general because the moments of inertia depend upon the internal degrees of freedom, if only because of vibrations, so that there is some coupling between the Euler angles, the angular momentum, and the internal environment.

Decoherence Originating from Scattering

There is something unrealistic about the example we considered, because nothing in the universe is completely isolated. There are always some particles present, even in intergalactic space, and there are always photons, if only from the cosmological thermal background and one may wonder what their effect is. This question was investigated by Joos and Zeh who have shown that the external environment is also a strong source of decoherence.[11] Their theory is given in the appendix and we shall limit ourselves presently to the leading ideas and consequences.

To understand how decoherence arises, one must resort to collision theory to find the phase effects originating from the collisions of a macroscopic object with the atoms, molecules, and photons in its external environment. It is known that, from the standpoint of quantum mechanics, the existence of scattering can be expressed by a nonzero phase shift between the wave functions describing the colliding partners (the object and one atom, for instance) before and after collision. Since the object is macroscopic, an elastic collision with a microscopic particle does not provoke a sizable recoil so that the position **x** of its center of mass is unchanged under the collision. It turns out that the collision phase shift depends upon this position and this property plays an important role in the generation of decoherence. Let us therefore explain its origin.

We shall denote by **k** the de Broglie wavenumber vector of the incident atom before the collision and by **k**′ its value after the collision. One can compare what happens when the center of mass of the macroscopic object has a position **x** or a position **x**′ by performing a translation over a distance **x**′ − **x**. It is known that the momentum operators are the generators of translation and let us apply this to the momentum **p** of the colliding atom ($\mathbf{p} = \hbar\mathbf{k}$). One can go from an ingoing atom-object wave function where the atom has a momentum **p** and the center of mass of the object is at **x** to another where the atom has the same momentum **p** and the center of mass is at **x**′ by multiplying the wave function by a factor $\exp(i\mathbf{k}\cdot(\mathbf{x}'-\mathbf{x}))$. When the same operation is done on the outgoing state, where the atom after collision now has the wave number **k**′, the overall multiplying factor is $\exp(i\mathbf{k}'\cdot(\mathbf{x}'-\mathbf{x}))$. This implies that the collision phase shifts must differ by a quantity $(\mathbf{k}'-\mathbf{k})\cdot(\mathbf{x}'-\mathbf{x})$ when one compares the two positions **x** and **x**′ for the center of mass of the object.

Let us now assume that the initial wave function for the center of mass of the object is $\psi(\mathbf{x})$. A unique collision with one atom, which is a random event, modifies the initial wave function by multiplying it by a factor $\exp[i(\mathbf{k}'-\mathbf{k})\cdot\mathbf{x}]$, the quantity $\mathbf{k}'-\mathbf{k}$ being random and its probability governed by quantum scattering. When many such collisions take place successively at random, the result can be described by saying that the modulus of the wave function does not change but its phase for different values of **x** becomes a random signal, uncorrelated for sufficiently different values of **x**. This is therefore a case where decoherence shows explicitly its character as a changing complicated phase. The outcome is that, when this random effective wave function is used to construct a density operator, the latter will be found to be practically diagonal in the position coordinates.

One can estimate the order of magnitude of the effect. To do so, start from an initial nondiagonal density operator with matrix elements $\rho(\mathbf{x}, \mathbf{x}')$ in position and the question is by what factor arising from decoherence this matrix element is reduced after a time t. One finds that, once again, it depends exponentially upon time and explicit calculations give for its value

$$\exp\{-\Gamma(\mathbf{x}' - \mathbf{x})^2 t\}.$$

The decoherence exponent Γ depends in a simple way upon a few physical quantities controlling the importance and the effects of scattering, namely, the average number n of external atoms (or molecules, or photons,...) per unit volume, the *total* collision cross section σ of these particles with the object (the theory does not assume that the collisions are necessarily elastic), their average velocity v (which is c for the photons), and the average de Broglie momentum k. It is given by

$$\Gamma = c_1 k^2 n \langle \sigma \rangle v,$$

where c_1 is a numerical constant of the order of 1 and the average of the cross section is taken over all the direction of the incident particles.

The macroscopic character of the object appears in the macroscopic value of the cross section, which is given essentially (for a given incident direction of the atom) by the area of the projection of the object onto a plane. For a ball of radius R, this is πR^2. The quantum nature of the phenomenon is shown by the occurrence of the de Broglie factor k^2, which can be written as $(p/\hbar)^2$.

If one considers again the case of a pendulum, assuming its cross section to be of the order of 1 square centimeter, in air under standard conditions for temperature and pressure, the decoherence factor after one nanosecond is now found to be of the order of $\exp(-10^{21})$. This is still much smaller than what we found earlier as the effect of the internal environment starting at zero temperature.

We might therefore summarize this by saying that the moon shows a sharp image in the sky because it receives light from the sun. The same result was also obtained in the previous chapter from the fact that the moon was produced by classical events and later submitted to classical conditions. In any case, everything goes in the same direction: the moon is there undoubtedly, even when nobody looks at it, and its outline is perfectly sharp. No doubt you noticed it.

Chiral Molecules

Another important example of decoherence by an external environment has been given by Jona-Lasinio and Claverie.[12] It is concerned with chiral molecules and it is an example where the collective system undergoing decoherence is not macroscopic since it is a molecule. A chiral molecule has two different shapes that are related to each other by a space inversion. In view of the invariance of electromagnetic interactions under space inversion, these two states have exactly the same energy and their wave functions ϕ_R and ϕ_L are related to each other by a parity transformation. The simplest example of that type is the ammonia molecule NH_3: it has the shape of a tetrahedron and if one thinks of the three hydrogen nuclei as sitting upon a horizontal plane, the nitrogen nucleus can stand either above or below the plane, the two states being related by a symmetry with respect to that plane. There are many other examples of chiral molecules, particularly among biomolecules such as DNA.

The quantum theory of chemistry rests upon the Born–Oppenheimer approximation. It starts as is well known by considering arbitrary given positions of the nuclei and then computing the wave functions and the energy eigenvalues of the atomic electrons when they are submitted to the Coulomb field of these nuclei. An energy eigenvalue $V(r)$ depends upon the various distances r separating the nuclei. One can then find the energy levels of the molecule by adding the mutual Coulomb repulsion of the nuclei to $V(r)$ and solving the Schrödinger equation for the nuclei. In the case of a chiral molecule, it is convenient to introduce among the variables describing the positions of nuclei one changing sign under a space inversion. This could be the height z of the nitrogen nucleus above the hydrogen plane in the case of the ammonia molecule.

If one keeps the distances between the hydrogen nuclei fixed while letting z vary, the potential $V(z)$ one obtains is shown in Figure 7.1. It has two strong minima at $z = \pm 1$ in convenient units, showing the locations where the nitrogen nucleus can sit and there is a strong barrier between them. This barrier represents the large amount of energy that would be needed to force a nitrogen atom to cross the plane of the hydrogen atoms. The potential also depends upon the distances between these atoms but this will be less important for our purpose.

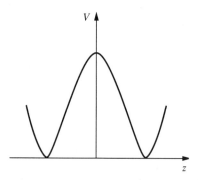

7.1. *The potential energy for a chiral molecule*

The symmetric form in z of the potential gives rise formally to a degenerate ground state with two different eigenfunctions ϕ_R and ϕ_L centered around $z = \pm 1$. When the Pauli principle is taken into account, the ground state does not remain degenerate but is split by exchange effects so that the wave functions for the actual ground state ψ_g and for the first excited state ψ_e are given to a very good approximation by

$$\psi_{g;e} = 2^{-1/2}\{\phi_R \pm \phi_L\}.$$

Their splitting in energy is essentially given by the matrix element of the potential $\langle \phi_R | V | \phi_L \rangle$, which is a small quantity because of the very small overlap between the two wave functions ϕ_R and ϕ_L.

One is therefore tempted to conclude that, in thermal equilibrium, there will be a statistical mixing of the two states g and e. A consequence of this would be that a medium containing the molecules would not have any optical activity and would react exactly in the same way (i.e., with the same refracting index) to a right-handed or a left-handed polarized light. This prediction is not however always in accordance with observation: even when the process for producing the molecules is parity-invariant, the fluctuations in the refractive index[13] show that the statistical state actually consists of right-handed and left-handed molecules in the case of compounds containing arsine AsH_3, whereas compounds of ammonia NH_3 behave as if consisting of symmetric and antisymmetric molecules. In other words, the density operator is diagonal in the $\{\phi_R, \phi_L\}$ basis for arsine and in the $\{\psi_g, \psi_e\}$ basis for ammonia.

The origin of this effect is once again decoherence. It is due in the present case to an interaction of a molecule with its environment, i.e., with the other molecules surrounding it, whatever their chemical nature may be. This can be shown by considering the effect of a very small perturbing potential $V_1(z)$ acting on the state of a molecule. It is convenient to choose it different from 0 only when the variable z is between two neighboring values a and b. Because of the very small energy splitting between the two states g and e and the fact that the effect of a perturbation is in first approximation inversely proportional to the energy difference, this is a situation where even a very small perturbation can produce a drastic change in the wave function. This means of course that one cannot use perturbation theory to find the effect of the external potential $V_1(z)$ but the problem is simple enough, because of the localization properties of the potential, to be investigated by other techniques in the theory of differential equations.

This has been done by Jona-Lasinio and Claverie and they found that the perturbed system has a ground state 0 and a first excited state 1, each one of them being localized in the vicinity of one of the two potential minima $z = \pm 1$ if the perturbation is not too small. Which one of them is the ground state depends on the sign of the perturbing potential and on its exact location. More precisely, if one denotes by $\chi_{0,1}(z)$ the two wave functions diagonalizing energy, their localization properties can be found by comparing their values at the positions of the two potential minima $z = \pm 1$. One finds them to be related by

$$\frac{\chi_0(1)}{\chi_0(-1)} = -\frac{\chi_1(1)}{\chi_1(-1)} \approx \exp\left\{-\frac{2}{\hbar}\int_a^b [2mV(z)]^{1/2}\, dz\right\}.$$

The energy splitting is also much increased, which may explain why the fluctuations are big enough to produce an optical activity. It is given by

$$\frac{E_1 - E_0}{E_e - E_g} \approx \exp\left\{\frac{2}{\hbar}\int_a^b [2mV(z)]^{1/2}\, dz\right\}.$$

These results are valid as long as the perturbing potential is much larger than

$$\Delta E \cdot \exp\left\{-\frac{2}{\hbar}\int_0^b [2mV(z)]^{1/2}\, dz\right\},$$

ΔE being the potential barrier at $z = 0$. There is also another condition on the range $b - a$, but it is practically always satisfied. These results mean in practice that the external perturbations can be very small and nevertheless very efficient in producing chirality effects, as long as they are large as compared with the very small unperturbed splitting $E_e - E_g$. The theory is quantitative since one can estimate the relevant parameters and particularly the order of magnitude of the actual external potentials by using other experimental data. The results seem to agree with observation. One can explain, for instance, why under standard conditions one observes an optical activity in compounds of arsine but not in compounds of ammonia because of the difference in their energy splittings.

Finally, we try to express most simply the nature of the effect. An isolated molecule has symmetric and antisymmetric states and their energy splitting is so small that they have the same population in thermal equilibrium. They have intrinsically no possible optical activity. When the molecules are in a real environment, two molecules with opposite values of z do not feel the same action of the environment at the same time, as is clear from their different shapes. This effect can be assimilated in the Feynman–Vernon formalism to a coupling depending upon the value of the parameter z or, in the scattering formalism introduced by Joos and Zeh, to a random phase shift depending again upon z. Whatever its precise mathematical origin (and it may have more than one) the interaction of the molecule with its environment will tend to diagonalize the reduced density operator according to the values of z, i.e., it will favor an equilibrium distribution containing two optically active populations. Polarization effects become then possible in that case as an effect of statistical fluctuations.

6. Back to Schrödinger's Cat

An obvious outcome of these various theories is the wide validity of decoherence for macroscopic objects (and also often for microscopic bodies). They may rely upon different mathematical techniques and also different detailed physical mechanisms, but the systematic occurrence of decoherence when there is dissipation points towards some kind of universality and a domain of validity far exceeding what one is able to compute explicitly. Decoherence takes place much more rapidly than dissipation and, though both effects behave exponentially in time, the characteristic time for decoherence is very much shorter than for dissipation.

If one leaves aside nondissipative systems, to be investigated in more detail in Chapter 10, one can then say that a macroscopic system will never be found to be in a quantum superposition of macroscopically different states. This is true in particular for the instruments that are used for making quantum measurements, where the experimental data are only so many classical properties that can be seen and registered. The essential consequence of decoherence has already been anticipated in Chapter 2 and it should have by now become clear: the data that can be obtained from a quantum measurement are described by a diagonal density operator, so that they cannot show any quantum interference effect. They are correctly described by ordinary probability calculus as so many different and mutually exclusive classical events.

Decoherence therefore provides an answer to one of the most difficult problems by which quantum measurement theory was plagued for many years. Its results can be combined with the ones already obtained in the previous chapter to provide a complete derivation of the classical behavior of dissipative nonchaotic and macroscopic objects, whether they occur in ordinary circumstances or they are used for performing quantum measurements. The previous theory of determinism, now completed by the understanding of dissipation, shows clearly the possibility of keeping a memory of past facts by a record. It can therefore be said that one has now a complete understanding of classical physics by considering it as an outgrowth of quantum mechanics.

The reader will have recognized that this is a clearcut answer to one basic aspect in the problem of Schrödinger's cat and a complete one for the problem of Wigner's friend. It says that, at any given time, the cat is either objectively dead or alive. There is no interference effect among the two cases. When we open the box and find the cat to be dead, our knowledge of dissipation effects allows us to make an autopsy and to say when she died. Of course, another question remains, which is to understand how one can go from this probabilistic standpoint according to which the cat is dead or alive to what is seen in reality where only one possibility is actually found to be true empirically.

All these results are so important that one should be careful to bring attention upon any remaining weakness in their proof or any uncertainty in their consequences. Three points are worth mentioning in this respect. The first one has to do with the fact that the separation of different possible measurement data into several clearcut events has only been shown to be true for the practical case where one observes only the collective properties of a measuring device. One might wonder what would happen if one were to make a quantum measurement upon the instruments rather than simply looking at them. Would decoherence appear again as a completely satisfactory answer for the problem of superpositions or could it be circumvented so that the problem we thought had been solved would reappear? This question will be considered in the next few sections. The second question is far less important and somewhat technical. It was found in various models that the density operator becomes diagonalized in the collective coordinates but not much investigation has been done of its diagonal properties in momentum or in the case when one uses quasi-projectors to express classically meaningful collective properties. It would in principle be necessary to give a satisfactory answer to these questions to conclude that the ordinary description of classical physics is wholly understood. The few specific models where one can get hints indicate that there is no real problem, but there is no complete answer, maybe because this is a somewhat boring problem from which nothing particularly remarkable is expected. Finally, there remains one very serious problem in the connection between classical and quantum physics. Everything that was obtained up to now was only a consequence of the first principles of quantum mechanics with no input of empirical information, though our efforts were directed towards a reconstruction of what is actually observed. All this is therefore pure theory. Since quantum theory is probabilistic, what is obtained can only deal with a possibility and it can say nothing about actuality. Empirically, a measurement gives only one datum; theoretically, we can only put all the possible data upon the same footing and we do not understand the uniqueness of actual facts. This is a very deep question that will have to be considered later on with all due attention.

CAN ONE CIRCUMVENT DECOHERENCE?

7. A Criticism of Decoherence

Bell's Argument against the Significance of Decoherence

After drawing the consequences of his work on decoherence in collaboration with Eliott Lieb, Klaus Hepp published an important paper proposing that decoherence provides a satisfactory answer to the old problem of quantum superpositions of macroscopically different states.[14] John Bell responded

DECOHERENCE

by a far-reaching criticism, which remained unanswered at that time.[15] He acknowledged the interest and the value of decoherence for *all practical purposes,* as he expressed it; he nevertheless denied the fundamental character of the answer, maintaining that the problem, as a question of principle, was still essentially the same as it had always been. This is obviously an extremely important point, since the challenge is to decide between a breakthrough or a stalemate in the progress of interpretation.

Bell's argument went as follows. When one starts from a pure state, it will always remain a pure state as long as it evolves according to the Schrödinger equation. Accordingly, even if the reduced density operator describing the collective observables becomes diagonalized, this is unessential: the full density operator still represents a pure state and it therefore contains the possibility of showing interference. True enough, no interference can be seen in a measurement of a collective observable, but there always exist more subtle observables that would be able to show it. By measuring such an observable, as was always assumed to be possible in principle, one will exhibit the existence of a surviving quantum superposition of macroscopically different states. The answer provided by decoherence is therefore valid "for all practical purposes", because one can only measure in practice a collective observable of a macroscopic object, but it is not a valid answer as far as basic principles are concerned.

One often answers the criticism by sticking to a strictly empirical point of view: Decoherence shows that all the measurements we can perform will never show a quantum interference and that is enough because no experiment will ever show us to be wrong. This is however a rather uncomfortable position. Let us consider an extreme example to illustrate it: An intergalactic civilization with very powerful technical means wants to play a trick on us. Its playful scientists have read Bell and they make surreptitiously a measurement of the type he proposed. Assume for instance that a man, in the year 1000, might have died of a cancer initiated by a unique irradiation originating from the decay of a uranium nucleus in his house of granite. He survived fortunately and he died only thirty years later from another cause. In the meantime, he had a child who was among the ancestors of Napoleon and also of Doctor Babbit, who now teaches quantum mechanics. The measurement made by the intergalactic jokers tests a wave function of the solar system. It contains a component of the nucleus decay products such that the old man died in the year 1000 and the measurement shows that this component is still present. The extraterrestrials show their results on TV news and Mister Babbit is asked to explain them. What will he say? Necessarily that there is no doubt that he himself exists and does not exist, that many books on history are both right and wrong. In a nutshell: facts don't exist as facts and truth is an empty notion. We already knew it from Schrödinger's cat and Wigner's friend who are apparently back again.

A Questionable Axiom

When looking at Bell's criticism, one immediately realizes that it relies upon an axiom, sometimes attributed to von Neumann though it is at most implicit in his works, namely, that every observable is measurable. Some authors introduce it as a rule in the formalism[16] but it should not be taken without some caution. We never met it, for instance, when trying to build up a consistent interpretation and one may legitimately question it. An apparently strong argument in favor of this axiom is provided by the ideal measurements, described in Chapter 2.4. Let us therefore examine this justification where it was shown how one could measure "in principle" an arbitrary observable A belonging to a system Q by coupling it with the momentum P of a "pointer". Is this really convincing? There are two ways of looking at it:

1. This argument assumes essentially that every hamiltonian (in the present case, the coupling hamiltonian between Q and the pointer) can be realized as a matter of principle. But one knows that there are only a finite number of particle types and that their hamiltonian always derives from four basic interactions (gravitational, weak, electromagnetic, and strong) so that this does not leave much freedom, certainly not enough to obtain all the couplings one can write with pen and paper.

    If one considers an abstract or pure quantum mechanics, i.e., a general "form" of physics, as von Neumann for instance did, this kind of assumption is legitimate. It is not so when one takes into account that matter is made of particles.

2. The pointer is after all just a one-parameter system, not a macroscopic one, so that its result has nothing to do with a real experimental datum.

    If however one pursues this argument with a counterargument, one can say that, in order to bring the pointer position to the rank of an observed phenomenon, one can use a truly macroscopic system to read it. This system will give a factual signature for the value of A. One is then back where one started: if this phenomenon is manifested by the value of a collective observable, the reading apparatus is again a truly macroscopic system in a superposition state, though still hiding it because of decoherence. A complete answer cannot therefore be found in that direction.

This discussion is an example of the endless arguments and counterarguments one can get from a formulation of quantum mechanics relying mainly upon words. It clears up the ground however on two points: (i) there is no cogent argument for or against von Neumann's principle at the level of elementary quantum mechanics. (ii) One should give up using only words and go to a quantitative analysis if one wants to clarify the question.

8. One Cannot Circumvent Decoherence

Let us make the problem more specific in the following way: a macroscopic apparatus M is initially in a state involving a quantum superposition of two macroscopically different states. This is what happens, at least formally, after it has been used to measure a quantum observable having two different eigenvalues. We shall assume that enough time has elapsed so that the reduced density operator of M has become diagonal by decoherence. Bell's idea is to be tested by measuring some noncollective observable A belonging to M in order to show that its complete density operator still represents a superposition. The measurement of A is made by another measuring device to be called M'.

We establish the following notations and conventions. The observable A has only two eigenvalues, say 0 and 1. The wave function of the first apparatus M is accordingly written as

$$\alpha \psi_0 + \beta \psi_1, \qquad (7.28)$$

the two wave functions being eigenstates of A with respective eigenvalues 0 and 1; they are time dependent and otherwise very complicated. When the second measuring apparatus M' shows that A has the value 0 (respectively 1), one pair of its conjugate collective observables (x, p) is found finally to be in a cell C_0 (respectively C_1) in phase space, according to the theory of macroscopic properties given in Chapter 6. The total number of *continuous* microscopic degrees of freedom of the apparatus M (including the microscopic ones) is denoted by N. The fact that these degrees of freedom are continuous will be found to be important because of the very large exponential behavior of the energy density. There are of course about three times as many continuous degrees of freedom as there are particles in M. There may be discrete degrees of freedom due to spin and they can be used in practice, for instance by using the corresponding magnetic properties, but one should not forget that each spin is carried by a particle and it is therefore associated with three continuous degrees of freedom. Similarly, the apparatus M' has N' continuous degrees of freedom.

Then come assumptions that will have to be justified later on though more conveniently stated beforehand:

1. The observable A can conveniently be taken to be the occupation number n_k of a specific energy eigenstate in the environment of M, or some observable essentially equivalent to it. This assumption is not however strictly necessary and it will be used only for the convenience of being explicit.
2. It was already noticed that the density of energy levels increases exponentially with the number of continuous degrees of freedom. With

the choice of observable made in (1), the probability amplitude β in equation (7.28) is exponentially small and can be written as

$$p = |\beta|^2 = K\exp(-CN^r). \qquad (7.29)$$

It will be shown later on that, for the electronic degrees of freedom in a metal, the exponent r is equal to $2/3$.

3. The two cells C_0 and C_1 in phase space are contiguous and they have scales (L, P) for x and p. The probability that a quantum fluctuation gives a wrong measurement result by showing a datum in C_1 for the value 0 of A or conversely is given under these conditions by

$$\varepsilon = (\hbar/LP)^q, \qquad (7.30)$$

and it was shown in Chapter 6 that q is equal to $1/2$. Note that, if the two cells C_0 and C_1 were at a definite distance D in phase space, ε could be exponentially small and the present assumption is essential because the algebraic character of ε will play a crucial role in the argument. The justification of this assumption is by the way the most tricky part of a complete discussion.

4. Finally, the product LP in equation (7.30) is proportional to some power of N', i.e.,

$$LP/\hbar = cN'^s. \qquad (7.31)$$

It will be shown in an example that one can take $s = 1/3$.

We can then use elementary arguments to show that N' is bounded from below by a large exponential in N as follows. The value of p contains interference terms coming from the initially superposed state of M. If one measures p, one will be able to show the existence of the superposition. In order to get p, one needs of course to perform many individual measurements under identical conditions. Most of the time, because of the smallness of p, these measurements will give a zero result ($A = n_k = 0$). The significant results $A = 1$ will occur very seldom. Let us just consider the probability for obtaining one such event, which is of course quite insufficient for knowing the value of p but anyway an indication that it is attainable. When such an event happens, one must of course be sure that it is a genuine measurement and not an erroneous one coming from a quantum fluctuation giving randomly a datum in C_1 when it should have been in C_0. The condition for this is $p \gg \varepsilon$. This is because, if the inequality were in the other direction, one would have to attribute the result to its most probable cause, i.e., to a fluctuation. The milder inequality $p > \varepsilon$, though quite insufficient to give a significant result, is again at least an indication that one may be approaching one.

Using equations (7.29–31), this condition gives a relation between N and N', namely

$$N' > K'\exp(CN^r/qs). \qquad (7.32)$$

DECOHERENCE

It is then clear that, since the apparatus M is macroscopic so that N is very large, say typically of the order of 10^{27}, the exact values of the constants K' and C do not matter. One finds in that case that N' must be roughly of the order of 10 to the power 10^{18}.

No object that big is conceivable in a universe where the total number of protons within the visible horizon is of the order of 10^{80}. It was furthermore already noticed at the beginning of the chapter that the working of such a monster would be incompatible with special or general relativity. The conclusions are therefore the following:

1. It is impossible to circumvent decoherence when it takes place in a sufficiently big object.
2. This is not only a practical matter. The principles forbidding it are not those of quantum mechanics properly speaking but of relativity, not even mentioning the finiteness of the universe.
3. Some observables cannot be measured, even as a matter of principle.
4. Decoherence is a fundamental answer to the inexistence of macroscopic superpositions and not only a practical one.

9. Justifying the Assumptions*

It remains now to justify the above assumptions and to improve the estimate of the constants. We shall reconsider the assumptions one by one. It will be found that they sometimes rely too much on incomplete arguments, if only because one does not have a sufficiently general theory of decoherence at one's disposal. Nevertheless, it appears that the problem under consideration can be approached in principle in a systematic way so as to escape the realm of philosophy.

The Choice of an Observable to Be Measured

It was said that a good observable to be measured might be the occupation number of a definite energy level for the environment (i.e., the projector upon the corresponding eigenvector). One might keep in mind the example of an oscillator environment with mutually incommensurable frequencies of the oscillators so that there is no degeneracy and a one-to-one correspondence holds between the energy eigenstates and a basis for the oscillator states.

One could also assume that some sort of Caldeira–Leggett formal description of the environment can be justified so that there are as many oscillators as there are energy eigenstates of the environment,[17] though each oscillator can be only in its ground state or its first excited state. This is probably valid for superconductors but not a general feature. One could also argue, by analogy with Zurek's model in Section 2, that what is really important is the number of different frequencies, which would give the same conclusions. This would lead us almost immediately to the conclusion that an occupation number is practically the only candidate. This is why it is selected as the best example, though one should not rely strongly upon this assumption and rather consider it as an example.

Let us however take this kind of model in order to discuss a bit more what should be the best observable measure in order to circumvent decoherence. Collective observables are of course excluded. One might think of planting a unique and specific foreign atom in the apparatus to be tested. Once the apparatus has acted to make a measurement, one would get a superposition of two states for the atom, among other features, and one might think of performing a measurement upon it. One can however construct a new reduced density operator involving not only the collective variables but also the atom. It is then easy to show that this larger density operator is still decoherent. The simplest way to show it consists in using the Joos–Zeh argument in Section 5 by taking into account the interaction of the atom with its neighbors and, through them, with the entire environment. The new reduced density operator can be used to describe any kind of measurement upon the atom and, since it is diagonal after a short time, it is clear that such a measurement cannot show any sign of coherence.

A similar reasoning applies if one tries to measure a collective excitation of the system, for instance, a phonon with a specific wave number. One can again lump together the degrees of freedom describing it with the collective degrees of freedom when one computes the reduced density operator and it would again be subject to decoherence. In order to show it, one can use directly the general theory by treating formally the excitation as a collective observable. As a matter of fact, Zurek's model of an environment made of a collection of two-state systems shows that, as long as one measures a subsystem while leaving out a large number of "atoms", decoherence still holds. In order to get something significant, one must therefore go as far as measuring practically simultaneously a whole set of commuting observables. The simplest way to do this is to measure directly an eigenstate of the environment energy, assuming no degeneracy as the case occurs when there is no overall constant of the motion.

We shall finally consider how a measurement of an occupation number could show the existence of interference, the corresponding level being denoted by k and the projector upon the occupation number being denoted by n_k. Expand the total wave function of M, say $\Psi^1(t)$, in the energy eigenstates of the environment

$$|\Psi^{(1)}\rangle = \sum_j c_j^{(1)}(t)|j\rangle, \qquad (7.33)$$

DECOHERENCE

where the collective state has not been written down explicitly for greater clarity. If the initial state is a superposition

$$|\Psi\rangle = a|\Psi^{(1)}\rangle + b|\Psi^{(2)}\rangle,$$

then the average value of n_k at time t is given by

$$\langle n_k \rangle = |a|^2 \left|c_k^{(1)}(t)\right|^2 + |b|^2 \left|c_k^{(2)}(t)\right|^2$$
$$+ \left\{ ab^* c_k^{(1)}(t) c_k^{(2)*}(t) + \text{complex conjugate} \right\},$$

clearly showing the existence of interference terms.

The Example of a Resonance Measurement

The main point of the following discussion has to do with the immediate vicinity of the two cells C_0 and C_1 entering in the assumptions of the previous section. A specific example is used for greater clarity. One may wonder whether a measurement of an energy eigenstate can be made, even in principle, in the case of a realistic system. Nothing distinguishes a definite energy eigenstate from its neighbors, except precisely for its energy. This means that one must select it by means of its energy, which amounts to a frequency marker. This cannot be much different from having it resonate with an external oscillating signal, almost exactly tuned to the frequency $E_k/\hbar$. Because of the closeness of the energy levels, the width $\Delta\omega$ of this signal must be smaller than $\Delta E/\hbar$, ΔE being the difference in energy with the nearest neighbor state. According to the fourth uncertainty relation, this means that the measurement will have to last a time longer than $\Delta\omega^{-1}$, i.e., much larger than the age of the universe for not too small a macroscopic system. We shall however disregard this obvious impossibility because this is only meant as an example and its peculiarities should not be given too much weight.

The resonating signal must be produced by some device D, which is a part of M'. It will be taken for granted that no natural signal is sharp enough so that it has to be produced by a properly devised man-made (or intelligence-made) system, which must be macroscopic. Two cases are possible:

Case 1. The device D chops off the frequencies of an ordinary signal except in a very narrow frequency band. The width of this band is at most of the order of the difference in energy of two neighboring levels. The device D is necessarily a classically working instrument and it performs its expected duty when its collective coordinates and momenta belong to some well-defined cell C_1 in classical phase space. There must be some sort of continuity between the frequency one obtains and the (collective) parameters of the device, or at least we shall assume it. This is by the way the weakest point in the present discussion and anyone wanting to find a flaw in it (and perhaps proposing a specific experiment) would have to propose a counterexample. After thinking from time to time about this for more than a year, I could find neither a counterexample nor an entirely general proof. Anyway, this assumption means that the properties of having the signal in a frequency band $\Delta\omega$ or outside it correspond to two cells C_0 and C_1 adjoining each other. This implies that the two properties "the signal fits the frequency of level k" and "it does not fit it" are associated with two quasi-projectors E and E' such that $[E, E']$ is of order ε (in trace norm) with the previous value of ε.

Case 2. The device D directly generates the signal by whatever means. It produces a signal resonating with level k and with no other. It is then conceivable that it does not work continuously and the two previous properties will become simply "D works" and "D does not work". The corresponding projectors E and E', according to Theorem 6.B, may then have an exponentially small commutator. However, D must now select by itself the level k. Since D is made of particles, it cannot produce naturally (say, by emission) a signal that would exactly fit the frequency $E_k/\hbar$ up to an error of order $\Delta E/\hbar$ because we would be back to Case 1. So, one must resort to a computational device generating at least the number E_k with an error ΔE, leaving to the imagination the task of converting the number in the computer into a testing device for the state k so defined. It turns out that this computer, or anything like it, will need at least $E_k/\Delta E$ real memories. (This last statement is in fact not trivial and we shall take it as granted for a moment in order to first complete the discussion, with the intent of coming back to it later on.)

We are therefore left with two possibilities. In Case 1, the two cells C_0 and C_1 are contiguous, as was assumed in the previous section. In Case 2, the apparatus M' has at least $E_k/\Delta E$ degrees of freedom and this lower bound on N' is essentially the same as the one we found in Section 8.

Numerical Estimates

We shall first relate the parameters (L, P) to the number of degrees of freedom N' in Case 1. The device D is made of matter; L and P are error bars; L cannot be larger than the size L_0 of a typical piece of D. The scale P is at least a fixed quantity. It is defined in most cases by the temperature T' of M' and of order $(kT'/m_e)^{1/2}$ in a metal (m_e being the electron mass), or a quantity compatible with a classical control if one were to let T' go to 0. A rough estimate for N' is therefore obtained by relating it directly with L by

$N' = \nu L^3$, ν being the number of conducting electrons per unit volume in the case of a metal. One has then

$$\varepsilon = [\hbar P^{-1}(\nu/N)^{1/3}]^{1/2}.$$

Finally, one needs only an estimate for the occupation probability p of a specific energy state. In view of the tremendously large number of eigenenergies, it must be of the order of $E_k/\Delta E$. This ratio is given, in the case of a metal, by

$$\frac{\Delta E}{E_k} = K_0 \exp\left\{-K'\frac{m_e \varepsilon_F v^{2/3}}{\hbar^2} N^{2/3}\right\},$$

where m_e is the electron mass, ε_F the Fermi energy, and v the volume per conducting electron. The constants K_0 and K' are known explicitly but unnecessary for our purpose. This gives, for instance, a constant C in equation (7.32) for copper of the order of 0.1, i.e., practically 1.

It is then easy to complete the calculation, which need not be done explicitly since the only interest of the present evaluation was to make sure that no large correction can modify the conclusions already made.

About Experiments Involving Computers

The second case considered above assumed that some computer could be used to generate the number E_k and the necessary memory in it was asserted to be necessarily very large, of the order at least of the (exponential) number of energy eigenstates in which **M** can be found. (This number will be denoted in the following by J.) This assertion remains to be justified. It turns out however that it brings us into a very interesting and rapidly developing field of research and we shall therefore consider it with a wider generality. Our aim is to show directly that any computer-generated measurement involves necessarily a tremendously big computer if it is to circumvent decoherence.

It will first be necessary to say what *algorithmic complexity* is. Computers deal with numbers, the simplest form of which is given by strings of bits (1 and 0). They also work according to a program, which can also be coded as a string of bits. The theory of information processing is much simpler if one does not bother about the limitations coming from the finiteness of actual computers and we shall therefore deal with a so-called universal computer able to write down infinite strings if necessary. This is not contrary to the goal we have in mind since it will be enough for our purpose to show that the smallest program capable of performing the necessary calculations consists of at least J bits, from which it will follow that an actual computer doing it needs memory of at least size J.

It can be proved that all universal computers are equivalent in the following sense: the programs defining the same calculation in two of them differ in length (in bits) by a finite quantity, which is at most the length of a compiler from one to the other. An example of our problem is to generate E_k or something similar as the output of a computer. If E_k needs to be written with at least J bits to separate it from neighboring values, this does not necessarily mean that the amount of memory to be used is as high as J. One might conceive of a program allowing the calculation of E_k and acting directly upon other components of M' to control their functioning without having to explicitly write down the actual value of E_k. If this program were much shorter than the length of E_k, the previous discussion would have to be entirely reconsidered. As an example, it is easy to write down a rather short program computing the first billion decimals of π. On the other hand, the smallest program computing a given "random" number plainly consists in writing down this number.

These considerations are made clearer by introducing the concept of *algorithmic complexity*.[18] The algorithmic complexity $I(q)$ of a binary string q is the length (in bits) of the shortest program which, when presented to a universal computer, causes the computer to compute q and then to halt. It would take too long to explain in detail how this notion can be applied to quantum calculations and we refer the interested reader to a basic article by Carlton Caves on the subject.[19] We shall accordingly consider a specific example. Suppose one wants to compute the wave function Ψ of a counter following a Stern–Gerlach detector, some time t after the measurement. The spatial wave function of the incoming atom will be denoted by ϕ. One assumes the hamiltonian to be known and, if one can use a simple enough model for it, the part of the program containing the necessary information about the hamiltonian can be rather short. If the initial wave function $\Psi(0)$ is simple (having a small algorithmic complexity), it may turn out that $\Psi(t)$ also has a small algorithmic complexity. One may convince oneself of this by considering the academic example of free electrons in a perfect crystal, if they constitute the active part of the counter and if the model is reliable. It seems therefore at first encounter as if the result we wanted to obtain does not hold.

Caves notices however that, because of the close vicinity of eigenenergies, the calculation is extremely sensitive to any small error in the initial data. In essence, the calculation is a chaotic process. The initial data must be controlled with much information, of order J, if the output is to be reliable. This means in the present case that the spatial atomic wave function ϕ must be known with this amount of information. There is however a basic result[20] according to which the average algorithmic complexity of a string obtained from a random process is equal to the information obtained from its knowledge, up to an additive finite constant. This implies in the present case that the initial wave function ϕ must be generated by a preparing device having an algorithmic complexity at least of order J. The problem we had of knowing $\Psi(t)$ with enough precision is thereby brought back to the preparing instrument and nothing is gained. Somewhere there must be a computer with a program needing at least memory of size J to be written down. The same holds true if, in place of asking for the full wave function $\Psi(t)$, one asks only for one of its components along some simple basis, (with small algorithmic complexity) such as one of the coefficients $c_k^{1,2}(t)$ in equation (7.33) whose previous knowledge is necessary in order to assert the existence of interference in the special experiment we described.

10. The Direction of Time

The present response to Bell's criticism can be used to shed some new light on an old problem of irreversible thermodynamics and also to show that the direction of time in logic must be the same as in thermodynamics.

Thermodynamics and Irreversibility

Since the beginnings of Boltzmann's kinetic theory, there have been controversies around the conflict opposing the time reversal invariance of classical dynamics and the unique direction of time required by the second principle of thermodynamics. A well-known example is the following. All the molecules of a gas are initially confined in one half of a cylinder. The confinement is realized by a thin membrane. The membrane is destroyed and the gas expands in the whole cylinder. There was therefore initially a partially ordered state and there is finally a completely disordered one, something expressed by thermodynamics as an irreversible increase in entropy.

Is it really always irreversible? This assumption was frequently criticized. Let us leave aside the membrane and let us assume nevertheless that the molecules were prepared initially in one half of the cylinder. This is somewhat unrealistic but theoretically conceivable. At a later time t, each molecule has well-defined position and velocity in classical physics. If one then prepares another cylinder where the molecules are exactly located at the same positions but the velocities are exactly opposite, the classical laws of dynamics predict that, after a time t, all the molecules will be located in one half of the cylinder. It is therefore conceivable that order might come out of disorder and entropy can in principle decrease.

There are two standard answers to this objection. The first one is empirical. It says that one can always prepare an ordered state, in the present case with the help of a membrane. On the contrary, it is impossible in practice to prepare all the molecules of a gas with precisely defined positions and velocities such that the enormous billiard game in which they participate results in bringing all of them to the same corner.

The other answer has some similarity with the first one, though it is more concerned with the theoretical formulation of the problem. It says that one cannot give the positions and the velocities of the molecules with infinite precision, but only within some error bounds. One must therefore consider a whole family of initial states and not only a unique one. This situation must be described in the framework of probability calculus, which can only predict average values. The reversal of time upon the situation realized at time t will reestablish an ordered state at time 0 only for a completely negligible set of positions and momenta, i.e., with zero probability. This is what is called the approach of *coarse graining,* meaning that physics can only deal with finite regions in phase space but not with individual points.

The same problem also exists in quantum mechanics where it becomes even more troublesome, if possible. One can define the entropy in terms of the density operator by

$$S = -k\text{Tr}\{\rho \log \rho\},$$

where k is the Boltzmann constant. The time evolution of the density operator under the Schrödinger equation, $\rho(t) = U(t)\rho U^\dagger(t)$, implies that the entropy is a constant because of the unitarity of the evolution operator $U(t)$. This is puzzling since it looks at first sight like a universal property in spite of the fact that quantum mechanics is by itself already a probabilistic theory.

Theorists try to obviate this difficulty by again using methods of coarse graining that we don't need to describe, except to mention a criticism one often hears, which is that the choice of some coarse graining rather than another is arbitrary and it is not specified by an objective necessity. Entropy would then depend upon our arbitrary choice.

Some people also say that one observes only collective quantities such as pressure, volume, temperature, energy, and so on. It is then possible to make use of methods analogous to the ones we met in the theory of decoherence by restricting the discussion to a reduced density operator. One then finds that the corresponding entropy increases. This looks like a satisfactory answer. It meets nevertheless the same old criticism: If one applies time reversal to the density operator $\rho(t)$ and prepares the quantum system with exactly this time-reversed density operator, one will recover the ordered state by a straightforward time evolution. The entropy one obtains from a reduced density operator will therefore also decrease. One is therefore left at the same point where Boltzmann and Ehrenfest stood long ago.

It is interesting to consider if something new can be added to this old problem by taking into account a feature of quantum mechanics that had been ignored up to now, namely that some measurements and therefore some preparations of a state are impossible, even as a matter of principle.

DECOHERENCE

Let us consider in greater detail how one can prepare a state that would be the time-reversal of $\rho(t)$. It will be simpler to consider the case when one has initially a pure state, which is then also the case for $\rho(t)$. The theory of decoherence shows that the wave function at that time is very subtle and precise; the phases of its probability amplitudes upon a basis of energy eigenstates for the environment are extremely sensitive to the slightest modification. In order to prepare the time-reversed state, one would have therefore to produce these amplitudes with an extreme accuracy. This is by many orders of magnitude much more difficult than the unique ideal measurement we discussed in Section 8. The measuring behemoth one met in that case would be like a speck of dust when compared with the preparing device that is now needed.

This means that the standard empirical answer can now be given a higher status: it stands upon first principles. One can totally exclude the possibility of preparing the exact time-reversed state of a given state, at least when the system has a large enough number of continuous degrees of freedom. It also means presumably that there is an objective limit to the conceivable fineness of a coarse graining, corresponding to the conceivable upper size of a device that can be used in a measurement or a preparation.

Logical Time and Thermodynamical Time

We can elaborate a bit more upon these results: The direction of time in thermodynamics can be defined as going from a state that is prepared towards another that cannot be prepared. This is similar to the usual idea according to which it is going from order to disorder, except that the notions of order and disorder are now given a meaning within the framework of quantum mechanics. The ordered states are easy to produce, theoretically easy to conceive, and they occur in reality. Disordered states are lost among a multitude of similar ones, they cannot be analyzed with precision, and there is no method allowing the production of one of them, except through the spontaneous evolution of an ordered state, in which case there is no way of reproducing it exactly otherwise.

This is most easily seen in the breaking of a quantum superposition where the decoherence effect shows much better what happens than the dissipation effect alone. One starts from a complete density operator $\rho(0)$; as a matter of fact, one does not know exactly what it is except that its trace over the microscopic degrees of freedom is a given reduced density operator $\rho_r(0)$. It can also happen that a few other things are known, such as the average energy of the internal environment, for instance. This ignorance does not really matter because any initial $\rho(0)$ will generate, by dynamical evolution, another state $\rho(t)$ such that, whatever $\rho(0)$, the reduced density operator $\rho_r(t)$ is the same. On the contrary, if one applies a time reversal operation to $\rho_r(t)$

and calls it $\rho'_r(0)$, there is a unique operator $\rho'(0)$, among all those having $\rho'_r(0)$ for their partial trace, that will generate the time-reversed operator of the reduced initial state $\rho_r(0)$ by dynamical evolution. It is furthermore impossible to prepare $\rho'(0)$, even as a matter of principle, when the system is big enough.

It may therefore be said that the really important thermodynamical direction of time is the one due to decoherence rather than dissipation, as Léon van Hove noticed first.[21] It corresponds to the formal choice of a retarded Green's function rather than an advanced one (the choice of $+i\varepsilon$ rather than $-i\varepsilon$ in the calculations of Section 3) when one solves the equations of motion.

If one then considers how we arrived at the impossibility of preparing some states, one sees that logic entered the proof in a significant way. It might also be recalled that the existence of a privileged direction of time in logic was already noticed. One can then ask whether there is a relation between the two directions of time in thermodynamics and in logic. In fact, they must be the same as one can see easily. Decoherence and dissipation follow the same direction of time. When decoherence is used to build up classical physics, one must perform a trace operation over the environment so as to concentrate strictly upon classical properties, while eliminating any detailed memory of the environment. This procedure can always be followed when going from an initial ordered state towards a final less-ordered one. The reasons are the same as before: this is the direction where a prediction can be made with confidence when taking dissipation into account. One can logically predict for instance that a pendulum initially swinging will finally hang motionless along its wire, but one cannot go the other way round. If one were trying to do it, one would have to assert the exact state of the motionless pendulum, including of course its environment with all the tiny phase details. If this were a pure state, it could be expressed completely by using a quantum projector, i.e., it would be described by a quantum property. However, since no preparing device can reproduce it, there is no means of linking this quantum property to a classically meaningful phenomenon, i.e., to associate it with an experimental fact. Accordingly, the logical implications to be drawn among classical properties in the framework of logic, when it is applied to a real physical situation, are restricted by decoherence in such a way that *the direction of time in logic must be the same as the direction of time in thermodynamics.*

It should be observed that this statement is not an intrinsic necessity of pure quantum mechanics. If the universe contained only two or three particles, there would be no such restriction and one would be allowed to choose arbitrarily the direction of time in logic. The existence of macroscopic systems offering the possibility of classical properties is therefore essential, for this question as well as many others in interpretation. Finally, it should be mentioned that Gell-Mann and Hartle have also indicated the possibility of linking the direction of time in logic with the cosmological direction of time. It would be interesting to extend the present considerations to this framework, but this has not yet been done.

Are Observables Observable?

The present analysis shows that the projector upon a specific energy eigenstate of the environment is an observable that cannot be measured. It is practically obvious that, among the multitude of observables that are conceivable in quantum mechanics, only a very small proportion are measurable and most others only have a mathematical status. Very few observables are measurable. This does not mean that an observable that cannot be measured definitely has no interest. It may be useful, for instance, to consider the probability of occupation of a given energy eigenstate of the environment in some theoretical discussions. This is necessary if one wants to prove that it is not measurable. One can also consider some logical properties mentioning it. One can even compute their probabilities, but they are only a tool of language without any empirical meaning. It may also be noticed that von Neumann's chain, which was mentioned in Chapter 2, where a measuring apparatus measures another, to be measured by a third one, and so on, is shown to be meaningless because the second apparatus is already an impossible monster.

APPENDIX: DECOHERENCE FROM AN EXTERNAL ENVIRONMENT

The calculations showing the effect of decoherence coming from an external environment will be given in more detail in this appendix. It is based upon original work by Joos and Zeh,[22] though we find a difference with them of a factor 8π in the decoherence exponent, which may be due to a slight error in their work. It has in any case no practical consequence.

One considers a macroscopic object. Choose a basis of states where the center-of-mass position X is diagonal. The other degrees of freedom, either collective or microscopic, are not explicitly written down so that a vector in this basis is simply denoted by $|x\rangle$. One wants to investigate the effect of decoherence upon the density operator $\rho(x, x')$. Consider an initial state $|x\rangle|\chi\rangle$ where χ represents a particle in the external environment, either an atom, a molecule, or a photon. The scattering of the particle by the object does not produce a recoil of the object when it is heavy enough. After a scattering, the particle is in a state $Sx|\chi\rangle$ where the operator S_x is the collision matrix for a particle hitting the object located at position x. The interaction therefore corresponds to the transition

$$|x\rangle|\chi\rangle \to |x\rangle S_x|\chi\rangle.$$

The density operator for the massive object after collision is obtained by taking a partial trace of the complete object-particle density operator over the states of the particle. One thus obtains

$$\rho(x, x') \to \rho(x, x') g(x, x'),$$

where

$$g(x, x') = \langle \chi | S_x^\dagger S_x | \chi \rangle.$$

Let us assume that the whole system is enclosed in a cubic volume with a large side L. The possible values of the particle wavenumbers k are therefore discrete and one can choose a basis such that

$$\langle k | k' \rangle = \delta_{kk'}. \tag{7A.1}$$

Denote by S the collision matrix when the object is at the origin of coordinates and its matrix elements can be written as $\langle k|S|k'\rangle = S(k, k')$. The operator S_x is obtained from S by a space translation along the vector x so that

$$\langle k | S_x | k' \rangle = S(k, k') e^{-i(k-k') \cdot x}.$$

We first consider the case when the object is hit by a beam of monokinetic particles with a fixed wavenumber k_0. This corresponds to $|\chi\rangle = |k_0\rangle$. The decoherence factor $g(x, x')$ is given in that case by

$$g(x, x') = \sum_k e^{-i(k-k_0)\cdot(x-x')} \cdot |S(k, k_0)|^2.$$

We separate out the contribution of the term $k = k_0$ from the sum. The quantity $|S(k_0, k_0)|^2$ is the probability for no interaction. Since the beam crosses an area L^2 along its way, this probability is related to the total cross section σ_T of the particle upon the object by

$$|S(k_0, k_0)|^2 = 1 - \frac{\sigma_T}{L^2}.$$

For $k \neq k_0$, one must take into account the conservation of energy in a scattering. Since the recoil is negligible, this gives simply $|k| = |k_0|$. The quantity $|S(k, k_0)|^2$ is related to the scattering cross section. To get this relation, let us introduce an element of solid angle $d\Omega$ in the direction where

DECOHERENCE

the particle is scattered. Since the normalization (7A.1) corresponds to one particle per volume L^3, one has

$$\sum_k |S(k, k_0)|^2 = \frac{1}{L^2} \frac{d\sigma}{d\Omega} d\Omega,$$

where the sum over the vectors k is restricted to the solid angle $d\Omega$. The quantity $d\sigma/d\Omega$ is the elastic differential elastic cross-section. Its integral σ_e over all directions is the total elastic cross section. It is equal to σ_T when there is no reaction between the particle and the object. For instance, a red object scatters red photons elastically and it absorbs blue photons, so that the total and elastic cross sections are not equal in that case. In general, one has $\sigma_T \geq \sigma_e$.

One thus gets a relation between the decoherence factor and the elastic cross section:

$$g(x, x') = 1 - \frac{\sigma_T}{L^2} + \frac{1}{L^2} \int e^{-i(k-k_0)\cdot(x-x')} \cdot \frac{d\sigma}{d\Omega} d\Omega. \qquad (7A.2)$$

It is convenient to split the decoherence factor into its real and its imaginary parts and to write it as

$$g(x, x') = 1 - \frac{K(x, x')}{L^2} - i\frac{\phi(x, x')}{L^2}.$$

The functions $K(x, x')$ and $\phi(x, x')$ may be rather complicated. This is due to the fact that the elastic cross section contains a lot of information about the object. It can describe, for instance, how the object can be seen from every angle, for all colors. Fortunately, one needs only a few of its properties, which are the following:

1. When $x = x'$, one has

$$K(x, x) = \sigma_T - \sigma_e, \qquad \phi = 0.$$

2. When the distance $|x - x'|$ is large compared with the wavelength, one has $|k|\cdot|x-x'| \gg 1$, the contribution of the integral in equation (7A.2) is negligible, and one has $K(x, x') \approx \sigma_T$.

3. When $|x - x'|$ is small compared with the wavelength, one can expand this integral to second order to get

$$K(x, x') \approx \sigma_T - \sigma_e$$
$$+ \sum_{i=1}^{3} \sum_{j=1}^{3} \frac{1}{2} (x - x')_i (x - x')_j \int (k - k_0)_i (k - k_0)_j \frac{d\sigma}{d\Omega} d\Omega.$$

From this one gets

$$K(x, x') \cong \sigma_T - \sigma_e + \kappa |x - x'|^2 k_0^2 \sigma_e,$$

κ being a constant of the order of 1.

Since a section L^2 of the quantization volume is very large, the decoherence factor due to the scattering of a unique particle is practically equal to 1. However, the accumulation of many scatterings influences it rapidly. If the beam of particle contains n particles per unit volume, the number of them crossing a section L^2 of the box during a time t is nL^2vt, v being the particle velocity. One then gets

$$g_t(x, x') = \left[1 - \frac{K + i\phi}{L^2}\right]^{nL^2vt} \cong \exp[-(K + i\phi)nvt].$$

When $|x - x'|$ is large with respect to the wavelength, this gives

$$|g_t(x, x')| \cong \exp(-nv\sigma_T t). \qquad (7A.3)$$

The quantity $\sigma_T nv$ is the total number of particles interacting with the object during the time t. The exponent in equation (7A.3) is therefore in general a large quantity, except for very small values of t or for a very small density of particles in the environment.

When $|x - x'|$ is small compared with the wavelength and when there is no absorption ($\sigma_T = \sigma_e$), one gets

$$|g_t(x, x')| \cong \exp(-\kappa |x - x'|^2 k_0^2 \sigma_e nvt).$$

Here again, this quantity becomes rapidly very small, except for very small values of the distance $|x - x'|$. One can therefore conclude that there is effectively a dynamical diagonalization of the density operator in the position variables.

These results remain valid when one considers a continuous distribution of wavevectors in place of a monochromatic beam. One needs only replace the quantities occurring in the decoherence exponents by their average values.

PROBLEM

The decoherence factor (7A.3) can be written as $\exp(-t/T)$, where T may be called the decoherence time. One considers as the object a sphere having a radius R, under the following conditions: (i) in air at normal temperature and pressure, (ii) in a perfect vacuum at the surface of the earth, in

DECOHERENCE

the full light of the sun, (iii) in an intergalactic vacuum, containing only the cosmological 3°K radiation, and (iv) in a laboratory vacuum with 10^6 particles per cm^3.

Joos and Zeh give the value of the decoherence times (in seconds) for these various cases as the following:

Object	Dust	Aggregate	Big Molecule
R(cm)	10^{-3}	10^{-5}	10^{-6}
(i)	10^{-36}	10^{-32}	10^{-30}
(ii)	10^{-21}	10^{-17}	10^{-13}
(iii)	10^{-6}	10^{+6}	10^{+12}
(iv)	10^{-23}	10^{-19}	10^{-17}

Comment on these values.

8

Measurement Theory

1. Reality and Theory. Facts and Phenomena

The two previous chapters have shown why classical physics is so good an approximation in every ordinary circumstance that it should be the most natural framework for man's visual representation of reality and, owing to its proper logic, the foundation of common sense. Two conditions were found to underly the applicability of classical physics: dynamics should be regular and decoherence should have already taken place, the second condition implying the existence of dissipation. It may be noticed moreover that the first condition, in so far as it refers to a description of the initial state in classical terms, assumes implicitly that the initial state is a state of fact, as it was called in Chapter 6, namely a situation where the quantum state is in agreement with a classically meaningful quasi-projector. The description of classical physics is moreover not restricted to definite conditions and it is easily extended to the case when ordinary probability calculus applies to it. The initial state is in that case a weighted sum of classical quasi-projectors. It might also be mentioned that the second condition for classical behavior is sufficient and not always necessary since one knows many cases of dissipationless systems behaving in a classical way (e.g., superconductors, light in the conditions of geometrical optics).

The classical properties, whose existence is thereby predicted by the theory, possess almost all the characters of the actual facts of experimental physics. They can be described in terms of classical concepts, registered, later remembered or recovered from a record; different properties are mutually exclusive and, when several of them can occur alternatively, one can use ordinary probability calculus to discuss them. They are acquired once and for all and no subtle experiment can ever shed a doubt about their existence, past or present (as far as decoherence cannot be circumvented).

The spirit of this approach may be worth stressing again: Contrary to what is usually done, we have never considered the existence of classical properties as being sufficiently granted by a direct observation of reality, but it was derived theoretically from the first principles. This procedure goes a very long way towards the goal of establishing the consistency of quantum mechanics but it also has a very important consequence or, if one wants to say so, a drawback. Since the whole construction relies upon a probabilistic theory, the classical properties it predicts can only be potential, things that can happen and not things that do happen.

Another way of saying this is that a last basic character of real facts is still missing, namely their uniqueness. One can never expect it in any case from a theory where even the language one must use is founded upon probabilities. One sometimes deems the existence of this ultimate and essential difference between theory and reality as a deep conflict between them. This question is certainly worth more than a short discussion and we shall come back to it later on; it is undoubtedly the touchstone of quantum mechanics. We shall however only draw a very simple consequence from it, which is to be careful not to use the same word for two different notions. Accordingly and from then on, the classical properties of macroscopic objects will be called *phenomena* when they are conceived as a consequence of the theory, whereas they will be called *facts* when they actually occur in reality. What we obtained in the last two chapters is therefore a theory of phenomena and we must now face the problem of measurements with their factual data.

2. An Introduction to Measurement Theory

We can now construct measurement theory in the same way as we did for classical physics, namely along a consistent deductive way originating from the basic principles. The task will be in fact much easier and one can say without excess that quantum measurement theory is not a difficult problem, the really hard one being classical physics. The present chapter will deal with this question and it consists of four parts, the first two being concerned with the practical applications of quantum mechanics and the other two with perhaps less necessary and somewhat more philosophical questions about the relation of quantum physics to reality.

The first part deals with a unique measurement. We first try to make clear what a measurement really is and one of the main conceptual keys is to distinguish between two notions about which there is often some confusion: the first one concerns what is given and shown by an apparatus, and it is by necessity a classical property, the *datum;* the second one is more properly the *result* of the experiment and it is a quantum property of the measured system stating the value of a quantum observable, as, for instance, the value of a spin component. A large part of measurement theory hinges upon a rather simple theorem asserting the logical equivalence of data and results. Its most useful consequence however is the usual Born rule for the probabilities of the various data and results. This is obviously a very important practical result but it has also another interesting aspect in the present approach, which is that it marks the threshold from where probabilities cease to have only a logical role and they can acquire at last an empirical meaning.

The second part of the chapter considers two successive measurements and its main outcome is the empirical rule of wave packet reduction, so that one may be said to have recovered the two essential rules of the Copenhagen interpretation concerning empirical physics. What is perhaps even more important is that reduction is nothing mysterious when one applies it in practice to the analysis of a long series of two successive measurements. It is just an economical rule simplifying the use of histories and relying on two common conventions: one is only interested in the data of the two measurements and one does not wish to know anything about the quantum properties of the measuring instruments. Reduction appears in these conditions as the consequence of a simple theorem in logic requiring no mysterious physical effect in the system that is measured twice successively. The only important physical effect entering in the result is the existence of decoherence, which takes place obviously in the measuring instruments and not in the measured system. The present theory being general, one obtains of course the reduction rule in general and not only in the case of ideal experiments.

In the third part of the chapter, we will come to grips with what is certainly the deepest problem of quantum mechanics, namely, how to deal with the existence of a unique actual fact as the datum of an individual measurement, when the theory can only cover the various possible outcomes as so many phenomena necessarily put on the same footing. This is an opposition between facts and phenomena, reality and theory, and something very essential indeed.

Our strategy will be to peel off the question so as to arrive at the root of the problem. We will accordingly develop the theory a bit more so as to make clear where the difficulty lies. This development will show that the logical interpretation of quantum mechanics is not completely averse to the unicity of facts and it can on the contrary rather well agree with it. The crux of the matter is that this agreement cannot hold for all times. If there exists a time (to be called the present one) where there is a unique set of facts in place of all possible phenomena, then it turns out that the past should also be unique; only the future has to remain entirely potential. The opposition between reality and theory is therefore not so sharp as some people would think and it is on the contrary rather remarkable that they agree about the most basic features of past and present, which is their uniqueness, as opposed to the future and its potentiality.

The real difficulty is that this conciliation cannot withstand the ever-changing character of the present time. In other words, the theory offers no mechanism, no possibility, of explaining the occurrence of a unique fact. Our next move will be to wonder whether this difficulty really belongs to physics or to the "meta" part of physics (some sort of philosophy of knowledge). A good reason for contemplating this possibility is that there exists a famous answer to the question, the so-called "many-world interpretation" introduced by Everett.[1] It starts from a supposedly unique given initial state of the universe and it leads to a multitude of presently existing realities, all different except in the incapacity to communicate with each other because of decoherence. The recent results concerning the impossibility of circumventing decoherence would then support the consistency of Everett's answer, except that it assumes that reality is not unique. One cannot however assert as certain the unicity of reality because Everett's construction shows that it could be multiple without our knowing it. The point is therefore that many people do not refuse Everett's answer because it is inconsistent but because of philosophical reasons consisting in a legitimate but a priori preference for a unique reality. In other words, if quantum mechanics were absolutely true and Everett were right, no experiment would be able to confirm or to reject it. The question of actuality is therefore not necessarily a matter of physics.

There exists another answer besides Everett's. It is pretty obvious once the problem has been sufficiently clarified and it even becomes possible at that point to recognize in it the gist of Bohr's opinion when he talked about wave reduction. The idea is that quantum mechanics has crossed an ultimate threshold in the history of our knowledge of the basic laws of physics, beyond which a mathematical theory must acknowledge its ultimate inability to account for all the aspects of reality. It must do so because otherwise no distinction would be left between theory and reality, something we shall try to argue as an absurdity. To say by the way that quantum mechanics has reached such an ultimate frontier in the field of knowledge means that what one is contemplating at its most extreme point of contact with reality is a witness to its feats and not at all to its failure, as quite a few people believed.

The last part of the chapter will abandon these eerie questions for a more technical but nevertheless important one: what can be said with certainty to be true in a world obeying quantum mechanics? A part of the answer is obviously that facts should be considered as true. A more subtle one is that a past fact can also be held to be true in view of a good record. Some other properties dealing with a microscopic observable can also be said to be true, with due regard for all the conditions this implies. This is where one rejoins Bohr's considerations about what can be said and cannot be said in view of the experimental setup, though one must disagree with his interdiction of talking sensibly about the microscopic world. Once again, he had been right when saying what is allowed and wrong when saying what is forbidden.

This discussion of truth (or what can be said to really happen) will be finally applied to the preparation of a state so as to give an explicit mathematical construction of the state of a system expressing how it has actually been prepared. The whole construction of interpretation will thus close upon its starting point and thus achieve consistency.

MEASUREMENT OF A SINGLE OBSERVABLE

The Framework

A measurement is a special kind of experiment where two physical systems interact. One of them is the measured system and it will be denoted by Q. The other one is the measuring apparatus to be denoted by M. The measured system can be either microscopic or macroscopic; it can be an atom, an electron, a pi meson, or anything like that. It can also be a subsystem of a larger system as for instance the spins of a certain specific kind of nuclei in the case of nuclear magnetic resonance. In any case, it is described by quantum mechanics and one is interested in the value of one of its observables to be called A.

The measurement device is more sophisticated and necessarily macroscopic. It can include various subdevices, coils or capacitors generating magnetic or electric fields, detectors, electronic equipment, recording devices, and so on. The recording of the data is important in the discussion; it can be simply the position of a pointer on the dial of a voltmeter, or a number registered in a memory of a computer, or a figure in a laboratory book or many things of that kind.

There are often natural objects that can act in the same way as a man-made measuring device. The decay of a uranium nucleus or the arrival of a hard cosmic ray in a rock can leave a track of crystal defects and this can be revealed long after by chemical means and, as far as questions of principle are concerned, it does not matter whether this revealing process will be realized some day or not, or whether the rock is on Earth or on some satellite of Uranus, whether a man has ever seen the rock or not. The existence of the defects is an actual fact and this is all what matters. Similarly, the data of experiments devised by man remain facts, even if past ones, even if the book where some of them were written has been destroyed and its memory forgotten. This means that the existence of measurements and their role in the foundations of the theory are in no essential way linked with man's activity or his role as an observer.

It was said that the measuring apparatus must be macroscopic. This is essential for the data to be described as phenomena, in the sense already mentioned. This is a fundamental condition for two reasons: because they must be described by classical physics with a very good approximation and because the various possible outcomes of a measurement should be described correctly by ordinary probability calculus. The first condition means that a datum is a property of collective observables and the second condition requires an efficient effect of decoherence and therefore the existence of an environment. Together they require a macroscopic measuring apparatus. It might also be recalled for completeness that the statement of a datum by

MEASUREMENT THEORY

classical logic cannot be perfectly clearcut, but this point has been already discussed at length and we shall leave aside the possibility of being mistaken with a probability of 10^{-20} or 10^{-30} or some other small number in our logic.

Another important distinction was mentioned in the introduction: What one sees on a dial or reads on a listing is a *datum,* i.e., a classical property expressing the position of a pointer or the content of a computer memory. But nobody is primarily interested in this kind of data, only in what it means: Nobody says that the pointer of a detector located behind a Stern–Gerlach device has shown a current of three milliamperes. One says that the counter indicates that an atom crossed the detector and, even more specifically, that the z-component of the spin atom is, for instance, $+2$. This statement is what we shall call the *result* of the measurement, namely a property of the measured system Q expressing a value for the measured observable. How one can pass from data to results is one of the major problems of the theory, even if a time-honored practice has been to ignore it.

3. What Is a Measurement?

Among all the interactions between a macroscopic object and various other systems, very few are actually measurements. The window of a bubble chamber is bombarded continuously by air molecules and none of these collisions ends in a measurement. One must therefore specify what is peculiar to the interactions giving rise to a measurement.

The Theoretical Description of a Measurement

It will be convenient to restrict the discussion to the case when A has a purely discrete spectrum with eigenvalues $a_1, a_2, \ldots, a_n, \ldots$. This assumption is not really a restriction. When an observable A having a continuous spectrum is measured, the apparatus cannot give an infinitely precise value for the result since the data cannot have an infinite precision in view of their classical and therefore approximate character. All that one can expect is to get the value of A within some maximal error bounds, a property that is associated with a projector E. One may therefore consider that the observable that is measured is really this projector, which has only two discrete eigenvalues 1 and 0. In other words, a measurement can always be considered as a collection of yes/no measurements answering the question: what interval contains the value of A?

Denote by $\mathbf{H}_Q$ the Hilbert space of the measured system Q, the observable A being associated with a self-adjoint operator in this space. The hamiltonian of the Q-system will be denoted by H_Q. We will similarly denote by $\mathbf{H}_M$ and H_M the Hilbert space of the apparatus M and its hamiltonian.

The full Hilbert space **H** of the complete system $Q + M$ is the tensor product of these two Hilbert spaces and the total hamiltonian can be written as

$$\mathbf{H} = H_Q + H_M + H_{\text{int}},$$

where the term H_{int} represents the interaction between the two systems Q and M.

The Initial State

We will assume that the systems Q and M do not interact beforehand. The initial density operator of Q will be denoted by ρ_Q. Similarly, the initial state of M is ρ_M and it will be sufficient to consider only the reduced density operator for the collective observables, although one sometimes needs also to specify global characteristics of the internal environment such as its temperature (in the case, for instance, of calorimetric measurements). Even if there are some initial microscopic correlations between Q and M, they disappear when one takes the reduced density operator and the overall initial state of the system $Q + M$ is given by the tensor product $\rho_Q \otimes \rho_M$.

The Duration of a Measurement

It will be assumed that the interaction takes place during a time interval $[t_m, t_m + \Delta t]$, the interaction beginning at time t_m and extending a time Δt. It is not however necessary to define these times in a too precise manner, as can be shown by considering the two main cases that can be distinguished, as already mentioned in Chapter 2. In the first case, a particle (e.g., an elementary particle or an atom) enters at time t_m into the detector (a bubble chamber, a wire chamber, a counter...). It is described by a wave packet penetrating the body of the detector at time t_m, though it is clear that, because of the finite extension of the wave packet, this time is not perfectly well defined and, moreover, not all the parts of the detector are equally sensitive. It may even happen that the particle travels a rather long way inside the detector before an interaction occurs, as happens for instance with the detection of neutrinos. All this means that neither the time at which the measurement begins nor its duration are precisely defined. In the second typical case, the time at which the measurement begins is controlled by the apparatus itself. If one uses, for instance, a laser so as to scatter light upon an atom beam (to measure the atom's velocity by a Doppler effect), the measurement begins when the laser starts to work. One might say that this time is controlled by the dynamics of the laser, which is implicitly contained in H_M. In any case, the measurement will be significant under the condition that the final state of the measuring device is only slightly affected by the precise value of t_m, which means that the probability distribution of the measured observable must be rather insensitive to its dependence upon time. This is however a practical limitation and not a matter of principle.

MEASUREMENT THEORY   331

The Observables Having a Role in a Measurement

Two observables, and not only one, will play a basic role in the discussion. The first one is of course the measured observable A. Its eigenvectors will be denoted by $|a_n; s\rangle_Q$, s being a degeneracy index.

The description of the measuring apparatus would be extremely complicated if one were to take all its details into account. We will therefore characterize it by only one observable, to be called B. This simplification is only a matter of convenience and one should not attach undue significance to it, since it would be easy, if desired, to describe M more elaborately in a specific case. The simplest way to proceed is to assume that, among all the devices comprising M, one is a mechanical clockwork similar to the mileage counter in a car, showing numbers indicating the data. It can be described in a way similar to what was done for the pendulum in Chapter 7. The clockwork can stop at a few final positions just as a pendulum comes to rest and we will denote its various possible indications by associating them with the values of an observable $B = 1, 2, \ldots, n, \ldots$.

Remark. One can define more rigorously the observable B as follows: Define a cell C_n in the phase space of the clockwork, containing the point $x = n$ corresponding to the nth stopping position and assigning an upper limit to the velocity. Associate a projector E_n with each cell of that kind and define the observable

$$B = 0.E_0 + 1.E_1 + 2.E_2 + \cdots + \lambda[I - (E_0 + E_1 + E_2 + \cdots)], \quad (8.1)$$

λ being a large negative number. The last term is introduced only as insurance for the theory against all the conceivable aberrant or uninteresting situations that could be due to a malfunction or a breakdown of the apparatus. Furthermore, since one is dealing with a realistic experiment, the number of possible data is finite and the sums in equation (8.1) are finite. The pedantic observable B can be used conveniently to characterize the data.

It will be assumed that the initial state of the apparatus corresponds to the position $B = 0$ when the pointer indicates the value 0 or, as we shall say, the *neutral position* of M. This can be expressed by the equation

$$E_0 \rho_M E_0 = \rho_M, \quad (8.2)$$

which has the form given in Chapter 6 to a phenomenological (classical) initial property, i.e., what was called a state of fact. In other words, the initial neutral position for the pointer is a state of fact. It should be added that equation (8.2) is only valid up to a very small but finite correction, but it will be simpler to forget about it since this point should have become clear by now.

Finally, to complete the question of notation, we will use an M-Hilbert space basis with vectors $|b_n, r\rangle_M$, where b_n denotes an eigenvalue of B (it will also be sometimes abbreviated as n) and r is a degeneracy index representing the many other characteristics of M. Finally, a state vector $|a_n, s\rangle_Q \otimes |b_q, r\rangle_M$ for $Q + M$ will be written globally as $|a_n, s, b_q, r\rangle$.

Characterizing a Measuring Apparatus

We shall now distinguish a measurement from an ordinary interaction. This relies upon two conditions. The first one is the possibility for the apparatus to show data, and this has just been discussed. It does not depend upon the possibility for a human mind to have access to the data but upon the spontaneous separation of the possible data by decoherence into distinct classical possibilities. The second condition is that the interaction between Q and M must be such that a specific value of A in the initial state should generate a specific value of B in the final state.

It was seen in Chapter 2 how this can be realized in the framework of von Neumann's measurement theory. We would like however to be more general and closer to real experiments. It will be assumed that, when the initial state of Q is an eigenstate of A with the eigenvalue a_n, after interaction the recording device goes from its initial neutral position to the position n. This assumption is not empirical but it relies upon dynamics. It does not assume that one can actually prepare a pure eigenstate of Q, but that this kind of transition is a formal consequence of the Schrödinger equation for the interacting system $Q + M$. As such, it can be expressed as a property of the complete evolution operator $U(t)$ now to be considered.

In view of the fact that the functioning of M is often controlled from the outside and the hamiltonian interaction can therefore be time-dependent, we shall write the evolution operator between times t_1 and t_2 as $U(t_2, t_1)$ rather than in the form $U(t_2 - t_1)$ that is valid only for a time-independent hamiltonian. We will have to write down the evolution operator in several cases and its form may depend upon the times to be considered. At a time t later than the initial time 0 and prior to the beginning of the interaction t_m, Q and M do not interact and one has simply

$$U(t, 0) = \exp(-iH_0 t/\hbar), \quad \text{where } H_0 = H_Q + H_M.$$

We shall mainly consider the case where the system Q survives to the measurement and its interaction with M ceases afterwards (if Q is, for instance, an atom, it does not remain finally stuck in the walls of the apparatus or in the bulk of the detector). The two systems Q and M cease to interact after the time $t_m + \Delta t$ and one can write the evolution operator after the

MEASUREMENT THEORY

measurement as

$$U(t, t_m + \Delta t) = \exp\{-iH_0(t - t_m - \Delta t)/\hbar\}.$$

The measurement interaction takes place between the times t_m and $t_m + \Delta t$ and the corresponding evolution operator, $U(t_m + \Delta t, t_m) = S$, will play an important role in the theory. The notation is inspired by the theory of collisions where a so-called S matrix relates the initial and final states of a colliding system (a measurement can also be considered in some sense as a sort of a collision between the measured system and the measuring one).

We can then write explicitly the characteristic property of a simple measurement in terms of this operator. If at time t_m, the system Q is in an eigenstate of A with the eigenvalue a_n (at least in a formal way, i.e., as a mathematical assumption) while the measuring system M is in its neutral position $B = b_0 = 0$, then, at the end of the interaction, the system Q is still in an eigenstate of A with the same eigenvalue and the state of M has become a linear superposition of eigenstates of B, all of them belonging to the eigenvalue b_n. One has therefore:

$$S|a_n, s, b_0, r\rangle = \sum_{r's'} c^{(n)}_{rr';ss'} |a_n, s', b_n, r'\rangle. \tag{8.3}$$

A more general condition is valid when the final datum remains a good signature of the initial value of A but the final state of Q is no longer an eigenstate of A (or at least not an eigenstate belonging to the initial value). The ability of the apparatus to measure A is again a property of the interaction operator S written as:

$$S|a_n, s, b_0, r\rangle = \sum_{r's'm} c^{(n,m)}_{rr';ss'} |a_m, s', b_n, r'\rangle. \tag{8.4}$$

The important point is again that there is a one-to-one correspondence between a_n in the initial state and b_n in the final state.

It may be noticed that these conditions do not consider the possibility of an inefficient measurement where, sometimes, M would not register and its pointer would remain in the neutral position though an interaction with Q occurs. In that sense, we are considering a perfectly efficient measuring device. Perfection is something rare in reality but it would be easy to modify the theory to take care of an imperfect efficiency of the measuring device and this question will be left aside.

An important property of the operator S is its unitary character, which is obvious since S is an evolution operator. This is expressed by the relations $SS^\dagger = S^\dagger S = I$. The group property of evolution operators, i.e.,

$$U(t_3, t_1) = U(t_3, t_2)U(t_2, t_1),$$

can then be used to obtain expressions of the evolution operator starting from time 0 in terms of the free evolution operator $U_0(t) = \exp(-iH_0t/\hbar)$ and in terms of S by:

$$U(t) = U_0(t), \qquad \text{for } t \leq t_m,$$
$$U(t) = U_0(t - t_m - \Delta t)SU_0(t_m), \quad \text{for } t \geq t_m + \Delta t.$$

4. The Main Theorems

Data and Results

The simplest part of measurement theory, though the most important one, consists of clarifying the relation between the experimental data and the results of the experiment, which assert the value of the measured observable at the beginning of the measurement. We want to understand better how a knowledge of a datum implies a knowledge of the result and it will be shown that the two are logically equivalent, all the other results of measurement theory following essentially from this equivalence. This basic problem is therefore a matter of logic. As such, it requires building a logic where the logical equivalence can be proved. We know from the no-contradiction theorem that the equivalence will also hold in every consistent logic involving both properties so that we can use the simplest logic of that kind.

The Logic of a Measurement

The simplest logic must include the various possible results and the corresponding data. More precisely, there is one possible result for each eigenvalue a_n of the measured observable A and, as a property, it means that "the value of A is equal to a_n at time t_m" (when the measurement begins). It is convenient to give a similar standard form to the data. This can be done by introducing a formal collective observable B with eigenvalues $\{b_n\}$, as in equation (8.1). A property stating a datum is then expressed as: "the value of B is equal to b_n at time $t_m + \Delta t$" (when the measurement is completed). These various properties (together with their negation if necessary) are enough to define a quantum logic L involving a complete family of histories referring to the two times t_m and $t_m + \Delta t$. The main question is to find whether it is consistent and what are the implications in it. This is what we shall now consider.

The Logical Equivalence of Data and Results

We must first make sure that the logic L is consistent and this amounts to proving the validity of its consistency conditions. As a matter of fact, it would be enough to consider a unique specific datum (together with its negation) and the corresponding result (together with its own negation) as the only properties to enter, so that one has to deal with the simplest kind of logic, already discussed in Chapter 5. The check of its consistency is made in Appendix A. The proof of the expected logical equivalence amounts to calculating the relevant conditional probability, also given in the same appendix. All these calculations are straightforward though a bit technical. Their main ingredients are equation (8.3) expressing that one is dealing with a perfect measurement and the unitarity of the transition operator S. Their outcome is the following theorem, which holds true for measurements of both types I and II:

Theorem 1. *The data and the corresponding results of an experiment are logically equivalent.*

The content of the theorem is so intuitive that it might look rather trivial and, in a tacit way, the whole Copenhagen interpretation was already relying upon it. This result is however necessary in order to give full significance to a measurement and to conclude that a quantum property necessarily follows from the observation of an empirical factual datum.

Probabilities

The second theorem deals with the explicit probabilities for the results of an experiment. It is again a rigorous statement of something one often considers as obvious by sheer habit. It also holds for measurements of both types I and II and it can be stated as follows.

Theorem 2. *The probabilities of the data are equal to the probabilities of the corresponding results.*

This theorem may look rather obvious once the logical equivalence between data and results has been accepted, though its formal proof is not quite trivial and is to be found in Appendix A. Its nontrivial character and its interest can be seen when one writes down explicitly the two kinds of probabilities that are asserted to be equal. The probability of a datum n is given by definition by

$$p(b_n) = \text{Tr}\{\rho E_n^B(t_m + \Delta t)\}, \tag{8.5}$$

where the projector belongs to the measuring device and is associated with the datum b_n:

$$E_n^B = \sum_r |b_n r\rangle\langle b_n r|.$$

The probability of the result does not depend at all upon the details of the apparatus and it is defined by

$$p(a_n) = \text{Tr}\{\rho_Q E_n^A(t_m)\}, \tag{8.6}$$

where the projector E_n^A belongs to the measured system and is associated with the result n. In other words,

$$E_n^A = \sum_s |a_n; s\rangle\langle a_n; s|.$$

When one thinks of how complicated a measuring apparatus can be and how different two experimental devices purporting to measure the same quantity may be, it is remarkable that there exists such a simple universal correspondence between them.

This theorem is a turning point in the construction of the theory. It was shown in Chapter 4 that the form for the probabilities is unique. If one finds therefore in a series of experiments that the various data occur at random, the only probability distribution one can assign to the events is given by equation (8.5). These probabilities for classically meaningful data correspond to definite events because of decoherence. One can then measure them empirically by looking at the frequencies of the various data in a long series of trials. The long way we followed starting from the principles of the theory at last reaches direct contact with empirical physics and the theory can be finally tested. Moreover, the probabilities that were only up to now useful theoretical tools for a logical language at last acquire an empirical meaning.

Since the use of ordinary probability calculus has been shown to be completely legitimate for the data, one can use its well-known methods to show that the probabilities can be well approximated by the frequencies in a long series of trials. This is meaningful and there is no inconsistency in using a basically probabilistic theory while nevertheless considering that each datum, though relevant from this theory, is something meaningful, registered, and known with a high level of certainty. So, we have finally shown that quantum mechanics provides a consistent framework for its own empirical verification. This is very important since it allows us to get rid of the Copenhagen pronouncements enjoining the exclusion of every reference to quantum mechanics when considering a measuring apparatus and describing the data by classical physics exclusively. We can now keep the best of both worlds by using the facilities of classical physics for describing the data and knowing that we have the right to do it, though we admit only one kind of physics.

The probability in equation (8.6) has the well-known form first written down by Max Born. This formula is however not an axiom but the outcome of a theory having its roots in the universal rule of interpretation.

Repetition

One can repeat a measurement by doing it again after getting the result. In the special case of a measurement of Type I where one assumes that the observable A is a constant of motion for the system Q (or that the two experiments can be made in a sufficiently short time), one can also easily prove the following theorem:

Theorem 3. *Under the above assumptions, a repeated measurement again gives the same result.*

WAVE FUNCTION REDUCTION

In the first part of the chapter we obtained the first empirical rule of the Copenhagen interpretation concerning the values of the probabilities. The second Copenhagen rule had to do with two successive measurements and it was expressed in terms of wave function reduction. This is what we are now going to consider, to see whether we can justify this rule and better understand it.

5. Two Successive Measurements

Theory

Consider an experimental setup where two observables A and A' are measured successively. Both observables belong to the same system Q; the first measurement is made by an apparatus M and the second one by another apparatus M'. As an example, think of an atom crossing successively two Stern–Gerlach devices which are not necessarily oriented along the same direction. We need only assume that the measured system Q is isolated between the two measurements.

We shall denote typical data of the two measurements by b_n and b'_k and the corresponding results by a_n and a'_k. We are interested in the joint probability of the two data or in the conditional probability for the second datum given the first one, $p(b_n \mid b'_k)$. It can be obtained experimentally from a series of joint measurements or from a series of measurements of the second observable in a subsample consisting of the events giving the datum b_n in the first measurement.

The theory is very similar to the one already made for a unique measurement. One considers now the physical system $Q + M + M'$. The time ordering of the measurements is ensured by taking the time t'_m for the beginning of the second measurement to be later than the time $t_m + \Delta t$ for the end of the first one. One introduces the simplest logic L involving both kinds of data and the corresponding results, together with their negation. Since a datum is a property holding at the end of a measurement and a result refers to its beginning, one has to deal with four-time histories.

The first point is to check the consistency of this logic, which is easily found to be valid so that all the probabilities are significant. One can also compute the conditional probabilities for each datum and its associated result and one again finds them to be logically equivalent. Finally, the quantity we are most interested in is the conditional probability for the second datum given the first one $p(b_n \mid b'_k)$ or, equivalently, the joint probability $p(b_n, b'_k)$ that is its product by $p(b_n) = p(a_n)$. This last quantity is given explicitly in terms of the projectors representing the two data by

$$p(b_n, b'_k) = \text{Tr}\{E_k^{B'}(t'_m + \Delta t)E_n^{B}(t_m + \Delta t)\rho(0)E_n^{B}(t_m + \Delta t)E_k^{B'}(t'_m + \Delta t)\}. \tag{8.7}$$

This is not very illuminating and one may wonder whether a simpler expression involving only the projector for the second result could not replace it, as suggested by the Copenhagen rule. One may directly consider the conditional probability $p(b_n \mid b'_k)$ rather than the joint probability $p(b_n, b'_k)$ and ask whether it can be written in a simple form such as

$$p(b_n \mid b'_k) = \text{Tr}[\rho_f(t_m + \Delta t)E_k^{A'}(t'_m)], \tag{8.8}$$

where an effective density operator $\rho_f(t_m + \Delta t)$ would describe the Q-system at the end of the first measurement and the projector represents directly the result of the second measurement. This formula is much simpler than equation (8.7) since it is written in the Hilbert space of the measured system alone and the unique projector entering is also much simpler than the projectors representing the very complicated measuring devices.

It is quite remarkable that such a simplification is possible and it follows from the next theorem, whose main ingredients are now the semidiagonal property of the second transition operator S' and its unitarity:

Theorem 4. *The conditional probability for the result (or the datum) of a second measurement given the datum of a first one can be written in the simple form given by equation (8.8). The formal operator $\rho_f(t_m + \Delta t)$ entering in it is given explicitly by*

$$\rho_f(t_m + \Delta t) = \frac{1}{p(a_n)}\text{Tr}_M\{E_n^{B}(t_m + \Delta t)\rho_Q \otimes \rho_M E_n^{B}(t_m + \Delta t)\}. \tag{8.9}$$

This is obviously the general form of the rule for wave function reduction. It is easily seen that equation (8.9) defines a positive operator with unit trace in the Hilbert space of Q. It evolves according to the Schrödinger equation for Q as long as this sytem is isolated, i.e., during the interval between the two measurements. The occurrence of the inverse of the probability $p(a_n)$ in front of equation (8.9) can be traced back to the fact that equation (8.8) gives a conditional probability.

Equation (8.9) looks more complicated than what is usually given in elementary textbooks. The reason is that it provides a general answer, whatever the complexity of the first apparatus and the intricacies of the first measurement. The formula works, for instance, in rather complex circumstances such as the following. A first measurement gives the momentum of a charged particle when it entered a bubble chamber. The particle did not stop in the chamber but one can nevertheless know its momentum by counting the density of bubbles left along its track. The chamber was also in a magnetic field so that the trajectory of the particle was modified. This kind of measurement is quite different from a measurement of Type I and it should not be surprising that one obtains a formula like equation (8.9), since it must make room for such intricacies, depending upon the details of the first apparatus and of the initial state.

Note that the exact form of wave function reduction must use a quantum quasi-projector to take into account the first datum. When insisting that the first apparatus should be strictly described by classical (and not semiclassical) physics, the Copenhagen interpretation was therefore depriving itself of the general expression of wave function reduction. These defects could not however be detected because they occurred in a case where the theory did not seriously try to make predictions (and where it would have been difficult anyway to do better explicitly). The present interpretation is clearly consistent and complete but this is only an advantage as far as matters of principle are concerned. There is no real change for all practical purposes because it is anyway very difficult to make precise predictions except when the first measurement is of Type I.

Simple Cases

We can recover the prescriptions of the Copenhagen interpretation in simple cases, as one should expect in view of their success with so many experiments.

Let us initially assume that the first measurement is *strongly* of Type I, by which we mean that any initial state $|a_n; s\rangle$ of the measured system remains exactly the same after the first measurement. This case occurs when the transition operator S is diagonal with respect to the degeneracy indices in the Hilbert space of the Q-system. Equation (8.4) becomes in that case a formula first given by Lüders:[2]

$$\rho_f(t_m + \Delta t) = \frac{E_n^A \rho_Q(t_m) E_n^A}{\text{Tr}_Q[E_n^A \rho_Q(t_m) E_n^A]}.$$

When furthermore the eigenvalue a_n is nondegenerate, one gets the well-known elementary formula

$$\rho_f(t_m + \Delta t) = |a_n\rangle\langle a_n|.$$

What Is the Meaning of Wave Function Reduction?

Wave function reduction has been obtained here as a consequence of the theory of histories and not as a new principle. As a matter of fact, it is only a convenient recipe and its use is not compulsory, since one could as well remain in the framework of the general theory by keeping track of the properties of both instruments, as was done in fact to prove the recipe. Accordingly, *wave function reduction is a convenience but not a necessity.*

Nothing requires that we restrict the description of physical reality to the unique system Q during the interval of time separating the two measurements or after a measurement. One can always include everything relevant to the experiment by using histories taking care of all the systems with which the measured system interacts. As a matter of fact, this is what one should do in the majority of real cases where a system is not prepared, strictly speaking, by a measurement. But the possibility of using reduction is limited to very special conditions, so the theory would be very poor if it could only deal with this rather exceptional case. On the contrary, histories are able to describe everything, at least in principle. The literature on quantum mechanics for a long time had a tendency to treat measurements as some sort of a paradigm, something more or less essential, not only for the understanding of experiments in a laboratory, but also for the understanding of Nature. It is much better not to so drastically invert the normal hierarchy so that measurement remains only a special case of interactions and not the other way around as a necessary paradigm of preparation.

Wave function reduction is a logical result having among its assumptions the existence of decoherence in the measuring apparatus, which is a true physical effect. These two notions should not be confused. Reduction only says how physical reality can be most conveniently described under restrictive and precise assumptions—namely, taking into account the result of a measurement and not paying any attention to the measuring device after the measurement. *Reduction is not in itself a physical effect but a convenient way of speaking.* The real physical effect conditioning its possibility is the existence of decoherence in the apparatus (allowing us to treat the data by ordinary probability calculus). It should be stressed that decoherence takes place in the measuring device and not in any way in the measured system to which reduction is applied. In that sense and contrary to what was assumed in the past, reduction is not a physical effect affecting the measured system. Its usefulness comes from a physical effect in the apparatus.

Finally, it is interesting to examine the status of reduction in the framework of logic. It happens to be something well known to logicians, which they call a *modus ponens*. This notion is most easily understood in the case of mathematical logic: The proof of a theorem in mathematics is a chain of implications starting from the basic axioms and ending with the statement of a theorem. However, as is well known, one can also directly combine several theorems to obtain new results without going back to the axioms. This is precisely an example of a modus ponens, namely, one leaves aside the long chain of implications leading from the axioms to the theorems and one starts directly from the theorems. This procedure, which is so well known that most of us have forgotten a large part of the proofs of the theorems we use in our professional life, can be justified rigorously by the methods of logic. One can also borrow another example from information theory, which is basically a logical calculus: a modus ponens is used when the result of an intermediate computation (a proposition) is stored in a memory and the calculation leading to it is erased.

Wave function reduction is similarly a recipe allowing the erasure of irrelevant information: one erases the first measuring apparatus from consideration, though it is still there; one also forgets much of the initial preparation of the system (sometimes even everything). One sticks only to the result. The justification of this simplification is much less trivial in the present case than it is in mathematics and information theory, but the result is the same, namely, a *modus ponens* or essentially a rule for oblivion.

ACTUAL FACTS

We now arrive at the most fascinating aspect of quantum mechanics, which is its relation with facts or, if one prefers, with reality. There is an obvious gap between quantum mechanics and reality: Quantum mechanics is probabilistic, which means that it can only envision a multitude of possibilities on the same footing, whereas actual facts, with their definite uniqueness, remain outside its reach. This essential question will now be analyzed according to the general strategy of the logical approach, which is to extract as much as possible from the theory itself before comparing it with reality.

What Is Satisfactory in the Present Theory

Let us examine where we now stand: It has been shown that most macroscopic objects can be described by classical physics, including its logical setup. As long as one considers only a purely classical world where all the significant interactions take place between macroscopic objects, one obtains a satisfactory representation of reality. It is perfectly valid when the initial state is itself a classical situation, which has been called a state of fact. One can then understand everything happening in the classical world, except for the existence of very small and negligible quantum fluctuations and some delicacies concerning the description of chaotic systems. There is no significant distinction in classical physics between potential phenomena and actual facts because of determinism (as long as one disregards chaos). The difficult questions arise therefore only when one considers the macroscopic effect of a quantum event.

An Essential Problem

The word "essential", though often used in a loose way, can profitably be applied to the problem facing us now. The firm classical ground upon which physics rested was shattered when the first quantum measurement was made and with the later discovery that quantum theory cannot give a complete account of reality. The theory can account for the random occurrence of measurement data as long as they are considered by probability calculus as potential phenomena. However it cannot answer the very simple and obvious question: how can an actual unique datum be realized at the end of an individual measurement?

The present situation of the theory is best described by a parable: An angel was born in heaven and nourished only with mathematics and logic for the strength of his mind. He was due however for a visit to the earthly world, so that he first had to learn its laws. An archangel taught him the principles of quantum mechanics as they are written in the book. The young angel understood everything easily and he knew therefore what to expect when arriving on Earth: he was to see a multitude of simultaneous phenomena piled up together, as they were generated from time immemorial by the accumulation of many little quantum happenings. How great was his surprise when he discovered in place of that a unique clearcut reality showing everywhere sharply defined features. There were no dreamy wandering and ambiguous pictures, except for the clouds in the sky, but only hard and distinct outlines. He never quite recovered from the experience.

6. Actual Facts and the Present Time

The Direction of Time

We recognized on several occasions the capacity of quantum mechanics to extract from its bosom what we believed to be primary common sense notions. What we had considered as being most obvious was rediscovered in a new light. It was a surprise to find that it was already present, though deeply hidden, in the principles of the theory. We are now going to meet another example of this kind with the different qualities of past and future.

Conventional quantum mechanics has not much to say about that, if only because of its invariance under time reversal, but we had already had clear signs that one can go further by using its logical aspects. It was shown in Chapter 5 that the logical structure of quantum mechanics does not share time-reversal invariance; it was also shown in Chapter 7 that the theory agrees with the existence of dissipation and the logical direction of time coincides with the direction in thermodynamics. It will now be shown that the theory can be reconciled with the uniqueness of facts, at least at a given (present) time.

An Academic Case

It is convenient to first consider an academic example. Let us treat a very simplified model of the universe, which is inspired by a laboratory. It is a closed world consisting only of a few measuring devices and a unique quantum system Q upon which measurements can be made automatically. The system Q is, for instance, an atom going from one apparatus to the next. All the measurements are ideal (i.e., of Type I), the measured observables being denoted by $A, A', A'', \ldots$ and the measuring devices by $M, M', M'', \ldots$. The observables will be assumed to have nondegenerate eigenvalues for simplicity. At a given initial time, all the pointers of the instruments are in their neutral position and the state of Q is assumed to be given. The quantum system Q interacts first with the apparatus M at a time t and the measurement gives datum b_n, then Q interacts with M' at a time t' to give a datum b'_m and so on.

Consider a time t_0 later than the first two measurements and earlier than the third one. The various indications that can be shown by the instruments at that time are (b_n, b'_m, b''_0), with known statistical weights p_{nm}, the neutral indication b''_0 indicating that the third apparatus has not yet registered and it is still in its neutral state. We shall now enforce the uniqueness of facts at time t_0 by asserting that there actually exists a unique set of such data.

Taking Present Facts into Account

If one takes into account what is actually observed at what we called the present time, there are no potential phenomena but only ascertained facts. Data are shown on the dials of M and M', while the dial of M'' still shows a neutral indication. Though the theory did not predict this uniqueness, it can nevertheless account for it consistently as follows. We know how to describe the present situation because we already met it in the case of a purely classical laboratory containing only macroscopic objects and no quantum system Q to be measured. This previous learning can be used to improve our understanding of the present situation and to draw its consequences.

We shall assume that the times t and t' when the first two measurements took place have been registered. This assumption does not raise any special difficulty since one knows how to describe a clock and how to make a record in the classical framework, as we understand it on the basis of quantum mechanics. One can then use the knowledge of the present facts, including the records, to reconstruct the past events by using logic. One can tell what was the "classical" state of all the instruments at any given time in the past, except during an interval of time when one of them was performing a measurement, and one can also reconstruct the properties of the Q system at the beginning of all the past measurements. All this is a direct consequence of the explicit constructions that were given in Chapters 6 and 7 and of the theory of measurement as already seen. The past appears therefore as unique.

As far as the future is concerned, we can only predict the probabilities for the various possible outcomes of the next measurement to come and the description of the future remains therefore completely potential and probabilistic.

The Structure of Time

This poor man's model of the universe indicates a remarkable result that can be formulated as follows. From a given present time where there are actual facts, one can logically reconstruct a unique past, whereas the future has to remain potential. It should however be stressed that this reconstruction applies only to the past classical events and to the results of past measurements. It does not allow us, for instance, to say what the state of the internal environment in an apparatus was.

This approach also makes clear an interesting point: past facts are not absolutely real; they only *were* real. One can never indicate a past fact by pointing a finger at it and saying "that". One must call for memory or use a record, a note, or a photograph. Nevertheless, the derivation of a unique past is possible because quantum mechanics allows for the existence of memory and records. It goes without saying that everybody always takes this structure of time for granted but, up till then, it was always tacitly added to our understanding of physics and never supposed to be a consequence of it.

Our ideal laboratory and the real universe differ mainly by the existence in the latter of various physical effects erasing the memory of many past facts or of many details, because of chaos, damping, or wearing. This will be considered as inessential and the structure of time that was found will be assumed to apply to reality. Notice that this point of view agrees much better with the ordinary process of gaining knowledge than the extreme theoretical point of view where one always starts from a given initial state (given by whom, from where?), the state undergoing a time evolution according to a gigantic Schrödinger equation. What we observe in reality is always something existing right now, even if we interpret it as a trace of an event in the past, whether it be a crater on the Moon, the composition of a star atmosphere, or the compared amounts of uranium and lead in a rock. The laws of physics and of logic, or much more simply common sense, allow us to reconstruct vast patches of the past from the present data and the present records. As for the future, it remains forever misty and the realm of possibilities. This is how all knowledge is obtained and one can also envision it in that way in the framework of physics, even at the basic level of quantum mechanics.

The Answer Is Insufficient

In spite of some progress, the problem of actuality is however not solved. The reason why is perfectly obvious: time is changing. We do not understand how a potential phenomenon among many similar ones becomes a unique actual fact and the theory offers no hint. It is intrinsically probabilistic and we have been able to squeeze it until all the conditional probabilities practically equal to 1 have been used. It cannot give more and it provides no mechanism and no explanation for the last problem remaining to be understood.

7. Everett's Answer

A tentative answer was proposed by Everett in the fifties and, as already mentioned, it has always attracted the attention of quite a few theorists. It is particularly significant, not so much for its own value (at least, according to the present author's viewpoint), but for the light it sheds upon the nature of the problem. The point is that Everett's contribution is not so much a theory but essentially a *representation* of reality. The word "representation" is used here as philosophers do, namely as some picture of reality one may have in mind, which is not completely empirical but involves a part of imagination or convention, a way of looking at reality originating essentially from one's own philosophical inclinations. It is not science because no experiment can show it to be wrong and it is not a theoretical truth because there can be no proof of it. It is not nonsense because one cannot prove it to be inconsistent.

Everett took quantum mechanics completely at face value and he pushed its consequences to their ultimate limits so that this led him to the best understanding of decoherence at that time. The fact that he obtained a solution of the problem of actuality without changing an iota to quantum mechanics is very important, even if one does not like his answer. The point is that one may dislike it but one cannot prove it to be wrong, nor can one prove its validity by reasoning or experiment if one wants to believe it. This means that, despite the keen interest shown by many people for the problem of actuality, it might very well be that it does not belong to physics but to our manner of thinking about physics, which is after all a matter of philosophy. We are therefore going to consider Everett's construction, not for its own sake, but as an indication that the problem it tries to solve is not a problem of physics.

Everett's Representation

In its modern form, as advocated by Gell-Mann and Hartle,[3] Everett's representation can be described as follows: There was a unique initial state of the universe, perhaps even a pure quantum state described by a gigantic wave function. It evolved according to a big Schrödinger equation. The content of the universe was probably initially very erratic, when the quantization of space was supposed to dominate; it later contained practically pure thermal radiation, according to the big bang models, so that there were not many classical phenomena to be mentioned and there was practically no fact to be seen, except for very global ones. Some time after, phenomena became possible. For the sake of illustration, it will be convenient to consider the birth of a galaxy as such a phenomenon. According to pure quantum mechanics, it is conceived as a phenomenon because the corresponding reduced density operator is diagonal in the collective variables of the galaxy, but that does not tell us where the galaxy is. There is only a probability for the whole galaxy to be in some place rather than another.

Two states of that kind, where the galaxy is supposed to be possibly in two different regions of space, have no possible relation and they completely ignore each other because of decoherence. When a dweller on the galaxy (supposed to be in one of the possible locations) tries to track down the past by using what he can see, he can only infer the past of the galaxy until he reaches a barrier of ignorance. This is because of dissipation and decoherence and we have already discussed this question in Chapter 7: one cannot reconstruct exactly an event too far in the past. Decoherence in particular implies that, when two classical objects (e.g., two observers) belong to two different histories of the universe, they must completely ignore each other.

Everett concluded from a similar argument that there is no contradiction between a universally valid quantum mechanics and the fact that someone on a specific branch of history sees everything classical as unique. His essential idea was to transform this into a positive assertion: all the histories of the universe are real. They exist for their own sake just as much as the one we can see. There is a reality not so different from our own where Caesar was not assassinated, another where Genghis Khan died in infancy, another where a famous big meteorite missed the Earth long ago and the person writing this book on quantum mechanics is a dinosaur. There is not a unique reality but many different noncommunicating zones.

One can recognize in this vision a modern revival of older philosophical conceptions—namely, Plato's representation, where ideas were more real than the apparent world around us, which is considered as only a shadow illusion on the wall of a cavern. There is some analogy with the present case, where all the potentialities make together the "true" reality, while our limited powers of measurement forbid us penetrating the intimate secrets of the quantum superpositions where the existence of other worlds remains hidden.

A Criticism

One cannot refute Everett's representation, but one can dispute. Our criticism will be based upon aesthetic grounds and also economy of thought. They are not compelling and all they can show is how ugly and how expensive this representation is. Each time there is a click of a Geiger counter anywhere in the world, it is an indication that two realities have become separated and this is their only difference. You continue to live in both of them and there is a tremendously large number of copies of yourselves with similar or different lives while their histories diverge more and more. There is undoubtedly something preposterous in such a vision. It obviously gives an undue importance to the little differences generated by quantum events, as if each one of them were vital to the universe. This is not at all what one can learn from physics, which shows on the contrary that the most important features of reality are classically deterministic and that, although there is much randomness in it, this is practically always due to classical chaos. Let us consider the first point. Despite the very high number of individual quantum effects occurring in the universe, a tremendous number of them give the same result in practice. Although the emission of a photon by an atom in the photosphere of the sun is a quantum event and the same is true for a nuclear reaction between two protons in the core of the sun or the absorption of a photon by a photographic plate, there are so many photons that they individually do not change the universe by an iota. They are completely absorbed in big collective effects that are almost perfectly described by a deterministic classical physics and the sun continues to shine and the earth to revolve around it.

This does not mean that no randomness is changing the face of reality but, once again, practically every significant consequence of randomness is of a classical nature. If a piece of rock contains one thousand alpha ray tracks originating from the decay of uranium nuclei, it could as well have contained one thousand and two according to quantum mechanics and nobody cares. But if it is true that dinosaurs were killed by the fall of a meteor chaotically perturbed in its trajectory, this is undoubtedly much more important, because we might never have been here otherwise. The erratic motion of water in a river, its effect upon a pebble, the details of the weather, the fact that a storm is taking place here rather than there, all this is governed partly by determinism and partly by classical chaos. Can one say that the fact that chaos is ultimately stochastic because of Heisenberg's uncertainty relations, as discussed in Chapter 10, is a sufficient reason for giving up its description by classical probabilities and replacing it by an infinity of universes, where each possible little event is perfectly realized? Certainly not and, once again, Everett's representation is unaesthetic because it gives a tremendous importance to what has none.

As far as an economy of thought is concerned, its rules have been given long ago by William of Ockham and his famous razor: never use many when one is enough (*multiplicitas non ponenda sine necessitate*). There was never anything in the history of thought so bluntly contrary to Ockham's rule than Everett's many worlds. Finally, the theory of decoherence has shown that the various outcomes of a measurement can be described by conventional probability calculus. Why then return to an antiquated view of quantum mechanics using again and again infinite superpositions? Why keep the skin of the grape when the juice has been drunk?

An Assessment

While acknowledging that Everett's representation and its later investigation by Bryce De Witt contained significant seeds of progress for the development of decoherence and despite the impossibility to disprove it, we feel it impossible to accept as a satisfactory answer to the problem of actuality. It violates too much common sense, it gives too much weight to what in practice has none and it adds nothing to our philosophical understanding of science. This is why we are going to look for another answer.

Hidden Variable and Physical Reduction Theories

For the sake of completeness, a few other answers that have been proposed at one time or another must be mentioned. They differ from the previous one in that they do not accept completely quantum mechanics and they introduce basic modifications in it. The first one assumes the existence of hidden variables. It will be described in more detail in the next chapter so that we shall not enter into its discussion presently, except for a remark remaining again at the level of aesthetics. Recent progress in interpretation has allowed one to resolve the old conflict between probabilism and determinism, even though determinism was found to be approximate. It would be strange under these conditions to assume that exact determinism, as described by hidden variables, could still hold while being located at a much lower inaccessible level, just to provide a cause for the result of a quantum measurement. This criticism is even stronger in the case of Bohm's theory, to be described in the next chapter. It keeps the wave function of quantum mechanics and the Schrödinger equation and it adds to that "real" coordinates of particles moving deterministically in a potential partly generated by the wave function. Such a theory cannot avoid meeting again all the problems of interpretation associated with the derivation of classical physics from quantum mechanics so that, for instance, the quantum effects correcting determinism must still be present and will have to be solved in the same manner. The op-

MEASUREMENT THEORY

position between an approximate determinism at our level and a perfect one where one cannot see it would be still more awkward.

One should also mention another proposal by Ghirardi, Rimini, and Weber.[4] They assumed that a real physical effect, up to now undetected, occurs here and there, randomly in space and time. It is very rare and a particle may have typically to wait for a billion years before being frozen in space by it. It is also rather local, taking place over distances of the order of a thousand angströms. As for its nature, it is simply an actual reduction of the wave packet for a particle, exactly as elementary quantum mechanics would have it in the Copenhagen framework. In other words, the wave function of a particle is suddenly reduced to its part lying in a small space region and it is increased in magnitude so that it remains normalized. This is exactly what would follow from a position measurement in the case of a realistic wave function reduction, except that there is no macroscopic apparatus making a measurement and, in some sense, the "measurement" is made randomly and leisurely by space itself.

Ghirardi et al. have shown that the cumulative effect of such localizations could be efficient enough to fix the position of a solid object containing some 10^{23} particles in about one tenth of a microsecond. The choice of the length and time parameters for the effect is more or less a matter of choice but not a very open one. This is because of the necessity of maintaining the existence of the many quantum effects in solid state physics and atomic physics that are predicted by ordinary quantum mechanics and agree with observation, so that the new effect cannot be too dominant. On the other hand, it should be efficient enough to "freeze" the position of a macroscopic system in a short enough time.

There is a difficulty with this theory. If each linear dimension of a solid substance is reduced by a factor of ten, the number of particles in it is reduced by a factor one thousand and the time necessary before a freezing of the system is multiplied by one thousand. One can then wonder what happens with small detectors. We saw in Chapter 2 how a photographic emulsion acts as a photon detector and it was mentioned that the number of interstitial silver ions producing the effect is of the order of a few billion. It is then easy to see that one will have to wait almost a year before the actualization of one emulsion grain among all others. The authors acknowledge this difficulty[5] and, to get rid of it, they must resort to what happens when the emulsion is developed or when an observer looks at it. The development involves many more particles that are frozen more rapidly. If an observer looks at the undeveloped emulsion under a microscope to discover where the reacting grains are, the big system to be frozen by the reduction effect in a short enough time is the whole brain of the observer. The idea is clever but it must share the criticism we addressed to hidden variables: it restates as a basic principle something one can successfully explain otherwise as a secondary effect of quantum mechanics, when it is properly understood. In the present case, all that Ghirardi, Rimini, and Weber try to explain has been already explained by decoherence and by the present theory of measurement, with a better efficiency and while avoiding their difficulties.

8. A Law of Physics Different from All Others

The Root of the Question

Facts exist. Nobody can explain that as a consequence of something more basic or, as the poet said, "There is no why for a rose, it blooms because it blooms" (Angelus Silesius). What kind of a problem is the actuality of facts? Probably not one within the reach of experiment, since experiments start from facts and how could one use experiments to find why facts exist? Is it even a problem of physics? One might suspect that it is not when one considers the background behind two rather popular answers. When people are willing to accept the existence of hidden variables, even if they are fundamentally inaccessible to experiment, what they want to obtain is not anything having to do with empirical physics but with ontology (i.e., the part of metaphysics dealing with "being"). They want to reach what is "behind" experimental physics so as to preserve their own representation of reality, which happens to demand some sort of causality. Everett's solution shows on the contrary that the problem disappears when one accepts quantum mechanics as it is at the expense of adopting another representation of reality that cannot be falsified by experiment. A representation does not say how to do physics but only how to think of it and a significant problem along these lines would be to find another representation recognizing the success of quantum mechanics while allowing the uniqueness of reality.

Could it be therefore that the problem of actuality does not after all belong to physics but it is only a matter of representation, which belongs to philosophy? If one does not like this word, is it only a problem offering no experimental grip at all and therefore simply a ground to dispute preconceptions?

Since one is questioning the nature of the representations suggested by physics, one may ask another question halfway towards the solution: Why can't we accept that things are what they are without asking physics to explain everything in full detail in terms of mathematical laws? Being modest and accepting this proposal by not asking science to be absolute looks like a very reasonable and cautious position, but there is a tricky consequence: if such a renouncement of knowledge were made necessary by reality itself, it would mean the existence of some limit to the extent of physics and, as such, it would be some sort of a new principle of physics. A principle of that kind would be very different from all previous physics and it would even go in the opposite direction of our present construction. After adding a logical structure to quantum mechanics so as to extract the existence of phenomena from the principles of the theory, one is now proposing stopping the quest on the verge of reaching actuality.

This proposal is not in fact completely new and it first appeared in another guise in Bohr's considerations on wave packet reduction. It is moreover not a renouncement nor a defeat but, on the contrary, a remarkable mark of accomplishment for the enterprise of physics and also probably a valuable indication about its nature. This is what we are now going to show. It was found that the principles of quantum mechanics contain in germ everything belonging to reality, except for its uniqueness. Uniqueness is, however, the essence of reality, as was often stressed by philosophers. Wittgenstein for instance says that, when looking at something real and before being able to give it a name, one can only indicate it by pointing a finger at it and saying "that." Without the uniqueness of "that," the whole idea of reality is lost. Conversely, the whole logical structure of quantum mechanics rests upon the use of probabilities. Reality and theory disagree therefore only on one point, but one which, in both cases, partakes of their essence.

Never before in science was such an exemplary conflict encountered. There is a remote chance that it might be avoided if at least some part of reality were outside the domain of quantum laws. Considering what is known presently, the only possible candidate is spacetime, because there is not yet a consistent quantum theory for it. If it were to behave classically, one could probably conceive new models ensuring datum uniqueness. Whatever it may be, we shall ignore in any case such far-fetched possibilities and proceed as if all the exit doors were completely closed.

When reality and theory disagree in their fundamental nature, one cannot avoid evoking another old problem belonging to the foundations of physics: why can reality be so well represented (or mimicked) by a mathematical construction? Einstein said it also in a famous sentence: the most inexplicable feature of reality is that it can be explained. The tentative answers to the problem of actuality that were already mentioned—Everett's representation, hidden variables, or realistic reduction—completely ignore this second fundamental question. They proceed as if it were an unquestionable law of nature that nature should be at all times, everywhere and in all its aspects, completely described by mathematically formulated rules. Although this has been found to be valid up to some point when physics was younger and covered less ground, why should it be an absolute rule? The point is of course to assume an absolute agreement between theory and reality and no longer just a satisfactory correspondence. To believe in such a kind of absoluteness is a very strong and frequent prejudice among physicists. It is however undoubtedly more questionable than some other preconceptions entering in the proposed answers to the problem of actuality, to which it is always tacitly added and never mentioned.

In order to see what the assumption of an absolute account of reality by a theory would imply, we shall accept it for a moment for the sake of the argument. Everything happening in reality is supposed to be exactly described by a great theory, which remains to be discovered but whose existence is nevertheless taken for granted. Accordingly, if an initial state of the universe is supposed to exist, everything that is going to happen later is already contained in it. Time becomes a perfectly unnecessary notion since an initial set of data is enough to determine everything (though not perhaps a unique reality). Time is only a parameter to be used for sewing the properties of the universe together. Freedom is of course an illusion and, in that sense, one is again facing the most extreme difficulties accompanying classical determinism. What is much worse, reality itself is an illusion or, more precisely, it becomes identical with its own mathematical image. The ever-changing, ever-creating reality is identified with the timeless stillness of mathematics. Although the present arguments are avowedly philosophical, they were worth stating, if only to make clear the real framework in which the searchers of total theories are trying to drive science. The exact opposite of this extremism is to push the program of physics far enough that it exhibits its own limits by reaching a point where mathematics is directly facing reality. Quantum mechanics has been able to achieve this feat. This is why we suggest that the actuality of facts is something that needs not be explained by a theory. This is also why we consider that, when one finds a gap between theory and reality only at their common extremities, this is not a failure but the mark of an unprecedented success for quantum mechanics, as compared with all the theories before it.

The Answer as a Rule of Physics

It might be said that this point of view is essentially a resignation. Maybe. But it is also an acknowledgment of the wonder of reality, which is the main motivation for being a physicist. This "solution" also raises new questions for the philosophy of science and particularly the theory of knowledge, but they are better left to Chapter 12 where these matters will be considered. For the time being, we shall only try to clearly state this point of view as some sort of a principle of physics or rather as a convenient representation in which one can work. As a law, it would be somewhat different from all the other ones and it might be stated as follows:

Rule 5. *Reality is unique. It evolves in such a way that the actualizations of different facts originating from identical conditions follow the statistical rules predicted by the theory.*

It may be recalled that, in Bohr's interpretation, the reduction of wave functions was also a rule different from all others. It appeared as a preliminary necessary assumption allowing one to assert the exact initial conditions of an experiment. Since one cannot discuss the result of an experiment if one does not know under what conditions it takes place, the rule of wave function reduction appeared as something that cannot be checked experimentally because its role was to tell us what the experiment is. It was believed that this rule could only be accepted or rejected with the whole theory of quantum mechanics itself.

Things have changed somewhat. The reduction of the wave function is no longer central, since it has been brought to the level of an unnecessary convenience, while its true origin has been found to be decoherence. One even knows that decoherence is not universal though it must be acting in a measuring device and there is no measurement otherwise. Bohr's essential point remains anyway. It has become much clearer since it is no longer linked with the strange technicalities of wave function reduction and it now comes from a direct comparison between theory and reality.

THE NOTION OF TRUTH

The existence of actual facts will now be shown to be a key to the notion of truth. What can be said to be true? Are there true properties of a microscopic quantum system? These nontrivial questions were first investigated by Heisenberg and answered by him in a rather restrictive way. It should also be remembered in this connection that the existence of a criterion for truth was mentioned in Chapter 5 as one of the basic notions of logic and no such criterion has yet been given. The main difficulty with the notion of truth in quantum mechanics comes from the existence of many different consistent logics, i.e., complementarity.

These questions are not essential and even rather superfluous when one is only interested in the practical applications of the theory. They come nevertheless to the forefront when one wants to consider rather subtle problems that plagued interpretation for a long time and introduced a doubt about its consistency, the best known example being due to Einstein, Podolsky, and Rosen (henceforth often abbreviated as EPR). The basic notions concerning truth will be given in the next section and they will be applied in the following one to the relation between a preparation process and the mathematical form of the state density operator. The discussion of the EPR problem as such will be postponed till the next chapter.

9. The Criteria of Truth

What Is Truth?

It should first be stated that the kind of truth one is interested in here is not whether the theory itself is true, a question that can be answered only as usual by comparing its predictions with experiment. We are assuming the theory to be correct and we ask what criterion for truth one can use in its framework. It is clear for instance that a fact should be considered as being true and one cannot conceive of physics without this foundation. When it comes however to distinguishing between the data and results in an experiment, a result is not by itself a fact, but only a logical consequence of a fact. Can it also be said to be true? Is there something else one can assert to be true, such as for instance the straight-line motion of a particle in a vacuum? These are the questions to which we want to find an answer.

Truth is of course a subject for which logicians show a great interest. They have given formal conditions for it, which are mainly technical. The purpose of these conditions is so that one can attribute a so-called "truth value" (i.e., the value "true" or "false") to some propositions, so as to make sure that no contradiction will ever arise. It will not be necessary to give all these conditions here in detail. They define how the truth values of different propositions can be combined with the logical operations (and, or, not) and the logical relations ($=$, if . . . then). For instance, one assumes that, if a is true and b is true, then "a and b" is true; if a is true and $a \Rightarrow b$, then b is true; and a few other rules in the same vein. In any case, they only tell us the conditions one should impose on truth and they do not tell us how to find it.

One may distinguish two main categories of truth, which occur when one is dealing with a formal language, particularly mathematics, or with an empirical science. The first case was already briefly described in Chapter 5, where it was indicated that truth originated from a convention attributing a truth value to a few basic axioms. This aspect of the question need not concern us here.

In an empirical science or, if one prefers, when one is dealing with reality, the situation is much more intricate. This is because truth now appears under two aspects: logically, it is still a formal quality which can be transmitted from one proposition to another by correct reasoning and, in reality, the origin of truth is to be found in facts. Logicians have found it quite a difficult problem to accommodate simultaneously these two aspects of truth. There are several different approaches to this question and it can be shown that physics can more or less follow an approach that was advocated by Wittgenstein, so that one can at least feel sure that no trivial error has been made in the logical background.[6]

The Criteria of Truth

According to all authorities, one can safely assume that a fact is true by definition. Wittgenstein added that a logical consequence of an actual fact is also true. The simplest example of this case is the true character of a past phenomenon as it can be derived from knowledge of an actual faithful document using classical determinism.

It might look at first sight as if this criterion for truth is exactly the same as what was proposed by Heisenberg, according to whom what is true is always the result of an experiment. It is not so, however, since Heisenberg made a questionable identification of the result of an experiment with the corresponding datum. Data are facts and as such they are true. A result is something else. It is not even a phenomenon and its status is somewhat slippery. We know that it is logically equivalent to a datum but we also know that there are many different consistent logics. This is where Theorem 1 and the no-contradiction theorem in Chapter 5 come to our help since they show that the result of an experiment is logically equivalent to a datum in every logic mentioning both of them.

But what about the logics not mentioning the result, a critic would say. Aren't they raising a risk that the result of an experiment does not satisfy the basic conditions for truth? This risk is real, as other examples dealing with properties not amounting to the strict result of an experiment will show later on.

One can fortunately avoid these pitfalls by using the following construction of truth proceeding in three steps:

1. First restrict the quantum logics to be considered to the consistent logics involving all the relevant present facts, also introducing their negations as usual.[7] They will be called *sensible* logics. A sensible logic may very well involve other kinds of quantum properties in addition to the facts, assuming of course that it remains consistent. Because of the freedom in the choice of these supplementary propositions, there are still many different complementary sensible logics. It should also be stressed that the facts to be included in a sensible logic are not restricted to the ones that happen to be known to an observer. They are in principle all the relevant facts, whether they are known to us or not. Hence the present approach, despite its logical character, is clearly objective. A fact will be taken to be true by definition and its negation to be false. Comparison with the logicians' approach shows that this postulate is consistent with all the conditions conventionally demanded from truth.[8]
2. Now extend the sensible logics by including all the classical phenomena that are linked to the present true and false ones by determinism. Determinism being a logical equivalence, we shall agree to extend the notion of truth to a phenomenon that is deterministically equivalent to a true present fact. This extension is also known to be logically consistent from the work of logicians. It justifies the use of documents (records or memorized past facts). It also presumably removes ambiguities coming from relativity, although this remains to be proved.
3. One now extends the notion of truth to every quantum property satisfying the following two criteria.[9] First it can be added to every sensible

logic to give a larger logic that is automatically consistent. Secondly, in this extended logic, it is logically equivalent to a fact.

Criterion 1 means that one can add such a proposition a (together with its negation) to the field of propositions of any sensible logic L. This addition increases the number of consistency conditions to be satisfied, but we require that these supplementary conditions are automatically satisfied in every sensible logic L, because of dynamics. This is a very important step because it allows us to overcome all the difficulties coming from the existence of complementary logics, since it is valid in all the logics one might use. It means that what is true is insensitive to complementarity. As for the second criterion, it may be considered as obvious.

Note also that step 2 is a special case of step 3. One may also suspect that the second criterion is an automatic consequence of the first one, but this has not yet been proved. It is also easy to prove that these conventions satisfy all the conditions for truth as they are used in formal logic but we shall not give the proof.

Examples

The most important example (and perhaps the only one except for those arising from determinism) is the result of a measurement. The corresponding data are assumed to be registered and the records provide facts, which are logically equivalent to the data as they are produced at the end of measurements. Because of the transitivity of logical equivalence (if $a = b$ and $b = c$, then $a = c$), it is enough to prove the logical equivalence of a result with a datum to make sure of its logical equivalence with the factual record of the datum.

Using the methods given in Appendix A, one can show that the result of a measurement can be added to every sensible logic and that the necessary further consistency conditions are automatically satisfied. This is a consequence of dynamics, i.e., of the semidiagonal character of the S-operator connecting the results and the data. The first criterion is therefore satisfied, and the second has already been obtained in Theorem 1. Accordingly, the result of an experiment is true with no ambiguity.

Reliable Properties

Along with true properties, it is sometimes convenient to introduce some other so-called reliable properties. They will be useful for clarifying the apparent paradox considered by Einstein, Podolsky, and Rosen and they also play a significant role when one wants to rigorously justify the calculation of systematic errors in a series of experiments.

MEASUREMENT THEORY

A property will be said to be *reliable* when it satisfies the two following conditions:

1. It is contained in one or several sensible logics or it can be added to them consistently. There exists however at least one sensible logic to which the property cannot be added without generating an inconsistency.
2. In the sensible logics where it enters, the property is the logical consequence of a fact.

Such a property is called reliable because of the no-contradiction theorem, since it can never lead to any contradiction of a fact. However, because of the second condition, there are sensible logics where it can be considered as nonsense. It cannot therefore avoid the arbitrariness originating from complementarity.

A simple example of a reliable property is found in an experiment where an ordinary source of light (a star for instance) emits a wave that is detected by a photomultiplier, which is located so far away that the beam of light is very dim and the photons are detected only from time to time. One can then say that, just before the measurement, a photon was approaching the detector. But one can also say that the initial spherical wave emitted by the star was still propagating in the whole space. Both properties can enter different sensible logics but they are obviously complementary. To say that there was a photon is a logical consequence of detection. To say that there was a full wave is also a logical consequence of emission. This is the best-known example of complementarity and it is clear that neither property is true (because otherwise it would have to be true that there was a full wave while one photon was approaching the detector). Both of them are however reliable according to our definition. The various properties expressing straight-line motion as they were discussed in Chapter 5 also belong to this category.

An Example

The above truth criteria were proposed in response to a criticism by d'Espagnat[10] who questioned a careless use of the word "true" in Griffiths's initial paper on histories and also an overly cautious use of the word "reliable" in the first papers by the present author on the logical approach. A simple example was used by everybody on that occasion. Its consistency conditions are easily checked as well as the implications to be used and the relevant calculations can be left as an elementary exercise. This example is the following. A spin-1/2 particle is initially in an eigenstate of the x component S_x of spin with the eigenvalue $+1/2$, as a consequence of its preparation process. One assumes that the particle is isolated and uncorrelated with anything else, in order not to mix this simple problem with the much more involved EPR problem. The z component of spin is measured at a time t_2 and the result of the measurement is found to be $S_z = +1/2$. One then considers a large logic L involving the properties of both the preparing and the measuring devices and also stating the value of S_z at some intermediate time $t_1 (0 < t_1 < t_2)$. The property stating that "$S_z = +1/2$ at time t_2" is the result of a measurement and it is therefore true, as already mentioned. It is easily shown that the property stating that "$S_z = 1/2$ at time t_1" is logically equivalent to it in the logic L and it is therefore a logical consequence of the measurement datum. It is however only reliable because there exist other sensible logics not involving it.

The last statement can be justified by exhibiting another logic complementary to L. Consider a large logic L' differing from the first one only by what it can assert at time t_1 where it states the values of S_x as being $\pm 1/2$. The proposition "$S_x = +1/2$ at time 0", expressing a property at the end of the preparation process is a logical consequence of the preparation datum, as shown by Theorem 5 to be given in the next section, and it implies the property stating that "$S_x = +1/2$ at time t_1" in the logic L'. The consistency of each logic L and L' together with their complementary character show that what was stated in each of them as a property occurring at time t_1 is reliable but not true.

Estimating Systematic Errors in Experimental Physics

Reliable properties are also useful because they provide a convenient setting for a consistent discussion of the systematic errors occurring in an experiment. Their estimate often requires a systematic census of all the effects that would be able to perturb the experimental device or to mimic a true measurement. For instance, the result one obtains could be due to another particle rather than the expected one (e.g., a cosmic ray), or the right kind of particle might have followed an unexpected road (e.g., a particle scattered upon the shielding rather than following a straight-line path in the experimental device). These various possibilities are often estimated by using probabilistic methods (Monte Carlo calculations), which rely upon a classical behavior of the particles.

The Monte Carlo techniques consist in looking for the various possibilities and of course none of them can be said to be true. They remain at the level of an assumption. They are however reliable since most of them rely upon the assumption of a straight-line motion of the particles and this is enough to compute reliably the corresponding probabilities and to get a significant estimate of the final errors.

The Link between Complementarity and the Presence of Measurement Devices

One of Bohr's important ideas was that complementarity is broken by the actual presence of a measuring device, which is enough to select the right properties to be asserted. For instance, the presence of a detecting antenna compels us to speak of an electromagnetic signal in terms of the electric field whereas the presence of a photomultiplier selects the language of photons. Bohr gave so much weight to this idea that he denied even a meaning to a quantum system by itself and he insisted that one should only speak of the whole system including the measuring instruments.

One does not need to adopt such a drastic standpoint and it is perfectly legitimate to consider a quantum system by itself, as long as it is isolated. One can nevertheless recover Bohr's essential results. What one can say to be true when a measurement has taken place is the result of the measurement. It asserts a property of a measured observable A, one for which there exists a basis where the transition operator S is semidiagonal and this is a dynamical property of the coupling between the measured system and the measuring system, holding only for a specific observable. The properties that can be added to a sensible logic with an automatic consistency are therefore well defined and they completely agree with Bohr's statement.

10. Up to What Point Can One Know the State?

What Is the State of a System?

A last question remains to be considered, which is to understand more clearly the state of a system, how one can effectively know it, and what kind of knowledge this is, whether it can be said to be true and in what sense. We know from Gleason's theorem that there exists a state of an isolated system and one can obtain the density operator of a part of it, whether isolated or not, by a partial trace. This is however a very abstract and nonconstructive assertion and we need better if we want to know the state effectively.

Measurement theory gives a straight answer. At the end of a measurement such as it was described in the first two parts and using the same notations, the Schrödinger equation tells us the state of the system $Q + M$ at the end of the measurement, say $\rho(t_m + \Delta t)$. According to Theorem 4, or rather to a closer examination of its proof, one can say that every future measurement, either of Q or M, following an event where the measured datum is n (corresponding to the result a_n for the measured observable A), will have a probability that can be obtained by using the density operator found in equation (8.9), i.e.,

$$\rho_n(t_m + \Delta t) = p_n^{-1} E_n^B \rho(t_m + \Delta t) E_n^B, \qquad (8.10)$$

p_n being the probability for the result a_n. One can then derive the density operator for the subsystem Q of $Q + M$ by taking a partial trace over the Hilbert space of M, which will give equation (8.10). This is plainly in accordance with Dirac's views according to which the knowledge of a state can be obtained from an analysis of the preparation process, particularly when the preparation is a prior measurement.

We shall now successively examine two questions. The first one is to find what kind of knowledge is provided by equation (8.10) and the second one is to generalize this formula to a case where the preparation of a state cannot be reduced to a simple and unique measurement. When asking what kind of knowledge one has obtained, we are taking into account the existence of controversies that are centered in this question, particularly the one arising from a criticism of Bohr's interpretation by Einstein, Podolsky and Rosen.[11] As a matter of fact, its examination will lead us to rediscover in a straightforward way the situation that was pointed out by these authors.

Is There a True Knowledge of a State?

We should first make it clear what is meant by knowledge. We have previously found two different kinds of true properties. The first one is primary—actual facts; the second one is logically derived and includes mainly past facts (as reconstructed from memory) and the quantum results of a measurement. The knowledge of a state is certainly not primary and it cannot in general even be expressed by a simple property (which means that the density operator is not in general proportional to a projector). There has been however so much discussion in the past of these questions, often considering the case of a pure state for clarity, that it may be worth the effort to consider the following question: Is the knowledge of a pure state true, in the sense already given? This question can be submitted to logic because the density operator associated with a pure state is a projector (with rank 1); it is therefore associated with a property and one may find whether it is true or not.

We can get an answer if we consider a finer description of a problem already investigated, namely a measurement of Type I where a system Q is measured by an apparatus M, the notation being the same as in sections 1 and 2. The initial state of Q is taken to be a pure state if not otherwise stated. One again considers a logic L for the system $Q + M$, involving the possible data and the results of the measurement, but one wants to know more about the whole system. One therefore includes in this logic histories mentioning various properties of the system Q before and after the measurement. We consider a case where the measurement gives a datum n and we want to know whether this is enough to completely determine the initial state one can take for the histories of Q after the measurement. If it is completely

MEASUREMENT THEORY

determined, the associated property will be true and one will be able to assert that the knowledge of the state of Q after the measurement is also true. The first question one must ask is of course whether the logic L is consistent or not. It will be convenient to consider that the first measurement is of the yes/no type, its result being that A has the value a_n. This is because the histories one may think of are very different according to the various outcomes of a measurement, so that it is better to consider the simplest case.

Since the question we are interested in gives a privileged role to the histories of Q after the measurement (when it is isolated again), it will be convenient to introduce two sublogics of L dealing only with the system Q when it is isolated, before and after the measurement, say L_1 and L_2. More precisely, L_1 is characterized by the initial state ρ_Q at time 0 (now a pure state); it involves properties of Q occurring in the histories of the larger logic L at a time between 0 and t_m. It also contains the result of the measurement stating that the value of A is equal to a_n at time t_m (as well as the negations of all these properties for logical completeness). The most important characteristic of the second logic L_2 is a specific initial density operator to be given by the pure state $|a_n\rangle$ at time $t_m + \Delta t$ at the end of the measurement. This is of course the state we expect and that we are testing for truth. The second logic involves otherwise some various properties of Q occurring at various times after the measurement (together with their negations).

One can first consider the case where the measured eigenvalue a_n is nondegenerate. An investigation of the consistency of the overall logic L yields the following theorem from which it follows that the knowledge of the state after the measurement is true in that case.

Theorem 5. *A logic describing the behavior of a measured system Q and a measuring apparatus M before and after a measurement of Type I for a nondegenerate observable A is consistent if and only if its two sublogics describing Q respectively before and after the measurement are consistent. The second logic must necessarily have the initial density operator $|a_n\rangle\langle a_n|$ at the end of the measurement, if a_n is the result of the measurement. The conclusions are valid whether or not the initial state of Q before the measurement is a pure state.*

The proof of this theorem is given in Appendix A as the best representative of all the theorems met in the present chapter, because its proof involves all the techniques that must be also used in the other proofs whereas most of them are simpler. The theorem was stated in the case when the initial state is not pure because it shows that there are cases when a true knowledge of a state can be completely obtained by a measurement.

The assumptions of the theorem are however very restrictive and we shall be particularly interested in the case where the measured observable has degenerate eigenvalues, the initial state of Q now being assumed to be pure. The most convenient description of the degenerate case consists in introducing a complete set of commuting observables for Q, one of them being A. There is no real gain in generality by having more than two such observables and we shall therefore consider this case, denoting by (A, A') the two commuting observables. We then reconsider the proof of Theorem 5 and find that two cases must be distinguished. First choose a basis of the Hilbert space made of tensor products of respective eigenvectors for A and A' and the first case occurs when A is uncorrelated with A' so that the initial state can be written as $|\psi_A\rangle \otimes |\phi_{A'}\rangle$. The other case covers all those where this form cannot be obtained for any A' and it will be called the correlated case.

It is possible to extend Theorem 5 to the uncorrelated case and all its conclusions remain upon replacing $|a_n\rangle$ by $|a_n\rangle \otimes |\phi_{A'}\rangle$. If the state $|\phi_{A'}\rangle$ is known truly, for instance, because of a prior measurement of a nondegenerate observable, the new state is also known truly. It turns out that no way can be found to extend Theorem 5 in the correlated case. Its most representative example turns out to be also the clearest case exhibiting the problems put forward by Einstein, Podolsky and Rosen (EPR) so that it can be asserted that the EPR situation appears as a canonical case in the logical approach to quantum mechanics.

In view of the importance of the EPR problem, which is worth a specific investigation, and for the sake of pursuing only one objective at a time, we shall not immediately enter into its discussion. We proceed as follows. In the next subsection, we investigate the simplest state involving completely correlated observables to show that it corresponds to a form of the density operator we shall meet again when dealing specifically with the EPR problem. This subsection is technical and not very interesting, except for the result, and it can be skipped without inconvenience. We shall then mention a key result in the logical analysis of the EPR problem so as to go on with our main problem at present, which is whether or not one can actually obtain a knowledge of a state. As for the EPR analysis itself, it will be postponed till the next chapter where it belongs.

*Correlated Observables**

We first recall how correlations are defined. Consider the simplest case when both observables A and A' have only two eigenvalues 0 and 1 and they provide a complete set of commuting observables. When they are not explicitly specified, these eigenvalues will be denoted respectively by a and a'. It will also be convenient to consider only the clearest case when the correlation is maximal though it would be easy to generalize the results to more general conditions if they were needed.

The correlation coefficient R of two observables A and A' in a given state is defined in general by

$$R = \frac{|\langle(A - \langle A\rangle)(A' - \langle A'\rangle)\rangle|}{\Delta A \Delta A'},$$

where for instance ΔA is the root mean square uncertainty of A. The two observables are said to be correlated when R is nonzero. In the present case and denoting by ρ the density operator, one has, for instance,

$$\langle(A - \langle A\rangle)(A' - \langle A'\rangle)\rangle = \text{Tr}\{EE'\rho\} - \text{Tr}\{E\rho\} \cdot \text{Tr}\{E'\rho\},$$

E and E' being the projectors associated with the eigenvalue 1 for the observables A and A', respectively. One can introduce explicitly the matrix elements of the density operator as

$$\langle a, a'|\rho|b, b'\rangle = \rho_{aa'bb'}.$$

The correlation is maximal when it is equal to 1, in which case one obviously has

$$f(\rho) \equiv \langle(A - \langle A\rangle)^2\rangle\langle(A' - \langle A'\rangle)^2\rangle - \langle(A - \langle A\rangle)(A' - \langle A'\rangle)\rangle^2 = 0.$$

One can make the function $f(\rho)$ more explicit by writing it in terms of the matrix elements of the density operator, finding easily after a calculation of matrix traces that

$$f(\rho) = \rho_{1111} - (\rho_{1111} + \rho_{1010})(\rho_{1111} + \rho_{0101}).$$

The matrix elements of ρ are constrained by the two conditions

$$\rho = \rho^\dagger, \qquad \text{Tr } \rho = 1.$$

The first condition shows that the matrix elements of ρ entering explicitly in $f(\rho)$ are real. Both conditions together show that these matrix elements are independent and their values determine the matrix element ρ_{0000}. It is then easy to find the density matrix making $f(\rho)$ maximal since this function is a second-order polynomial. One thus gets

$$\rho_{1111} = \rho_{0000} = -\rho_{0101} = -\rho_{1010} = 1/4. \tag{8.11}$$

All the other matrix elements of ρ are 0, as one can show using the condition $\text{Tr } \rho^2 \leq \text{Tr } \rho$, expressing that the eigenvalues of ρ are bounded by 1.

It is easy to show that this is a pure state and one has $\rho = |\psi\rangle\langle\psi|$, with

$$|\psi\rangle = 2^{-1/2}\{|A = 1\rangle \otimes |A' = 0\rangle - |A = 0\rangle \otimes |A' = 1\rangle\},$$

an expression familiar in the literature on the EPR problem, an example being given by a pure state of two spin-1/2 particles with zero total spin. This will be met again in the next chapter.

What Can Be Known about a State?

One of the most clearcut results of the analysis of the EPR problem to be given in the next chapter is that one cannot find a *true* property asserting the state at the end of a measurement in a correlated case (a maximal correlation occurring in the EPR example). A property of that kind is only reliable, i.e., in some way arbitrary and in no case to be known with certainty. Since it is meant to assert the state of a system after a measurement, one must conclude that it is impossible to know a state with certainty except under rather special conditions.

We are therefore led to the following conclusion. We know that there should exist a state for each isolated system in the universe but, most often, we cannot know what it is with certainty, except in a few cases that are carefully devised by physicists with all their tricks so as to deal with sufficiently simple systems. It should be stressed that this conclusion is not simply the result of laziness and economy precluding measurements that are too difficult, but an intrinsic limitation. We shall nevertheless now proceed to give an explicit construction of the state of a system. This attempt could be considered as contradictory in view of what has just been obtained. The difference is however that we do not claim to obtain *the* state of a system but *one* state, taking all the facts and only the facts into account. There are many equivalent states fitting this condition but it turns out that (at least most probably) they cannot be distinguished from each other by real experiments.[12]

11. Explicit States

A Few Warnings

We have already heard from time to time a few warnings indicating serious difficulties against a perfect knowledge of a system state:

1. A classical property expressing a phenomenon can be expressed equally well by every member of a family of equivalent quasi-projectors and it is not defined by a unique well-defined projector. This is a source of ambiguity and, though these differences are very small, they cannot be completely ignored.
2. It is impossible to know exactly the internal state of a macroscopic object (i.e., the state of its environment), even as a matter of principle.
3. The systems one observes usually have suffered various interactions, though none of them was a measurement. These interactions modified the state, but how could one take that into account?
4. When a measured system consists of two strongly correlated parts, it is found that the knowledge of the final state cannot be completely true.

MEASUREMENT THEORY

The difficulties arising from each problem are rather different and each of them carries its own lesson. The first one is basic, when it asks how large the difference between two density operators should be for them to be considered as physically different. The second one is also a serious matter of principle: how can one say what is the state of a system (including the environment) if one cannot answer the question by performing detailed measurements? The third difficulty shows that, in most cases, one must include a system in which one is interested in a much larger system involving all the objects with which it came to interact at one time or another. The lessons of the last point have already been drawn. Altogether, these various remarks tend to shed doubt upon the conventional belief in the indisputable existence of a unique state.

What Has a Physical Meaning?

Physicists often use expressions such as "this has a physical meaning" or "that does not mean anything physically". They are usually expressed in a rather loose way, but it will be useful to see whether one can give them a more precise content.

It will be said that *a property or a proposition has a physical meaning* when one can give it a truth value, at least in principle. This means that one can conceive, at least in principle, of an experimental device to check whether the property mentioned is true or false. Clearly, this definition in practice covers only the properties expressing a phenomenon or the possible results of a conceivable experiment, though it remains to say what one means by "in principle" and "conceivable".

One can think of two different possibilities, if one wants to avoid anthropocentrism, i.e., not restricting oneself to what man can do at present. The first states that a measurement can be conceived if it can be realized in the present universe. The second one is more idealistic and it considers the finiteness of our universe to be inessential. One can accept every experiment that is consistent with all the laws of physics, including relativity. The point of view according to which every observable can be measured, whatever it is, has been shown to be untenable. The difference between the two first points of view is inessential as long as one does not try to give explicit orders of magnitude.

Equivalent States

The existence of a limitation upon the physical meaning of properties implies a similar limitation upon the significance of a difference between two density operators. Consider two mathematically different density operators ρ and ρ' and ask whether they can be considered as physically different. It will be convenient to introduce the family of projectors or quasi-projectors $\{E_k\}$ for all the properties of the system having a physical meaning. One can define the corresponding differences in their probabilities

$$\text{Tr}\{|\rho - \rho'|E_k\} = p_k. \tag{8.12}$$

Let us also denote by p the upper bound of the quantities $\{p_k\}$.

Probabilities that are too small have no physical meaning. This is a direct consequence of the discussion in Chapter 7 concerning the possibility of circumventing decoherence, where it was shown that the determination of some explicit very small probabilities was inconceivable because it would need impossible measuring devices. We shall not try to give precise numbers for the corresponding confidence threshold, which moreover would depend upon the properties one considers. In any case, this means that the conditions of equivalence (8.12) between two different density operators allow the existence of mathematically distinct state operators that are not physically distinct. A mathematician would notice that these conditions (8.12) are seminorms so that they define correctly a notion of vicinity between two operators, but this is not really important. Conversely, one has also obtained a formal criterion for two state operators to be physically different: they should differ by the probabilities they assign to at least one measurable property, this probability being sufficiently different from 0 to be measured.

Introducing the Universe

Two difficulties we mentioned at the beginning were concerned with the approximate character of the phenomena and the difficulty of assigning a state to the internal environment of a macroscopic system. Both of them can be alleviated by using physically equivalent density operators as they have just been defined. Another kind of difficulty had to do with the past interactions of a physical system with various other systems and it will now be considered.

When one wants to define the state of an isolated system, one must almost always, in principle, take into account in its preparation all the systems with which it interacted in the past. These systems however had interactions with other systems and so on, so that one does not see where to stop. The simplest thing to do is therefore to consider the whole universe as the only really isolated system. One might try to make the system smaller, for example, by restricting it to be the solar system (but what about cosmic rays?) or the laboratory, or a few instruments in it. The choice to be made depends upon the approximation one is ready to accept. Whatever it may be, the reasonings would not be essentially different in these various cases.

If one is able to construct (at least theoretically) the state of the "universe", this will be enough to define explicitly the state of every part of it that is momentarily isolated by taking a partial trace. One is therefore led to follow a line of thought advocated by Gell-Mann and Hartle,[13] who stressed the interest of inserting the interpretation of quantum mechanics in a cosmological framework in order to get a completely consistent theory. The next problem is hence to define the state of the universe up to an error having no "physical meaning".

The State of the Universe

Assume for definiteness that the universe had a beginning at some time 0. We will avoid however a discussion of quantum cosmology and assume that there existed a state of the universe where only matter and radiation were appreciably quantum, not space itself, rather soon after time 0, the corresponding density operator being denoted by $\rho(0)$. It will also be assumed that one can define a universal time as can be done in the homogeneous and isotropic models of spacetime currently used.[14] It might be objected that this procedure rejects the problems one wanted to solve very far back in time without changing their nature. This is a relevant criticism but it should also be said that the uncertain knowledge of these boundary conditions is the only serious limitation we shall meet and it will turn out to have very little influence upon the actual definition of the state of a real limited system.

Many facts occurred in a multitude of places and times since the beginning of the universe. Let us denote by $\{E_k(t)\}$ the family of all the quasi-projectors for the actual facts occurring at some time t anywhere in the universe. It may be noticed that these quasi-projectors are defined up to an equivalence involving small but nonvanishing errors and, according to the discussion in Section 8, these facts should be responsible for the present state of the universe. One could also follow Gell-Mann and Hartle who prefer to rely upon Everett's approach and to consider "our" actual present as a branch of the histories of the universe separated from all the other ones by decoherence. Whatever it may be, the result for the present state of the universe is exactly the same in both approaches so that, for the sake of definiteness, we shall stay within the framework advocated previously. We can then define explicitly the state of the universe at a time t as being the outcome of all the past and present facts in the following way. According to Theorem B of Chapter 6, the projectors of the various facts occurring at the same time t' commute. Let us therefore lump them together by introducing their well-defined product, which is also a projector:

$$F(t') = \prod_k E_k(t'). \tag{8.13}$$

One can then recapitulate all the facts occurring up to a time t by introducing the operator (again a projector):

$$G(t) = T\left\{\prod_0^t F(t')\right\},$$

where the symbol T denotes a time ordering of the product in such a way that a projector $F(t')$ is on the left of $F(t'')$ if $t' > t''$. The operator $G(t)$ accounts for all the accumulated facts resulting in the present behavior of the universe. The state of the universe at time t is then given by

$$\rho(t) = \frac{G(t)\rho(0)G^\dagger(t)}{\text{Tr}[G(t)\rho(0)G^\dagger(t)]}. \tag{8.14}$$

If one considers equation (8.14) in more detail, one may notice that many facts persist in a deterministic way so that many projectors $E_k(t')$ relative to different times and describing a deterministic evolution multiply each other. They are redundant and their product is simply given by the first projector expressing the initial fact when it took place. This remark is also valid for the classical interactions between classical systems.

A drawback of the formula is that it does not constrain much the internal and external environments. This insufficiency can be at least partly removed. Some factual data concerning local temperatures, for instance (whether the temperature is measured or not), can be added to the definition of the state by a slight change. This is in any case necessary when the time t approaches 0, when the universe is supposed to contain essentially thermal radiation. It would also be necessary to take into account continuous classical systems, i.e., fluids, though it is not quite clear how to do it. These difficulties do not seem however to be a matter of principle but rather technicalities in a domain where one cannot become technical. It would also be necessary in principle to make sure that the arbitrariness in the projectors describing a fact is not large enough to introduce changes in the state operator (8.13) that would be be physically significant. We shall however stop here this discussion about the sex of angels.

Theory and Practice

A sensible physicist, and there are many of that kind, might rebel against the outcome of the present analysis. Do you mean, he would say, that in order to know the state of a system before doing an experiment, I have to know the state of the universe in all its gory details? And for what? Just to take a trace over it and then to forget most of it and keep only a token? Is this reasonable?

He would be right. The point is however that the present analysis was not directed towards practical applications. It was aimed at understanding how to relate the mathematical concept of state to the reality of facts and to verify the objective character of the theory. This is why rather qualitative and general arguments were good enough, whereas a careful mathematical investigation would have been necessary if the result had been of practical importance.

The practice of physics fortunately does not require a total knowledge of past and present events. The whole art of physics is, on the contrary, to deal with carefully devised experimental conditions where this is unnecessary. A physicist deals nevertheless frequently with a system for which he does not know in detail all the facts entering in its preparation. In that case, he usually resorts to information theory so as to make the best of what he knows. Information theory makes perfect sense when it relies upon an underlying objective basic theory and what was said here shows that this foundation exists. Its practical use is briefly described in Appendix B.

APPENDIX A: THE THEOREMS OF MEASUREMENT THEORY

The theorems given in the present chapter as well as the logical considerations behind the notions of true or reliable properties have all more or less the same pattern, though each one of them is slightly different. They might have been organized differently from the text if our goal had been to follow the easiest mathematical way, but then it would not have been very convenient from the standpoint of the physical ideas. This is why I have chosen to provide in the present Appendix the necessary techniques without entering into the details of all the proofs. The emphasis has been put on Theorem 5, in which all the main difficulties arise, as well as on the implication leading from a datum to the corresponding result.

1. Preliminary Notions

We first consider the traces occurring in the consistency conditions and in the probabilities of measurement theory.

Notations and Assumptions

All the projectors or quasi-projectors will be denoted by the same letter E. When necessary, it will be indicated whether the projector refers to the system Q or M by a subscript, for instance E_Q. When the observable to which a projector refers has to be made explicit, this will be indicated by a superscript and the corresponding eigenvalue by a subscript; for instance, E_n^A denotes the projector associated with the eigenvalue a_n of A and E_n^B a projector for the observable B associated with the eigenvalue b_n of B. Moreover, for clarity, the time at which the measurement begins will be denoted by t_m and the time when it comes to an end by t'_m (i.e., $t_m + \Delta t$ in the notation of the text). It is convenient to express the semidiagonal property of the operator $S = U(t'_m, t_m)$ in the case of a measurement of Type I by

$$E_{n'}^A E_{n''}^B S E_0^B E_n^A = S E_0^B E_n^A \delta_{nn'} \delta_{nn''} \tag{8A.1}$$

We will also use its unitarity: $SS^\dagger = S^\dagger S = I$. An explicit form for the evolution operators is given by:

$$U(t) = U_0(t), \qquad \text{when } 0 \leq t \leq t_m$$
$$= U_0(t - t'_m) S U_0(t_m) \qquad t \geq t'_m.$$

What Must Be Analyzed

We will have to write down some consistency conditions and to compute some probabilities. The consistency conditions will be written in the form advocated by Griffiths, i.e., as

$$\text{Re}[\text{Tr}\{E_{p-1}(t_{p-1})\ldots E_k(t_k)\ldots E_1(t_1)\rho E_1'(t_1)\ldots E_k'(t_k)\ldots E_p(t_p)\}] = 0,$$
$$\tag{8A.2}$$

where the two projectors associated with the time t_k are mutually exclusive and k goes from 1 to p, p being the total number of intermediate times.

Nota. As a matter of fact, the projectors occurring in equation (8A.2) with an index not equal to k should involve an arbitrary sum upon elementary projectors (see Appendix IV.A). One may however reduce the analysis to yes/no measurements for most applications so that the present form is usually sufficient. In any case, one can also check afterwards when necessary that all the results are correct, which is easier to see than to write down briefly.

As a rule, we will use only sensible logics where all the facts are taken into account. It means that we must systematically introduce in all the traces an initial projector $E_0^B(t_m)$ expressing that the apparatus is in its neutral state at the beginning of the measurement (or the negation of this property for the sake of logical completeness), as well as a projector $E_n^B(t'_m)$ for some data b_n. We will also consider only logics where the result of the measurement is mentioned, as expressed by some projector $E_{n'}^A(t_m)$. Under these conditions,

MEASUREMENT THEORY 371

the consistency conditions take the form of equation (8A.2) and they involve traces of the following type:

$$T = \text{Tr}\{\ldots E_n^M(t'_m)E_{n'}^A(t_m)E_0^M(t_m)\ldots \rho \ldots E_0^M(t_m)E_{n''}^A(t_m)E_{n'''}^M(t'_m)\ldots\}. \tag{8A.3}$$

In practice, we will never have to treat cases where the indices n, n', n'' and n''' are all different, but one may nevertheless work with the trace (8A.3), which covers all the cases to be met. The projector $E_0^M(t_m)$ has been written twice without introducing its negation for the following reasons: (i) It does not matter whether one uses the property stating that the apparatus is initially in its neutral position at the time 0 or at the time t_m. This is because the systems Q and M do not interact before the time t_m so that M remains isolated during that period. The projector $E_0^M(t)$ is therefore conserved by time evolution when M is isolated, in view of Theorem VI.C expressing the deterministic behavior of M. So, one can as well state this initial neutral property at the beginning of the measurement or use it at time 0 by bringing the corresponding projector at time $t = 0$ into contact with the density operator. (ii) The fact that the apparatus is initially in its neutral position can be expressed by

$$E_0^B \rho_M E_0^B = \rho_M, \tag{8A.4}$$

so that $E_0^B \rho_M \overline{E}_0^B = 0$. Accordingly, the consistency conditions involving this projector are automatically satisfied.

Finally, the projectors that have not been explicitly written down in the trace (8A.3) may represent two histories of the system Q, one of them prior to the measurement or the other after it.

An explicit calculation of the trace (8A.3) with the help of the properties of the evolution operator and the S operator is a necessary preliminary for all the proofs and it will be done.

The Structure of a Trace

We proceed as follows.

1. First make the time dependence of projectors explicit by writing the trace (8A.3) as

$$E_{k+1}(t_{k+1})E_k(t_k)E_{k-1}(t_{k-1}) = \cdots E_{k+1}U(t_{k+1},t_k)E_kU(t_k,t_{k-1})E_{k-1}\cdots. \tag{8A.5}$$

2. A general form of a trace is then given by

$$T = \text{Tr}\{F_Q E_n^B S E_{n'}^A E_0^B F_Q' \rho F'' _Q E_0^B E_{n''}^A S^\dagger E_{n'''}^M\}.$$

Here F_Q represents a product of all the projectors and the intermediate free evolution operators entering in the description of the properties of system Q after the measurement. These operators have been brought together using the cyclic invariance of a trace. equation (8A.5) shows that F_Q is simply given by a product of time-dependent projectors, the origin of time being now taken at the time t'_m. Similarly, the operators F'_Q and F''_Q are products of time-dependent projectors with origin at time 0, each of them referring to some history of Q before the measurement. The occurrence of the factors S and $S^\dagger$ originates from the intermediate evolution operator $U(t'_m, t_m)$ and its inverse.

3. One can use the semidiagonal property of the transition operator S, as expressed by equation (8A.1) to obtain a simpler expression for the trace. First define an operator Σ, which maps the subspace of the M Hilbert space associated with the eigenvalue b_0 of B on the subspace associated with the eigenvalue b_n. This operator therefore acts in practice only upon the degeneracy indices and we will write

$$\Sigma_{rr'} = \langle b_n, r, a_n | S | b_0, r', a_n \rangle.$$

Physically, it is clear that this operator represents the very complicated processes occurring while a measuring device is working, from its initial interaction with the system Q to its final setdown showing the data.

Using equation (8A.1), we then get

$$T = \text{Tr}\{F_Q E_n^B E_n^A \Sigma E_n^A E_0^B F'_Q \rho F''_Q E_0^B E_{n''}^A \Sigma^\dagger E_{n'''}^A E_n^B\} \delta_{nn'} \delta_{n''n'''} \delta_{nn''}.$$

4. We therefore have a trace where the operators relevant to one or the other of the two systems Q and M have been separated. The trace can then be factored as

$$T = \text{Tr}_Q\{F_Q E_n^A F'_Q \rho_Q F''_Q E_n^A\} \text{Tr}_M\{E_n^B \Sigma E_0^B \rho_M E_0^B \Sigma^\dagger E_n^B\} \delta_{nn'} \delta_{n''n'''} \delta_{nn''}.$$

5. Finally, the first trace can be computed explicitly by using a basis in the Q Hilbert space made of the eigenvectors of A. When they are nondegenerate, one gets

$$\text{Tr}_Q\{F_Q E_n^A F'_Q \rho_Q F''_Q E_n^A\} = \text{Tr}_Q\{E_n^A F'_Q \rho_Q F''_Q\} \text{Tr}_Q\{F_Q E_n^A\},$$

so that:

Lemma. *When the eigenvalues of A are nondegenerate, the traces (8A.3) occurring in the consistency conditions and the probabilities can be factored as*

$$\begin{aligned} T = &\text{Tr}_Q\{E_n^A F'_Q \rho_Q F''_Q\} \text{Tr}_Q\{F_Q E_n^A\} \\ &\times \text{Tr}_M\{E_n^B \Sigma E_0^B \rho_M E_0^B \Sigma^\dagger E_n^B\} \delta_{nn'} \delta_{n''n'''} \delta_{nn''}. \end{aligned} \quad (8A.6)$$

MEASUREMENT THEORY

One sees that the two traces over Q refer respectively to the history of Q before the measurement with an origin of time at time 0 and to the history of Q after the measurement with an origin of time at time t'_m. In the trace relative to M, the occurrence of the factors Σ and $\Sigma^\dagger$ expresses the effect of the interaction.

2. The Proof of Theorem 5

According to this theorem, a logic L containing the data, the results, and some histories of the measured system Q before and after the measurement is consistent if and only if two logics L_1 and L_2 belonging to Q are consistent. It assumes that the eigenvalues of A are nondegenerate. The logic L_2 has an initial density operator at time t'_m that is given by the projector E_n^A.

The proof consists in finding under what conditions the real part of a trace (8A.6) vanishes so that the consistency conditions are satisfied. In the traces occurring in a consistency condition, every projector appears twice (in two different places), except for a unique time t_k where the two projectors are explicitly different. We shall distinguish two cases according to the fact that these special projectors belong either to the system M or Q. If they belong to M, among the factors occurring in the trace (8A.6) is the equality $F'_Q = F''_Q$, and the two traces relative to Q are both real and positive. This is obvious for the first one, which can be written as

$$\mathrm{Tr}_Q\{E_n^A F'_Q \rho_Q F''_Q\} = \langle a_n | F'_Q \rho_Q F'_Q | a_n \rangle.$$

As for the second one, i.e., $\mathrm{Tr}_Q\{F_Q E_n^A\}$, it is a typical history probability for the system Q with an initial state operator E_n^A and it is therefore also positive. The real part of the full trace therefore vanishes because the trace itself vanishes when $n \neq n'$, as shown by the Kronecker symbols in the quantity (8A.6). It might be noticed that, since the properties under consideration refer to a macroscopic apparatus that is described by classical logic and therefore represented by quasi-projectors, the trace vanishes only up to an error of order ε. This is where the possible violations of determinism enter in the behavior of the apparatus and the corresponding errors, even if very small, cannot be avoided.

When the two projectors that are different belong to the history of Q before the measurement, the trace over M is positive as well as the trace over Q relative to the later history. One is then brought back to a consistency condition for L_1. The same result is obtained when the two different projectors refer to the later history of Q and one gets a consistency condition for L_2.

3. Data Imply Results

The property a expressing the experimental datum is represented by the projector $E_n^B(t'_m)$ and its probability is given by

$$p(a) = \text{Tr}\{\rho E_n^B(t'_m)\}.$$

The property b expressing the result is represented by the projector $E_n^A(t_m)$ and its probability is given by

$$p(b) = \text{Tr}\{\rho E_n^A(t_m)\}.$$

The probability of the property "a and b" is given by

$$p(a.b) = \text{Tr}\{E_n^A(t_m)\rho E_n^A(t_m)E_n^B(t'_m)\}.$$

The proof of the two implications $a \Rightarrow b$ and $b \Rightarrow a$ will be obtained if one can show that

$$p(a.b) = p(a) = p(b).$$

Let us prove that $p(a.b)$ is equal to $p(b)$. Using the fact that M is initially in its neutral state, as given by equation (8A.4), and the cyclic invariance of the trace, one gets

$$p(a.b) = \text{Tr}\{\rho E_0^B E_n^A(t_m)E_n^B(t'_m)E_n^A(t_m)E_0^B\}.$$

Writing down explicitly the time evolution, this can be written as

$$p(a.b) = \text{Tr}\{\rho E_0^B U_0^{-1}(t_m)E_n^A S E_n^B S^\dagger E_n^A U_0(t_m)E_0^B\}.$$

The deterministic stability of the apparatus in its neutral state (Theorem C of Chapter 6) gives

$$E_0^B = U_0(t_m)E_0^B U_0^{-1}(t_m), \tag{8A.7}$$

from which one gets

$$p(a.b) = \text{Tr}\{\rho U_0^{-1}(t_m)E_0^B E_n^A S E_n^B S^\dagger E_n^A E_0^B U_0(t_m)\}.$$

Using the semidiagonal form of the operator S as given by equation (8A.1), this probability can also be written as

$$p(a.b) = \sum_{n'n''} \text{Tr}\{\rho U_0^{-1}(t_m)E_0^B E_n^A S E_{n'}^B E_{n''}^A S^\dagger E_n^A E_0^B U_0(t_m)\}.$$

Using the two decompositions of the identity operator

$$\sum_{n'} E_{n'}^B = I_M, \quad \sum_{n''} E_{n''}^A = I_Q,$$

one gets:

$$p(a.b) = \text{Tr}\{\rho U_0^{-1}(t_m)E_0^B E_n^A S S^\dagger E_n^A E_0^B U_0(t_m)\}.$$

MEASUREMENT THEORY

The unitarity of the operator S gives now

$$p(a.b) = \text{Tr}\{\rho U_0^{-1}(t_m) E_0^B E_n^A E_0^B U_0(t_m)\}.$$

Using again equations (8A.4) and (8A.7), the desired proof is obtained. The reverse implication from a result to a datum is proved similarly.

APPENDIX B: THE DENSITY OPERATOR AND INFORMATION THEORY

In the vast majority of practical applications, one cannot perform enough measurements when preparing the system Q to be measured later (or these measurements are not precise enough) to obtain a complete knowledge of the initial density operator. The explicit expression (8.34) of this operator in terms of the objective state of the whole "universe" is of course of no help in practice since it assumes a perfect knowledge of all the relevant facts. So, one must use at best the information one has obtained in order to get a reasonable approximation for the actual density operator.

The information consists most of the time in the knowledge of some average values for a number of observables $A_1, A_2, \ldots, A_q$, i.e., some numbers

$$\text{Tr}(\rho A_j) = m_j, \tag{8B.1}$$

together with the normalization condition

$$\text{Tr}\,\rho = 1. \tag{8B.2}$$

An example is given by a beam of charged particles, the average velocity of which is known by a classical measurement (for instance, by measuring the electric current carried by the beam). This is an average over a large number of particles, but it is enough to give information in the form (8B.1). One also quite often knows the average value of the energy. When the particles cross a device selecting the ones with a specific value of some observable (for instance a spin component with the help of a Stern–Gerlach device), one can take $A_j = E$ in equation (8B.1), where E is the projector associated with this value and the right-hand side m_j of equation (8B.1) is equal to 1. Furthermore, the knowledge of error estimates for the various quantities can be expressed by writing the square of the corresponding quantum uncertainty in the form (8B.1).

Information theory defines the information for a system having discrete classical random states and a probability distribution $\{w_1, w_2, \ldots\}$ by the quantity

$$I = \sum_\alpha w_\alpha \log w_\alpha, \tag{8B.3}$$

which is defined up to an additive constant. In the case of quantum mechanics, information can be written in terms of the density operator in the form $I = \text{Tr}(\rho \log \rho)$.

One can then look for a density operator satisfying the conditions (8B.1) and (8B.2) and minimizing the information (8B.3) so as to express at best our knowledge of the conditions and our complete ignorance of everything else. Conditions (8B.1) and (8B.2) are therefore considered as constraints. This problem of minimization can be solved by using the technique of Lagrange multipliers. Introduce as many parameters $\lambda_1, \ldots, \lambda_q$ as there are data (8B.1) together with a Lagrange parameter μ to take into account the normalization condition (8B.2). It is then easily found that the density operator giving the required minimal information has the form

$$\rho = \exp\left(\mu + \sum_{j=1}^{q} \lambda_j A_j\right).$$

The parameters $\{\lambda_j\}$ and μ are implicitly determined by the conditions (8B.1) and (8B.2).

As an example, let us assume that the only information concerns a projector E giving the result of a selection at the beginning of the experiment, the rank of this projector being finite and equal to N. The constraints are given in that case by

$$\text{Tr}(\rho E) = 1, \qquad \text{Tr}\,\rho = 1, \tag{8B.4}$$

from which one gets $\rho = \exp(\mu + \lambda E)$. If one expands the exponential as a series and takes into account the relation $E^2 = E$, one gets

$$\exp(\mu + \lambda E) = e^{\mu}\{1 + (e^{\lambda} - 1)E\}.$$

One can then find the quantities $\exp(\mu)$ and $\exp(\lambda)$, in view of the conditions (8B.4), using the relation $\text{Tr}\,E = N$. If the dimension of the Hilbert space is equal to n (which is possibly infinite), one gets

$$\{n + N(e^{\lambda} - 1)\}e^{\mu} = 1, \qquad Ne^{\lambda + \mu} = 1,$$

from which one gets immediately

$$\rho = \frac{E}{N} = \frac{E}{\text{Tr}\,E}.$$

This is precisely the form systematically used by Griffiths for an initial state, as was mentioned in Chapter 3.

As another example, let us consider a macroscopic system. The information says that its collective observables are inside a cell in phase space

MEASUREMENT THEORY

having a quasi-projector F. One also knows the average energy of the environment. Using the same methods, one finds in that case

$$\rho = \frac{1}{K} F e^{-\beta H_E},$$

where the normalization coefficient K is given by

$$K = \text{Tr}\, F \cdot \text{Tr}_E(e^{-\beta H_E}).$$

This final expression, when the parameter β is expressed in terms of the temperature by $\beta = 1/kT$, can be used in many semiclassical problems.

9

Questioning Quantum Mechanics

Serious doubts have been raised at one time or another about the idea that a wave function can provide complete knowledge for the actual state of a physical system. The most famous criticism of that kind was put forward in 1935 by Einstein, Podolsky, and Rosen and the first part of the present chapter discusses this. These doubts are often associated with a feeling that other significant data could exist, completing or replacing the wave function. These putative data are usually called hidden variables. An example of a hidden variable theory due to David Bohm will be briefly described to give an idea about this trend of research. A very general test for the existence of hidden variables has been given by John Bell and his basic results, together with their comparison with experiments, will be the last topic to be covered.

THE EINSTEIN–PODOLSKY–ROSEN EXPERIMENT

Among the many changes in our representation of nature afforded by quantum mechanics, three at least are often resented as particularly unpleasant: the impossibility to give full credence to some very familiar and deeply intuitive notions such as the position of a particle or its trajectory; the irreducible character of probabilities ("God does not play dice"), through which nature appears to be in some sense erratic; finally, what can be felt as a breaking of the laws of physics occurring when a fact actually takes place. It is tempting to assume that these three difficulties are only three different manifestations of some incompleteness of the theory, which would come from its neglect of a more realistic but yet inaccessible foundation of physics. Maybe after all a particle is really a particle as the atomists of older times thought of it and it has a well-defined position; maybe probabilities are needed because our experiments are too coarse; and perhaps the actual outcome of a measurement is determined anyway by some hidden unknown quantity. At least, this is what quite a few people would like to believe.

No believer in these ideas would however go so far as to give up the main results of present-day quantum mechanics. Its success, the precision of its predictions, and the overall consistency of the formalism are too good to be thrown out. So the aim of these people is only to reexpress everything in a different framework somewhat more akin to our usual and deeply felt habits of thinking. If possible, they would also like to discover some corrections to ordinary quantum mechanics, even if very slight, in order to compare their ideas with experiments. These various trends can be used to define the main steps of the present chapter. We will first criticize quantum mechanics from this point of view, after which a tentative explicit theory will be described and, finally, we will consider what experimental criteria might show the existence of these hidden features of physics.

Of course, one can also hold the opposite view that our habits of thinking originate mainly from an aquaintance with our surrounding macroscopic world and there is nothing sacred about them. The most appropriate way of looking at the foundations of physics would be not to cast them into a preconceived mental frame but rather to learn directly from a study of Nature what the right modes of thinking are. This standpoint, which is so exactly opposite to the tendencies of hidden variables searchers, was already stressed by Francis Bacon and it is clearly dominant in the present book. Nevertheless, for the subject at hand, when it comes to considering either a careful criticism of the present theory, a vigorous attempt at building up alternative ones, or a search for possible experimental criteria for some basic notions, it does not matter what philosophical views one stands for and these questions fully belong to physics.

Einstein's Approach

When looking at quantum mechanics after it had been formulated by Bohr and Heisenberg, Einstein recognized that it had to contain a large amount of truth but he thought that something more was still needed, particularly for explaining the existence of facts. He was not alone in these views, which were also shared by Schrödinger and Louis de Broglie.

Whereas de Broglie attempted to build up an alternative theory, Einstein wanted to proceed with more method. Since no contradiction with experiment could be found, the only starting point for a deeper investigation was to find some sort of a crack or a contradiction in the theory itself. This idea led him to propose the ideal experiments mentioned in Chapter 2, in order to exhibit such a contradiction. His lack of success led him to pause in this attempt for a few years.

Later on, in 1935, in collaboration with Boris Podolsky and Nathan Rosen, he proposed a new ideal experiment, the germ of which is probably due to Podolsky. The idea was not to find an internal inconsistency within the theory but, more modestly and perhaps more deeply, to show that it does not offer a complete description of physical reality, that something more is conceivable, which is suggested by the theory itself though not completely covered by it. If this were so, one might expect that a better theory is possible which can overcome these deficiencies.

1. The Background

Elements of Reality

Nothing can replace a direct reading of the original paper by Einstein, Podolsky and Rosen,[1] but we shall try to convey their meaning while taking into account the developments that were to follow so as to shed some light upon the most delicate issues.

EPR, to abbreviate these authors, founded their considerations directly upon a theory of knowledge, stating that "objective reality is independent from any theory" and "any serious discussion of a physical theory must take into account the description of reality that is given by the theory as well as the concepts with which the theory operates." In other words, one must judge a theory by looking at both its formalism and its connection with reality. They stressed that the concepts should be expected to correspond with physical reality so that one can represent reality by means of them. One might recognize in these introductory sentences some lessons coming from the theory of relativity and their spirit is similar to what was advocated in Chapter 2 as to what an interpretation should be. It may also be noticed that these innocent-looking sentences are politely but definitely directed against Bohr's interpretation, which does not succeed in representing reality by means of the concepts with which the theory operates since it has to call for the extraneous concepts of classical physics and contains a conflict between the Schrödinger equation and the collapse of the wave function. Then, EPR try to define what a complete theory is. This is not the kind of completeness that was stressed in Chapter 2 where it meant an ability to give specific predictions whatever the experimental setup. It is something much nearer to the foundations of knowledge and it proceeds from the basic notion of an *element of reality*.

As a side remark, it may be worth mentioning how difficult it is to distinguish reality from its codification by our brains or its description, either by words or by mathematical concepts, not to mention many habits and prejudices generated over the millennia of human existence. So, when EPR try to isolate some specific elements from the bulk of reality, they are facing one of the most difficult problems in the theory of knowledge. Accordingly, the step they are taking at this point must be watched very carefully and it will be found later on to be essential. They assert that "If, without perturbing a physical system in any manner, one can predict with certainty the value of some physical quantity, then there exists an element of reality corresponding to that quantity." They also add that something is known with certainty when the corresponding probability is equal to 1, so that this notion has a meaning even if probabilities are intrinsic. This definition is sometimes considered to be rather obscure and far-fetched, or even suspected to be put forward in an ad hoc manner for the sake of the argument following it. This is not so and one can recognize in it both Einstein's unique experience of physics and his familiarity with philosophy, particularly including Spinoza's modes of a substance (Spinoza being probably the philosopher Einstein appreciated most, for very good reasons).

How are we to understand this definition? There have been many papers and even books about it and we shall not try to enter in their controversies but only to give some indications. An element of reality is something that can be conceived with precision and can also be known with certainty. Seen in that way, it is clearly a building block of knowledge. It is clear that most concepts that are used in classical physics are associated directly with an element of reality, whether it be a position, a shape, an orientation, a color, or a velocity, because they can in principle be known by looking carefully at the system, or they can be measured without perturbing the system. In that sense, what we called previously a phenomenon, namely a property having a clearcut classical meaning, is an element of reality. But EPR do not say that the elements of reality consist only of the classical phenomena. They try to take a position at a higher and perhaps more controversial level. An element of reality is something you can observe while you remain a pure outsider, an observer and a thinker who is not directly involved in this world. Said otherwise, an element of reality is something one can conceive legitimately without becoming involved with the system, something objective not involving the observer as a subject or an actor. Of course, in order to substantiate this notion, EPR had to give examples of nontrivial elements of reality that are not simply phenomena.

Once elements of reality are assumed to exist, EPR can define a theory as being complete when every element of reality has a counterpart in it. One might say more loosely that a theory is complete when it can account for everything one can think of legitimately, observe while remaining an outsider. Their main argument is then to show that quantum mechanics is not complete in this respect so that there are some notions one can think of legitimately that are not covered by the theory. Of course, nothing could be more opposed to the logical approach that is used in this book, where what one can think of legitimately is supposed on the contrary to be governed by the theory itself. This clearcut opposition is very convenient because it implies that EPR's point of view is much more easily and clearly compared with the logical interpretation than with Bohr's interpretation at which it was aiming.

An Example

EPR had to give an example of an element of reality that is not simply a classical phenomenon. To do so, they considered a particle that is described by quantum mechanics with the one-dimensional wave function $\psi = \exp(ip_0 x/\hbar)$. The momentum observable is an element of reality in that case since its value is equal to p_0, with probability 1, and this is known without perturbing the system by any measurement. On the contrary, the position of the particle can only be known with the help of a specific measurement, which will strongly perturb the state of the system so that it is not an element of reality. EPR expressed this by saying that, in quantum mechanics, "when the momentum is known, the position has no physical reality."

Using this example, one can translate EPR's elements of reality into a more formal though less philosophical framework. The present element of reality is simply a special case of a property. An element of reality, when considered as a property, is found to be associated with a projector E leaving the state vector unchanged. In terms of a density operator ρ, its existence would correspond to the validity of the relation $E\rho E = \rho$. As a matter of fact, a pure state $|\psi\rangle$ is always associated with many elements of reality, though they refer generally to very abstract and quite uninteresting observables, the simplest one being the observable $|\psi\rangle\langle\psi|$. Conversely, most mixed states cannot be associated with any element of reality. So, it looks as if the concept of an element of reality, as it was defined, is rather inadequate to what it was supposed to be, namely something well suited to reality. On the contrary, it is most often concerned with mathematical oddities when it is not purely and simply an empty concept in most cases.

The Main Example

EPR then passed to another example that was to become famous. It involves a strongly correlated system, already met in Chapter 8 when we noticed that the recipe for wave packet collapse is sometimes rather subtle from the standpoint of logic. As a matter of fact, the collapse of the wave function plays a central part in EPR's discussion so that it is quite interesting to notice that EPR hit the mark since they had undoubtably found a weak point in Bohr's interpretation. Perhaps even more remarkable is Bohr's answer[2] where he shows no hint of having realized the existence of a difficulty. It looks as if the increasingly dogmatic character of his interpretation had hidden from him in the present case the possibility of making an improvement.

We shall describe the formal aspects of the first ideal EPR experiment with perhaps more detail than needed, because it will make the later discussion easier. Consider two particles, taking them again for simplicity to be in a one-dimensional space. Let X_1 and X_2 be their position observables, P_1 and P_2 the corresponding momenta. Consider the following wave function

$$\psi(x_1, x_2) = \delta(x_1 + x_2). \tag{9.1}$$

It cannot of course be normalized but this is inessential as will be shown later on. One can look at this state by using different representations for the wave function. As it has been written, it can be considered to give the projection of a state vector $|\psi\rangle$ upon a basis of eigenvectors for X_1 and X_2, which reads $\psi(x_1, x_2) = \langle x_1, x_2 | \psi \rangle$. If one introduces other commuting observables representing the center-of-mass position and the relative momentum,

$$X = X_1 + X_2, \quad P = P_1 - P_2,$$

one can also use the wave function

$$\langle x, p | \psi \rangle = \delta(x)\delta(p), \tag{9.2}$$

where x and p are the eigenvalues of X and P, respectively. Finally, if one wants to consider the two commuting observables P_1 and X_2, one can use

$$\langle p_1, x_2 | \psi \rangle = \exp[i p_1 x_2 / \hbar].$$

2. Analyzing the EPR Experiment

We now consider what quantum mechanics can say in the case of the EPR experiment and whether the corresponding assertions are satisfactory from the standpoint of consistency. This will be done twice for a comparison, first according to the analysis made by EPR themselves and secondly from the standpoint of the logical interpretation.

The EPR Analysis

EPR's reasoning proceeded as follows. They considered the simultaneous measurement of the observables P_1 and X_2. According to the Copenhagen interpretation, there is a wave packet reduction when the measurement of the momentum P_1 of particle 1 gives the result p_1. The new wave function for the whole system made of particles 1 and 2 is obtained by applying the projector $|p_1\rangle\langle p_1|$ associated with the result upon the initial state vector $|\psi\rangle$. Hence it is given, up to an inessential normalization factor, by

$$\phi(x_1, x_2) = \exp(i p_1 x_1 / \hbar) \cdot \exp(i p_1 x_2 / \hbar).$$

This is an eigenfunction[3] of $P_2 = (\hbar/i)\partial/\partial x_2$ with the eigenvalue p_1 so that, according to their definition of an element of reality, EPR could assert that the value of P_2 must be equal to p_1 since it is known with probability 1 without any disturbance of particle 2. Nevertheless, quantum mechanics tells us that, when one measures the position X_2 of this particle, one cannot also take for granted the value of its momentum P_2 so that the theory is unable in the present case to encompass the knowledge of an element of reality.

So, at least in the present case, there exists an element of reality having no counterpart in the theory. According to the definition of completeness by EPR, this means that quantum mechanics is incomplete. Their conclusion is clear: "Whereas we have shown that the wave function does not give a complete description of physical reality, we leave open the question to know whether there exists such a description or not. We believe however that such a theory is possible."

Einstein versus Einstein

It should be noticed that, since their elements of reality do not follow directly from a measurement but from a reasoning, EPR have in fact brought the discussion into the domain of logic. Their conclusion might therefore be reformulated in another way by saying something like: "Quantum mechanics, as described by the Copenhagen interpretation, leads to logical inconsistencies, which is why it must be considered to be incomplete. We expect however that a more elaborate theory will get rid of these inconsistencies."

It is interesting to compare this statement made in 1935 with a later one also made by Einstein in 1949 when he said: "This theory (quantum mechanics) is until now the only one which unites the corpuscular and ondulatory dual characters of matter in a logically satisfactory fashion; and the (testable) relations which are contained in it, are, within the natural limits fixed by the undeterminacy-relation, complete. The formal relations which are given in this theory—i.e., its entire mathematical formalism—will probably have to be contained, in the form of logical inferences, in every useful future theory."[4]

In this later statement, Einstein maintains his earlier point of view, though in a weaker form. He now agrees that quantum mechanics is complete but this is restricted to what can be actually tested. It is not clear whether or not he still thinks that other notions, which cannot be tested by experiments, are meaningful from the standpoint of a theory of knowledge. He is however more specific concerning the kind of future theory he is still expecting when he says that quantum mechanics should essentially remain while becoming a consequence of it through straightforward logical inferences. As far as one can see, this means that he was expecting in 1949 that a logically consistent theory was to be discovered and that the usual practical rules of measurement theory would follow from it as so many theorems.

It is perhaps not too preposterous to apply this prediction to the present logical interpretation, which has been found to be logically consistent and from which the practical rules of the Copenhagen formulation are obtained as so many theorems. One might accordingly expect that a complete account of an EPR experiment is possible in its framework, showing no hint of a paradox though providing a clear understanding of what an "element of reality" really is. This is what we shall now consider.

Orientation

We shall therefore now investigate in some detail what can be said about an EPR experiment from the standpoint of the logical interpretation of quantum mechanics. The interest of this problem is twofold. First, because this is a very old question and it would be worth getting a clearcut answer to it. The second reason is that the EPR experiment was also met in the last chapter as a special case occurring in a deductive measurement theory. One cannot therefore agree with Bohr when he responded by acknowledging nothing as essentially new in the argument by EPR. On the contrary, they had pointed out the existence of a significant problem in the logic of interpretation, which had remained hidden behind its too dogmatic formulation.

Note that, since this was the only nontrivial problem we met in developing a consistent interpretation in the previous chapter (even if it was inessential), we can be sure that no further pitfall awaits us if this one can be solved. So it remains the last hurdle to obtaining a completely consistent interpretation. Of course, its solution cannot be trivial and some preliminary orientation will be useful.

The problem will be treated in several steps. In the next subsection, it will be recalled why there is no logical meaning in asserting that a particle has a definite position and momentum. This means that the elements of reality have no more existence in the logical interpretation than they had in the Copenhagen approach. This more or less obvious point will then leave room for more questions concerning the exact status of the "elements of reality". It turns out that the example chosen by Einstein et al. is not appropriate to a complete discussion of this question. This is because the wave function they used is very unstable under the effect of wave packet spreading so that they could only consider simultaneous measurements. This is a serious drawback and the possibility of performing the two measurements one after another at different times will turn out to be quite illuminating, as one may expect from the key role of time in quantum logics.

A more convenient version of the EPR experiment has been proposed by David Bohm. In its simpler form, it deals with a system of two spin-$1/2$ particles in a state of zero total spin. This is also precisely the example we met in last chapter when investigating formally the various logical cases for two successive measurements. Since the spin states of the particles are unaffected by evolution when no magnetic field is present and there is nothing like a spreading of wave packets in that case, the measurements can be made at different times. This case will be described in the next section as well as the case of two photons, which is very similar.

The analysis of these two problems according to the logical interpretation will then be given in Section 4. It essentially hinges upon the notion of truth, as defined in Chapter 8. It was then shown that a system may have true properties, which are defined irrespective of a choice of logic, as long as the real facts are duly taken into account. It was also mentioned that there exist other so-called "reliable" properties, which have a meaning in some logics, but not in all of them. Taking a reliable property for granted can never lead to a strict logical contradiction and it may very well logically agree with the facts within the logic in which it is formulated. There is however no objective reason for using a logic where a reliable property holds rather than another where this property does not make sense and this remains entirely a matter of choice. To make one choice rather than another is arbitrary. Accordingly, the reliable properties are themselves arbitrary. They belong to our freedom of imagination and our freedom of speech but they cannot be supposed to be true according to the ordinary standards of truth.

The analysis we shall have to perform will by necessity be rather technical since a basic feature of the logical interpretation is to disentangle all kinds of verbal confusion by a straightforward calculation. The results are nevertheless quite simple since it will be found that the EPR elements of reality are only reliable properties and they are never true properties. This means that they have nothing to do with reality since they are arbitrary forms of speech, if only because their statement involves an arbitrary choice of logic. We shall then consider some consequences of this result. After considering a side issue in Section 5, namely the case of relativistic particles for which one sometimes assumed that an EPR device would allow superluminal signals, we shall conclude this survey in Section 6 by considering the question of the separability of quantum mechanics, which has also been the subject of much discussion for a long time.

The Original EPR Experiment

To begin with, we consider the original EPR example and what can be said about the simultaneous knowledge of X_1 and P_1. Let us assume that the two observables P_1 and X_2 are actually measured. This is essentially a measurement similar to the ones we discussed in Chapter 8 and is well understood. One introduces the main properties entering in its discussion and, as long as time does not enter explicitly, one can simply follow EPR and assume that the wave function (9.1) represents the initial state of the system, both measurements taking place at that initial time 0 and ending at a time $t = \Delta t$.

The quantities P_1 and X_2 are measured with the help of two instruments M_1 and M_2, which may be assumed to lie in different regions of space. Each apparatus records a datum registering a fact. The two propositions expressing these facts will be denoted by f_1 and f_2. Two other properties express the results of the measurements and they may be denoted by a_1 and a_2, or in words:

a_1 = "the momentum P_1 of particle 1 is equal to p_1",
a_2 = "the position coordinate X_2 of particle 2 is equal to x_2".

For more precision, one could have introduced small error bounds around these values but this would not play any significant role in the discussion and it may be left aside, just as we have not used a normalized wave function but a delta function. As for the EPR element of reality, it is expressed by the following proposition:

$b =$ "the momentum P_2 of particle 2 is equal to p_1".

As long as one introduces only the data and the results of the measurements and one leaves aside the element of reality, this is exactly what was considered in Chapter 8 so that one has the usual implications between the data and the results

$$f_1 \Rightarrow a_1, \qquad f_2 \Rightarrow a_2.$$

As for the element of reality, it is impossible to add it to any logic for the whole system, since this would amount to introducing together the two simultaneous properties a_2 and b associated with two noncommuting observables. Accordingly, one finds that the proposition expressing the element of reality has no logical content. It is in contradiction with the universal rule of interpretation so that it is empty, even if the words expressing it look like they have a meaning in our fuzzy ordinary language.

3. Bohm's Version of the EPR Experiment

One can conceive many different versions of the EPR experiment. The essential features to be kept are the commutativity of the two measured observables and the strong correlation of one experimental result with an element of reality, this correlation occurring in the initial state. The simplest case was discussed in Section 8.6 where the corresponding density operator was found to be given by equation (8.11), which is exactly what one finds for a pure spin-zero state for two spin-$1/2$ particles. This was also the example put forward in 1951 by David Bohm. It is particularly convenient since it allows one to find out the exact status of the EPR elements of reality by a simple logical discussion.

Two Spin-$1/2$ Particles

More precisely, we consider the following system. A spin-zero particle decays into two spin-$1/2$ particles. The quantum numbers of the initial particle and the two decay products are such that the two decay products are produced in a state of zero total spin. It happens for instance when both particles are produced in a state with zero orbital angular momentum. Two detectors M_1 and M_2 are used to measure respectively the spin component of particle

1 along the z direction and the spin component of particle 2 along some unit vector $\mathbf{n}$. The spin state of the two particles can be written as

$$|0\rangle = 2^{-1/2}\{|s_z^1 = 1/2\rangle|s_z^2 = -1/2\rangle - |s_z^1 = -1/2\rangle|s_z^2 = 1/2\rangle\}, \quad (9.3)$$

where $|s_z^1 = 1/2\rangle$ denotes a state of particle 1 with a spin component along the z axis equal to $+\hbar/2$. This expression does not depend upon the choice of the quantization axis and one could as well have written it as

$$|0\rangle = 2^{-1/2}\{|s_n^1 = 1/2\rangle|s_n^2 = -1/2\rangle - |s_n^1 = -1/2\rangle|s_n^2 = 1/2\rangle\},$$

with $s_n = \mathbf{s} \cdot \mathbf{n}$.

The Case of Two Photons*

Rather than using two spin-1/2 particles, one can as well consider photons. This is essentially because a photon has only two independent states of spin though its spin is 1, because of its zero mass. We indicate briefly how the formalism has to be modified in that case, although this is a rather technical point and it can be skipped without inconvenience. The interest of this example is that it allows an easier treatment of relativistic effects and, moreover, this is also how the experiment has actually been performed with the highest precision. This last feature is however not as essential as it would seem because the problem at hand does not really hinge upon the statistics of the measurements but upon what one is allowed to assert once the result of an individual measurement is known.

An experiment with photons can use an atom of cesium for instance, which has three energy levels—E_0 (the ground state), E_1, and E_2, satisfying the following conditions. When the highest level E_2 is excited (by a tunable laser), it emits a photon 1 while decaying to the level E_1, thereafter emitting rapidly a second photon 2 and falling back to the ground state. Furthermore, the total angular momentum of the two photons is 0 because the ground state has the spin and parity quantum numbers 0^+ and the two other levels E_1 and E_2 have the quantum numbers 1^- and 0^+, respectively. If the two photons are detected in opposite directions, the components of their space wave function having a zero total orbital angular momentum will be the only ones to contribute and the total spin must be 0.

To describe the spin state of a photon, consider a photon having momentum $\mathbf{p}$ directed along a unit vector $\mathbf{n}$. Because of the zero mass of the photon, the possible eigenvalues of the spin projection $\mathbf{s} \cdot \mathbf{n}$ along $\mathbf{n}$ can only be ± 1, the absence of the eigenvalue 0 reflecting the transverse character of the electromagnetic field. The observable $\mathbf{s} \cdot \mathbf{n}$ is called the helicity of the photon. Every spin state of the photon can then be written as a linear combination of the two helicity eigenstates $|\mathbf{p}, \pm 1\rangle$. A state of two photons with their momenta **p** and **p**′ along opposite directions (i.e., **n** = −**n**′) and having a total spin 0 can be shown to be given by

$$2^{-1/2}\{|\mathbf{p},+1\rangle \otimes |\mathbf{p}',+1\rangle - |\mathbf{p},-1\rangle \otimes |\mathbf{p}',-1\rangle\}.$$

If, rather than the helicity **s** · **n**′ of the second photon, one uses the opposite observable **s** · **n**, i.e., one chooses to measure the two spin components along the direction of the momentum of photon 1, one gets

$$2^{-1/2}\{|\mathbf{p},\mathbf{s}\cdot\mathbf{n}=+1\rangle \otimes |\mathbf{p}',\mathbf{s}\cdot\mathbf{n}=-1\rangle - |\mathbf{p},\mathbf{s}\cdot\mathbf{n}=-1\rangle \otimes |\mathbf{p}',\mathbf{s}\cdot\mathbf{n}=+1\rangle\}, \tag{9.4}$$

which is exactly similar to equation (9.3). This shows why the discussion of an EPR experiment with photons will be essentially the same as in the case of two spin-1/2 particles.

In the case of spin-1/2 particles, the experiment involves the measurement of each spin along different directions. Because of the peculiar spin properties of a zero-mass particle, the relevant measurements will be somewhat different (although formally similar) in the case of photons. One will then detect a photon after going through a polarization analyzer. The discussion of the experiment therefore requires some understanding of polarization, which we are now going to consider.

A coherent monochromatic completely polarized beam of radiation with its wavenumber vector **k** directed along the z direction is classically described by an oscillating electric field, which can be written as

$$\mathbf{E}(z,t) = \mathrm{Re}\{\mathbf{E}_0 \exp[i(kz - \omega t)]\}, \tag{9.5}$$

where $\mathbf{E}_0$ is a complex vector in the xy plane. The radiation intensity that is carried by the component of the field along a direction **n** in the transverse xy plane is proportional to the time average of the quantity $|\mathbf{E}\cdot\mathbf{n}|^2$, i.e., to

$$\sum_{\alpha;\beta} n_\alpha E_{0\alpha} E_{0\beta}^* n_\beta \sum_{\alpha;\beta} = n_\alpha R_{\alpha\beta} n_\beta, \tag{9.6}$$

where the components (α,β) of the vectors are defined with fixed axes x and y. This shows that the complex vector $\mathbf{E}_0$ defines a so-called (2×2)-polarization matrix R having rank 1, i.e., only one nonzero eigenvalue.

When the field is quantized, the intensity in the **n** direction becomes proportional to the number of photons which can be measured by a detector standing behind an analyzer selecting a linear polarization along the **n** direction. It turns out that the polarization matrix (9.6) is proportional to the density operator for the spin states of the photon in a basis depending upon the choice of axes in the transverse plane. As usual, a pure state is described by a density matrix having rank 1 and a unit trace. Using the expression (9.6), a photon having a linear polarization along the x axis (i.e., when the vector $\mathbf{E}_0$ is directed along the x axis), has a density matrix with elements

$$\rho_{xx} = 1, \qquad \rho_{xy} = \rho_{yy} = 0. \tag{9.7}$$

Since any self-adjoint 2×2 matrix can always be written formally as a real linear combination of the unit matrix and the three Pauli matrices, one can also write the matrix (9.7) as

$$\rho = \frac{1}{2}(I + \sigma_3).$$

Similarly, the density matrix for a photon which is linearly polarized along the y direction is given by

$$\rho = \frac{1}{2}(I - \sigma_3).$$

The case of a completely polarized wave having an elliptic or a circular polarization will be seen in a moment.

Before that, let us consider the case of a mixed state, which corresponds to an arbitrary density matrix. It can always be written in the form

$$\rho = \frac{1}{2}\{I + \sigma \cdot \mathbf{P}\},$$

where the first term ensures the value 1 for the trace and the Pauli matrices σ play only a formal role. The three-dimensional Stokes vector $\mathbf{P}$ characterizes the polarization of the beam. Since the eigenvalues of ρ are respectively equal to $1/2(1 + |\mathbf{P}|)$ and $1/2(1 - |\mathbf{P}|)$ and since they must lie between 0 and 1, the length of $\mathbf{P}$ must be smaller than 1. The extremity of this vector therefore lies inside the unit sphere. When $|\mathbf{P}|$ is equal to 1, one eigenvalue of ρ is equal to 1 and this corresponds to a pure state for the photons and to a completely polarized wave, which can then be written in the form (9.5). Taking $\mathbf{E}_0$ along a real direction $\mathbf{n}$ in the xy plane, one finds that the Stokes vector associated with a photon linearly polarized along the $\mathbf{n}$ direction has the components

$$P_1 = 2n_x n_y, \qquad P_2 = 0, \qquad P_3 = n_x^2 - n_y^2.$$

A circularly polarized wave corresponds to a vector $\mathbf{E}_0$ with xy components proportional to $(1, \pm i)$ and a Stokes vector

$$P_1 = P_3 = 0, \qquad P_2 = \pm 1.$$

A Stokes vector with $|\mathbf{P}|$ smaller than 1 corresponds to a partially polarized radiation and, when $|\mathbf{P}|$ is 0, the wave is unpolarized.

The link with the helicity states can then be obtained by writing down the density matrix ρ for a completely left-handed polarized wave in terms of the helicity states as

$$\rho = |\mathbf{p}, +1\rangle\langle\mathbf{p}, +1|,$$

with a similar expression for a right-handed circularly polarized wave.

Finally, one can describe the effect of an analyzer. Let the polarization selected by the analyzer be given by a Stokes vector $\mathbf{P}$ with unit length; then the matrix

$$E = \frac{1}{2}\{I + \sigma \cdot \mathbf{P}\}$$

is a projector. From a formal standpoint, one gets exactly the same results when considering the measurement of the component of a spin-1/2 particle along a direction $\mathbf{n}$ in space and when one detects a photon behind an analyzer with $\mathbf{P} = \mathbf{n}$. This means, together with equation (9.3), that the discussion of the EPR experiment is exactly the same in both cases so that one can restrict the discussion to the case of two spin-1/2 particles.

4. Truth and Reliability in the EPR Experiment

The questions that were raised by Einstein et al. find a clearcut answer within the framework of the logical interpretation. As was already stated, this answer consists mainly in recognizing that the elements of reality represent only reliable properties but not true properties, so that they are in fact arbitrary, depending upon a special choice in the description corresponding to a special choice of logic. They have therefore no objective character and there is no reason to follow Einstein et al. when they say that the theory is incomplete if it does not take them into account.

We will now give a more convenient statement of the EPR problem in the case of spin-1/2 particles, which will allow us to introduce time explicitly so that the discussion can be more complete. According to the definition of a reliable property, this discussion will consist essentially in exhibiting explicitly several different consistent logics, all of them including the experimental facts. In one of them, the element of reality will be formulable and it will turn out to be the logical consequence of a fact (a measurement datum). In another logic, it will be found impossible to express it, both results together showing the reliable character of the element of reality.

The EPR Experiment with Spins

Let us follow Bohm by considering two spin-1/2 particles 1 and 2 that have been produced in a total spin 0 state. Two measurements of spin are then performed. The first one measures the spin component of particle 1 along a direction with unit vector **a** at a time t_1. The exact duration Δt of this measurement will be unimportant in the discussion so that it can be assumed to be very short. In a later measurement, the spin component of particle 2 along a direction **b** is measured at time t_2.

If one were to follow EPR's approach, one would say that the first measurement produces a reduction of the wave function. If one assumes that the result of the first measurement is $\mathbf{s}^{(1)} \cdot \mathbf{a} = +1/2$, reduction gives a state vector at the end of the measurement which is obtained by applying the projector

$$|\mathbf{s}^{(1)} \cdot \mathbf{a} = +1/2\rangle\langle\mathbf{s}^{(1)} \cdot \mathbf{a} = +1/2| \otimes I^{(2)}$$

to the initial state vector, giving the reduced state vector

$$|\mathbf{s}^{(1)} \cdot \mathbf{a} = +1/2\rangle \otimes |\mathbf{s}^{(2)} \cdot \mathbf{a} = -1/2\rangle.$$

One would then conclude that the spin component of particle 2 along **a** at the end of the first measurement is $-1/2$ with a probability 1, without having to perturb particle 2. Accordingly, the property $\mathbf{s}^{(2)} \cdot \mathbf{a} = -1/2$ is an element of reality during the time interval separating the two measurements.

When the two directions **a** and **b** are parallel, nothing peculiar happens since the second measurement only confirms this element of reality. When the two directions differ and if, for instance, the second measurement gives the result $\mathbf{s}^{(2)} \cdot \mathbf{b} = +1/2$, one gets the same kind of paradox as before. The theory of measurement given in Chapter 8 shows that it is true that $\mathbf{s}^{(2)} \cdot \mathbf{b} = +1/2$ at time t_2 whereas assuming a knowledge of the element of reality would lead one to assert that $\mathbf{s}^{(2)} \cdot \mathbf{a} = -1/2$ at that time, both statements being obviously incompatible.

About Errors

From the standpoint of practical physics, there is something puzzling in the marked logical difference between the two cases we have met, when the two directions **a** and **b** are parallel or they are not. In the first case, stating the element of reality seems to give rise to no problem whereas it gives a logical inconsistency in the second case. However, two realistic directions that are defined by two different instruments are generally not strictly parallel but only approximately so. So, actually where is the frontier between the two cases and when can one say that the two directions are parallel enough for the element of reality to be valid?

This question has no easy answer within the framework of the Copenhagen interpretation. In the logical interpretation, the answer is simple and it comes from the unavoidable errors coming with the statement of a physical phenomenon. Let ε be the typical logical error occurring in the statement of each experimental data. No difference can be made between two statements that are logically equivalent up to an error ε. But the two properties stating that

$$\mathbf{s}^{(2)} \cdot \mathbf{a} = -1/2 \quad \text{and} \quad \mathbf{s}^{(2)} \cdot \mathbf{b} = +1/2 \tag{9.8}$$

are associated with two projectors E and E' in the spin Hilbert space and it is easily shown that

$$\text{Tr}\,|E - E'| \leq 4\sin\theta/2,$$

where θ is the angle between the two directions $\mathbf{a}$ and $\mathbf{b}$. Accordingly, there is no significant logical difference between the two properties (9.8) when the angle between these two directions is small enough and more precisely when $\theta < \varepsilon/2$.

The Answer to the EPR Paradox

To prove that the property stating that "$\mathbf{s}^{(2)} \cdot \mathbf{a} = -1/2$" is reliable but not true (and therefore arbitrary), consider the global system consisting of the two particles together with the two measurement devices. Construct two consistent logics, both being sensible, i.e., including the data from the two measurements. To one of these logics, say L, one also adds the statement of the element of reality, i.e., $\mathbf{s}^{(2)} \cdot \mathbf{a} = -1/2$ at some time t such that $t_1 + \Delta t \leq t < t_2$, the duration Δt of the first measurement having been reintroduced here to avoid any ambiguity. This is what EPR would have asserted as an element of reality and it is a reliable statement since it will be shown that the logic L is consistent and the element of reality is a logical consequence of the first datum. So as far as one is prepared to be satisfied with reliability without asking for truth, EPR were right. They logically had the right to state that they know this element of reality and they did not run into the risk of self-contradiction when doing so. The condition $t < t_2$ with a strict inequality must however be stressed because, were one to take $t = t_2$, it would turn out that the logic L is inconsistent. This means that the element of reality has no meaning in any logic when stated at the same time as the second measurement. There is therefore no risk of a paradox.

Even when the shadow of a paradox is dissipated, it remains to prove that the element of reality is restricted to be only reliable. To do so, introduce a second sensible logic L' where now $\mathbf{s}^{(2)} \cdot \mathbf{b} = +1/2$ at the same time t as before. This can be considered as some sort of an anticipation. One says that, before the second measurement took place, the spin state of particle 2 was already what this measurement would give. This has of course no realistic meaning for somebody who does not believe in prophecy or who is not himself a prophet but, nevertheless, the logic L' so obtained is perfectly consistent. Of course, the element of reality cannot be added to the logic L' while preserving consistency. The existence of this logic L' is enough to make sure that there exist logics where the element of reality cannot be envisioned once all facts are known. From the other properties already established, one can conclude that the element of reality is reliable but not true.

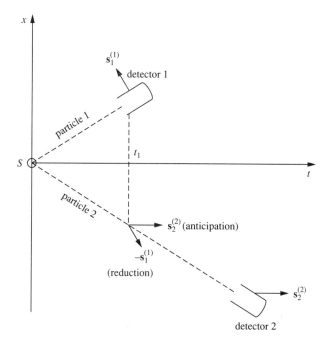

9.1. *The Einstein–Podolsky–Rosen–Bohm experiment.* Two different and complementary logics are shown, both of them including the two experimental results. They differ in their treatment of what happens to the nonmeasured particle at the time of the first measurement. The first logic relies upon wave packet reduction and the other one anticipates the result of the second measurement.

*Formal Aspects**

We shall now make the previous argument more explicit, though not giving all the calculations in detail. As usual, they amount to computing a few traces and how this is to be done was already explained in Chapter 8 and particularly in its Appendix.

It is convenient to use a shorthand notation where for instance $[1, \mathbf{a}, -, t]$ denotes the property according to which the spin component of particle 1 along the direction **a** has the value $-1/2$ at time t. We will have to consider the following properties:

1. The results of the two measurements, namely:

$$[1, \mathbf{a}, +, t_1], \quad [2, \mathbf{b}, +, t_2]. \quad (9.9)$$

2. Two properties concerning the actually measured particle at the end of each measurement, namely:

$$[1, \mathbf{a}, +, t_1 + \Delta t], \quad [2, \mathbf{b}, +, t_2 + \Delta t].$$

3. A property of the two particles together, which is what one would consider when stating wave packet reduction in the conventional approach, i.e.,

$$[1, \mathbf{a}, +, t_1 + \Delta t] \cdot [2, \mathbf{a}, -, t_1 + \Delta t],$$

the dot denoting as usual the logical operator "and."

4. The state of the two particles at the end of the second measurement, namely,

$$[1, \mathbf{a}, +, t_2 + \Delta t] \cdot [2, \mathbf{b}, +, t_2 + \Delta t].$$

We will only consider the case where $t_1 + \Delta t < t_2$. The case when $t_1 = t_2$ or when the two measurements overlap is not essentially different from what was found in Section 2 concerning the initial simultaneous version of the EPR experiment. One needs only to replace everywhere in this case what was then said about "momentum" by the same statement for the value of $\mathbf{s} \cdot \mathbf{a}$ and what was said about "position" by the value of $\mathbf{s} \cdot \mathbf{b}$, keeping the same name as before for the particles.

We shall restrict our considerations to sensible logics (i.e., consistent logics including the data from each measurement). As mentioned in Chapter 8, one can always add to them the results (9.9) of each measurement, the consistency conditions being automatically satisfied and these results being true in the sense given earlier. Each of them is logically equivalent to the corresponding data. This entails that another property is implied by a fact if it is implied by a result. This remark can be used to get rid of an explicit reference to the measuring devices and one can work entirely with logics involving only different spin properties at various times. Accordingly, all the calculations are reduced to the logics for the spin system, which must include the experimental results as possible properties because of the restriction to sensible logics.

The logic L may involve the following properties:

- The result of the first measurement and the final state of particle 1 as resulting from this measurement, i.e.,

$$[1, \mathbf{a}, +, t_1], \quad [1, \mathbf{a}, +, t_1 + \Delta t].$$

- Any statement concerning the spin of particle 1, either before or after the first measurement, as one might expect from the result of this measurement, i.e.,

$$[1, \mathbf{a}, +, t] \quad \text{for any } t, \quad 0 < t < t_1 \text{ or } t \geq t_1 + \Delta t.$$

- Any extension of this kind of statement to a property of particle 2, as would be expected from conventional wave packet reduction, i.e., any EPR element of reality

$$[2, \mathbf{a}, -, t] \quad \text{for some } t, \quad \text{such that } 0 < t < t_2. \tag{9.10}$$

- The results of the second measurement together with the corresponding state of particle 2 if one wishes to mention it:

$$[2, \mathbf{b}, +, t_2], \quad [2, \mathbf{b}, +, t_2 + \Delta t].$$

Of course, the negations of all these properties must also be included in the field of propositions.

The second (anticipating) logic L' may contain:

- What comes from the second measurement, i.e., the properties

$$[2, \mathbf{b}, +, t_2], \quad [2, \mathbf{b}, +, t_2 + \Delta t].$$

- The eventual consequences concerning what happened to particle 2 between the two measurements, i.e.,

$$[2, \mathbf{b}, +, t] \quad \text{for } t, \quad \text{such that } t_1 < t < t_2;$$

one cannot extend this before $t = t_1$ because it would be logically inconsistent with the result of the first measurement.

- One might also assume that the first measurement determined the properties of the first particle thereafter, i.e., use the properties

$$[1, \mathbf{a}, +, t] \quad \text{for } t \geq t_1 + \Delta t,$$

It might be noticed that the first logic L relies most strongly upon the result of the first measurement and it draws all allowable consequences from it, whereas the second logic L' puts as much weight as possible upon the result of the second measurement, which is why it has to be anticipative. A convenient informal way of looking at these logics would be to attribute them respectively to two different experimentalists, each one hating the other and wanting to make the most of what he himself has obtained, tolerating the existence of the other only to the point where ignoring what he has obtained would violate the laws of logical consistency.

Straightforward calculations, patterned after the example of Appendix 7.A, show that both logics L and L' are consistent and they are of course complementary. Furthermore, the element of reality (9.10) is implied in logic L by the result of the first measurement whereas it cannot be added to the logic L' without breaking consistency. This is enough to prove that the element of reality is reliable but not true.

5. The Relativistic Case

Relativity was sometimes associated with an EPR experiment to suggest the possibility of information signals propagating faster than light. One might of course anticipate that these considerations were based upon a logical flaw, as will now be shown. We consider the case of two photons produced in a state of total spin 0 and detected behind a polarization analyzer in two different faraway galaxies, assuming that decoherence has not broken the quantum superposition.

We cannot classify the various cases as was done before assuming that the two measurements follow each other in time or are simultaneous (or at least partly overlapping in time). In a definite inertial reference system, it can be said that the measurement on particle 1 takes place near a point $\mathbf{x}_1$ at a time t_1 (the finite duration of the measurement being clearly irrelevant in the present case), whereas particle 2 is measured near a point $\mathbf{x}_2$ at a time t_2. We distinguish the cases when the spacetime points $(\mathbf{x}_1, t_1)$ and $(\mathbf{x}_2, t_2)$ are spacelike separated or timelike separated.

The previous logical discussion can still be carried out in the chosen inertial frame. The main difference comes from the explicit form of the evolution operators entering in the explicit time-dependent form of the projectors. When a detector is located in a finite space region, which can be treated as small, the photon propagator for its travel from the source located at the origin of space to the detection region near $\mathbf{x}_1$ is not given as before by the nonrelativistic expression for a massive particle

$$\langle \mathbf{x}_1 | U(t_1) | 0 \rangle = (2\pi i t_1/m)^{-3/2} \exp\{-m\mathbf{x}_1^2/2it_1\}$$

(with units where $\hbar = 1$). This has to be replaced by the relativistic retarded propagator for a zero-mass particle, i.e.,

$$\langle \mathbf{x}_1 | U(t_1) | 0 \rangle = \frac{1}{|\mathbf{x}_1|} \delta(t_1 - |\mathbf{x}_1|/c).$$

Without entering into the details, we indicate the main results one obtains in the two cases:

1. When the two detection events are timelike separated, we assume for definiteness that t_1 is earlier than t_2 (whatever the inertial reference system). We recover exactly in that case the conditions prevailing

for two nonrelativistic spin-1/2 particles when the two measurements occur successively. The first measurement made at $(\mathbf{x}_1, t_1)$ indicates for instance that the photon number 1 has its polarization in the x direction (as defined by an analyzer) and this is a true statement. We can also assert that photon number 2 is polarized (along the y direction, which is what comes out of the zero-spin total wave function) before being detected. This is however an EPR element of reality, which is only reliable and not true just as before.

2. When the two detection events are spacelike separated, we find the same conclusions as in the case of two nonrelativistic spin-1/2 particles when the two measurements are performed simultaneously. This does not come as a surprise since there always exists a reference system where the two measurements are actually seen as simultaneous. The logical conclusion is then that the EPR element of reality cannot enter in any sensible logic so that it cannot even be formulated in a consistent way. It is logically empty and therefore cannot carry any amount of sensible information.

In other words, the EPR element of reality is only something that can occur in the mind of somebody in Galaxy 1 when he sees the result of the experiment. He can choose to believe it or not and nobody can prove him wrong since his belief is reliable, at least when the other photon detection takes place absolutely later than the one he sees. The choice is left to his own freedom of arbitrariness as a thinker or, more precisely, as a logician. What he believes does not belong to reality if one agrees that only truth can be the touchstone of reality.

On the other hand, if the other photon is detected at a spacelike distance from the first measurement, a judge can wait until the news of this faraway measurement arrives. He might then tell our thinker that he was completely wrong, that he had no right to believe what he believed because he was committing, even if unknowingly, the sin of logical inconsistency. One should therefore never utter an element of reality when seeing only one piece of evidence, because one cannot be sure to be exculpated later on the grounds of reliability. The judgement can be much worse and might be a sentence for having made an unquestionable logical error or, in other words, a blunder. This is the worst one can receive for an incautious use of the elements of reality, but it is bad enough to banish them from serious consideration.

6. Separability

Another aspect of the EPR experiment was particularly stressed by later writers, even if it was noticed previously. It has to do with the global character of wave packet reduction, which was believed to occur once and for all in whole space without a finite propagation velocity nor a limitation in distance. Thus, for instance, when a polarized photon is detected in Galaxy 1, it was assumed that the polarization state of the second photon was immediately defined at whatever distance it might be.

Some people did not take the problem very seriously and they more or less felt that it was only a matter of language. As someone said, the same thing happens to a man when his wife gives birth to a child when she is far away. He immediately becomes a father whatever the distance. These skeptical people however did not write much on the subject while others published long and learned papers about what they called the nonlocality or *nonseparability* of quantum mechanics.

Nonlocality means that one immediately becomes a father, even when the maternity hospital is in the Andromeda nebula, or the spin state of particle 2 immediately becomes a knowledgeable element of reality when a spin component of particle 1 has been measured. It stresses the fact that this connection takes place whatever the distance. Nonseparability means the same thing, except one insists upon the impossibility of considering a particle independently of the other one, as long as they are strongly correlated in view of a common event in the past. According to another trivial example, a man is a husband and cannot legally get married again because of a past event when he became strongly correlated with his wife. It was sometimes said that this kind of limitation in the independent character of a particle implies that the reality of that particle is not contained in itself but inseparable from what exists outside it, however far. If this were true, it would link realism, i.e., the possibility of getting a true knowledge of physical reality, to this question of separability.[5] One would not be able to assert that a particle behaves freely, without any outside action, since its properties might be changed under the effect of something happening to another particle much farther away. This was after all in some sense an essential part of the argument by EPR.

One can be more precise and borrow from d'Espagnat the definition of separability as follows.[6] A theory is separable if, according to it, when "a physical system remains for some time mechanically (as well as electromagnetically and so on) isolated from all other systems, the evolution of its properties for that time cannot be influenced by operations being made upon another system." The EPR analysis or more simply the assumption of wave packet reduction implies that quantum mechanics is nonseparable and this result was considered as well established for a long time. This is not so obvious however because one should say more precisely what are the "properties" one is talking about and what it means for them to be "influenced". If one accepts that the properties are what have been defined as such in Chapter 3, one sees that there are quite a few types of them. Some of them can be mentioned without any justification for believing them or not. They only have a probability. Some are reliable. They are the consequence of a known fact but, nevertheless, they are generated arbitrarily by my freedom of speech and they do not universally fit the facts. Other properties do fit the facts whatever the conditions; they are valid beyond the arbitrariness of complementarity; they are true.

It seems appropriate to restrict the kind of properties considered when dealing with physics (and not only with the talk made by physicists), to the true properties. But when this obvious restriction is made, the whole discussion of the EPR experiment given previously shows that the true properties of a momentarily isolated system are not sensitive to any kind of interaction occurring in another system. So one can say confidently that, as far as the true properties are concerned, quantum mechanics is separable.

HIDDEN VARIABLES

7. Hidden Variable Theories

The Underlying Conception of Physics

The criticisms by Einstein, Schrödinger, de Broglie, and others are indicative of a discomfort with quantum mechanics which cannot be easily rejected as unfounded. Their objections were directed against various aspects of the theory, but it might be said that the main criticism comes from a refusal to accept the existence of probabilities that are intrinsic or inescapable or, if one prefers, the existence of an absolute randomness in physics. Since in any case all the experiments confirm the random character of microscopic systems, many people have tried to assimilate this to something one can conceive more easily, namely a randomness coming from our ignorance of some causes, which are inaccessible but nevertheless real. It amounts to bringing back the classical version of probability calculus where the intrinsic properties of the physical systems are assumed to exist and to be well defined, though unknown and perhaps in some way unknowable. In one form or another, the philosophy of hidden variables was always behind all these considerations.

In this kind of representation of physical reality, one assumes that a particle is actually a particle. At any time, it has a well-defined position and a unique velocity, though they cannot be determined in practice. One can only determine their statistical distributions. Furthermore, one must accept that these distributions are correctly described by ordinary quantum mechanics because of the tremendous experimental confirmations. One therefore assumes that quantum mechanics is an essentially correct theory, at least when considered as a formal calculation tool, but it does not model a hidden reality and, in that sense, it remains incomplete. These hidden parameters also provide in principle the efficient cause needed to decide why a measurement apparatus will ultimately give a specific result rather than another one. They offer a classical substratum from which quantum mechanics would be erected as a phenomenological theory.

It is only fair to the reader to confess that the present author does not share this philosophy, since it is better to confess one's prejudices rather than hiding them under the pretense of a complete objectivity. The reasons for this attitude are however only aesthetic. An interpretation where all the classical concepts rely upon a deeper quantum basis, including its logical aspects, has an elegance and consistency that cannot be matched by a return to antiquated ideas. It seems difficult to accept that the deep and unexpected mathematical properties one found for the projectors expressing classical properties or the decoherence effect do not contain a large part of truth. They rely upon so few assumptions and recover so many well-known features of reality that had remained unexplained before them that they have the kind of beauty Dirac coined as the mark of truth. Their success in easily getting rid of so many difficulties and objections that were accumulated during half a century cannot but reinforce that feeling. In the same vein, to retrieve classical determinism from a purely probabilistic foundation is also a rather striking justification of this probabilism.

There would by now be something awkward in trying to lay the foundations of physics upon an exact deterministic substrate when it has been found that the ordinary, eye-catching determinism we see around us, of which our minds have become impregnated, is only something approximate. This criticism is quite serious since, as far as the presently existing hidden variables theories are concerned, they would have to resort to the same explanation for large scale determinism insofar as they accept ordinary quantum mechanics as being effective. There would be an absolutely well-defined position and momentum for all particles but this would be beyond any check and, nevertheless, the collective position and momenta of ordinary objects would have to behave deterministically only in an approximate way.

Bohm's Theory

This being said, we should now describe, even if rapidly, what these hidden variable theories are. There exist essentially, at least to the author's knowledge, two main types of hidden variable theories. The first one is due to David Bohm[7] and belongs to a category that was introduced by Louis de Broglie under the name of pilot-wave theory. The second type is due to Edward Nelson[8] and assumes that particles diffuse, even in empty space, according to an abnormal diffusion law, the origin of which remains mysterious. Both are very clever but we shall mention here only Bohm's theory because the ideas of Nelson would bring us too far into the study of stochastic processes.[9]

Bohm's theory will be considered here only for the case of a unique particle under the action of an external potential. The mathematical description of this physical particle utilizes several mathematical quantities—namely, a field and the coordinates of an ordinary particle. The field is nothing but the usual Schrödinger wave function $\psi(x)$ and the particle itself is described as usual by its position and momentum. These quantities evolve in time according to a dynamics where the field obeys the usual Schrödinger equation

$$i\hbar \frac{\partial \psi}{\partial t} = -\frac{\hbar^2}{2m}\Delta\psi + V(\mathbf{x})\psi.$$

Before specifying the dynamics for the position and momentum of the particle, it will be useful to rewrite the Schrödinger equation in a slightly different form. Exhibiting the modulus and the phase of the wave function, one can write

$$\psi(\mathbf{x},t) = R(\mathbf{x},t)\exp\{iS(\mathbf{x},t)/\hbar\}.$$

When separating the real and the imaginary parts in the Schrödinger equation, one gets several equations for the functions $R(\mathbf{x},t)$ and $S(\mathbf{x},t)$

$$\frac{\partial S}{\partial t} + \frac{1}{2m}(\nabla S)^2 + V + Q = 0,$$

$$\frac{\partial \rho}{\partial t} + \nabla \cdot (\mathbf{v}\rho) = 0, \quad (9.11)$$

where $\rho = R^2$,

$$\mathbf{p} = m\mathbf{v} = \nabla S, \quad (9.12)$$

$$Q = -\frac{\hbar^2}{2m}\frac{\Delta R}{R}. \quad (9.13)$$

We can then formulate the dynamics for the particle degrees of freedom. The quantity $Q(\mathbf{x},t)$, which is given by the wave function, is considered to be a "quantum" potential acting upon the particle located at position $\mathbf{x}$. Since no experiment can tell us this position exactly, we describe it by probability calculus where its distribution is assumed to be given by $\rho(\mathbf{x},t)$. As for the momentum of the particle, it is assumed to be exactly given by equation

(9.12), also depending upon the wave function. Then one postulates that the motion of the particle is simply given by Newton's law, under the combined effects of the external and the quantum potentials, i.e.,

$$m\frac{d^2\mathbf{x}}{dt^2} = -\nabla(V + Q).$$

It is then clear that equation (9.11) is the Hamilton–Jacobi equation for the action S associated with this law of motion and it is easily checked that these conventions are mathematically consistent.

When these equations are solved numerically, they give a very nice illustration of the quantum effects. For instance, in the case of a two-slit interference experiment where the initial state consists in sending a bunch of particles against the screen, the quantum potential is responsible for diffraction through the slits and also for the repartition of the particles in an interference pattern on the other side of the screen. Some critics compare the gain of knowledge one obtains from these calculations to the use of false colors for a better viewing of complicated mathematical data. The tenants of the theory see of course much more in it than this minor role.

Spin is best treated as being directly described by quantum mechanics so that no "hidden" proper angular momentum for the particles has to be assumed. The Pauli principle is then entirely contained in the symmetry properties of the wave function. The quantum potential depends upon the positions of all particles. In the case of fermions, one can see from equation (9.13) that the quantum potential becomes infinite when the wave function vanishes upon a nodal surface. There is accordingly an infinite repulsion between two fermions, forbidding them from being at the same place.

The hidden variables theories are much less developed than conventional quantum mechanics. They have difficulties when one tries to extend them to a relativistic framework because there is no way to account simultaneously for the field and the particle characteristics of a given species of particles.

8. Bell's Inequalities

The question of hidden variables was greatly clarified by John Bell who found an empirical criterion for their existence.[10] It deals directly with a central question, which is whether hidden variables obeying classical probability calculus can determine the outcome of a measurement.

The Assumptions

One can best explain the basic ideas in the case of two nonrelativistic spin-1/2 particles, though the case of photons is essentially similar. The setup to be considered is once again an EPR–Bohm experiment where the two particles 1 and 2 are produced in a total spin 0 state. One assumes that the system made up of these two particles can be described by hidden variables collectively denoted by λ, their exact number remaining unspecified and inessential. The classical probability density for these quantities is denoted by $\rho(\lambda)$. The spins of the two particles are measured simultaneously. The component α of the spin of particle 1 along a direction **a** is measured (with the convention $\alpha = \pm 1$, the value 1 being written in place of $1/2$ by a change of scale in order to simplify the mathematical formulas). The component β of the spin of particle 2 along another direction **b** is also measured. One assumes that the results α and β depend only upon the hidden parameters, including those belonging to the measuring devices.

The essential assumption is that the result given by one measuring device does not depend upon the orientation of the other. This is another version of localizability where one assumes that the result of each measurement depends only upon the hidden variables relevant to the immediate vicinity of the apparatus and not at all upon what takes place in a remote region of space. The meaning of this restriction is best understood by considering what would happen if it were not true: the hidden variables in the second measurement device would be influenced by the presence and the dynamics of the first device, when the latter is interacting with particle 1. This would imply the existence of an interaction at a distance between the two instruments and their environment (including the particles), i.e., an interaction propagating faster than light. The assumption can therefore be considered as quite reasonable and the main hypothesis remains the existence of hidden variables.

Bell's criterion involves the mean value of the product $\alpha\beta$ as obtained after a large number of individual measurements. This mean value will be denoted by $P(\mathbf{a}, \mathbf{b}) = \langle \alpha\beta \rangle$. It depends upon the directions along which the spins are measured. We denote by $\langle \alpha(\lambda, \mathbf{a}) \rangle$ the average value of α. By assumption, it depends only upon the direction **a** along which the spin component of particle 1 is measured but not upon the direction **b**. It may depend however upon the value of the hidden variables λ, at least the ones that do not depend upon the directions **a** and **b**. The average is therefore taken over the hidden variables which depend upon the directions **a** and **b**. The mean value $\langle \beta(\lambda, \mathbf{b}) \rangle$ is defined in the same way. We can then write

$$P(\mathbf{a}, \mathbf{b}) = \int d\lambda \rho(\lambda) \langle \alpha(\lambda, \mathbf{a}) \rangle \langle \beta(\lambda, \mathbf{b}) \rangle. \tag{9.14}$$

The hidden variables occurring in the integral do not depend upon the directions **a** and **b**. This expression therefore assumes that one can take the mean values of α and β over all the other hidden variables and they are uncorrelated. In other words, it is equivalent to taking the average of the product or the product of the averages. This is because the existence of a correlation would mean that an individual measurement result α could depend upon some hidden variables related to **b**. This equation (9.14) has been analyzed in great detail by Clauser, Horne, Shimony, and Holt.[11] Its complete discussion would take us too far afield and we shall now proceed, referring the interested reader to their paper.

*Derivation of Bell's Inequality**

The results α and β can only take the values ± 1, from which one gets the inequalities

$$|\langle \alpha(\lambda, \mathbf{a}) \rangle| \leq 1, \quad |\langle \beta(\lambda, \mathbf{b}) \rangle| \leq 1. \tag{9.15}$$

Let us then consider three directions **a**, **b**, and **c** where the unit vector **c** is different from **b**. Using equation (9.14), one has

$$P(\mathbf{a}, \mathbf{b}) - P(\mathbf{a}, \mathbf{c}) = \int d\lambda \rho(\lambda) [\langle \alpha(\lambda, \mathbf{a}) \rangle \langle \beta(\lambda, \mathbf{b}) \rangle - \langle \alpha(\lambda, \mathbf{a}) \rangle \langle \beta(\lambda, \mathbf{c}) \rangle],$$

the integration being now taken over all the hidden variables. Introducing another direction **a'** and adding and subtracting the same quantity, we can also write this as

$$P(\mathbf{a}, \mathbf{b}) - P(\mathbf{a}, \mathbf{c}) = \int d\lambda \rho(\lambda) \langle \alpha(\lambda, \mathbf{a}) \rangle \langle \beta(\lambda, \mathbf{b}) \rangle [1 \pm \langle \alpha(\lambda, \mathbf{a'}) \rangle \langle \beta(\lambda, \mathbf{c}) \rangle]$$

$$- \int d\lambda \rho(\lambda) \langle \alpha(\lambda, \mathbf{a}) \rangle \langle \beta(\lambda, \mathbf{c}) \rangle [1 \pm \langle \alpha(\lambda, \mathbf{a'}) \rangle \langle \beta(\lambda, \mathbf{b}) \rangle].$$

But in view of the inequalities (9.15), one has

$$P(\mathbf{a}, \mathbf{b}) - P(\mathbf{a}, \mathbf{c})| \leq \int d\lambda \rho(\lambda) [1 \pm \langle \alpha(\lambda, \mathbf{a'}) \rangle \langle \beta(\lambda, \mathbf{c}) \rangle]$$

$$+ \int d\lambda \rho(\lambda) [1 \pm \langle \alpha(\lambda, \mathbf{a'}) \rangle \langle \beta(\lambda, \mathbf{b}) \rangle],$$

from which one gets

$$|P(\mathbf{a}, \mathbf{b}) - P(\mathbf{a}, \mathbf{c})| \leq 2 \pm [P(\mathbf{a'}, \mathbf{c}) + P(\mathbf{a'}, \mathbf{b})]. \tag{9.16}$$

This inequality does not assume that the total spin is 0 and it is due to Clauser, Horne, Shimony, and Holt. When the total spin is 0, the two components of the spins of both particles along the same direction are necessarily exactly opposite so that one has $P(\mathbf{a}, \mathbf{a}) = -1$. Putting this value in the inequality (9.16) and taking $\mathbf{a}' = \mathbf{c}$, one gets Bell's inequality

$$|P(\mathbf{a}, \mathbf{b}) - P(\mathbf{a}, \mathbf{c})| \leq 1 + P(\mathbf{b}, \mathbf{c}). \tag{9.17}$$

Bell's Inequality as a Criterion

Bell's inequality (9.17) imposes some conditions upon the correlations one should obtain when measuring the spin components of the two particles in different directions. They give a test for the existence of hidden variables as compared with conventional quantum mechanics, if it turns out that quantum mechanics, which predicts specific values for these correlations $P(\mathbf{a}, \mathbf{b})$, does not satisfy this inequality. As a matter of fact, the quantum prediction is $P(\mathbf{a}, \mathbf{b}) = -\mathbf{a} \cdot \mathbf{b}$. The violation of Bell's inequalities occurs for instance when one takes the three vectors $(\mathbf{a}, \mathbf{b}, \mathbf{c})$ to be in the same plane, the angles between $\mathbf{b}$ and $\mathbf{a}$ being $\pi/4$, between $\mathbf{c}$ and $\mathbf{a}$ being $3\pi/4$ and between $\mathbf{b}$ and $\mathbf{c}$ being then $\pi/2$. One then gets

$$P(\mathbf{a}, \mathbf{b}) = -P(\mathbf{a}, \mathbf{c}) = 2^{-1/2} \quad \text{and} \quad P(\mathbf{b}, \mathbf{c}) = 0.$$

Bell's inequality is clearly violated since it can then be written as $2^{1/2} \leq 1$.

Similar results are obtained in the case of two photons emitted in a total spin zero state and detected along two opposite directions. The directions $(\mathbf{a}, \mathbf{b}, \mathbf{c})$ are then replaced by convenient orientations for the polarizers. This correspondence is obtained from the polarization formalism that was given in Section 3 and it will not be described here in detail since it is only a formal matter.

Experimental Results and Extensions

Experiments endeavoring to test Bell's inequality have has been done for both protons and photons. The case of photons is much more clearcut because control of a photon polarization is much easier by optical means than what one can obtain from the spin measurements for protons. The photons were produced in a total spin zero state as obtained from the cascade decay of a caesium atom, as explained in Section 3. The results of the experiments[12] agree perfectly with the predictions of quantum mechanics. We then conclude that there do not exist hidden variables or, at least, that they are terribly well hidden. This might be because there exist interactions propagating faster than light between the measurement devices. Or it could be that one was unlucky when working with spin because spin is well described by quantum mechanics whereas hidden variables only control the position of particles. Or it could be that hidden variables are sensitive to quantum ingredients, as in Bohm's theory, so that they might not be properly localized. The lack of a sign from Heaven cannot discourage a true believer and the quest still goes on.

The most interesting recent result consists in finding some cases,[13] involving more than two spins, where the Bell inequalities are always violated by quantum mechanics and not only for some special choices for the angles in the simplest case. The realization of these experiments looks however rather difficult.

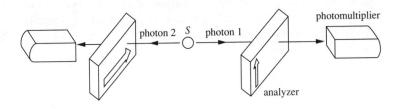

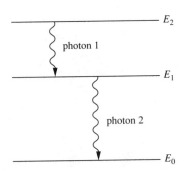

9.2. *Aspect's experiment*

PROBLEM

Consider two spin-1/2 particles, 1 and 2. Show that the projector upon the state with total spin 0 is the 4×4 matrix

$$\frac{1}{16}[I + \sigma^{(1)} \cdot \sigma^{(2)}]$$

and the projectors upon the states $| \mathbf{s}^{(1)} \cdot \mathbf{a} = \pm 1/2 \rangle$ are given by

$$\frac{1}{4}[I^{(1)} \pm \sigma^{(1)} \cdot \mathbf{a}] \otimes I^{(2)}.$$

Considering the various histories for which $s^{(1)} \cdot a = \pm 1/2$ at time t_1 and $s^{(2)} \cdot b = \pm 1/2$ at time t_2, respectively, compute the probability distribution for the measurement results in an EPR–Bohm experiment by computing the relevant traces.

Consider the logic L involving the possible results together with the EPR elements of reality $s^{(2)} \cdot a = \pm 1/2$ at time t_1. Show that it is consistent and gives rise to the implication $[s^{(1)} \cdot a = +1/2 \text{ at time } t_1] \Rightarrow [s^{(2)} \cdot a = -1/2 \text{ at time } t_1]$.

Consider the logic L' involving the possible results together with the anticipating statement $s^{(2)} \cdot b = \pm 1/2$ at time t_1. Show that the logic is consistent and it gives rise to the implication $[s^{(2)} \cdot b = +1/2 \text{ at time } t_2] \Rightarrow [s^{(2)} \cdot b = +1/2 \text{ at time } t_1]$.

10

Nonclassical Macroscopic Systems

In the early Copenhagen interpretation, it was assumed that a macroscopic system must behave classically. The present chapter considers the exceptions to this rule. In the first part, this is discussed in the case of superconducting devices, after recalling the necessary background in superconductivity theory. The relevant experiments will be described in the next chapter. In a second part, one considers the remarkable case of chaotic systems. In principle, they do not obey the equations of motion of classical mechanics after a finite time but, nevertheless, their statistical properties agree with them.

The conditions insuring a classical behavior of a macroscopic quantum system have already been given in Chapters 6 and 7. This behavior is of course apparent only in the collective degrees of freedom of an object since the microscopic degrees of freedom for the matter of the object remain quantized, an atom in this matter still having discrete energy levels, for instance. The conditions we have found are sufficient, which means that the system should behave classically when all of them are satisfied. There are three conditions: (i) The dynamics is regular for the time interval during which the system is considered, (ii) There is dissipation so that decoherence can take place, and (iii) The initial state is a state of fact. The last condition means that there exists a quasi-projector F associated with a regular cell in the collective phase space so that the density operator satisfies the relation

$$F\rho F = \rho,$$

with a good enough approximation (in trace norm).

When one of these conditions is not satisfied, or several of them, it may happen that the system behaves in a nonclassical way, but this is not automatic. The quantum behavior can take various forms. For instance, a harmonic nondissipative oscillator which is initially in an energy eigenstate $|n\rangle$ with a very high excitation number $n \gg 1$ violates both conditions (ii) and (iii). It behaves like a random classical system with a motion to be given by

$$\mathbf{x}(t) = \left(\frac{n\hbar}{2m\omega}\right)^{1/2} \sin(\omega t + \phi)$$

with a random phase ϕ. The randomness is what remains of the underlying quantum effects in the present case (this point will be left as an exercise).

Another example to be worked out later on in detail involves a superconducting device exhibiting a macroscopic tunnel effect. As for chaotic systems, it will be found that they cease to obey classical physics after a finite time necessary for the full establishment of chaos.

NONCLASSICAL SUPERCONDUCTORS

Superconductors play an important role in the present framework because they are among the very few systems for which dissipation may be negligible and therefore decoherence can become a slow effect. As an introduction, we shall first recall the theoretical background necessary for understanding the relevant experiments.[1]

1. Superconductors

Elementary Theory of Superconductivity

There are several kinds of superconductors but, for our present needs, we shall consider only the simplest case. The superconductivity effect is due to the interaction between the electrons and the vibrations of the crystal lattice in a metal. It can be understood as follows. An electron located at a given point in space exerts an electrostatic force upon the neighboring atomic ions in the crystal lattice. This force produces a slight deformation of the lattice in the neighborhood of the electron. Because of elasticity, however, the deformation does not remain local but it extends at some distance, just as locally pushing a piano string deforms it over a large distance. The bulk of the crystal is electrically neutral but, when the ions are displaced, an electric polarization appears, which in turn can exert an attractive force upon other electrons. As a result, there exists in a conductor an electron-electron interaction, which adds to their Coulomb interaction. When deriving it from quantum mechanics, it can be interpreted as an exchange of phonons between the electrons, but we shall not need to enter into these aspects.

The interaction is very weak but it has a long range because of the large distance over which an elastic deformation can extend. The Coulomb force is much bigger but it is screened by the ions so that it decreases exponentially over a finite distance (the Debye distance). Therefore, at least at zero temperature, the attractive interaction due to phonons can predominate and produce bound states of two electrons, the so-called Cooper pairs. As a matter of fact, this effect can only happen for electrons whose momenta are sufficiently near the Fermi surface, because otherwise their kinetic energy would be frozen and the relative wave functions of the two electrons would not be localized. It can be shown that, at least in ordinary superconductors, a Cooper pair is made of two electrons having a total spin 0 and essentially opposite momenta. Its radius is of the order of 10^{-4} cm and its binding energy e_0 of the order of 10^{-3} eV. The interaction between an electron and the lattice is responsible for the existence of Cooper pairs, but it also contributes to the electrical resistivity of the metal. Hence good superconductors can be found among poor conductors such as, for instance, lead.

When the temperature is low enough so that $kT \ll e_0$, the Cooper pairs are stable and they are not destroyed by thermal motion. There is an upper (critical) temperature above which superconductivity vanishes. Below that, the Cooper pairs can be treated like freely moving particles with spin 0, charge $2e$, and mass $2m$, e and m being respectively the charge and mass of an electron.

Since the Cooper pairs are spin zero particles, they obey Bose statistics and therefore all of them will find themselves in the same quantum state near zero temperature. This means that, if one denotes by $\mathbf{x}_j$ the center-of-mass position of a pair, its wave function is $\psi(\mathbf{x}_j)$ and the wave function for the set of all Cooper pairs (assuming that there are N of them) is given by the product

$$\Psi(\mathbf{x}_1, \ldots, \mathbf{x}_N) = \prod_{j=1}^{N} \psi(\mathbf{x}_j).$$

The hamiltonian for these pairs having charge $2e$ and mass $2m$, in the presence of a magnetic field with a vector potential $\mathbf{A}(\mathbf{x})$, is given by the well-known expression

$$H = \sum_{j=1}^{N} \frac{(\mathbf{p}_j - 2e\mathbf{A}(\mathbf{x}_j))^2}{4m}.$$

It will be convenient to exhibit the modulus and phase of the wave function by writing

$$\psi = \rho^{1/2} e^{i\vartheta}, \qquad (10.1)$$

where ρ is proportional to the volume density of the pairs. This is a constant quantity in the bulk of the superconductor since it contributes to the electric charge density, which has to be a constant because otherwise strong electric forces would be created and would restore homogeneity.

It will also be useful to write down the probability current of a pair, which is given in the presence of a magnetic field by

$$\mathbf{J}(\mathbf{x}) = \frac{1}{2 \cdot 2m}\{[(\mathbf{p} - q\mathbf{A})\psi]^* \psi + \psi^*(\mathbf{p} - q\mathbf{A})\psi\}, \qquad (10.2)$$

where $q = 2e$. The Schrödinger equation

$$i\hbar \frac{\partial \psi}{\partial t} = \frac{1}{2 \cdot 2m}(\mathbf{p} - q\mathbf{A})^2 \psi$$

implies the conservation of probability in the form

$$\frac{\partial \rho}{\partial t} + \nabla \cdot \mathbf{J} = 0,$$

which can also be interpreted as the conservation of electric charge for the pairs, the corresponding charge density being $qN\rho$ and the electric current density $\mathbf{j}(\mathbf{x})$ being equal to $qN\mathbf{J}(\mathbf{x})$. One can also express the probability current density in terms of the phase and modulus of the wave function by inserting equation (10.1) into equation (10.2), which gives

$$\mathbf{J}(\mathbf{x}) = \frac{\hbar}{2m}\left\{\nabla\vartheta - \frac{q\mathbf{A}}{\hbar}\right\}\rho.$$

The electric current density is therefore

$$\mathbf{j}(\mathbf{x}) = \frac{\hbar q}{2m}\left\{\nabla\vartheta(x) - \frac{q\mathbf{A}(x)}{\hbar}\right\}n, \tag{10.3}$$

where $n = N\rho$ is the volume density of pairs.

When an electric current is established in a superconducting circuit at zero temperature, it persists indefinitely and this is why these materials are called superconductors. The origin of the effect can be understood by going back to the basic reasons for the existence of electric resistance. In ordinary conductors, resistivity is due to collisions between the conducting electrons and anything that can alter the perfection of the crystal lattice: impurities, vibrations, crystal defects, dislocations, grain boundaries, and so on. But the binding energy of a Cooper pair is somewhat higher than its thermal kinetic energy, at low enough temperature. The pair cannot then be broken because there is not enough energy available, nor can it be scattered by a defect because this would imply that it becomes extracted from the Bose condensate and that also needs a significant excitation energy. As a result, the pairs can conserve a collective common velocity resulting in an overall electric current, which is not dissipated.

This very brief survey of basic superconductivity theory can be used to identify some collective variables, namely the electric current density $\mathbf{j}(\mathbf{x})$ and the pair density $n(\mathbf{x})$. Equation (10.3) shows that their expression involves the phase $\theta(\mathbf{x})$ of the individual wave function, which is therefore also a collective variable. One might wonder why these collective variables appear here directly as pure numbers and not as usual as quantum observables associated with an operator. The answer is that, at least at a finite temperature, the number N of Cooper pairs is not fixed but fluctuates and is therefore itself a quantum observable better described by quantum field theory. The same is true for $n(\mathbf{x})$ and $\mathbf{j}(\mathbf{x})$. The phase $\theta(\mathbf{x})$ then also becomes a quantum observable, which is canonically conjugate to $n(\mathbf{x})$. The quantities we found are the average values of these field observables.

The Meissner Effect

The conservation of charge in a stationary superconductor is given by the condition $\nabla \cdot \mathbf{j} = 0$. Choosing a gauge for the vector potential such that

$$\nabla \cdot \mathbf{A} = 0, \tag{10.4}$$

the expression (10.3) for the electric current in terms of the phase gives

$$\nabla^2 \vartheta = 0. \tag{10.5}$$

If the superconductor is an infinite wire, this equation implies that the phase ϑ is constant inside the wire. Equation (10.3) for the current density then becomes an important constitutive relation between the current and the vector potential first written down by London and replacing Ohm's law in the case of superconductors:

$$\mathbf{j}(\mathbf{x}) = -\frac{q^2 n}{2m} \mathbf{A}(\mathbf{x}). \tag{10.6}$$

An important consequence of London's law is the existence of the Meissner effect, namely, the vanishing of a magnetic field inside a superconductor. As a matter of fact, this property is as much a characteristic of superconductivity as the persistence of an electric current. It can be derived as follows. As a consequence of Maxwell's equations, the vector potential and the current density are related in the present gauge by $\nabla^2 \mathbf{A} = -\mathbf{j}$. London's law (10.6) then gives an equation for the potential inside the superconductor, which reads

$$\nabla^2 \mathbf{A} = \lambda^2 \mathbf{A}, \tag{10.7}$$

where the quantity λ is the inverse of a length a, the penetration depth, which is given by

$$\lambda^2 = \frac{1}{a^2} = \frac{nq^2}{2m}.$$

The penetration depth is 2.10^{-5} cm for lead and this is a typical order of magnitude so that one can consider it as being very small when compared to the size of a wire. At this large scale, the boundary of the wire can be represented by a plane $x = 0$ so that, in its neighborhood, the main variation of the vector potential depends only upon the distance x to the boundary and equation (10.7) becomes

$$\frac{d^2 \mathbf{A}}{dx^2} = \lambda^2 \mathbf{A}.$$

This gives $\mathbf{A}(x) = \mathbf{A}_0 \exp(-x/a)$ for the field inside the superconductor. The potential and also the magnetic field therefore vanish very rapidly at a short distance from the superconductor boundary. It should be noticed that, in view of London's law, this implies that a superconducting current is a surface current flowing in a boundary region having a width of the order of the penetration depth.

From a more advanced standpoint, it might be noticed that these results have been obtained in the special gauge (10.4) and this is particularly true of London's law (10.6), which is not gauge invariant. This lack of gauge invariance is a deep feature of superconductivity, which can be characterized as a spontaneous breaking of gauge invariance symmetry.[2]

A Superconducting Ring

A superconducting ring can carry a current for a very long time (the experiment has been done with currents persisting several years). This current generates a magnetic field and a flux Φ_i through the ring. If there are other sources of magnetic fields in the surroundings, the total magnetic flux across the ring will be denoted by Φ, including the effect of the external sources together with the flux generated by the ring itself.

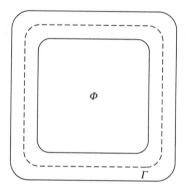

10.1. *A superconducting ring*

Because of the geometry of the ring, we can no longer use the reasoning which led to equation (10.5) for the constancy of the phase. We expect, however, that the existence of the Meissner effect is insensitive to this change in geometry and this can be checked by theory and experiment. It implies that the current must be 0 well inside the superconductor and particularly along a closed curve Γ inside the ring as shown in Figure 10.1. From equation (10.3) giving the current in terms of the phase and the vector potential, one then gets

$$\nabla \vartheta - \frac{q\mathbf{A}}{\hbar} = 0 \qquad (10.8)$$

NONCLASSICAL MACROSCOPIC SYSTEMS

along Γ. Taking the circulation of the gradient $\nabla\vartheta$ along Γ, one finds that the integral $\int_\Gamma \nabla\vartheta \cdot d\mathbf{s}$ is equal to the change in phase of the wave function ψ when a complete turn is made around the ring. But such a turn cannot change the value of the physical quantities associated with the wave function and the phase can therefore only change by a multiple of 2π. Taking the circulation of both terms in equation (10.8) along Γ, one gets

$$q\int_\Gamma \mathbf{A} \cdot d\mathbf{s} = q\Phi = k \cdot 2\pi\hbar, \qquad (10.9)$$

where k is an integer. So one finds that the magnetic flux across the ring can take only discrete values and vary by finite jumps, which are integral multiples of the elementary flux

$$\Phi_0 = \frac{h}{|q|} = \frac{h}{2|e|} = 2 \cdot 10^{-7} \text{gauss-cm}^2. \qquad (10.10)$$

The Josephson Junction

A Josephson junction is essentially a very thin insulating layer located between two superconducting materials. It is often made of a layer of oxide. The junction therefore also appears as a condenser and some electric charges can accumulate across its boundaries so that there is a voltage across it. We denote by U_1 the potential energy of a Cooper pair on one face of the junction and by U_2 its value on the other face.

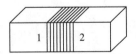

10.2. *A Josephson junction*

The Cooper pairs can cross the junction by a tunnel effect, if it is thin enough, and this can give rise to a macroscopic current. To get an elementary theory of this remarkable effect, we shall denote by ψ_1 the value of the wave function immediately on the left of the junction and by ψ_2 its value on the right. The tunnel effect can then be described in a phenomenological way as a coupling between these two quantities, which amounts to writing down the Schrödinger equation in the form

$$i\hbar\frac{d\psi_1}{dt} = U_1\psi_1 + K\psi_2, \qquad i\hbar\frac{d\psi_2}{dt} = K\psi_1 + U_2\psi_2. \qquad (10.11)$$

The coefficient K represents the coupling between the wave functions on both sides of the junction, representing the tunnel effect by its transmission amplitude. The coefficient K will be assumed to be a real number for simplicity, though this is unessential. An explicit calculation using the usual theory of the tunnel effect for crossing a potential barrier (representing here the higher potential energy in the insulator) shows that this coefficient decreases exponentially with the width of the junction. In equation (10.11), we have neglected the kinetic energy terms, which are quite small in view of the small velocity of the pairs.

When a current crosses the junction, the relation (10.3) between the current and the phase shows that the phase cannot be a constant. We shall therefore denote by ϑ_1 and ϑ_2 its values on the left and the right of the junction. Since the current carries charges, the charge densities ρ_1 and ρ_2 on the boundaries of the junction can vary with time. Since the junction also behaves like a capacity, let us denote by V the voltage across it. Conveniently choosing the origin of energies, we insert in equation (10.11) the following values for the parameters and also define a phase difference by putting:

$$U_1 = qV/2, \quad U_2 = -qV/2,$$
$$\psi_1 = \rho_1^{1/2} e^{i\vartheta_1}, \quad \psi_2 = \rho_2^{1/2} e^{i\vartheta_2},$$
$$\delta = \vartheta_2 - \vartheta_1.$$

Separating out the real and imaginary parts in equation (10.11), we get

$$\frac{d\rho_1}{dt} = \frac{2K}{\hbar}(\rho_1 \rho_2)^{1/2} \sin\delta,$$
$$\frac{d\rho_2}{dt} = -\frac{2K}{\hbar}(\rho_1 \rho_2)^{1/2} \sin\delta,$$
$$\frac{d\vartheta_1}{dt} = \frac{K}{\hbar}(\rho_2/\rho_1)^{1/2} \cos\delta - \frac{qV}{2\hbar},$$
$$\frac{d\vartheta_2}{dt} = \frac{K}{\hbar}(\rho_1/\rho_2)^{1/2} \cos\delta + \frac{qV}{2\hbar}.$$

(10.12)

The charge densities differ from their equilibrium value in the immediate vicinity of the junction and the current I across it is proportional to $d\rho_1/dt = -d\rho_2/dt$, which gives

$$I = I_0 \sin\delta, \tag{10.13}$$

where the critical current of the junction I_0 depends only upon the coefficient K and the average density of Cooper pairs, i.e., upon the junction itself.

The phase difference δ can be obtained by subtracting the third equation (10.12) from the fourth, which gives

$$\frac{d\delta}{dt} = \frac{qV}{\hbar}. \tag{10.14}$$

Equations (10.13) and (10.14) are the basic phenomenological relations governing Josephson junctions.

It may be noticed that, though the current is controlled by a tunnel effect and therefore by quantum mechanics, this does not mean that the system does not obey classical physics. The same is true for practically all macroscopic systems when the equations of motion and the parameters entering in them can be obtained from a quantum analysis. The present collective variables are the current I (or the charge Q across the junction) and the voltage V and their dynamical relations have all the characteristics of a deterministic classical system.

SQUIDs

A superconducting quantum interference device (SQUID) is essentially a superconducting ring containing a Josephson junction. There are many variants of this basic configuration, using for instance several junctions or several rings; the ring itself is not necessarily circular. The theory is always essentially the same and we shall only consider the case of a ring containing two junctions to be denoted by a and b. Two external contacts at points c_1 and c_2 allow measurement of the voltage across the junction or the injection of an external electric current into the ring. External magnets can also produce an external magnetic flux through the ring.

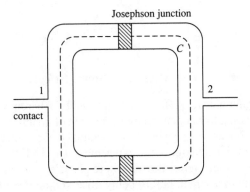

10.3. *Schematic view of a SQUID*

Taking the circulation of equation (10.8), which relates the vector potential to the phase, along a circle Γ interior to the ring yields the relation

$$\int_\Gamma \nabla \vartheta \cdot d\mathbf{s} = \frac{q}{\hbar} \int_\Gamma \mathbf{A} \cdot d\mathbf{s} = \frac{q}{\hbar} \Phi,$$

where the total magnetic flux through the ring appears. The integral on the left-hand side is equal to the change occurring in the phase ϑ under a complete turn around the ring. It is equal to the algebraic sum of the phase discontinuities across each junction and it must be a multiple of 2π since the wave function only depends on the space variables. We therefore have

$$\delta_a - \delta_b = \frac{q\Phi}{\hbar} + 2k\pi, \tag{10.15}$$

where δ_a (respectively δ_b) is the phase change across junction a (respectively b) when going from the half-part 1 of the superconductor to the half-part 2.

The current I_a crossing the first junction a from the contact c_1 to c_2 is given by $I_a = I_0 \sin \delta_a$, together with an analogous expression for the current across the second junction b. The total current from one contact to the other is then given by $I = I_a + I_b$. Using equation (10.15) and denoting by δ_0 the quantity $(\delta_a + \delta_b)/2$, this can be written as

$$I = I_0 \sin \delta_0 \cos \frac{e\Phi}{\hbar},$$

where we have reintroduced the charge $e = q/2$ of the electron. One then finds that the maximum of the current crossing the ring from outside is given by

$$I_{\max} = I_0 \left| \cos \frac{e\Phi}{\hbar} \right|. \tag{10.16}$$

2. The Aharonov–Bohm Effect

A very interesting physical effect was proposed by Aharonov and Bohm.[3] It sheds an interesting light upon the interpretation of quantum mechanics. As often happens, it was first proposed as a gedanken experiment before being realized. In its initial ideal form, the experiment was the following. Consider a rather ordinary interference device where particles coming from a source S cross two slits in a screen before being observed on the other side of the screen. The experiment is made with charged particles—for instance, electrons. The originality of the experiment consists in putting a long magnetic coil near the screen, between the two slits. An isolating casing prevents the electrons from penetrating inside the coil.

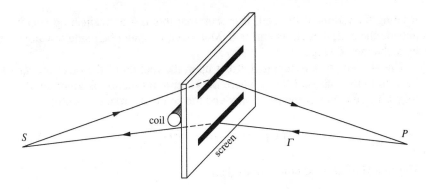

10.4. *The ideal Aharonov–Bohm experiment*

One might naively expect that no effect will take place and the interference fringes will be unaffected by the presence of the coil, whether or not a current is going through it. This is because one knows that the magnetic field vanishes outside the coil. The electrons are therefore always in a region with zero magnetic field and they feel no force, so that their classical trajectories are unaffected. According to conventional wisdom, the magnetic field has a physical reality and this is not the case for the vector potential. If this were true, the electron would never see a field and therefore would feel no effect. Bohm and Aharonov showed however that the position of the interference fringes must be sensitive to the existence of the field *inside* the coil where the electrons never enter.

To understand the effect, remember that the hamiltonian of a charged particle in the presence of a magnetic field is given by

$$H = \frac{1}{2m}\left[\frac{\hbar}{i}\nabla - e\mathbf{A}(\mathbf{x})\right]^2 = \frac{1}{2m}(\mathbf{p} - e\mathbf{A})^2,$$

in terms of the vector potential.

An important point is that, though there is no magnetic field outside the coil, there is a nonvanishing vector potential. This can be seen by taking a z axis along the axis of the coil and denoting by B the magnetic field inside it. A possible choice for the vector potential is then given in cylindrical coordinates by

$$A_z = A_r = 0;$$
$$A_\phi = Br \quad \text{inside the coil } (r \leq R),$$
$$= BR \quad \text{outside } (r \geq R),$$

R being the radius of the coil. It is seen that this is a nonvanishing vector outside the coil, though its curl is 0. Moreover, it cannot be made to vanish by a change of gauge.

The absence of a dynamical effect upon the motion of the electron can be seen in the classical limit, where the Hamilton function is given by the following function of the position $\mathbf{x}$ and the conjugate momentum $\mathbf{p}$:

$$h(\mathbf{x}, \mathbf{p}) = \frac{1}{2m}(\mathbf{p} - e\mathbf{A}(\mathbf{x}))^2. \tag{10.17}$$

The first Hamilton equation gives then

$$\frac{d\mathbf{x}}{dt} = \frac{\partial h}{\partial \mathbf{p}} = \frac{1}{m}(\mathbf{p} - e\mathbf{A}(\mathbf{x})), \quad \text{or} \tag{10.18}$$

$$m\frac{d\mathbf{x}}{dt} = \mathbf{p} - e\mathbf{A}(\mathbf{x}), \tag{10.19}$$

showing that the kinetic momentum $m(d\mathbf{x}/dt)$ does not coincide with the Lagrange momentum $\mathbf{p}$. The energy (10.17) remains given by $E = \frac{1}{2}m\mathbf{v}^2$, and it is insensitive to the magnetic field. This is because a magnetic force $e\mathbf{v} \times \mathbf{B}$ is orthogonal to the velocity and exerts no work. The second Hamilton equation gives

$$\frac{d\mathbf{p}}{dt} = -\frac{\partial h}{\partial \mathbf{x}} = \frac{e}{m}(\mathbf{v} \cdot \nabla)\mathbf{A}(\mathbf{x}). \tag{10.20}$$

The Newton equation of motion is obtained by combining equations (10.18) and (10.20) to give

$$m\frac{d^2\mathbf{x}}{dt} = \frac{d\mathbf{p}}{dt} = e(\mathbf{v} \cdot \nabla)\mathbf{A}(\mathbf{x}) = e(\mathbf{v} \times \nabla \times \mathbf{A}) - e\mathbf{A}(\nabla \cdot \mathbf{v})$$
$$= e(\mathbf{v} \times \nabla \times \mathbf{A}) = e\mathbf{v} \times \mathbf{B},$$

using a well-known formula for the double vector product. One therefore recovers the motion of a particle submitted to a Lorentz force, the force vanishing where there is no magnetic field.

A puzzling aspect of the Aharonov–Bohm effect is that one might contemplate another approach to the problem by restricting the position of the

NONCLASSICAL MACROSCOPIC SYSTEMS

electron to be strictly outside the coil where it cannot enter and where it does not feel any magnetic force, using the simpler hamiltonian

$$H = \frac{1}{2m}\left[\frac{\hbar}{i}\nabla\right]^2 = \frac{1}{2m}\mathbf{p}^2, \tag{10.21}$$

and imposing that the wave function vanishes on the isolating casing as a boundary condition. This would give exactly the same classical motion and there would be no sensitivity of the interference fringes upon the magnetic field in the coil. The comparison between the two formulations will serve as a guide in the discussion.

In both cases, semiclassical physics gives a relation between the momentum and the phase $S(x)/\hbar$ of the wave function, which was mentioned in the preceding chapter, i.e.,

$$\mathbf{p} = \nabla S. \tag{10.22}$$

One can then compute as usual the phase difference between two piecewise straightline trajectories L and L' where the first starts from the source S, follows a straight line and crosses the upper slit, then follows a straight line again to reach a given point P on the other side of the screen; the second trajectory L' also goes from S to P but crosses the lower slit. Let Γ denote the closed path starting from S and going to P and back to S by following L first and then L' in the reverse direction. The phase difference $\Delta\phi$ along the two paths L and L' can be expressed as a contour integral along Γ so that

$$\Delta\phi = \int_\Gamma \frac{1}{\hbar} \nabla S \cdot d\mathbf{x}.$$

If one replaces the phase gradient by the momentum, using equation (10.22), one gets in the case of the naive hamiltonian (10.21), for which $\mathbf{p} = m\mathbf{v}$, the phase difference:

$$\Delta\phi = \int_\Gamma \frac{1}{\hbar} m\mathbf{v} \cdot d\mathbf{x} = k(L - L'),$$

where we used the fact that the velocity always goes along the trajectory and it has a constant magnitude, which can be expressed in terms of the quantum de Broglie wave number $\mathbf{k}$. Here, L and L' denote respectively the total length of each path. The result is the usual interference formula and there is no effect of the magnetic field upon the fringe pattern.

Compare this with the case where one uses the correct relation (10.19) between the momentum, the velocity, and the vector potential. One finds that another term must be added to the previous phase difference. It is given by the circulation of the vector potential around Γ, which can be expressed in terms of the magnetic flux by

$$\Delta\phi_{magnetic} = -\int_\Gamma \frac{e}{\hbar}\mathbf{A}\cdot d\mathbf{x} = -\frac{e}{\hbar}\Phi,$$

Φ being the flux through a portion of a surface bounded by Γ. The point is now that this surface necessarily crosses the coil so that the flux is nonzero. As a result, we find that the interference pattern must be sensitive to the value of the magnetic field and the fringes move when the field is changing.

When the experiment was proposed, it gave rise to some discussion. Most people thought that the effect would exist because of the universality of gauge invariance. The experiment was first realized by using a SQUID as follows.[4] Insert a small coil inside the ring of a SQUID. Equation (10.16), which gives the maximal current along the ring in terms of the magnetic flux, shows the characteristics of an Aharonov–Bohm effect. The Cooper pairs move everywhere in a region where the magnetic field is 0 but the maximal current depends upon the flux, i.e., upon the existence of a magnetic field outside. To measure the maximal current, one can vary the phase by changing the impedance of the external circuit injecting the current through the contacts. Note that, if one were to use a complete superconducting ring without junctions, the magnetic flux would assume only integral values by equation (10.9) and no interference effect would be seen.

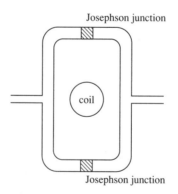

10.5. *The Aharonov–Bohm experiment with a SQUID*

The effect has been entirely confirmed and this was the first example of a macroscopic system showing quantum interference effects and therefore a quantum behavior. There exist, however, even more striking experiments with a SQUID, as we shall see now.

3. The Basis of Leggett's Experiments

Anthony Leggett[5] proposed some remarkable experiments using a SQUID in order to prove that a macroscopic system undoubtedly can behave in a nonclassical way under well-chosen conditions. His initial motivation was to criticize the Copenhagen interpretation where one assumes the impossibility of such a situation. This is because the Copenhagen interpretation uses simultaneously quantum physics and classical physics and it must therefore have some way of stating which one is to be chosen in each case. Bohr and Heisenberg had said that a macroscopic system was to be described by classical physics and this was the point Leggett was scrutinizing, while he had also in view a possibly deeper criticism of conventional quantum mechanics.

The Hamiltonian of a SQUID

As a preliminary, we will write down the equations for the dynamics of a SQUID in a hamiltonian form. We shall consider a SQUID with one junction (the equations are practically the same as the ones already discussed for two junctions), an external magnetic field being eventually created by external coils. We first consider the case when no current is injected into the ring from outside.

It will be found that the dynamical equations have a classical hamiltonian form when one chooses the total magnetic flux across the ring as the collective coordinate. This is a regular dynamics. One may expect that it is, as usual, a consequence of some background quantum dynamics where, at least to first order in Planck's constant, the quantum hamiltonian is simply the operator having the classical Hamilton function for its Weyl symbol. The hamiltonian is therefore well defined, at least to that order, if one can find a classical Hamilton function. One might also contemplate a derivation of this hamiltonian from first principles but this would be much more difficult in view of the actual complexity of the system.

As a superconductor, the system is nondissipative and, if dissipation is strictly 0, it cannot show decoherence. If one is therefore able to prepare it in a state that is not a state of fact, one can expect to observe a quantum behavior. As a matter of fact, the system is slightly dissipative, if only because the Josephson junctions have some resistance and some corrections will have to be applied using the general theory of decoherence. This means that the system can also be used as an experimental benchmark for the existence of decoherence and for its detailed theory, which makes it even more interesting.

With these applications in mind, we shall first consider the system in its classical version. The notation will be the following. Let Φ denote the total flux through the ring and Φ_{ext} the flux produced from outside. The difference $\Phi - \Phi_{ext}$ is due to the self-inductance L of the ring. We have seen that a Josephson junction behaves like a capacity C and we will denote by I_0 its critical current. The unit (10.10) of magnetic field is still denoted by Φ_0.

The classical equations of motion are Hamilton equations, the corresponding quantum hamiltonian being given by

$$H = \frac{1}{2C}\left[\frac{\hbar}{i}\frac{\partial}{\partial \Phi}\right]^2 + U(\Phi). \tag{10.23}$$

This is quite similar to the hamiltonian of a particle with position coordinate Φ and formal mass C. The potential energy is given by

$$U(\Phi) = \frac{(\Phi - \Phi_{ext})^2}{2L} - \frac{I_0 \Phi_0}{2\pi}\cos\left(2\pi\frac{\Phi}{\Phi_0}\right).$$

When a current I_{ex} is injected from outside, the potential energy becomes

$$U(\Phi) = \frac{(\Phi - \Phi_{ext})^2}{2L} - \frac{I_0 \Phi_0}{2\pi}\cos\left(2\pi\frac{\Phi}{\Phi_0}\right) - I_{ex}\Phi. \tag{10.24}$$

Justification of Leggett's Hamiltonian

The derivation of these results in their classical version is a simple exercise in electric circuit theory. It rests upon the following relations:

1. The current I_J crossing the junction is given in terms of the phase discontinuity across the junction by equation (10.13), namely, $I_J = I_0 \sin\delta$.

2. The phase δ is given in terms of the magnetic flux by an equation analogous to equation (10.15), except that there is now only one junction. This gives $\delta = 2\pi\frac{\Phi}{\Phi_0} + 2k\pi$.

3. The flux due to self-induction is related to the current I_r in the ring by $\Phi - \Phi_{ext} = LI_r$.

4. The time variation of the phase δ is related to the voltage V across the junction by equation (10.14), namely

$$\frac{d\delta}{dt} = \frac{qV}{\hbar} = -2\pi\frac{V}{\Phi_0}.$$

5. Finally, one must consider the electric charge Q across the junction. The junction itself has a capacity C so that one has as usual $Q = CV$. The change of Q with time is given by the balance of currents. One must be careful about sign conventions (so that I_0 is positive, for instance) and when this is done, one gets

$$\frac{dQ}{dt} = I_r + I_J - I_{ex}.$$

NONCLASSICAL MACROSCOPIC SYSTEMS

Eliminating all the variables to keep only the flux, one obtains a differential equation of motion very similar to a Newton equation, namely

$$C\frac{d^2\Phi}{dt^2} = -\frac{\partial}{\partial \Phi}U(\Phi),$$

where the potential is given by equation (10.23).

Discussion

The potential $U(\Phi)$ depends upon the four parameters L, I_0, Φ_{ext} and I_{ex}. If one uses a SQUID which is made up of two parallel wires connected at one extremity by a Josephson junction and by a superconducting bridge at the other, one can let the value of L vary by changing the position of the bridge so that in practice there is one fixed parameter I_0 and the three other ones can vary. The four of them can be measured by calibration experiments so that the theory contains no free parameter.

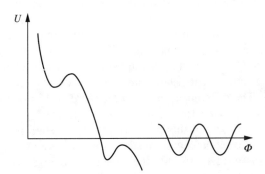

10.6. *Various forms of Leggett's potential*

The potential contains a linear term depending upon L, Φ_{ext} and I_{ex}, a quadratic term depending only upon L, and an oscillating term, which is intrinsic to the junction. The possible shapes of such a potential are rather flexible and one sees in Figure 10.6 two examples showing respectively a double-well potential and a potential allowing the existence of an unstable bound state, which can decay by a tunnel effect. The experiments one can perform to show a manifest quantum behavior of a SQUID are therefore quite characteristic, at least in principle. It should not be forgotten however that the corrections coming from small decoherence effects should be considered. Their theory is much more involved[6] and we shall come back to them when discussing the actual experiments in the next chapter.

CHAOTIC SYSTEMS

We will now consider quite a different problem, though still concerned with the quantum behavior of a macroscopic system. It has to do with the following question: Is it correct to rely upon the equations of classical dynamics when describing a macroscopic chaotic system? One of course assumes that what is undoubtedly reliable is its description by quantum mechanics. The question arises because this is a case where the dynamics is not regular after a finite length of time so that one cannot be sure that classical physics applies. Contrary to what was found in the case of SQUID's, it will turn out that this investigation does not lead to a convenient experimental test for the fine points of the new interpretation. This is because, once chaos is fully established, one must necessarily resort to statistical techniques in order to describe the experimental properties of the system, even at a classical level. Accordingly, the statistical description becomes common to both the classical and the quantum approaches. The relevant question one may ask is therefore whether the statistical properties are the same in both cases or not.

Although there is no complete proof but only some indications coming from a few simplifying assumptions, one can get an answer that is presumably general. It states that the statistical properties of the system are essentially the same whether one computes them by quantum or by classical dynamics. The differences between them are of the order of the corrections to classical physics (the "epsilons" in Chapter 6), probably too small to be checked. In view of this negative result, we shall not try to discuss in much detail the actual chaotic systems from a more physical standpoint. This question is in fact presently the subject of intense investigations and anything we might say would presumably be rapidly outdated.

4. Classical and Quantum Statistics of Chaotic Systems

As was mentioned in Chapter 6, the connection between quantum and classical physics becomes a nontrivial problem in the case of chaotic systems. It remains valid for sufficiently short times since the dynamics is then regular, but it breaks down after a finite time. This can be shown by using the criterion for the regularity of the dynamics, which consists in the conservation of regularity (i.e., in keeping a finite area to volume ratio) for a cell in classical phase space. When chaos is fully developed, an initially regular cell becomes highly distorted, consisting of complicated filaments having a local scale of the order of Planck's constant or smaller and regularity is lost.

For anybody who takes for granted that quantum mechanics is the fundamental theory and classical mechanics is only a practical approximation, this means that the classical description of a chaotic system becomes unphysical when chaos develops down to so small a scale. The equations of classical mechanics and the theorems one derives from them are then reduced to the role of toys with no real connection with physics. From a practical standpoint however, it does not seem that there are real cases where chaos actually goes that far down in scale and, in most cases, it stops before reaching it because of friction. As an example of this, one might quote fully developed turbulence. There is therefore little to expect from the standpoint of experimental physics in this comparison between classical and quantum chaotic systems. Even if it were not so, we shall now show why one would expect anyway that their statistical predictions would be the same.

Statistical Description

Let us consider a macroscopic physical system having a chaotic dynamics. It is initially prepared in a state of fact so that the initial state has a clearcut meaning from the standpoint of both classical and quantum physics. One observes it nondestructively (i.e., without perturbing it) from time to time, which is possible because of its macroscopic character. For instance, one might think of observing a macroscopic ball moving on a Sinai billiard without friction. Because of the quantum limitations, an observation can only determine the position and momentum of the ball within finite error bounds, large enough when compared with the uncertainty limits. This is particularly the case for the initial state, which is known therefore classically to lie in some classical box C_0 in phase space. When chaos is fully developed, the image of this cell has become very complicated and all that one can determine by a measurement is whether the values of the coordinates lie in some error box. If one looks at one copy of the system, this information is very poor and it tells practically nothing about the dynamics. One gets a much better knowledge of the system by observing many identically prepared copies of it and making statistics of different series of observations. An important question is then to find whether the statistics of the successive measurements can distinguish between quantum mechanics and classical physics.

Let us assume for convenience that the initial state is described by a regular cell C_0 in phase space, for instance, a rectangular cell corresponding to prescribed errors for the position coordinates and the momenta. Classically, this is described by an initial probability density $f_0(x, p)$, which is equal to V^{-1} in C_0 (V being the volume of the cell) and to 0 outside. In quantum mechanics, one may use a quasi-projector E_0 associated with the cell together with the initial value of the density operator

$$\rho_0 = \frac{1}{V} E_0. \tag{10.25}$$

We shall also assume that the observations consist in measuring nondestructively the position and momentum within prescribed errors. For instance, it can be assumed that one has covered phase space completely by a collection of rectangular cells $\{C_r\}$ and the measurement consists of finding in which cell the system is at some time. We shall denote by L and P the scales of position and momentum represented by the sides of these ($2n$-dimensional) rectangular cells. The measurements are assumed to be really performed by some convenient devices. They are measurements of a macroscopic system and, as such, they can be conceived either from the standpoint of classical physics or of quantum mechanics, in which case it will be convenient to consider them as so many yes/no experiments telling in what cell the system is.

It is convenient to take the configuration space to be bounded. The energy is also bounded since one starts from a finite cell C_0. Then the useful region of phase space is bounded. It is also convenient to assume that all the cells have the same volume V and that C_0 is one of them, in which case there is a finite number of covering cells. We associate a quasi-projector E_r with each cell C_r. If it is used to express the result of the measurement in the framework of quantum logic, the corresponding probability of error ε will be of order $(\hbar/LP)^{1/2}$. It can be considered either as the mathematical arbitrariness of the definition of the quasi-projector or as the risk of error when making a reasoning where the result associated with the cell C_r is taken for granted. Since the collection of cells $\{C_r\}$ covers all the available phase space, it is always possible to define the quasi-projectors in such a way that one has

$$\sum_r E_r = E,$$

where E is the projector onto the available phase space. However, since we work in any case only in that domain, there is no inconvenience in replacing E by the identity operator in the right-hand side of this equation.

We now restrict the category of chaotic systems to be considered. To do so, define a regular metric on phase space and consider how the distance between a classical trajectory starting from a given point (q, p) and another trajectory starting at a small distance δ from the point (q, p) will increase. It will be assumed that, whatever (q, p), there is at least one neighboring trajectory whose distance increases exponentially with time and behaves like $\delta \exp(t/\tau)$. It will also be assumed that the time of increase τ is essentially the same whatever the starting point (q, p) (i.e., it is everywhere of the same order of magnitude). The quantity τ^{-1} is what is called a Liapunov exponent and one therefore assumes the existence of a positive Liapunov exponent at each point and that they all are of the same order of magnitude.

According to Theorem C in Chapter 6, which describes the dynamical evolution of quasi-projectors and its correspondence with the motion of classical cells, the dynamics of the system must be regular for a time of the order of τ or less. We shall assume that the measurements always occur with the same time interval Δt between them, this interval being of the order of τ. We denote by D_r the cell which is the transform of C_r when it undergoes a classical motion during the time Δt. More precisely, Δt will be taken small enough for the two cells C_r and D_r to be regular and large enough so that they are significantly different whatever r, i.e., the evolution is effective enough during that time. It will be assumed that this is possible and that the dynamical and logical errors coming from Theorem C for these cells remain of the order of ε.

Quantum Evolution

The system being initially in the state (10.25), the probability $w_r(\Delta t)$ for finding the result associated with the cell C_r at time Δt, i.e., during the first measurement, is given by

$$w_r(\Delta t) = \text{Tr}\{\rho_0 U^{-1}(\Delta t) E_r U(\Delta t)\}.$$

Write $F_r = U^{-1}(\Delta t) E_r U(\Delta t)$. According to Theorem C, F_r is a quasi-projector, which is associated with the cell D_r and involves an error of order ε. To compute the probabilities for the results of the second measurement at time $2\Delta t$, use the results of Chapter 8 concerning the empirical use of wave packet reduction. If the first measurement indicated that the system is in cell C_r, one can predict the probabilities for the second measurement using the operator $\rho_r = (1/V)E_r$ to act as a density operator at time Δt. To consider a series of measurements on an ensemble of identically prepared systems rather than an individual system, one can take the operator

$$\rho(\Delta t) = \frac{1}{V} \sum_r w_r(\Delta t) E_r.$$

as the effective density operator at time Δt. It allows one to compute the statistics of the collection of measurements to occur at time $2\Delta t$.

After a time $n\Delta t$, it is clear that the operator giving the probabilities for the next series of measurements is the following effective operator:

$$\rho_{\text{eff}}(n\Delta t) = \frac{1}{V}\sum_r w_r(n\Delta t)E_r.$$

This effective density operator, which loses at each step all information concerning a scale finer than the cell's graining, can be compared with the density operator obtained by solving the Schrödinger equation. One can show that their difference is of order $n\varepsilon$ in trace norm, i.e., in practice very small. This may be important for applications to irreversible thermodynamics, but this point need not concern us here. It also confirms the possibility of nonperturbing measurements.

Let us simply try to find a relation between the probabilities $w_r(n\Delta t)$ and $w_r((n+1)\Delta t)$ at two successive times. One has

$$w_r((n+1)\Delta t) = \text{Tr}\{\rho_{\text{eff}}(n\Delta t)U^{-1}(\Delta t)E_r U(\Delta t)\} = \text{Tr}\{\rho_{\text{eff}}(n\Delta t)F_r\}$$
$$= \frac{1}{V}\sum_s w_s(n\Delta t)\text{Tr}\{E_s F_r\}. \tag{10.26}$$

According to Theorem B in Chapter 6, the product $E_s F_r$ is a quasi-projector associated with the intersection of the cells C_s and D_r and the trace occurring in equation (10.26) is given by

$$\text{Tr}\{E_s F_r\} = \mu(C_s \cap F_r), \tag{10.27}$$

where $\mu(C)$ denotes the volume of a cell C. The relative errors when writing equations (10.26) and (10.27) are of order ε. With the same kind of error at each step, one can therefore write

$$w_r((n+1)\Delta t) = \sum_s w_s(n\Delta t)P_{sr}, \tag{10.28}$$

where $P_{sr} = \frac{1}{V}\mu(C_s \cap F_r)$. The quantities P_{sr} have the following properties: (i) They are nonnegative, and (ii) Since the cells $\{C_r\}$ cover the available phase space and $\{C_r\}$ and $\{D_r\}$ all have the same volume, one gets

$$\sum_r P_{sr} = \sum_s P_{sr} = 1.$$

Note that, in spite of the exponential time increase of the relative distance between neighboring classical trajectories, the errors in the evaluation of the probabilities increase only linearly with time.

An evolution equation such as equation (10.28) for a set of probabilities $\{w_r(t)\}$, when the coefficients obey the two conditions (i) and (ii), is a so-called Markov process. These processes have been studied in great detail

NONCLASSICAL MACROSCOPIC SYSTEMS

and their properties are well known. It can be shown, for instance, that all the probabilities $\{w_r(n\Delta t)\}$ tend towards the same limit when the time $n\Delta t$ tends to infinity. The limit is therefore a microcanonical statistical distribution. This is very interesting but we shall not elaborate on it.

Classical Description

In classical statistical physics, the system is described by a probability density distribution $f(x, p, t)$, initially confined in the initial cell. It evolves under classical motion according to the classical mapping g_t of phase space so that

$$f(x, p, t) = f(g_t^{-1}(x, p), 0). \tag{10.29}$$

The probability of finding the system in the cell C_r at time t is then given by

$$p_r(t) = \int_{C_r} f(x, p, t) \, dx \, dp.$$

However, the explicit form (10.29) for the distribution is useless in practice for a time t which is large when compared to τ. This is because of the chaotic character of the mapping g_t, which makes it impossible to compute explicitly. One is therefore led to look again for the correlations among successive measurements over a time interval Δt. The statistics for a large number of copies can again be expressed in terms of the frequencies $w_r(n\Delta t)$ for which the system is observed to be in the various cells $\{C_r\}$ at these times. The statistics is the same as what is given by the probability distribution

$$f_0(x, p, n\Delta t) = \sum_r w_r(n\Delta t) \frac{1}{V} \chi_r(x, p),$$

where $\chi_r(x, p)$ is the characteristic function for the cell C_r. Conversely, one can compute the distribution expected at time $(n + 1)\Delta t$ in terms of the distribution at time $n\Delta t$ by writing

$$w_r((n+1)\Delta t) = \int_{C_r} f_0[g_{\Delta t}^{-1}(x, p), n\Delta t] \, dx \, dp$$

$$= \sum_s w_s[n\Delta t] \int_{C_r} \frac{1}{V} \chi_s\{g_{\Delta t}^{-1}[x, p]\} \, dx \, dp. \tag{10.30}$$

The integral in the second equation is, up to a factor V^{-1}, the volume of the intersection between the cells C_r and D_s and one recognizes the quantity

$$\int_{C_r} \frac{1}{V} \chi_s\{g_{\Delta t}^{-1}[x, p]\} \, dx \, dp = P_{sr}.$$

Since the initial classical and quantum probabilities are the same, one finds that they also evolve over time in the same way, in view of the formal identity between equations (10.30) and (10.28). Their only difference takes place in the approximations that have been made writing equation (10.28), so that it corresponds to a relative error of order ε at each step.

So, to conclude, looking at the correlations in the statistical evolution of a chaotic system, one obtains the same predictions from classical and from quantum mechanics, up to small corrections that are practically outside the range of experiment. Therefore, although the description of a chaotic system by classical physics is fundamentally meaningless when chaos is developed, this has no practically observable consequence. Of course, one should remember that this result has been obtained by making simplifying assumptions and we cannot claim it to be completely general. More complete conclusions will probably come out of the presently very active investigations of quantum chaotic systems.

11

Experiments

This chapter is devoted to the experimental aspects of interpretation. Up till now, no experiment has been found that lies outside the framework of interpretation, so that the theory always provides unambiguous predictions which are in agreement with observation. This state of affairs supports the contention that the interpretation is complete, in the empirical sense given in Chapter 2.

There has been a strong renewal of interest in the foundations of quantum mechanics during the last decade or so. Many clever or difficult experiments, often both, have been realized for testing the Copenhagen interpretation, investigating the conditions where its predictions are unprecise, and looking for faults in its assumptions. The last category refers particularly to the experiments made with SQUIDs, which were suggested by Leggett for showing the existence of nonclassical macroscopic systems. Their success alone would be enough to confirm the necessity of revising the older version of the Copenhagen interpretation. On the contrary, no experiment has shown a disagreement with the modern version.

In view of the large number of significant experiments, it was thought better not to try covering all of them in the present chapter but to restrict attention to the ones offering either a difficulty, a lack of prediction, or an ambiguity in the framework of the older interpretation. This also includes some experiments for which specific predictions had been made, though more from authority than from a full-fledged analysis, so that a more complete understanding was needed. The issue of delayed-choice experiments, for instance, had been explicitly predicted by Bohr, but he gave no precise reasons for his statements. They therefore provide an interesting test for the deductive character of the new interpretation.

The experiments to be discussed here have been arranged in several families according to the type of theoretical notions upon which they mainly rely. The first category includes the experiments for which the notion of consistent histories is particularly well suited. These are the decay of a particle, continuous measurements on a unique atom, and also the not-yet-observed Zeno effect. A second category includes all kinds of interference experiments, the insistance being put upon the subtler ones where clever devices were sometimes believed to test the passage of a particle along a specific path while preserving the existence of interferences; delayed-choice experiments are also put in that family. Finally, the last category contains the experiments showing quantum behavior of a macroscopic system and also a quantitative observation of a decoherence effect. Although some experiments require specific theoretical developments for their interpretation, we have tried not to excessively insist on the relevant calculations when possible and to stress rather the main issues and the essential results.

EXPERIMENTS SHOWING HISTORIES

Some experiments are particularly suggestive of the realization of a Griffiths history, sometimes taking place practically under the eyes of the observer. We begin with these since they give support to one of the basic features of the interpretation and show its intimate link with empirical physics.

1. The Decay of a Particle

There is no simpler experiment than looking at the decay of a sample of particles or of an individual one. This includes of course the decay of an atom with the emission of one or several photons. This question was already considered in Chapter 5 but it merits being discussed again, if only to prepare the discussion for more sophisticated experiments of the same type.

The Formal Description of a Decay

Let us consider an isolated system, which is initially an individual particle that can decay into several other particles. It can be an unstable "elementary" particle such as a free neutron, a π-meson, a K-meson, a hyperon Λ or Σ, or many other ones whose list fills up the tables of particle properties. The decay can be due to a strong, an electromagnetic, or a weak interaction. The initial particle can also be an atomic nucleus, in which case the three interactions manifest themselves respectively by alpha, gamma, and beta radioactivities. It can also be an excited atom, which usually decays by emitting a photon.

A basic rule of interpretation is that every mentioned property must be expressed within the basic formalism. So when saying that there is a particle at the beginning, one must use a projector in Hilbert space corresponding to this property. This projector will be denoted by E_p, where the index p stands for "particle". One also needs to express that the particle has decayed and the simplest way to do it is to say that the decay products are there at the time to which one refers. This property can also be expressed by a projector.

To give an example, consider the decay of a free neutron. It decays by the effect of weak interactions into a proton, an electron, and an antineutrino.

EXPERIMENTS

This can be written in a familar way as the reaction

$$n \to p + e + \bar{\nu}.$$

In order to describe the decay, one needs at least to consider a Hilbert space containing two subspaces—a Hilbert space $\mathbf{H}_n$ representing the states of the neutron and also the tensor product

$$\mathbf{H}_p \otimes \mathbf{H}_e \otimes \mathbf{H}_{\bar{\nu}} \qquad (11.1)$$

representing the decay products. As a matter of fact, a complete theory of the decay resorts to a much wider field-theoretic Hilbert space, but, whatever the context, the complete Hilbert space $\mathbf{H}$ always contains the two we mentioned as subspaces.

There is a projector projecting $\mathbf{H}$ upon $\mathbf{H}_n$ and it may be taken as the projector E_p. There is also a projector E_d, where the index d stands for "decayed", projecting onto the space (11.1) and expressing that the decay products have been produced. Of course, one could also consider several more precise events where the energies of the decay products and the directions along which they have been emitted are restricted, giving their spin states, and so on. It is not necessary to elaborate upon these obvious aspects.

The Exponential Decay Law

The elementary theory of decay rests upons the conditional probability $p(t, t + \Delta t)$ for a particle still intact at time t to have decayed at a later time $t + \Delta t$. It was mentioned in Chapter 5 that these conditional probabilities raise problems for the Copenhagen interpretation, since the existence of the undecayed particle at time t is not asserted by a measurement. It looks obvious but it is anyway questionable as resting on a retrodiction. This difficulty has been solved long ago and much more elaborate theories have been devised, giving directly the probability $p(t)$ for the particle to survive at time t as well as a good description of the wave function for the decay products.[1] They show in particular that the final state wave function is an outgoing wave which cannot regenerate the initial particle, at least when the time t satisfies the condition

$$t \gg (\hbar/E), \qquad (11.2)$$

where E is the kinetic energy available for the decay products. This condition will play a significant role later on.

If one introduces histories relative to the two times t and $t + \Delta t$ together with the properties stating that the particle is either intact or decayed, the conditional probability $p(t, t+\Delta t)$ will have a meaning if the corresponding logic is consistent. The corresponding consistency conditions are found to be valid if the condition (11.2) is satisfied by the time interval Δt, i.e.,

$$\Delta t \gg (\hbar/E). \qquad (11.3)$$

This result is interesting and suggests that a minimal time is necessary for the particle to decay (this is after all the fourth uncertainty relation between time and energy). This point will also be considered later.

It was shown in Chapter 5 that the probability $p(t)$ for the survival of the particle satisfies the relation

$$p(t) - p(t + \Delta t) = f(\Delta t)p(t), \qquad (11.4)$$

where the function $f(\Delta t)$ can be computed and is simply given in terms of the lifetime τ of the particle by

$$f(\Delta t) = \Delta t/\tau, \qquad (11.5)$$

when the condition (11.3) is satisfied with Δt replacing t, as well as the condition $\Delta t \ll \tau$. The lifetime can be computed by Fermi's golden rule.

One thus gets the familiar exponential decay law $p(t) = \exp(-t/\tau)$, which may have to be corrected for values of the time much larger than τ. These corrections are due to the fact that, because of the instability of the decaying particle, its energy is not perfectly well defined and its behavior at very large times depends on the details of its preparation. This is only significant in practice for resonances, i.e., particles with a very short lifetime decaying via strong interactions. It is better in that case to reconsider the process within the framework of scattering theory so that both preparation and decay are treated together. In most other cases of practical interest, the exponential decay law is valid.

We also recall the form of the decay probability function $f(\Delta t)$ for a very short value of the time interval Δt. Schrödinger's equation tells us that the change in the wave function must be linear in Δt so that the corresponding probability must be quadratic in Δt. It can be shown rigorously that the inequality $f(\Delta t) \leq \Delta t/\tau$ is always satisfied and the quantity $f(\Delta t)$ tends to 0 like Δt^2 when Δt tends to 0, whereas it is given by equation (11.5) when condition (11.3) is satisfied. This remark will play an important role in the discussion of the Zeno effect.

2. Repeated Measurements

Many theoretical papers have been devoted to the study of systematically repeated measurements.[2] Their most interesting features do not yet correspond to a practically realized experiment but their relation with the standpoint of interpretation is nevertheless worth a brief discussion.

The Theory

We shall consider how to describe measurements that are repeated iteratively with the same time interval between them. This question will be investigated from the standpoint of histories, which is the most convenient one. Let us first consider a simple case where a measuring device M tests the state of an unstable particle, i.e., it checks whether the particle is intact or has decayed at successive times $\Delta t, 2\Delta t, 3\Delta t, \ldots$. In the original theory of histories by Griffiths, where the measuring apparatus was not explicitly taken into account, one would naturally introduce a complete family of histories referring to these various times with properties associated with the two projectors E_p and E_d.

One can proceed as before, using a two-times history (at times t and $t+\Delta t$) to describe a decay. One again finds that a large family of Griffiths histories referring to the times $(\Delta t, 2\Delta t, 3\Delta t, \ldots)$ is consistent, at least when the condition (11.3) is fulfilled by the time interval. However, when the interval Δt becomes too small so that this condition is no longer satisfied, everything goes wrong. The decay state is no longer described by strictly outgoing waves; it looks formally as if a decayed particle could be regenerated from its decay products and, in any case, the consistency conditions are not satisfied by the family of histories. This clearly shows that Griffiths' original construction does not give a complete theory since it cannot describe this kind of experiment. We must therefore proceed to a more careful investigation. Once again, we must take into account the actual presence of the measuring instrument and it is not enough to state the results of a series of measurements only in terms of the measured particle. We must introduce the experimental *data*, i.e., the properties of the measuring apparatus that are effectively recorded, using sensible logics where these facts enter.

So to begin we define a logic involving only the facts. We do not mention anything about the particle itself but, at the times $\Delta t, 2\Delta t, 3\Delta t, \ldots$, we only state that the measuring device shows a macroscopic signal indicating that the particle is either intact or decayed. One can then show that this logic is *consistent*. This result might look surprising since it seems to conflict with what was found previously for the Griffiths family involving only the particle. The calculations upon which it relies are a bit heavy and we shall therefore only give a few indications about its key points. The S-matrix describing each interaction between the particle and the apparatus, as in Chapter 8, is again partly diagonal, since this is nothing but the definition of a good measuring instrument. Given a datum at a time $n\Delta t$, this diagonal feature implies that only the properties of the particle corresponding to this datum ("intact" or "decayed"), both just before and just after the measurement, must be satisfied. Of course, one must also assume that the measurement itself takes a time much shorter than Δt. Then, as usual, one must take decoherence into account. Let us assume for definiteness that the measuring device can only make sure that the initial particle has decayed by detecting one of its decay products. The decoherence effect taking place in the bulk of the apparatus (its environment) breaks the phase relations between the various components of the wave function for the decay products. As a consequence, the very fine phase matching that would be necessary to regenerate a particle at time $(n + 1)\Delta t$ from its decayed state at time $n\Delta t$ becomes completely lost. This implies first that there must be an ordering in time among the events entering the histories, or the probability is zero: once a particle has been found to be decayed, one cannot find it later back in its initial state. A second consequence is that the consistency conditions are now automatically satisfied because they always involve a product of two orthogonal quasi-projectors for the apparatus.

This is a remarkable example of one of Bohr's statements: one cannot (always) assert something uniquely in terms of the measured system but one must take into account the whole system consisting of the measured and measuring systems and stick to facts, i.e., rely only upon the experimental data.[3] It also shows how nontrivial the justification and the limitation of some sweeping statements made by Bohr may be and how measurement theory is really nontrivial when one wants to understand it in full detail. It should finally be added that, using the sensible logic dealing with the data, it is found that the result of each measurement is true, the proof being the same as in Chapter 8.

Quantum Jumps

There are practical limits to the frequency in repeating actual measurements. Let us however forget it for the sake of the argument so that the time interval Δt will be made arbitrarily small. It results from what has been said that every change one can observe in the system, whatever the means of observation, will always be a total transformation going from an intact particle to a completely decayed one during a time interval Δt. It is impossible to catch a particle *when* it decaying, like a fruit opening and ejecting its seeds. The idea of some short part of a history occurring while the decay is taking place has no logical meaning, even if a minimal amount of time is necessary for a spontaneous free decay to be complete. This brutal and discontinuous transition from a state to a radically different one has been called a *quantum jump*. It is all one can ever observe in a series of measurements. This was precisely the idea which led Heisenberg to his conception of matrix mechanics and it now reappears unchanged in the framework of a consistent interpretation.

3. The Zeno Effect

The Zeno effect is essentially a freezing of the state of a system when it is submitted to a very rapid series of repeated measurements. It has not been actually observed but there is no proof that it is impossible and it would be a quite interesting experiment from the standpoint of interpretation. This is why we are now going to discuss it briefly.

EXPERIMENTS

The name of the effect has little to do with physics. Zeno of Elea was a Greek philosopher, an elder contemporary of Socrates and a follower of Parmenides, who had a metaphysical doctrine according to which the Being does not move. This was considered as absurd by many people, either as contradicting common sense or because they held the views of another philosopher, Heraclites, for whom everything is ever changing. Zeno invented then a nice paradox according to which Achilles, the most rapid racer of all Greeks, could never catch up to a turtle. There is something analogous to that paradox in the so-called Zeno effect where one cannot see a change when looking too intently for it.

An Example

Let us consider a spin-1/2 atom or a neutron in a magnetic field **B** parallel to the x axis. The component of the spin along the z direction is measured repeatedly at times $\Delta t, 2\Delta t$, and so on. The evolution operator $U(\Delta t)$ is easily written down. If one denotes by γ the gyromagnetic ratio of the particle and by $\omega = \gamma B$ the precession frequency, $U(\Delta t)$ represents a rotation with an angle $\omega \Delta t$ around the x axis so that

$$U(\Delta t) = \exp\left\{\frac{1}{2} i\omega \Delta t \sigma_x\right\} = \cos(\omega \Delta t/2) + i\sigma_x \sin(\omega \Delta t/2).$$

Notice that, when $\omega \Delta t \ll 1$, this is approximately given by

$$U(\Delta t) = 1 + i(\omega \Delta t/2)\sigma_x.$$

The condition $\omega \Delta t \ll 1$ can be understood as follows. The atom has two spin energy levels in the magnetic field, which are separated by a gap $E = \hbar\omega$. The condition can therefore be written as $\Delta t \ll \hbar/E$, which is a case where a quantum jump has a very small probability, proportional to Δt^2 rather than to Δt. This small value of the transition amplitude is ultimately responsible for the Zeno effect.

The Zeno Effect

When the measurements are done, one can proceed as before and build up a sensible logic to describe the whole series of measurement data. It is found once again that the results of these measurements (for which $s_z = \hbar/2$ or $-\hbar/2$) can be inserted automatically in the logic and they are true properties. Assume that the particle is initially in a state where $s_z = \hbar/2$. We consider first the case where there is a unique measurement of s_z at a time t for which ωt is not necessarily small. The probability of finding the result $s_z = -\hbar/2$ is then given by

$$|\langle -1/2|U(t)| + 1/2\rangle|^2 = \sin^2(\omega t/2). \tag{11.6}$$

Let us now assume that repeated measurements of s_z take place with a time interval Δt small enough so that $\omega \Delta t \ll 1$. Let $p(t)$ be the probability for the whole series of observations where one finds the result $s_z = +\hbar/2$ from time Δt up to the time $t = n\Delta t$. Let then $q(t, \Delta t)$ be the probability of observing the value $s_z = -\hbar/2$ for the first time at $(n+1)\Delta t$. One has

$$q(t, \Delta t) = |\langle -1/2|U(\Delta t)| + 1/2\rangle|^2 \, p(t) = (\omega \Delta t/2)^2 p(t).$$

The dependency in Δt^2 should be noticed. If one proceeds as was done in the calculation of the exponential decay law, one gets

$$p(t) = \exp\{-n(\omega \Delta t/2)^2\} = \exp\{-(\omega^2 \Delta t/4)t\}.$$

This result should be compared with equation (11.6). It shows that, when the atom (or the neutron) is observed intently, it is found to remain much longer in its initial state than one would have expected from its behavior when it remains isolated. This is what is called the Zeno effect. It becomes particularly striking if one lets Δt tend to 0 (assuming the measurements to be possible), since then the theory predicts that one would indefinitely see the spin of the atom frozen in its initial state, whereas it had on the whole enough time to precess by a large angle. The Achilles atom, though able to spin much more rapidly than the atom Turtle, would be a dreadfully slow dancer in a waltz if he were constantly watched under the fire of repeated snapshots.

Discussion

Let us come back for a moment to the decay of a repeatedly observed atom. According to equation (11.4), the probability for the atom to be found in its excited, undecayed state at times $\Delta t, 2\Delta t, \ldots$ up to the time $t = n\Delta t$ and then in its decayed ground state at time $(n+1)\Delta t$ is given by

$$p(t) = \exp\left\{-\frac{f(\Delta t)}{\Delta t} t\right\}.$$

Here again, this probability tends towards 1 when Δt becomes very small, since $f(\Delta t)$ behaves like Δt^2. If one were to observe each atom of a fluorescent lamp with the most extreme care, the lamp would not emit light!

Of course, at least in the case of the fluorescent lamp, even a unique measurement of that sort is inconceivable, even in principle, according to the discussion given in Chapter 7. One should therefore not take this example seriously. However, in the case of a unique atom, this kind of measurement can be considered more carefully, though the prospect of realizing the experiment is not very promising. The energy gap E between different atomic levels is typically of the order of an electron-volt so that $\hbar/E$ is of the order of 4×10^{-15} seconds, much outside the present possibilities for repetitive measurements.

The interest in the Zeno effect was due for a long time to the conviction that it was a remarkable manifestation of wave function collapse. Its observation was therefore expected to show a clearcut proof of this important physical concept. It was thought that each measurement provokes a collapse bringing the state vector systematically back to its initial position, forbidding it to turn freely in the Hilbert space. This conviction was however wrong. The collapse of a wave function is not a physical effect and the present discussion did not have to mention it, the framework of histories being sufficient. In that sense, the Zeno effect will certainly lose a large part of its interest, as far as the foundations of physics are concerned and its observation would not prove the physical existence of wave function reduction.

Has the Zeno Effect Been Observed?

It is extremely difficult, if not impossible, to realize experimental conditions where the Zeno effect would be seen directly. This is because of the very demanding constraints upon the repetition time. It may be therefore interesting to discuss another physical effect, which was sometimes assimilated in the Zeno effect: line narrowing in nuclear magnetic resonance (NMR).

We recall the principle of NMR. A sample of matter in a solid, liquid, or gaseous form, whether chemically homogeneous or not, is submitted to a magnetic field $\mathbf{B}_0$ directed along the z direction. The nuclear spins have an average precession around this axis with a frequency $\omega_0 = \mu B_0/\hbar$, where μ is the magnetic moment of a nucleus. The corresponding energy levels are associated with the quantized values of the spin projection along the z axis and they are separated by an energy gap $\hbar\omega_0$. A microwave produces a magnetic oscillating field in the xy plane with a frequency ω so that it can be absorbed resonantly when ω differs little from ω_0. The absorption can be measured and this is the principle of NMR since the absorbed energy indicates the number of resonating nuclei while the resonance frequency gives a faithful signature of the nucleus.

The effect in which we are interested is the width of the resonance line. Its value is dominated by the following effect. A given nucleus not only sees the externally imposed magnetic field but also a local field produced by the neighboring atoms. When the medium is not ferromagnetic or antiferromagnetic, there are fluctuations in this local field from one site to another because of the random orientation of the neighboring spins. Let us denote by ΔB the uncertainty in the z component of this random part of the field and by ω_p the frequency $\mu\Delta B/\hbar$. When the nuclei are embedded in a solid matrix, the local field fluctuates slowly in time and the spin energy levels follow adiabatically these changes so that their gap varies from one nucleus to another between the values $\hbar(\omega_0 \pm \omega_p)$. The consequence is an uncertainty in the frequency resonance of order ω_p, which is the linewidth.

In a liquid or gaseous sample where the nuclei under study are a minority, one can observe a rather startling effect. The resonance lines can be quite appreciably narrower than in a solid, and they become even narrower when the pressure increases. In this case the local fields vary very rapidly because the spins of the neighboring atoms are carried along by their brownian motion with rather high velocities. The time for a change of the local field under the effect of brownian motion is proportional to the mean free path, which is itself inversely proportional to the volume density of atoms and therefore to the pressure, at a given temperature in a gas. For instance, with a thermal velocity of 300 meters per second and an intermolecular distance of the order of 10^{-7} cm, the fluctuations of the local field take place in a time of the order of 10^{-12} seconds. This can be taken as a typical order of magnitude for the correlation time τ_c during which the internal field varies by a quantity of order ΔB. Under such rapid changes, a nuclear spin does not have enough time to adiabatically follow the field and its energy varies within somewhat smaller bounds, which results in a line narrowing. One finds experimentally that the narrowing effect takes place in a most marked way when the fluctuations are so rapid that one has

$$\omega_p \tau_c \ll 1, \qquad (11.7)$$

the linewidth being then essentially given by

$$\Delta \omega = \omega_p^2 \tau_c. \qquad (11.8)$$

There are detailed theories of this effect and they are in very good agreement with observation.[4] They treat the local field as a random classical quantity having an average quadratic fluctuation ΔB and a correlation time τ_c. One can better understand the result if one assumes for simplicity that the local field is always along the z direction and randomly takes the two values $\pm \Delta B$ during successive time intervals τ_c. The perturbation exerted by the field lasts too short a time to be adiabatically followed by the energy levels when the condition (11.7) is satisfied. Rather than a full adiabatic change of order $\mu \Delta B$, there is only a perturbation of the wave function. This is given by time-dependent perturbation theory as it applies for short times, and the result is a reduction of the energy variation by a factor $\mu \Delta B \tau_c / \hbar$, i.e., $\omega_p \tau_c$. This results in a linewidth given by equation (11.8). Note that the condition for the correlation time τ_c and the amount of reduction occurring under the effect of a rapidly changing perturbation are similar to what entered in the discussion of the Zeno effect. Likewise the reduction factor for the width is the same as the freezing factor in the Zeno effect.

The question therefore arises whether the narrowing of NMR linewidths could not be a subtle manifestation of the Zeno effect, in which case this effect would have been observed. This assumption was however far from obvious since there is in that case nothing analogous to a measurement process. A closer study indicates that the two effects have indeed some similarities in so far as their explanations rest upon the same theoretical ideas but, nevertheless, they should be considered as substantially different. This is what the following analysis will try to make clear.

An Attempt at a Comparison*

The theories of line narrowing do not directly go back to the foundations of quantum mechanics and their treatment of the local field is phenomenological. To build up a theory digging deeper into the basis of the theory would probably require a serious effort with the only hope of practical results identical to what is already known. This would be too much work for too little outcome. This is why we will only attempt a partial investigation using analogies with other similar calculations.

The spin of a nucleus interacts with an environment consisting of all the nearby atoms. Let us assume for simplicity that the nucleus is fixed in space. The value of the local magnetic field can then be considered as a special collective observable. It behaves like a classical quantity because of decoherence so that one falls back almost immediately upon the assumptions underlying Abragam's theory. This rather intuitive procedure was however not checked by a complete calculation and it remains speculative. If one however accepts the result, it can be seen that the spin interacts with a classically behaving system, just as it happens in a measurement. A decoherence effect takes place in both cases within the macroscopic system which is interacting with the microscopic spin. There are however two important differences. There is nothing similar, in the case of line narrowing, to the enforcement of the spin state as coming from a factual datum. Even more important is the absence of any manifestation of a fact among a sample of potential phenomena. The root of the effect is elsewhere and it is due to the stochastic character of the external local classical field. If one were to interchange the roles of space and time for illustration, the narrowing of spectral lines would be much more similar to the Anderson localization of an electron in a random potential than to the Zeno effect. This would imply that the Zeno effect has not yet been seen. If it were seen, it would in any case not mean that one has "seen" a collapse of a wave function.

4. Observing a Unique Atom

After having spent some time upon measurements that have not been done, it will be better to go back to reality by reviewing real and beautiful experiments. One of the most impressive is the observation of a unique isolated atom,[5] which has become possible rather recently. The very existence of these experiments is the best reason one can give to show that quantum mechanics refers to individual systems and not only to statistical ensembles. Their interest is of course much greater. They can be used in practice to measure the value of some radiative parameters for atoms that were only poorly known beforehand. What is however most interesting from our standpoint is that they provide a perfect example of an experiment where an observer can see an ongoing history under his own eyes. They also show that the properties of an atom that may be said to be true, or real, are not always limited to a unique and isolated occurrence as happens for an ordinary measurement, but they also can go on forever repeating themselves.

Ion Traps

These accomplishments were made possible with the help of remarkable devices—ion traps, which capture almost indefinitely a unique atomic ion inside a bounded region of space.[6] We shall describe here the trap which was developed by W. Paul, though other types exist.

11.1. *An ion trap*

A Paul trap is at first sight an almost closed metallic box having a peculiar shape. It looks like a cylindrical can with a reentrant curved border, which is generated by the rotation of a hyperbola around one of its symmetry axes, say the z axis. It is a so-called equilateral hyperbola with its asymptotes making a 90 degree angle. The lid and the bottom of the can are also generated by the rotation of a similar complementary hyperbola (see Figure 11.1). The border, the lid, and the bottom are therefore portions of hyperboloids. The interest of this shape is due to its peculiar electrostatic properties. When the lid and the bottom are brought to an electric potential $-U$ and the border to the potential U, an electric field is produced in the box. It is easy to compute it and to find its effect upon the motion of a positively charged ion in the trap. One finds that the projection of the ion position upon the xy plane has an oscillatory motion so that this horizontal motion is bounded. The projection upon the vertical symmetry axis is however given by a combination of increasing and decreasing time exponentials. It is therefore unbounded and the motion is completely unstable. So, this is not yet the solution for trapping an ion.

This defect can be compensated by adding an alternating-current potential to the static one so that U is replaced everywhere by $U + V \cos \omega t$. The equations of motion become more complicated, but they are of a well-known type: they are Mathieu equations. Without entering into their derivation or into their mathematical study, it will be enough to mention that they depend upon the two following parameters

$$a = \frac{4eU}{mr^2\omega^2}, \qquad q = \frac{2eV}{mr^2\omega^2},$$

where e and m are respectively the charge and the mass of the ion and $2r$ is equal to both the minimal diameter of the border and the smallest height of the box, i.e., the distance between the apex of the generating hyperbola.

The Mathieu equations have remarkable properties. They have two types of solutions, which are either stable or unstable. In the case of a stable solution, the ion oscillates with a bounded amplitude, both in the horizontal and the vertical directions. In the case of unstable solutions, one or several coordinates increase exponentially with time. Another physical example of a Mathieu motion is given by a swing undergoing an oscillatory push, which can give rise to a stable or an unstable motion depending upon the frequency and the force of the push: the swing either only wiggles or it swings more and more. In the present case, one can choose the values of the two parameters a and q so that the ion remains confined in the trap. The Paul trap is very efficient and an ion can be kept prisoner in it for a very long time (the present longest trapping time is ten months). More recently, it was also found possible to confine a *neutral* atom by laser beams and interesting new experiments will presumably come out of it.[7]

A Basic Experiment

As an example, we shall consider the experiment by Nagourney et al. An ionized barium atom Ba$^+$ is confined in a trap. A laser (showing a blue color) can resonate between the ground state g (an $S_{1/2}$ state) and an excited $P_{1/2}$ state e. This produces rapid transitions between these two states. More properly, from a quantum point of view, one can show that the ion is in a so-called dressed state in the presence of laser light, which is a linear superposition of g and e. Though one might say that the excited state falls back repeatedly upon the ground state while emitting stimulated photons, the emitted photon is inseparable from the laser coherent light and it cannot be seen, even as a matter of principle. It can also happen however that this decay emits a fluorescence photon going in a direction other than the laser photons. This is a spontaneous emission and the photon can be detected.

The emission of fluorescence is substantial: one can obtain from it a photographic picture of the ion as a small blue source and one can even see it with the naked eye. One can also detect the fluorescence photons, with the help of photomultipliers, as their production proceeds. In a typical run, for instance, one sees an average of 1600 photons per second. The number of emitted photons is even higher because the solid angle covered by the detectors is not large.

The experiment can still be refined by sending into the cavity other (red) laser radiation. Its frequency is tuned to the transition between the excited state e and a third state $f(D_{5/2})$. The corresponding transition is forbidden so that it takes place only rarely in spite of the incitement due to the red laser. Furthermore, the transition from this state f to the ground state g is also forbidden so that the f state is metastable with a large lifetime of 30 seconds. Figure 11.2 shows what a fluorescence signal looks like when the two lasers are used together. For a rather long time, there is a fluorescence signal consisting in many photons that are seen by the photodetectors. The number of photons is so high that, in view of the finite response time of the detectors, the signal appears as almost constant with only slight fluctuations. Then, suddenly, with the random character of a quantum process, the fluorescence signal stops. Nothing, except maybe some small fluctuations coming from outside or a spontaneous firing of a photomultiplier, is seen for a long time. Then, the fluorescence signal reappears and everything occurs again as before. The time t during which fluorescence lasted and the time t' during which it stopped behave in a random way from one sequence to another. They are essentially distributed according to a Poisson process and the distribution of the values of t' is an exponential corresponding to the 30-seconds lifetime for the state f.

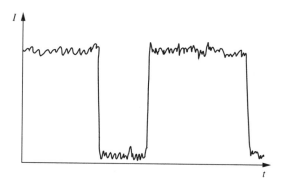

11.2. *A fluorescence signal emitted by a unique atom*

So, "everything happens" as if the ion were continuously going from the state e to g and back under the effect of the blue laser, spontaneously emit-

EXPERIMENTS   **447**

emitting fluorescence photons, then being excited with a quantum jump into the f state by the red laser, remaining unseen in that state as long as the state lives, and finally falling back to the ground state. Then, it is again caught into swinging to and from the state e and everything can start anew.

Histories and Reality

This description of what actually happens during the experiment is so simple and obvious that it looks convincing without further ado. It raises however serious difficulties for the Copenhagen interpretation because, as it stands, it is full of forbidden retrodictions. In the Copenhagen framework, one assumes that there is a reduction of the wave function when a photon is detected. This implies that the atom has fallen back to the ground state but is it sure that it was in its excited state just before? It is difficult to ensure it, because reduction tells us what happens after a measurement but it does not allow a retrodiction for what there was beforehand. Furthermore, the Schrödinger equation represented the atom wave function as a linear superposition of the ground state and the excited state, i.e., as a dressed atom, so that the question is particularly nontrivial in the present case. Was there also a quantum jump when a photon was emitted and not detected? This should also be a part of the story. What does it mean to say that the atom was lying in its metastable state while no signal was seen, if retrodiction is forbidden? The oscillations of the atom between its ground state and its excited state are so rapid in the presence of the blue laser that one can justifiably wonder when reduction is actually taking place. Is it when the photon is completely detected and registered or when it is emitted? After all, if reduction is a physical effect, either it needs a finite time or it takes place at a precise time. How and when? These questions do not hinder the possibility of computing the statistics of the signal but they are rather frustrating when it seems that there is a very clear picture of what happens that one is forbidden to believe in and ordered to keep one's nose upon formal computations. On the contrary, one is asked to believe in a reduction process in the favor of which there is no objective hint and which becomes the more obscure when more questions are asked about it.

These difficulties disappear in the logical interpretation. What was described is simply a history belonging to a sensible logic including all the data together with their true consequences. The histories where one states that, at a specific time, the ion is in one of the states g, e, or f are perfectly consistent. This is exactly the same kind of consistency as for the decay of a particle, up to trivial modifications. All that is said about the atom or the emitted photons consists of true properties or, if one prefers, real properties. The statistics of the signal follow directly from the probabilities of the corresponding histories. The undetected photons can also be considered by consistent histories and one can compute the corresponding probabilities.

As for the metastable state, its discussion is the same as in the case of an unstable particle, already considered in Section 1.

The complete theories of this kind of experiment are too technical to be reported here and the further details given by them bring more understanding about the fine points in the experiments than about the interpretation of quantum mechanics itself. Their overall point of view is however quite interesting because they rely more or less explicitly upon the concept of histories, or at least, all of them are expressed in the picture suggested by histories.[8] This shows how strongly the history framework compels recognition when really difficult problems of interpretation occur. As a matter of fact, these theoretical works introduced explicitly the concept of history, though not the idea of consistency, somewhat before Griffiths published his own work in 1984. It was first known as a "degrouping of photons".[9] Its relation to wave function reduction was then elucidated[10] and a complete analysis of photon emission by a unique atom, in agreement with the history point of view, was given as early as 1982.[11]

INTERFERENCES

Ordinary interferences have already been discussed in Chapter 5. The only nontrivial question, as far as interpretation is concerned, was to determine whether one can state that a particle (photon, neutron,...) went along a unique arm of the interferometer before being detected. The answer was negative and this assertion is deprived of any logical meaning. Nothing more then needs to be added concerning this simple case. What is to be considered now will be concerned with more subtle experiments, among which a specific example will be given. We will also treat the delayed-choice experiments.

5. The Badurek–Rauch–Tuppinger Experiment[12]

This is a neutron experiment, realized with great care. Its principle was suggested by J. P. Vigier and coworkers.[13] Some preliminaries will be useful before describing it.

The Spin Flipper

The experiment uses polarized neutrons and a device, the spin flipper, which can reverse (flip) the spin of a neutron. We shall describe how this works. One can produce beams of polarized neutrons by letting them go through an inhomogeneous magnetic field and afterward selecting a separated outgoing beam. This is essentially a Stern–Gerlach device. One also knows that two orthogonal spin states of a neutron, with for instance $s_z = 1/2$ and $s_z = -1/2$, do not interfere. When a neutron is polarized so that its spin state is $s_z = 1/2$, one can make it flip to the state $s_z = -1/2$ by a spin flipper. The principle of this device is very similar to NMR. A relatively high constant magnetic field $\mathbf{B}_0$ is directed along the z axis and, in a cavity, a microwave radiation produces a magnetic field $\mathbf{B}_1$ oscillating in the plane xy normal to the main field. The oscillating field can be considered as the linear combination of two fields turning in opposite directions and only one of them acts effectively on the neutron because it resonates with its precession. For simplicity, it will be assumed that the field $\mathbf{B}_1$ is rotating in the xy plane with a frequency ω, i.e., we will consider only one circularly polarized component of the microwave.

The part of the hamiltonian involving the spin degrees of freedom is given by

$$H = -(\mu B_0/2)\sigma_z - (\mu B_1/2)\sigma_x \cos\omega t - (\mu B_1/2)\sigma_y \sin\omega t. \quad (11.9)$$

An initial state $|+1/2\rangle$, corresponding to a spin component $s_z = +1/2$, becomes after its evolution during a time t a superposition $a_+(t)|+1/2\rangle + a_-(t)|-1/2\rangle$. It is easy to solve the Schrödinger equation, which gives

$$a_+(t) = \cos\frac{\Omega t}{2} - i\left(\frac{\omega - \omega_0}{\Omega}\right)\sin\frac{\Omega t}{2}, \quad a_-(t) = -\frac{\omega_1}{\Omega}\sin\frac{\Omega t}{2},$$

where $\mu B_0 = \hbar\omega_0$, $\mu B_1 = \hbar\omega_1$, and $\Omega^2 = (\omega - \omega_0)^2 + \omega_1^2$. Under the conditions of exact resonance—when $\omega = \omega_0$ and therefore $\omega_1/\Omega = 1$, one finds that the initial spin state becomes the oppositely polarized spin state $|-1/2\rangle$ after a time t such that $\Omega t = \pi/2$. One then says that the spin has flipped. This effect can be realized if the neutrons are monochromatic by choosing conveniently the length of the cavity where the microwave is acting. It may be noticed that a slight change in the external field $\mathbf{B}_0$ leads to a slight change in the resonance frequency.

The BRT experiment uses two spin flippers working at two slightly different frequencies ω and ω', which are sufficiently different for the two resonance lines to be clearly distinct in spite of their finite width. The two spin flippers are positioned along the two possible trajectories of the neutron in the interferometer.

The Experimental Device

The interferometer, where the neutrons enter one by one, involves several parallel diffracting plates. They have been cut from a single silicon monocrystal so that their crystal lattices are perfectly parallel. The neutron wave is thereby split into two parallel beams, as shown in Figure 11.3. The two spin flippers are positioned along the two beams. We will not discuss here the output device which allows one to observe the interferences, because it is after all an inessential feature. The experiment also involves a few refinements or peculiarities that will also be neglected; we shall only concentrate upon its most significant aspects.

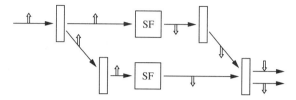

11.3. *The BRT experiment.* Each polarized neutron beam crosses a spin-flipper (SF) with a different resonance frequency.

The initial argument by Vigier et al. was to say that a spin flip absorbs energy since the energies of the two spin states differ by a quantity μB_0, μ being the neutron magnetic moment. As a matter of fact, either an absorption or a production of energy occurs according to the relative orientation of the initial spin and the magnetic field, but we shall consider only the case when there is absorption. The energy is taken from the absorption of a microwave photon and it can take place only in one of the two spin flippers. The photon has therefore either the frequency ω or ω'. If one were able to know what is this frequency, one could tell through which way the neutron went. Accordingly, this will be the main question to be considered: can it be said that the absorbed photon has a definite frequency?

Before doing any theory, it should be mentioned that the experimental results clearly show the existence of interferences. Therefore, if it were also possible to assert which way the neutron went, this would bring a drastic change in our conceptions of quantum mechanics. This is what is to be investigated.

Some Elementary Quantum Electrodynamics

Since one is asking nontrivial questions about photons, a little bit of quantum electrodynamics will prove useful. This can be done rather easily. Let us consider the microwave in the first cavity. It can be described classically by a vector potential $\mathbf{A}(x) \cos \omega t$ oscillating at the frequency ω. The vector potential, in the radiation gauge where its divergence is 0, satisfies the wave equation

$$(\Delta + k^2)\mathbf{A}(\mathbf{x}) = 0, \qquad (11.10)$$

where $k = \omega/c$. Boundary conditions expressing the physical properties of the cavity walls and its shape give a unique solution.

This electromagnetic field is in a unique mode. From the standpoint of dynamics, it is accordingly a system with only one degree of freedom. The dynamics and the time dependence can be exhibited by using the relevant solution $\mathbf{A}_0(\mathbf{x})$ of equation (11.10) satisfying the boundary conditions, which is defined uniquely up to a normalization factor. The time-dependent field is therefore written as $\mathbf{A}(\mathbf{x}, t) = \xi(t)\mathbf{A}_0(\mathbf{x})$. The coefficient $\xi(t)$ is the unique

EXPERIMENTS

coordinate describing the dynamics of the mode. The time-dependent wave equation

$$\left[\frac{1}{c^2}\frac{\partial^2}{\partial t^2} - \Delta\right]\mathbf{A}(\mathbf{x},t) = 0,$$

gives immediately, together with equation (11.10), the equation of motion

$$\frac{d^2\xi(t)}{dt^2} + \omega^2\xi(t) = 0.$$

This shows that, from the standpoint of dynamics, a wave in a cavity in a well-defined mode is essentially equivalent to a one-dimensional harmonic oscillator.

As long as only one mode is relevant, quantum electrodynamics is not more complicated than the theory of the harmonic oscillator. One considers the variable ξ as a quantum observable. Using it together with the canonically conjugate momentum observable, one can construct creation and annihilation operators as in Chapter 7. The energy levels of the oscillator are $(n + 1/2)\hbar\omega$, where $n = 0, 1, 2, \ldots$, and, if one changes the arbitrary origin of energy by taking it at the value $(1/2)\hbar\omega$, the energy eigenstates can be recognized as the states containing respectively 0, 1, or 2 photons and so on. The quantum state of radiation in this mode, containing exactly n photons, coincides with the nth excited state of the oscillator and it will be denoted by $|n\rangle$.

The most general state of the quantized electromagnetic field in this mode is given by a linear combination

$$|\Psi\rangle = \sum_n c_n |n\rangle.$$

At this point, one can introduce an important result by Glauber.[14] An actual microwave is generated by some electric and electronic devices working under classically meaningful conditions where a classical electric current generates the wave, which is transmitted to the cavity by a wave guide. Then, in these conditions, the quantum state of the electromagnetic field is necessarily a *coherent state*. The coherent states of an oscillator have been discussed in Chapter 7. A coherent state is a normalized eigenstate $|\Phi\rangle$ of the annihilation operator such that $a|\Phi\rangle = \alpha|\Phi\rangle$, where α is a complex number. If one goes back to the definition of the annihilation operator in terms of the position and momentum operators, one finds using Maxwell's equations that the number α is given, up to a multiplicative constant, by the combination $B_{1x} + iB_{1y}$ of the space components of the corresponding classical magnetic field.

What Happens in a Spin Flipper

We can now describe in more detail what occurs in a spin flipper when a neutron enters in the initial spin state $|+1/2\rangle$ and crosses it. When saying that the neutron crosses the spin flipper, one only means at this point that one half of the neutron wave packet crosses spin-flipper cavity and one is only interested in logical considerations, which will require a computation of a few traces. No interpretation has yet entered. The hamiltonian coupling the neutron at position **x** with the electromagnetic field is given by

$$H = -\mu \cdot (\mathbf{B}_0(\mathbf{x}) + \mathbf{B}_1(\mathbf{x}, t)), \qquad (11.11)$$

where the second term represents the coupling with the microwave in the cavity. The initial state of the wave is a coherent monomode state $|\Phi\rangle$ and the average number of photons in the cavity is a large quantity to be given by $\langle n \rangle = \langle \Phi | a^\dagger a | \Phi \rangle = |\alpha|^2$, where $\alpha \gg 1$. It should be noticed that the number of photons in the wave is not sharply defined and its uncertainty is given by $\Delta n = |\alpha|$, which is also a large quantity. When this is taken into account, it looks rather questionable whether a change of one unit in this number, when the neutron spin flips, could be observed. This is the question one has to discuss in detail.

The working of the spin flipper can be obtained from the hamiltonian (11.11) where the wave is now quantized in order to have a complete description. A more convenient form of this hamiltonian is obtained by expressing the field $\mathbf{B}_1(t)$ in terms of the time-dependent amplitude $\xi(t)$. One obtains, after a little algebra,

$$H = -(1/2)\hbar\omega_0\sigma_z - (1/2)\hbar\omega_1\{\sigma^{(+)}a\exp(i\omega t) + \sigma^{(-)}a^\dagger\exp(-i\omega t)\}, \qquad (11.12)$$

this expression being valid when the position of the neutron is inside the cavity. The average quantity $\langle \Phi(t)|H|\Phi(t)\rangle$ coincides with the hamiltonian (11.9) where the wave was described classically.

Interpreting the BRT Experiment

The interpretation of the experiment proceeds exactly in the same way as in the elementary case, which was given in Chapter 5. Introduce again as many projectors $E_2^{(n)}(t_2)$ as there are neutron detectors behind the system to find the effect of interferences, each projector stating that the neutron has its position in the active region of a detector at some time t_2. This time is controlled by the wave packet and it is the same as in Chapter 5. One also introduces two projectors $E_1^{(1,2)}(t_1)$ expressing that the neutron is located on the upper (respectively, lower) trajectory at a convenient time t_1, for instance in a space region V_1 (V_2) in front of the first (second) spin flipper. The basic question is again to find whether or not the corresponding logic is consistent and this amounts to a check of the following set of consistency conditions

$$\mathrm{Re}\,T^{(n)} \equiv \mathrm{Re}\,\mathrm{Tr}\{E_1^{(1)}(t_1)\rho E_1^{(2)}(t_1)E_2^n(t_2)\} = 0.$$

Some significant differences with the elementary case of Chapter 5 should now be stressed:

1. The physical system now consists of the neutron together with the quantized waves in the two cavities. The initial state operator ρ also takes this into account, its expression involving the neutron state as well as the two coherent states of the waves so that

$$\rho = |+1/2\rangle\langle+1/2| \otimes |\Phi(\omega)\rangle\langle\Phi(\omega)| \otimes |\Phi'(\omega')\rangle\langle\Phi'(\omega')|.$$

2. The traces are of course taken over the Hilbert spaces of the two electromagnetic modes as well as over the neutron Hilbert space.

3. The hamiltonian entering in the time propagators $U(t)$ is the sum of the hamiltonian for the free neutron, of the electromagnetic free hamiltonian

$$H(\text{radiation}) = \hbar\omega a^\dagger a + \hbar\omega' a'^\dagger a',$$

and of the coupling hamiltonian (11.12).

On the other hand, the semiclassical treatment of the neutron wave packets remains the same, if one neglects the small difference between the kinetic energies of the neutron in the two beams after crossing the spin flippers. This difference is due to the slight energy difference between two photons having the respective frequencies ω and ω', which later alters the neutron kinetic energy when it leaves the external magnetic field $\mathbf{B}_0$. It would be easy to take this explicitly into account and this has been done by Badurek et al., but it only gives rise to a negligible shift in the interference fringes, at least when one uses an ordinary interferometer. In the actual BRT experiment, things are a bit different because the interferometer is itself slightly different, but this is an inessential detail and it can be neglected without trouble.

In spite of these more complicated conditions, it is still possible to perform rather easily the calculation of the consistency traces in view of a few remarks:

1. These traces can be written as the diagonal matrix elements of a product of projectors

$$T^{(n)} = \langle i|E_1^{(2)}(t_1)E_2^{(n)}(t_2)E_1^{(1)}(t_1)|i\rangle,$$

where the overall initial state $|i\rangle$ is given by

$$|i\rangle = |\phi\rangle \otimes |+1/2\rangle \otimes |\Phi(\omega)\rangle \otimes |\Phi'(\omega')\rangle,$$

the state vector $|\phi\rangle$ representing the initial spatial wave packet of the neutron.

2. The propagators $U(t)$ act independently upon the states of the electromagnetic waves, except when the neutron crosses a spin flipper and an interaction with a wave takes place. This occurs somewhere between the

times t_1 and t_2. One can then write

$$T^{(n)} = \langle \phi | E_1^{(2)}(t_1)$$
$$\times \langle +1/2 | \langle \Phi(\omega), t_1 | \langle \Phi'(\omega'), t_1 | E_2^{(n)}(t_2) | + 1/2 \rangle | \Phi(\omega), t_1 \rangle | \Phi'(\omega'), t_1 \rangle$$
$$\times E_1^{(1)}(t_1) | \phi \rangle,$$

where the unnecessary symbols for a tensor product have been suppressed. The two wave functions $|\phi^{(1,2)}(t_1)\rangle = E_1^{(1,2)}(t_1)|\phi\rangle$ represent respectively the upper (lower) half of the wave packet at time t_1, if the time t_1 has been chosen so that the full wave packet is entirely contained in the region $V_1 \cup V_2$ (covering the front regions of both spin flippers) at that time, which is exactly what was done in the elementary case.

3. Introducing the state of the system at time t_1,

$$|\Psi^{(1,2)}(t_1)\rangle = |\phi^{(1,2)}(t_1)\rangle \otimes |+1/2\rangle \otimes |\Phi(\omega), t_1\rangle \otimes |\Phi'(\omega'), t_1\rangle,$$

one sees that a typical trace will contain a factor such as

$$F = \langle \Psi^{(2)}(t_1) | E_2^{(n)}(t_2) | \Psi^{(1)}(t_1) \rangle$$
$$= \langle \Psi^{(2)} | U^\dagger(t_2 - t_1) E_2^{(n)} U(t_2 - t_1) | \Psi^{(1)} \rangle,$$

which has to be evaluated more explicitly. The propagator $U(t_2 - t_1)$ acts upon the state vector $|\Psi^{(1)}\rangle$ representing one half of the wave packet, on the upper neutron trajectory with spin $+1/2$ in front of the spin flipper, while each cavity contains its own quantized electromagnetic wave. This propagator describes the motion of the neutron together with the interaction of its spin with the wave. Its effect upon the wave function in coordinate space is the same as in the elementary theory so that one should only pay attention to the spin, the waves, and their coupling.

4. In the region where the field $\mathbf{B}_0$ is acting (notice that this region contains the cavities), it is convenient to go to a rotating frame by using a unitary transformation generated by the operator $W(t) = \exp\{-(i/2)\omega_0 t \sigma_z\}$. This is a well-known trick in NMR theory and the interaction with a wave is then given simply by

$$H = -(1/2)\hbar\omega_1\{\sigma^{(+)}a\exp\{i(\omega - \omega_0)t\} + \sigma^{(-)}a^\dagger \exp\{-i(\omega - \omega_0)t\}\}, \tag{11.13}$$

where the two matrices $\sigma^\pm$ are the combinations $1/2(\sigma_x \pm i\sigma_y)$ of Pauli matrices, respectively raising and lowering the spin.

5. Let us write the hamiltonian (11.13) more compactly as

$$H = A(t)a + A^\dagger(t)a^\dagger \tag{11.14}$$

to find how it acts on the state of the associated microwave $|\Phi\rangle$. The first term gives $A(t)a|\Phi\rangle = A(t)\alpha|\Phi\rangle$. Write the second term as $a^\dagger|\Phi\rangle = \alpha^*|\Phi\rangle + |\Psi\rangle$. Using the definition of the coherent state as a normalized eigenvector of the annihilation operator, one easily finds that the vector $|\Psi\rangle$ is normalized and orthogonal to $|\Phi\rangle$. It is therefore much smaller in norm than the first term $a^*|\Phi\rangle$, which has the norm $|\alpha|^2$. Notice that only the second term contributes when the cavity is not working so that it contains no photon. In that case, the second term in the hamiltonian (11.14) represents a spontaneous spin flip with emission of a photon, the probability of which is extremely small in view of the small magnetic moment of the neutron.

More generally, one can use the following equivalent form of the coupling hamiltonian

$$H = -(1/2)\hbar\omega_1[\sigma^{(+)}\alpha \exp\{i(\omega - \omega_0)t\} + \sigma^{(-)}\alpha^* \exp\{-i(\omega - \omega_0)t\}]$$
$$\times |\Phi\rangle\langle\Phi| - (1/2)\hbar\omega_1\sigma^{(-)}\exp\{-i(\omega - \omega_0)t\}|\Psi\rangle\langle\Phi|.$$

The first term is exactly the coupling (11.9) with a classical wave and the second one is of relative order $\langle n\rangle^{-1/2}$ as compared to it. In view of its smallness, it can be treated by perturbation theory when computing the full propagator during the crossing of a spin flipper. The leading term reproduces the classical result exactly. The first-order correction gives a zero contribution to the trace because of the orthogonality of $|\Phi\rangle$ and $|\Psi\rangle$. One must therefore go to higher order in perturbation theory to find nonzero corrections. They are of order $\langle n\rangle^{-2}$ in probability and do not always correspond to a spin flip. They represent in fact only some very small quantum corrections to the working of a spin flipper, negligibly perturbing its performance. The rest of the calculation proceeds in the same way as in the elementary case and the traces occurring in the consistency conditions are exactly the same as before.

Conclusions

One can therefore assert that at most only one spin flip out of $\langle n\rangle^2$ is, in principle at least, detectable and, even so, it would come from a random quantum fluctuation in the spin flipper. The assumption according to which it might be in principle detectable was therefore wrong. A logic in which one would try to assert that the neutron went along a specific path is therefore necessarily inconsistent.

To conclude the discussion, one may also compare this result with two imaginary experiments:

1. Consider the academic case where the cavity contains initially *exactly* n photons rather than a coherent microwave. Then, Vigier's argument would work insofar as the traces occurring in the consistency conditions can be shown to vanish in case the final spin state flipped. They remain the same in the absence of a spin flip. There is therefore consistency only when one can make sure that there has been a spin flip. However, the cavity does not act any more as a spin flipper and

the spin flip has to be selected by some polarization analyzer (some sort of a Stern–Gerlach device) selecting the spin state of the outgoing neutrons. In that case, one finds that there are no interferences at all. All that has been learned in this case is that, even if one does not have enough information to assert which way the neutron went, this information is nevertheless contained in the photon wave function and it might be obtained nondestructively by a further photon measurement. But there are no interferences, so that the result is of no interest.

2. In the BRT experiment, one might also conceive of adding a photon measuring device distinguishing between the two states $|\Phi\rangle$ and $|\Psi\rangle$ at the end of each individual experiment. The result is however again rather poor. Most measurements will give the results $|\Phi\rangle$ and $|\Phi'\rangle$ together with interferences and there is no possibility of knowing which way the neutron went in that case. A very small fraction of events will give the result $|\Psi\rangle$ or $|\Psi'\rangle$, which provides information about the path followed by the neutron. However, this sample will show no interferences. Of course, the possibility of making this measurement upon the photon state is most probably impossible, even in principle.

The overall conclusion is therefore that interferences are incompatible with the possibility of asserting which way a particle went.

6. Delayed-Choice Experiments

The Idea

The experiments involving a delayed choice are in a category by themselves. Some physicists would consider that the question they try to answer is trivial, if not empty. Others would maintain that these experiments probe reality almost independently of the present background of quantum mechanics. Without committing oneself, one may try at least to understand the point of view of their tenants.

This is again basically an interference experiment, to be made with an interferometer having two well-separated arms. Let us suppose that two people are looking at it and they are discussing it, to see how one can arrive at the idea:

A: Tell me again the conclusions of quantum mechanics.

B: The photon does not go through only one arm of this apparatus, but along both of them together.

A: And what happens when there are photon detectors in the arms?

B: Then, the photon goes through only one arm. This also occurs when the detector only signals the passage of a photon and lets it go through.

A: Do you think that there is a time when the photon is travelling along both arms, when there is no detector?

B: I guess so.

A: What would happen if I were quick enough to insert detectors at the end of the arms before the photon gets out?

B: Well, let me think...

A: As I see it, the photon would be caught when it is already in some sense ubiquitous and present in both arms and then, I would oblige it to make itself known in only one place. So, it would suddenly have to be in one arm or to be in the other. If it manifests itself in Arm 1, it would have to disappear in some sense from Arm 2 if it is a real particle. In view of the distance between the two arms, this would imply action at a distance, which most people think impossible. Therefore, I would tend to think that the photon will continue to go along Arm 2 if your assumption of ubiquity is correct and if it was already there beforehand. It seems to me we might know through which arm the photon went and nevertheless see interferences.

B: I don't quite follow you, but in any case your experiment is impossible. You will never be quick enough to learn that a photon is in the interferometer and to put your detectors in its way.

A: Why not? Suppose I have already put the detectors in the apparatus and they need only to be activated once the photon has entered the interferometer. This is not impossible in principle. The geometry has only to be well chosen, with long bent arms, so that a signal that is sent when the photon enters the interferometer has enough time to reach the detectors before the photon arrives to them, which is possible if it follows a longer path than the signal.

B: I agree. But the electronic device activating the detector is a deterministic part of the apparatus and, according to quantum mechanics, its presence is enough to make the photon behave as if the detectors were already activated when it enters.

A: I do not quite understand that but I can beat it. Let *me* decide whether or not the detectors will be activated as soon as I know that it entered the interferometer. Would you say that my decision is deterministic and the photon will know in advance what I am going to decide?

B: I have no opinion about that and I am sure that your wife (or husband) can predict what you will choose but, anyway, although I know you to be quick-minded, you will not have enough time to make your decision.

A: It does not matter. I can play dice or, even better, in order to have enough time, I can activate the detectors by a random signal.

B: As far as I can see, Bohr maintained that quantum mechanics is objective and therefore only subject to what the apparatus actually is, no matter how it is activated or not. So, nothing will change in what I told you first.

A: And I am telling you that you cannot believe that a photon can follow two different paths at the same time and also believe in wave function reduction and in the impossibility of action at a distance. At least, my experiment will select the issues. Either one detects the photon and one sees nevertheless interferences. Then, it will mean that wave packet reduction is a physical process needing some time and incompatible with action at a distance. Or one does not see interferences. Then, it will mean that there is wave packet reduction but also action at a distance.

So, whatever is found, my experiment will have a conclusion: on one hand, wave function reduction is a local physical process or, on the other hand, there is some internal inconsistency in quantum mechanics, since it would imply action at a distance. How do you answer that?

B: You forgot a third possibility: wave function reduction might not be a physical process.

In one way or another, one may arrive at the idea of a delayed-choice experiment, even if the present author must confess that he does not really understand what is at stake. The experiment consists in deciding randomly to activate detectors in an interferometer at a time when the wave packet is already split among the two arms according to the usual quantum representation. What will come out of the result is not too clear. The experiment was conceived when wave function reduction was thought to be necessarily a physical process so that the locality of this process was the point to be investigated. Of course, as any reasoning where one uses the method of elimination after making a list of all the possible assumptions, it becomes unclear when another possibility is added, such as the unphysical character of wave function reduction in the present case. As it is, the experiment has been done and it should therefore be investigated according to the new interpretation.

EXPERIMENTS

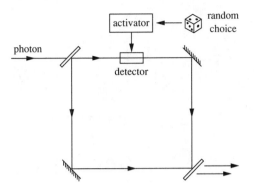

11.4. *An ideal delayed-choice experiment*

The idea of this kind of experiment apparently goes back to a remark by K. von Weizsäcker.[15] A trace of it also occurs in a single sentence by Bohr,[16] who stated that it does not matter whether our plans to manipulate the instruments have been fixed in advance or if one delays the decision till a time when the particle is on its way from one apparatus to another. These considerations have been renewed by John Wheeler[17] from whom we borrow these references. Some experiments of that type have now been performed[18] and their results confirm Bohr's point of view. Bohr's statement was however the expression of a conviction with no detailed proof and, after all, he never proved anything in this domain but always guessed what an answer should be from an overall intuitive sense of consistency. It will therefore be interesting to provide such a proof, if only as an exercise to check the completeness of the interpretation.

An Experimental Device

There are three main technical difficulties in the practical realization of the experiment: to make sure that there is only one photon in the interferometer at a given time, to check the presence of the photon in one arm without destroying it, and finally to make the random choice in a sufficiently short time.

In the experiments by Alley, Jakubowitz, and Wikes, the photon is produced in the following way. A very short laser pulse excites an emitting diode having a very low efficiency. The diode emits then a light signal that is reduced by a factor 10^{-16} in intensity as compared to the laser signal. The average number of photons entering the interferometer is much smaller than 1 so that, when the later events show that a photon was produced, the probability to have simultaneously another photon is very low. Furthermore, the very brief duration of the laser pulse is essentially maintained by the secondary emission so that the photon wave packet can be significantly shorter than the length of an interferometer arm.

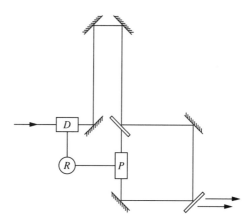

11.5. *A realistic delayed-choice experiment*

It would be very difficult to have an effective detector that would not destroy the photon. So there is a polarizer in front of the interferometer so that the photon entering it is polarized in a definite direction, say the *x* direction. In place of a detector, one uses a polarization rotator (a quarter-wave plate), which rotates the direction of polarization by 90 degrees, bringing it, for instance, in the *y* direction. It is located in one arm of the interferometer.

This is of course not enough since one needs to have a device that can be activated or not. So, in place of an inert quarter-wave plate, one uses a Pockells cell. It is essentially made of a substance showing a Kerr effect. When no electric field acts on it, it is optically neutral. When a field is applied, it becomes birefringent and behaves like a quarter-wave plate. So one can control the behavior of the cell and make it either neutral or active by the action of an external electronic signal.

To produce the signal, one detects the entering laser pulse and the resulting signal is sent through a generator of random bits. The signal coming from this randomizer is sent to the Pockells cell to activate it or not, according to the bit 1 or 0. One needs of course a very rapid electronic device and a geometrical arrangement allowing the electronic signals to go all the way through before the time when the photon wave packet reaches the cell.

The intuitive discussion of the events by unformalized histories is practically the same as before, except for obvious slight modifications: The sample of events corresponding to the case when the cell is activated shows no interference pattern. This can be considered as meaning that the polarization was rotated in the arm where the cell is located and therefore the photon had to be in that arm. It does however not exclude the possibility that the photon crossed both arms simultaneously. But one can also put a polarization analyzer in front of the photomultipliers detecting possible inter-

EXPERIMENTS

ferences, for instance by selecting a linear polarization along the y direction. A photomultiplier located farther away then registers a photon only in one case out of two statistically. This might be said to imply that the photon goes either through one arm or through the other, the two cases occurring with equal probability. One has therefore obtained the kind of situation the basic experiment was supposed to test, namely a photon going only through one arm whereas it was expected to be in both arms when the activation was decided.

Theory

For simplicity, we shall discuss only the ideal experiment that was first described. The actual experiment can be treated exactly along the same lines. So we assume that there exists in the first arm a nondestructive detector D. It is able to register a photon and let it go without a change. The data at one's disposal are the following ones: (i) a signal registering the entry of a photon into the interferometer, (ii) a record of the random bit that was sent by the randomizer, (iii) the signal by the detector D if it registers the photon, and (iv) the data given by a set of photomultipliers behind the interferometer. As a matter of fact, only one detector located at the theoretical position of a dark fringe would be enough.

It is of course impossible to model the decision of a human being. Even the working of an ordinary randomizer is rather difficult to describe and it might lead us into tricky questions concerning the description of classical randomness. We might rely for that upon the theory of K-systems given in Chapter 10 but it will be better to avoid it. So we rely upon a theoretically simpler device, which is much easier to describe by quantum mechanics. We assume that the laser entry signal is used to excite an atom A having a very short lifetime. The atom will decay randomly according to quantum mechanics, emitting a fluorescence photon. This photon activates the detector D, which records two kinds of data, namely its own activation and whether it has registered the photon in the interferometer. We don't need to worry whether this ideal device can be realized, since it retains obviously all the essential features of the basic experiment and this is enough for a discussion of principles.

Taking into account the finite extension of the wave packet and working as usual with its semiclassical propagation, one can introduce two significant times t and t' for histories. At time t, the wave packet is completely contained in the union of two space regions R_1 and R_2, one in each arm of the interferometer and the first one well in front of the detector D. We shall also assume that the detector D becomes activated at time t, if and only if atom A has already emitted a fluorescence photon by decaying. Let V_n be as usual a collection of space regions behind the interferometer, in each of which a photomultiplier is located. At time t', the wave packet is completely contained in the union of these regions. At the origin of time, the photon is just in front of the interferometer and Atom A has just become excited. The quantum system to be discussed is made of the photon and of Atom A. The data include two signals from D stating whether it has been activated at time t and, if so, whether it has registered the photon. To this must be added the various possible signals from the photomultipliers. As usual, the data are used to build up sensible logics. The relation between data and results is however by now sufficiently elucidated and we shall work directly with the properties of the system $A+$ photon without explicitly mentioning the macroscopic measurement apparata.

Let us now introduce the following histories:

1. Atom A is still excited at time t (or the detector D is inert, which is equivalent from the standpoint of logic). The photon is detected at time t' by a photomultiplier with label n.
2. Atom A has decayed at time t (so that the detector D is active). The detector D registers the photon and, at time t', the photon is again registered by a photomultiplier n.
3. Atom A has decayed at time t and D does not register the photon, which is still detected by a photomultiplier at time t'.[19]

The logic one can build from this family of histories is easily shown to be consistent. As a matter of fact, it consists only of data.

One can then add it to two other properties stating respectively that the photon is in the region V_1 (respectively V_2) at time t, which is expressed by the relevant projectors. Here something more subtle than usual will occur, which leads us to define two different logics L and L': In logic L, the properties of localization in the arms are inserted in the histories (b) and (c), but not in the histories (a). In logic L', they are also inserted in the histories (a). To say it more simply, in logic L one tries to assert in which arm the photon was before being detected by the detector D located in Arm 1 or was signalled by non-detection to be in Arm 2, but one avoids stating anything of that sort when the detector is not activated. This is consistent with Bohr's view of the experiment. In logic L', one dares to say in which arm the photon was, even when the detector is inert and cannot confirm it by a data. The consistency of L' would of course be a surprise.

As a matter of fact, the logic L' is obviously inconsistent because one set of consistency conditions is exactly the one occurring in the discussion of elementary interferences. As for the logic L, it is consistent, the verification being left as an exercise. One also finds in logic L an implication stating that

D has registered $\Rightarrow$ The photon is in the first arm at time t;
D has not registered $\Rightarrow$ The photon is in the second arm at time t.

The properties concerning the photons, as they are deduced, are only reliable.

So, (and this is a conclusion), nothing new is obtained and the intuitive picture one tried to test by the delayed-choice experiments is an illogical mental representation, forbidden as a matter of principle by the universal rule of interpretation. Once again, one recovers a conclusion that was already advanced by Bohr, except that it can now be proved.

LEGGETT'S EXPERIMENTS

7. The Experiments with SQUIDs

The Challenge

We have already had several occasions to recall Leggett's approach and his criticism of the Copenhagen assumption according to which the classical behavior of a macroscopic object is taken for granted. The relevant theoretical background was given in Chapter 10. We shall now briefly describe the relevant experiments. Recall that three conditions were found sufficient for a system to behave classically: a regular dynamics, the existence of dissipation, and an initial state corresponding to a classical state of fact. A SQUID does not satisfy the second condition because of the absence of dissipation in a superconductor so that a quantum behavior is possible in this case. This is not however an automatic feature since most superconducting devices behave classically. It is also necessary to prepare the system in a nonclassical state and this condition is the most drastic one.

Experiments

According to the quantum theory given in the preceding chapter, a SQUID is essentially a system with only one degree of freedom, which is conveniently described by the magnetic flux Φ across the ring of the SQUID. The dynamics is formally the same as for a nonrelativistic particle, the junction capacity C playing the role of a mass. This "particle" is submitted to a potential $U(\Phi)$, which is given by equation (10.24). The parameters entering in the definition of this potential are the main physical characteristics of the SQUID, namely the ring inductance L and the critical current I_0 of the junction, together with externally imposed conditions: the magnetic flux Φ_{ext} of external origin and the injected current I_{ext}.

Two shapes of potential are particularly interesting and they are shown on Figure 11.6. The first one is a symmetric double well and the second one is a potential allowing the existence of a metastable bound state, which can decay by a tunnel effect. Both are obviously well suited to the manifestation of quantum properties. The symmetric double well is well known for its remarkable effects, as, for instance, the existence of oscillations of a particle from one well to the other in spite of the potential barrier classically forbidding them. This case has been studied in much detail[20] but the realization of the experiment is particularly difficult because of the need for a well-controlled external magnetic flux and the sensitivity of the effects to even small decoherence corrections. We shall therefore only consider the second case, which has been the subject of many precise and cleverly devised experiments.[21]

A typical experimental setup is shown in Figure 11.7. The SQUID is made of a superconducting "ring" having the shape of a hairpin and consisting of two parallel wires; they are connected on one end to a Josephson junction device and, on the other end, a superconducting bridge closes the ring. External contacts are welded on the bridge to allow the injection of an external current. The narrowness of the ring reduces the sensitivity to unwanted external magnetic fields. All the physical parameters defining the potential $U(\Phi)$ can be obtained from separate calibration measurements, so that the theory contains no arbitrary parameter.

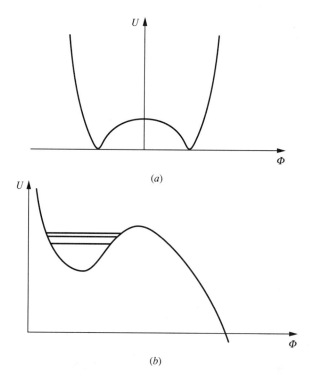

11.6. *Two potentials of importance in Leggett's experiments*

EXPERIMENTS

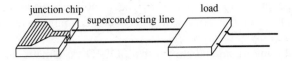

11.7. *An experimental setup for showing a macroscopic tunnel effect*

The shape of the second potential in Figure (11.6) shows the existence of a potential well allowing the existence of a few bound states. Their energies can be computed. They are metastable since the "particle" can cross the barrier on the right by a tunnel effect. The existence of the excited states has been verified as well as the value of their energy gap by perturbing the system under the action of a microwave. This is a clearcut first manifestation of a quantum behavior.

When the tunnel effect takes place, the initial "ground state" in the well turns into a "free" state of the representative particle. From this standpoint, one sees that the "particle" gains kinetic energy by "falling" along the potential. This can actually be shown by measuring the voltage V across the junction, the measurements being nonperturbing because of the macroscopic character of the system. The theory given in Chapter 10 shows that the voltage is proportional to the "velocity" $d\Phi/dt$. As long as the system remains in its bound state, the particle bounces semiclassically in its well, the average quantity $\langle d\Phi/dt \rangle$ is 0 and this is what the measurement gives. Accordingly, the bound state signals itself by a zero voltage. The crossing of the barrier and the following increase in kinetic energy is signalled by the occurrence of a voltage, which increases rapidly, at least during a finite time. So one knows when and how the crossing of the barrier occurs, which is another check of the quantum behavior.

At a very low temperature, the predicted effect is seen with the exact characteristics predicted by the theory. When the temperature is increased, though still remaining below the critical value for superconductivity, the quantum effects enter in competition with a classical one. The representative particle is then in a state of thermal equilibrium in the well and the thermal fluctuations in its kinetic energy can allow it to pass classically above the barrier. Here again, the experimental results agree quite well with the theory.

Decoherence

A very interesting aspect of these experiments consists in the possibility of verifying the theory of decoherence. This is very important since, most often, decoherence is so rapid and so efficient that one can only contemplate a situation where it has already fully taken place. This is rather frustrating because one always sees the classical behavior of the macroscopic systems and many people find it so obvious by sheer habit that they remain doubtful about a quantum transition where decoherence might have taken place, though too rapidly to be seen.

The conditions are much better with a SQUID. As a matter of fact, there is always a small resistance coming from the Josephson junction and from the imaginary part of the impedance in the entry circuit. It implies that one can never completely neglect dissipation and decoherence and they remain always sizable, though not dominant. The complete theory was done by Caldeira and Leggett.[22] The tunnel decay rate depends sensitively upon the resistance so that the agreement between theory and experiment is also evidence for decoherence.

When one plays with the position of the bridge while having an external impedance showing an effective resistance by having a real part, another effect occurs. The hairpin behaves like a conducting line. When it is long enough, the wave packet originating from the tunnel effect has enough time to be produced before reaching the bridge and there is practically no effect of the external resistance (this is a typical retardation effect). When the hairpin is too short, there is not enough place for the wave packet, which would be sensitive to the resistance while being produced, and the decay rate is highly reduced. This is by the way a general feature: instantaneous dissipation reduces drastically the rate of a tunnel effect so that most classical systems that could show one in principle according to elementary quantum mechanics are completely frozen in practice. The agreement between theory and experiment is very good[23] and it provides a clearcut verification for the existence of decoherence.

12

Summary and Outlook

This concluding chapter consists of two parts. The first one gives a summary of the main results. The second one is an excursion outside physics and deals with a few problems of a more philosophical character, such as the objective or the realistic features of the theory. These questions have been the subject of much concern since the advent of quantum mechanics and no advance in the understanding of interpretation would be complete without at least an attempt to recognize whether these old problems have become rather different or a little clearer.

THE RULES OF INTERPRETATION

In this first part, we will recall the main steps and the main results of interpretation, as they were obtained throughout the book. They have been already described, but their significance may perhaps be better appreciated now, when they are disentangled from their technicalities. The overall structure of the theory also becomes clearer when it is reduced to its essential lineaments.

1. The Basic Principles

Quantum mechanics relies upon a few basic principles defining the mathematical framework of the theory and the laws of dynamics.

The Mathematical Framework

The mathematical framework, in which both dynamics and logic must take place, is given by the following rule:

Rule 1. *The theory of an individual isolated physical system is entirely formulated in terms of a specific Hilbert space and a specific algebra of operators, together with the mathematical notions associated with them.*

These associated notions include vectors, bases, scalar products, and norm, as well as operators together with the associated notions of adjointness, unitarity, spectrum, eigenvalues, traces, and various kinds of norms and topologies. One specifies from the beginning a family of self-adjoint operators, closely associated with the description of the system. It includes first of all the operators associated with the momenta of particles and also the operators representing their positions, at least in the nonrelativistic case. At a higher level of the theory, these basic operators are replaced by quantized fields, which are much more difficult to interpret. A slighty different version of the theory was initiated by Dirac, where the observable quantities are primary and the Hilbert space framework is a consequence of them. It can be expressed in terms of C^*-algebras but the difference between the two approaches is essentially irrelevant for the needs of interpretation.

What is most important about the first rule is that it assumes that everything that might be said about the physical system should take place in its mathematical framework. This includes in particular the understanding of empirical properties and the whole of interpretation.

Dynamics

In nonrelativistic quantum dynamics, time is a continuous real parameter. It differs from all the other physical quantities, which are associated with operators. The dynamics of a physical system is given by

Rule 2. *The vectors ψ in the Hilbert space evolve in time according to the Schrödinger equation*

$$i\hbar \frac{\partial \psi}{\partial t} = H\psi,$$

where H is a self-adjoint operator, which is called the hamiltonian of the system.

The spectrum of this operator is assumed to be bounded from below, i.e., with no arbitrarily large negative eigenvalues. The operator H is the dynamical variable representing energy. The unitary operator

$$U(t) = \exp(-iHt/\hbar)$$

is called the evolution operator.

Another rule makes more precise the notion of noninteracting physical systems:

Rule 3. *A physical system S can be said to consist of two noninteracting systems S' and S'' when the Hilbert space of S is the tensor product of* the two Hilbert spaces associated with S' and S'' and its hamiltonian H is the sum of their hamiltonians H' and H'' or, more precisely,

$$H = H' \otimes I'' + I' \otimes H''.$$

These general rules must be completed in every specific case by specifying the Hilbert space, the basic operators or dynamical variables, and the hamiltonian. This structural information is the outcome of a research in particle physics covering three quarters of a century and not yet completely achieved. Some of its results might be stated as so many further principles and the interpretation may also sometimes rely upon them, as far as one needs to know what is the physical system to be described, i.e., what is to be interpreted.

The main outcome of this research is probably that every physical system is a collection of particles. They are best described by quantum field theory, which in particular takes into account relativistic invariance and the creation or annihilation of particles. One also knows how to go from a quantum field formulation to the description of nonrelativistic individual particles. When this is done, interpretation can take two different aspects. The first one gives a precise content to the properties of the fields and the particles in their embedding spacetime. The second one is nearer to what is empirically seen. It consists mainly in giving a theory of phenomena, i.e., the classically behaving properties of most macroscopic objects, including the instruments that are used in experimental physics. This is the part of interpretation that has made a significant progress recently and the only one that is discussed in this book.

2. Properties

The Criteria of an Interpretation

An *interpretation* consists in expressing the realistic and practical aspects of physics (namely, the physical objects, their behavior, and their empirical use) directly in terms of the basic concepts of the theory.

An interpretation is said to be *consistent* when it can be shown to be free from any internal self-contradiction. It is *complete* when it provides specific predictions (i.e., predictions about experimental data) for every possible experimental situation.

Properties

Physical quantities are called *observables*, even though not all of them can be observed. They are the primary physical notions in which one can be interested, the ones that can be represented by numbers. In quantum mechanics, an observable is defined as being associated in a one-to-one way with a self-adjoint operator in Hilbert space. The whole language of quantum physics can be expressed in terms of observables and their values at various times.

This can be made precise by the notion of *property*. A property asserts that the value of a given observable A belongs to a given set of real numbers D and it is therefore something very similar to the result of a measurement. No measurement needs however to be mentioned, at least to begin with, because this would assume that one understands clearly what a measuring instrument is and what it can measure. Despite the evidence of experimental physics, this assumption cannot be granted from the start because an instrument is a special physical system and, as such, it is submitted to quantum mechanics. It is moreover in most cases a very complicated system and some interpretation is already needed to understand what it is and how it works. The properties are accordingly considered as only formal assumptions during the construction of interpretation. Every property is associated with a well-defined projector in Hilbert space, to be given by

$$E = \sum_{a \in D} \left\{ \sum_r |a; r\rangle\langle a; r| \right\}.$$

When two properties are associated with the same projector E, they differ only trivially by a simple calculation: when, for instance, one of them states that the value of an observable A is in the domain $[-1, +1]$ and the other one states that the value of A^2 is in the domain $[0, 1]$. In that sense, a property is more properly defined by the corresponding projector than by its expression in words.

Probabilities and States

Probabilities do not enter the theory as something one can measure by a series of experiments but rather as what mathematicians call a measure, i.e., a priori numbers one can associate with hypothetical properties. This means that probabilities appear first of all as an essential ingredient of the language of physics, long before they can be associated with any kind of measurement. Probabilities are intimately related with the *state* of a system, which is defined as a datum from which one can derive a definite probability for every conceivable property.

The conditions to be imposed upon the probabilities of properties are of two kinds: (i) the ones respecting the meaning of properties, which are intimately related to logic, and (ii) the ones every probability should obey. The first conditions assert that the probability depends only upon the property to be considered and nothing else. In practice, it depends only upon the corresponding projector. The second class of conditions require first two rather trivial requirements—probabilities are nonnegative and they are normalized. The normalization condition means that a trivial property associated with the identity operator must have a unit probability. The most significant condition is additivity, which is also strongly linked to the logical meaning of the words expressing a property. It says that if two properties assert respectively that the same observable A is in a set D' or a set D'' and if these intervals are disjoint, then the probability $p(D' \cup D'')$ is the sum of $p(D')$ and $p(D'')$.

A key theorem due to Gleason expresses the powerful consequences of these straightforward assumptions. It states that there must exist a positive *density operator* ρ with unit trace, such that the probability of a property associated with a projector E cannot be anything other than

$$p = \text{Tr}\{\rho E\}.$$

This expression coincides with the probability rule first assumed by Max Born for the position of a particle in a given region of space, when the state of the particle can be described by a wave function. It does not appear presently as an axiom but rather as a direct consequence of the basic rules.

Time-Dependent Properties

By using the Schrödinger equation, one can choose the time at which a property holds and the time at which the state is given as being different. This corresponds to the intuitive notions of measurement (of a property) and preparation (of a state), to become clearer later on. One usually considers a time-dependent property stating that "the value of an observable A is in a domain D at a time t" and it is associated with the Heisenberg time-dependent projector

$$E(t) = U^{-1}(t) E U(t),$$

its probability being given by

$$p = \text{Tr}\{\rho E(t)\}. \tag{12.1}$$

3. The Logical Framework

Although the notion of property goes back to von Neumann (under the name of "predicate"), the discovery of a logical framework permitting the interpretation of quantum mechanics while respecting all the conventional rules of logic is only recent. Logic plays a central role in a modern interpretation, where it works as a guideline and an axiomatic foundation.

Logic

Logic is the science of reasoning and consistency. Consistency relies upon a few universal formal rules going back to Aristotle and the stoics. Logic should be moreover always applied in practice to a specific domain of knowledge and one then simply speaks of "a" logic when it is restricted in that way in its field of application. More specifically, the complete definition of a logic requires the following ingredients:

1. A field of propositions.
2. Three logical operators (not, and, or), allowing the combination of the propositions according to the rules of formal logic. One should also be given two relations among the propositions, the first one expressing a logical equivalence ($=$) and the second one a logical implication ($\Rightarrow$), which can be expressed by the locution "if ..., then ...". One has a *consistent logic* when all the axioms of formal logic are satisfied by these three operators and these two relations.
3. One also needs a criterion of truth for deciding when a proposition is true. It is particularly tricky in the case of quantum mechanics and it cannot be properly stated before obtaining a complete interpretation.

The most common logic that is used in physics is classical. It can be formalized in such a way that its basic propositions refer to the coordinates (q, p) of a point in the phase space of a classical system and they express that "(q, p) is in a domain D of phase space at a time t". The operators "not, and, or" are then associated with the corresponding operations in set theory (complement of a set, intersection, and union of two sets). Logical equivalence amounts to the identity of two sets in phase space and implication can be expressed as the inclusion of a cell in another one. The connection between two properties holding at different times is derived from the known classical motion: *determinism* is for instance a logical equivalence between a property associated with a domain D at time t and another one associated with a domain D' at a time t', when D' is the image of D by classical motion during the time $t' - t$.

Quantum Logics

Whereas classical physics rests upon a unique logic, quantum physics is more complicated, noncommutativity giving rise to complementarity, i.e., the existence of many different logics describing the same system. A quantum logic can be used to describe a physical system and to draw valid implications among its various properties. The field of propositions of a quantum

SUMMARY AND OUTLOOK

logic rests in an essential way upon the notion of Griffiths histories, which goes as follows in the simplest case:

1. One assumes an initial state for the system to be considered. The state is given at time 0 by a density operator ρ. The system must be isolated, at least as long as one is discussing it.
2. One considers an ordered sequence of times such that $0 \leq t_1 < t_2 < \cdots < t_n$. A *history* is then a sequence of properties holding at these times and it may be associated with a family of time-dependent projectors $\{E_k(t_k)\}$. One can be more precise by choosing a sequence of observables $A_1, A_2, \ldots, A_n$ and a corresponding sequence of real sets $D_1, D_2, \ldots, D_n$. No special commutation properties are assumed among the observables but several commuting observables can be used together at the same time, as, for instance, the various coordinates of position. The history states several properties according to which "the value of A_k is in the set D_k at time t_k," for $k = 1, 2, \ldots, n$. It is sometimes conveniently compared with a motion picture of the system while a unique property would be something like a snapshot.

A quantum logic envisions a family of mutually exclusive histories as so many conceivable events. This family can be obtained for instance by dividing the spectrum of each observable A_k into a complete collection of disjoint sets $\{D_k^r\}$ and the basic histories represent the possible occurrence of all these properties at the various reference times. Every history is therefore considered as a special proposition referring to n times together. It is often convenient to represent it graphically by a "rectangular box" in an n-dimensional space where the values of each observable are used for the coordinates. The field of propositions of the quantum logic is obtained by combining the histories with the logical operator "or", which can be done by considering unions of several rectangular boxes. These domains can then be combined by taking complement, intersection, or union, which completely define the three logical operators "not, and, or" in the field of propositions.

The Probabilities of Histories

One can associate a definite probability with every proposition belonging to the field of a special type of quantum logic. The conditions to be imposed upon these probabilities are again some logical conditions purporting to preserve the intuitive meaning of the underlying language, and the general axioms of probability calculus. The logical conditions are the following:

1. The probability of a history depends only upon the properties entering in it, i.e., upon the projectors $\{E_k(t_k)\}$.

2. A one-time history is an ordinary property and its probability is accordingly given by equation (12.1).
3. A history is said to contain a tautology when the property occurring at a time $t_{k+1}(k \geq 1)$ only repeats the property already occurring at the previous time t_k; i.e., when the projector $E_{k+1}(t_{k+1})$ is identical to $E_k(t_k)$. The probability is then assumed to be the same as for the case of a history with $(n - 1)$ reference times where the property at time t_{k+1} has been deleted.
4. There is also a tautology when the first property is trivial in view of the initial state so that $E_1(t_1)\rho$ does not differ from ρ. One can then again delete the first property to get an $(n - 1)$-times history.
5. A history is said to be nonsense when the property at some time t_{k+1} negates the one occurring at the immediately previous time t_k, i.e., when the two projectors $E_{k+1}(t_{k+1})$ and $E_k(t_k)$ commute and their product is 0. The probability is then assumed to be 0.

One can then define the probability of a general proposition as the sum of the probabilities for the alternative histories entering in it. The basic axioms of probability theory are in the present case:

6. Probabilities are nonnegative.
7. The probability of the trivial history (in which all the projectors coincide with the identity operator) is equal to 1.
8. Probabilities are additive. This means that when two propositions a and b are such that the proposition "a and b" is empty, the probability $p(a \text{ or } b)$ is the sum of $p(a)$ and $p(b)$.

These assumptions determine uniquely the probabilities of histories. This has been proved in full generality for $n = 2$ but it remains a conjecture for larger values of n, except when all the observables are either positions or momenta, in which case the result follows from the use of Feynman path integrals. This probability for a history is given in terms of the corresponding projectors by

$$p = \text{Tr}\{E_n(t_n)\ldots E_2(t_2)E_1(t_1)\rho E_1(t_1)E_2(t_2)\ldots E_n(t_n)\}. \qquad (12.2)$$

Consistency Conditions

It turns out that the additivity conditions are not all necessarily satisfied by the expression (12.2) and some necessary and sufficient *consistency conditions* must be satisfied to insure complete additivity. They are very important because they restrict the kind of propositions that can enter in quantum mechanics if the essential rules of logic are to be preserved.

SUMMARY AND OUTLOOK

The consistency conditions have been discussed in Chapter 4 and it will be enough to recall presently only the simple case when $n = 2$ and when the spectra of the two relevant observables A_1 and A_2 are both simply divided into two complementary sets. There is then only one consistency condition, which is

$$\operatorname{Re} \operatorname{Tr}\{E_1(t_1)\rho \overline{E}(t_1) E_2(t_2)\} = 0.$$

The main point is that these conditions can be checked by a straightforward computation.

Consistent Logics

When the field of propositions of a quantum logic satisfies the corresponding consistency conditions, it is possible to explicitly define in it the relations of logical equivalence and implication. A proposition a is said to imply a proposition b when the conditional probability $p(b|a)$ is 1, i.e.,

$$p(b|a) = \frac{p(a \text{ and } b)}{p(a)} = 1.$$

The two propositions (a, b) are considered as logically equivalent when $a \Rightarrow b$ together with $b \Rightarrow a$. The soundness of this construction is shown by the fact that a consistent quantum logic satisfies all the conventional axioms of ordinary formal logic.

It is also often very convenient to use approximate logics where the consistency conditions are satisfied only up to a small relative error and the conditions of implication can also be approximate. If for instance ε is a very small number, one can say that the proposition a implies the proposition b with a probability of error ε if one has $p(b|a) \geq 1 - \varepsilon$. This notion of approximation in logic is essential because it provides the necessary amount of freedom to reconcile quantum and classical logic or, more generally, quantum and classical physics, in spite of the strong differences between the two theories.

One can sometimes relate different logics by a few convenient changes:

1. A given logic can often be extended in many different ways: by using smaller intervals for the values of the observables so that the corresponding properties are refined; by adding more intermediate times so that the histories become more precise; by adding a time later than t_n, which extends the description towards the future. These procedures are of course only valid when the necessary consistency conditions generated by them are satisfied.

2. A logic can also be simplified by taking a coarser graining for the values of the observables or by suppressing altogether a reference time.
3. Two logics describing two noninteracting systems can be joined together into a single logic describing the overall system. The only condition is that the logical directions of time be the same for both systems.

Many different consistent logics can be used to describe one and the same physical system. This multiplicity is the essence of *complementarity*. Given two different consistent logics L and L', two cases are possible: (i) It may happen that there exists a larger consistent logic L'' containing both of them, in which case they are consistent with each other; and (ii) There is no such larger logic, in which case the two logics are said to be complementary, or foreign to each other. Finally, a simple *no-contradiction theorem* asserts the impossibility of self-contradiction. An implication $a \Rightarrow b$ is valid whatever the consistent logic containing both propositions a and b if the implication holds in one of them.

The Universal Rule of Interpretation

It is possible to lay the foundation of the whole interpretation of quantum mechanics upon a unique rule of a logical character:

Rule 4. *Every description of a physical system should be expressed in terms of properties belonging to a common consistent logic. A valid reasoning relating these properties should consist of implications holding in that logic.*

The power of this rule, which contains in germ most of classical physics and measurement theory, is due to the versatility of the concept of histories and of the embedding logics. As a matter of fact, histories are able to express everything belonging to the domain of empirical physics as well as everything one may assume about it or deduce from it. The importance of this rule in the present approach cannot be overstated. Whereas the previous rules, which have been used with a tremendous success for many years, can be said to tell us how to apply correctly quantum mechanics, the present one can show us how to *understand* it.

4. The Foundations of Classical Physics

Bohr had assumed the simultaneous existence of two different kinds of physics, a classical one and a quantum one, their only relation being a limiting correspondence principle. Contrary to this, the classical behavior of

SUMMARY AND OUTLOOK

the majority of macroscopic objects can now be considered as a direct consequence of quantum mechanics. The justification of this approach needs more than the usual derivation of Hamilton's equations from quantum dynamics by means of Ehrenfest's theorem. This is because some classical concepts are more a matter of logic than simply dynamics, as, for instance, when one considers determinism or what kind of certainty can be attributed to classical properties.

The understanding of classical physics shows several new features and a few notions that were often considered to be identical but which now must be carefully distinguished. One must distinguish between a macroscopic system (consisting of many particles) and a classical one (obeying classical physics, even if only approximately). One must also distinguish between a physical system (completely defined by its constituent particles and their interactions) and an object (corresponding to relatively few states of its constituent particles). A system made of several objects is analogous to a classical system, which ignores the atomic structure of matter. Finally, one must also distinguish between phenomena and facts. Phenomena are classically meaningful properties of a system of objects and they are theoretical and potential (i.e., hypothetical—they can be considered as part of an assumption). On the contrary, facts are actual and they are distinguished from phenomena by their reality and their uniqueness. The first step in the construction of classical physics concentrates upon the properties of a macroscopic object that can be considered just as well to be classical or quantum properties. In other words, it aims at expressing the classical logic of common sense as a special case of a quantum logic.

The Description of an Object

A macroscopic system in a state representing a specific object is best described by using *collective observables*. These are essentially the observables associated with the dynamical variables that are used in classical mechanics. There are also many other microscopic observables describing the matter of the object. A set of collective and microscopic coordinates constitutes a complete set of commuting observables for the whole object. It is often convenient to consider them as representing formally two different dynamical systems, the collective subsystem and the *environment*, which is parametrized by the microscopic coordinates and represents the matter inside the object and outside it. These two dynamical systems are coupled and their coupling is responsible for their energy exchanges, i.e., dissipation effects. It should be stressed that one does not know a general method for constructing the collective variables from the first principles and one can only conjecture the criteria for defining unambiguously what a collective observable is and to what extent it is collective. This unsolved problem remains one of the most important loopholes in the construction of a completely satisfactory interpretation.

The Principle of Correspondence

The relation between classical and quantum physics was most often understood as a *correspondence*. The classical dynamical behavior appeared as an approximation of quantum dynamics under conditions when the value of Planck's constant is relatively small. In the present interpretation, the correspondence is extended to the logical description of classical and quantum physics. A classical property is nothing but a special quantum property and it must be associated with a quantum projector according to the universal rule of interpretation or else it has no meaning.

Classical Dynamical Variables

Using collective and microscopic observables, one can always write the hamiltonian of a collection of macroscopic objects as a sum

$$H = H_C + H_E + H_{\text{int}}, \tag{12.3}$$

where the first term depends only upon the collective observables, the second one upon the microscopic observables, and the last term represents their coupling. This coupling is responsible for the energy exchanges between the two subsystems, i.e., dissipation and friction. The collective Hilbert space can be used to formulate classical physics in a quantum framework. It is a space of square-integrable functions defined upon a classical configuration space, which is parametrized by the values of the collective coordinates. The correspondence with classical physics holds in principle for every configuration space but it is simpler when the configuration space is euclidean, which is the only case to be considered presently.

The collective coordinate observables may be denoted by X_k ($k = 1, 2, \ldots, n$; n being the number of collective degrees of freedom). This is a complete set of commuting observables in the collective Hilbert space and the canonically conjugate momenta are denoted by P_k. The values of these observables are denoted by x_k and p_k, with the simplified notation $x = \{x_k\}$ for the set of all coordinates and similarly $p = \{p_k\}$. Given an arbitrary collective observable A, i.e., a self-adjoint operator in the collective Hilbert space, it can be associated with a *classical dynamical variable*, i.e., a function $a(x, p)$ of the position and momenta. This correspondence is unique under simple and general conditions that were given in Chapter 5 and it is expressed by the Wigner–Weyl formula:

$$a(x, p) = \int \langle x'|A|x''\rangle \delta\left(x - \frac{x' + x''}{2}\right) \exp\{ip \cdot (x'' - x')/\hbar\} \frac{dx\,dp}{h^n}.$$

The classical elementary variables x_k and p_k are then associated respectively with the observables X_k and P_k. This formula can be inverted by a Fourier transform to give the matrix elements of the observable in the coordinate representation from a knowledge of $a(x, p)$ and therefore the operator itself from its matrix elements. There are however some restrictions upon the observables for the formalism to be well under control and, in particular, the classical dynamical variables must be smooth functions. The *Hamilton function* is the classical dynamical variable $h(x, p)$ associated with the collective hamiltonian H_C. One can use it to write down the classical Hamilton equations of motion and this defines the classical motion in phase space.

Classical Properties

A classical property asserts that the classical collective variables (x, p) represent a point belonging to a given cell C in phase space. The simplest ones consist in giving the position coordinates and momenta within some prescribed error bounds, in which case the cell C is simply a rectangular box. A classical property allowing a correspondence with a quantum property must satisfy the conditions that the cell C is big enough (its volume being large when measured in units of Planck's constant) and its boundary is smooth enough. The condition of smoothness as well as an estimate of the errors involved in the correspondence rely upon simple geometric notions. Definite reference scales (L, P) are used for measuring the coordinates and momenta. They are representative of the dimensions of the cell, at least when its shape is simple enough, and they have been defined in general in Chapter 6. One also defines a *classicality parameter* ε for the cell, which is given by

$$\varepsilon = \left(\frac{\hbar}{LP}\right)^{1/2}.$$

One only considers the cells for which the classicality parameter is a small number. It is also convenient to introduce a metric in phase space, so that the distance between neighboring points with coordinates differing by (dx, dp) is defined by

$$ds^2 = \frac{dx^2}{L^2} + \frac{dp^2}{P^2}.$$

This metric can be used to define the area of a cell boundary and an *effective classicality parameter* η, which is the product of ε by the ratio between the area of the cell boundary and its volume in nondimensional variables x/L and p/P. The bulkier a cell is, the smaller this parameter. It may also become of the order of 1 or larger when the boundary is very distorted. A *regular* cell C is finally defined as a domain in one piece with no hole, big enough for ε to be small and bulky enough for η also to be small.

Quasi-Projectors

Because of the errors occurring in the correspondence between classical and quantum properties, a classical property cannot be associated with a unique projector in the collective Hilbert space but only to a class of operators sharing many properties in common with projectors. They are called *quasi-projectors*. Given a regular cell C, a quasi-projector can first be defined roughly as a self-adjoint operator F having only discrete eigenvalues, all of which in the interval $[0, 1]$ and most of them being close to either 1 or 0. The total number of eigenvalues close to 1 (i.e., the trace of F) is essentially the number $N(C)$ of quasi-classical states in C, which is the quantity

$$N(C) = \int_C dq\, dp/h^n.$$

The fact that it is almost a projector (i.e., almost all its eigenvalues are concentrated in the neighborhood of 1 or 0) is expressed by the relation

$$\mathrm{Tr}\{F - F^2\} = N(C) \cdot O(\eta).$$

The relation between the cell and a quasi-projector is insured by the fact that the classical dynamical variable $f(x, p)$ associated with this operator is the characteristic function of the cell C (equal to 1 in C and to 0 outside), except for a regularization consisting in rounding off the discontinuity on the boundary. The regularization introduces a slight freedom in the definition of the operator. Two quasi-projectors F and F' are said to be equivalent when

$$Tr|F - F'| = N(C) \cdot O(\eta).$$

Notice the occurrence of the so-called trace norm, which is the best measure for the proximity of two projectors where logic is concerned.

The outcome of this construction is that one can associate a quasi-projector (or rather a class of equivalent quasi-projectors) with a regular cell. Conversely, a quasi-projector can be used like a projector so as to

SUMMARY AND OUTLOOK

express the classical property as a quantum property. This correspondence is well defined, up to an error of order η. One could also say that, when a classical property is associated with a regular cell, it is just as meaningful when understood as a quantum property. One is then perfectly allowed to recognize the existence of classical physics while remaining in the framework of the universal rule of interpretation.

The Quantum-Classical Correspondence and Dynamics

The quantum-classical correspondence is preserved by dynamics, at least for a regular cell and during a finite time, which can be more or less long and is very long for most cases of practical interest. More precisely, classical dynamics relies upon the following result. Let C_0 be a regular initial cell and F_0 a quasi-projector associated with it. Let C_t be the cell generated from C_0 by the classical motion during a time t and F_t be a quasi-projector associated with it. The quasi-projector is well defined, up to equivalence, as long as the transformed cell C_t remains regular under classical motion. One must also assume that the intermediate cells $C_{t'} (0 \leq t' \leq t)$, through which the motion goes between time 0 and time t, are themselves regular. All this can be made quantitative by introducing a *dynamical classicality parameter* $\zeta(t)$, which is essentially the upper bound of the effective classicality parameters $\eta(t')$ for all the intermediate cells. The dynamics is said to be regular for the cell C_0 and the time t if $\zeta(t)$ is much smaller than 1.

The dynamical correspondence is expressed by the relation

$$\mathrm{Tr}|F_0 - U^{-1}(t)F_t U(t)| = N(C_0) \cdot O(\zeta(t)),$$

which is a generalization of Ehrenfest's theorem with explicit errors. It shows that the time evolution of the projectors under the effect of the Schrödinger dynamics agrees with the classical motion of the corresponding cells, at least as long as regularity is preserved. Every dynamical evolution is regular for a short enough time. The time of regularity is often very long for usual objects and particularly for the kind of instruments that are used in a laboratory, as well as most technical and natural systems. It is infinite for the preferred guinea pig of physics, the harmonic oscillator. In the case of a chaotic system, it is strictly finite and one has

$$t < \tau \log(LP/\hbar),$$

where τ is a (Liapunov) time characterizing the exponential rate of the growing distance between two neighboring classical trajectories in a chaotic motion.

5. Classical Logic

One then arrives at a turning point of the whole program of interpretation, which is to prove common sense, or at least to find in what case it can be used confidently in spite of the quantum character of the laws of nature. As far as physics is concerned, common sense consists in using ordinary logic (as it is inherent in ordinary language) and applying it to reality with the help of some knowledge of classical physics or even just some general feeling of it. It most frequently relies on the everyday representation of physical reality originating in our familiarity with ordinary empirical facts.

The validity of common sense can be proved by using as a foundation the rules of quantum mechanics. This is not a game of tricks since one also finds when common sense should not be applied, either because of its irrelevance to the situation or because of a logical inconsistency. It looks therefore as if one might consider common sense as originating from something deeper, something only quantum mechanics can disclose. This new approach to the meaning of physics turns the tables in such a way that many old problems are substantially changed, including many questions that used to be a plague for interpretation. Quantum mechanics was regarded for a long time as very strange, particularly because of its conflicts with common sense. Many of its criticisms or refutations were due to this utter strangeness. One can now proceed the other way round, by justifying (and simultaneously limiting) common sense as founded upon a firmer ground. This might be a very significant step, which would have its most important consequences in the domains of epistemology and philosophy.

The Validity of Classical Logic

Classical logic is essentially a learned version of common sense. For a system consisting only of macroscopic objects, it can be proved to be a consequence of quantum logic under the following conditions:

1. One only takes care of the collective degrees of freedom of the objects (i.e., the ones retained by classical physics).

2. Their coupling with the microscopic degrees of freedom (and therefore dissipation) is neglected in a first step.

3. Dynamics is assumed to be regular for an initial cell C_0 during a time t, with a dynamical classicality parameter $\zeta(t)$ much smaller than 1.

4. The properties to be mentioned in the propositions are associated with the cells $C_{t'}(0 \leq t' \leq t)$, which are obtained from C_0 by classical motion at a few discrete times.

5. The initial state ρ is assumed to be a *state of fact* compatible with the classical property represented by the cell C_0. This means more precisely that

SUMMARY AND OUTLOOK

it satisfies the following relation in terms of a quasi-projector F_0 associated with C_0:

$$\text{Tr}\,|F_0 \rho F_0 - \rho| = O(\zeta(t)).$$

Condition 4 defines a field of propositions for a quantum logic and the classical properties are expressed in quantum terms by using quasi-projectors. Classical logic under these conditions becomes a special kind of quantum logic and one can use the universal rule of interpretation to make sure of its validity. This logic is found to be consistent, up to errors of the order of $\zeta(t)$ and, by working out its implications, one finds that the usual reasonings of classical logic can be justified. Their statements are of course subject to some unavoidable errors, which means that their conclusions could happen sometimes to be empirically wrong because of the rare occurrence of large quantum fluctuations. The probability for such an error is at most of the order of $\zeta(t)$.

Among the implications holding in this logic, the most important ones concern *determinism*. It means that the classical property stating that "(q, p) is in the cell C_0 at time 0" is logically equivalent to the classical property according to which "(q, p) is in the cell C_t at time t." This well-known logical equivalence now also holds in the framework of quantum logic and it is valid up to an error in probability that is of the order of at most $\zeta(t)$. Determinism is again basically a logical equivalence between classically meaningful properties occurring at different times. One can then say that the observation of determinism at a large scale is in no way contradictory with quantum mechanics. One can predict on the contrary when determinism holds and prove it, while getting an estimate for the (most often very small) probability of its violation by large quantum fluctuations. These conclusions follow from assuming the universal validity of quantum mechanics.

The drastic opposition that was believed to exist between classical determinism and quantum probabilism was therefore in fact mainly an illusion. The two are perfectly consistent with each other, except for a slight change of attitude, which is not a change in physics. One should be prepared to accept that determinism can be violated in a random way with a very small (and computable) probability and also that it can be asserted only for sufficiently regular, i.e., sufficiently large-scale situations. It might be that a car can jump from one parking lot to another by a quantum fluctuation. Who cares? The probability is so small that it will never be seen. Conversely, I cannot claim that the position of my car is defined up to a billionth of an Angström when it goes at sixty miles per hour. This is the price quantum mechanics is asking when it recovers determinism and it is quite a cheap one when considering that common sense, classical predictions, and a renewed confidence in the validity of memory are offered as a bargain.

Finally, one recovers classical physics and practically all its certainty. This means that one can consider a fact as given and its record as faithful. One can also describe the surrounding reality by ordinary language, while considering that it corresponds faithfully to more basic notions expressed by the mathematics of Hilbert space theory. The main risk of inconsistency that could arise from the older Copenhagen interpretation has therefore been conquered and one no longer relies on two different kinds of physical laws, but only on one kind, the quantum ones. One can nevertheless express the result of an experiment (a fact) in classical terms, as Bohr would have it. There is no risk of self-contradiction and many interdictions that were pronounced by the Copenhagen interpretation to avoid the main pitfalls become unnecessary, since one knows quantitatively when classical physics applies and when it does not.

6. Decoherence

The Decoherence Effect

Decoherence is probably one of the most efficient effects in physics. It is a consequence of the coupling of a system (e.g., the collective subsystem in the case of a macroscopic object) with an environment involving a very large number of degrees of freedom. From the standpoint of mathematics, the effect is a tendency of two vector states representing the environment to become orthogonal after a finite time, when they are coupled to two different collective states. Intuitively, it is a consequence of the very high degree of infiniteness of the environment Hilbert space, as due to the large number N of microscopic degrees of freedom (which is reflected by a density of the energy levels increasing exponentially with a power of N, in the case of continuous degrees of freedom).

There are good reasons to believe that it is a universal effect, at least when the coupling with the environment is not too small. The presently available theories are however more limited in their scope because of the difficulty of a complete mathematical treatment of the problem. It is very difficult to control destructive interferences among wave functions of the environment depending upon a very large number of variables.

The Main Manifestation of Decoherence

The practical manifestation of the effect is best seen in the collective or reduced density operator, which is the density operator describing precisely the collective properties of an object. It is obtained from the complete density operator by performing a partial trace over the environment:

$$\rho_c = \text{Tr}_E \rho.$$

SUMMARY AND OUTLOOK

Decoherence produces a spontaneous diagonalization of this operator. It tends to become diagonal, the nondiagonal matrix elements vanishing exponentially with time. As a matter of fact, the effect is in most cases so rapid that one cannot catch it while it is at work and one only sees its result, which is a diagonal reduced density operator. Such a density operator is a mixed state where all remnants of quantum interference effects (between macroscopically different states) have disappeared. This is precisely the kind of density operator allowing the description of collective properties by probability calculus as so many distinct events. The decoherence effect therefore provides the long-sought disappearance of quantum superpositions among macroscopically different states.

Although there are not yet completely general theorems about this very important result, there is a fair evidence from many models that it proceeds as follows.

1. In the formal case of discrete collective degrees of freedom, the collective Hilbert space basis in which the diagonalization takes place is essentially unique and it is defined by the coupling with the environment. In the case of continuous degrees of freedom, diagonalization takes place in general in the coordinate basis, but it is soon followed by diagonalization in the momentum basis. One may conjecture that it occurs in any basis where different classical properties are defined by a set of quasi-projectors.

2. In the coordinates basis where $\rho_c(x, x'; t) = \langle x | \rho_c(t) | x' \rangle$, the nondiagonal matrix elements ($x \neq x'$) decrease exponentially with time and the nonvanishing diagonal matrix elements are concentrated around values of the position x satisfying a classical equation of motion with friction forces in the semiclassical limit. This association between decoherence and dissipation comes from the fact that both are due primarily to the coupling between the collective subsystem and the environment, the decoherence effect being however much more rapid than the dissipation effect. In an example where the collective subsystem is a harmonic oscillator interacting with an environment at very low temperature, one finds that

$$\rho_c(x, x'; t) \propto \exp\left\{ -\frac{1}{4} \frac{m\omega(x-x')^2}{\hbar} \frac{t}{\tau} \right\}, \qquad (12.4)$$

whereas the dominant matrix elements concentrate around solutions of the equation of motion

$$\frac{d^2 x}{dt^2} + \gamma \frac{dx}{dt} + \omega^2 x = 0. \qquad (12.5)$$

3. The decoherence time τ occurring in equation (12.4) is the inverse γ^{-1} of the friction coefficient. It occurs in the expression of the friction force in equation (12.5) and its exact value comes from the detailed form of the

coupling hamiltonian H_{int} in equation (12.3). This connection between the two effects is quite general so that the only cases where decoherence may not occur are the few ones where dissipation is negligible (superconductors, superfluids, and radiation providing the main examples).

4. The "environment" plays its role only because of its large number of degrees of freedom and of its interaction with the collective subsystem. It may be internal (in which case it stands for the matter in an object made of many atoms) or external (the air molecules or the photons of light surrounding an object). Both kinds of environments contribute to decoherence in a very efficient way, almost always in the case of macroscopic objects and even rather frequently in the case of microscopic ones.

7. Can One Circumvent Decoherence?

Not all the aspects of decoherence are yet completely understood, particularly because of the crude mathematical methods that are presently available. It remains to compare it carefully with semiclassical physics (if only to include the effect of dissipation in the statement of determinism) and with classical logic (to find what classically meaningful properties become clearly separated, as distinct events, by spontaneous diagonalization). Its consequences for the understanding of some long-disputed questions are also still partially unclear, though they are obviously important.

Among these questions is the status of an axiom of quantum mechanics that was put forward long ago by von Neumann. It stated that every observable can be measured, at least in principle. If this axiom were true, it would imply that every state vector can be prepared, at least in principle. Its validity looks however rather doubtful when one considers the huge family of self-adjoint operators in a Hilbert space or specific examples.

This is of course known not to be a necessary axiom of quantum mechanics and the question can only be whether or not it holds as a general result of the theory. This decision is very important for the following reason. Even when the reduced density operator has become diagonal, the full density operator still represents a quantum superposition if there was one initially (this being a consequence of the linear Schrödinger equation). One can therefore always conceive of an observable that would show the existence of this superposition if it could be measured. The possibility of this measurement would then imply that decoherence is only a practical answer to the basic problem of quantum superpositions, not a fundamental one.

One must of course make clear on what assumptions the validity of the "axiom" will be decided. Von Neumann was considering some sort of "pure" quantum mechanics, i.e., a somewhat formal frame essentially disjoint from

SUMMARY AND OUTLOOK

the rest of physics. At this level of generality, it was not unreasonable to assume that all kinds of hamiltonians and measurements are possible, at least "in principle". A different answer results when one takes into account other laws of physics, including the fact that matter is made of particles with a well-defined basic hamiltonian, together with the validity of special relativity and, accessorially, of general relativity.

An argument was given in Chapter 7 to show that von Neumann's assumption is not consistent with the rest of physics. It concerned a measurement intending to disprove decoherence, where an apparatus M initially in a pure state involving a superposition of macroscopically different states becomes decoherent as far as its collective degrees of freedom are concerned. Another apparatus M' measures a cleverly chosen observable A belonging to M and this is repeated many times exactly under the same conditions. The resulting statistics for the values of the observable A are then supposed to show the existence of interference effects.

The argument was based on the idea that the measurement by M' can only be useful if the probability p for getting a significant result (in a yes-no measurement for a value of A) is larger than the probability ε that a large quantum fluctuation in M' will produce the same final phenomenon as one would get from a positive measurement. It therefore relied upon decoherence theory and the known estimates for the errors limiting the validity of classical physics. It cannot yet be considered as an indisputable proof, but its result is anyway quite suggestive. Leaving aside irrelevant constants, one finds that the total number of (microscopic and continuous) degrees of freedom N and N' for M and M' must satisfy a condition such as $N' > \exp(N^{2/3})$, i.e., N' should be of the order of 10 to the power 10^{18} when N is of the order of some 10^{27} (a typical cat...). Remembering that the observed universe does not contain more than some 10^{81} protons, such a measurement is seen to be impossible and, moreover, the working of such a tremendously huge device is incompatible with special relativity and its existence is furthermore incompatible with general relativity (it would collapse immediately into a black hole).

Some Consequences and Perspectives

This result has various other consequences, if taken at its face value:

1. Not all the observables can be measured, but only a few of them. This means that the majority of the properties one could think of, even in the case of a relatively small system, can never be confirmed or disproved by any kind of measurement. They belong to the domain of conversational physics, not of real physics.

2. Decoherence cannot be circumvented once it has occurred, as long as the environment is large enough. It therefore affords a sensible answer to the problem of macroscopic superpositions.
3. Not all the conceivable states can be prepared. In particular, one cannot prepare the time-reversed state of the actual state of a macroscopic assembly of particles when some evolution has already occurred. This remark opens an interesting connection between decoherence and the foundations of irreversible thermodynamics.
4. The directions of time in logic, in decoherence and in thermodynamics should be the same.

8. The Theory of Phenomena

The outcome of the investigation of the behavior of macroscopic objects is a theory of phenomena, i.e., the classical properties of these objects. Its main point is that the existence of phenomena need not be added to the basic axioms of the theory and, on the contrary, it follows directly from them. This does not mean of course that the existence of real facts follows from the theory, which would be ridiculous and would amount to identifying reality with a mathematical construction. It means that the theory, when one draws systematically some of its consequences, predicts the main features of real phenomena. This is why one can say that quantum mechanics provides the means of its own interpretation. These phenomena (or classically meaningful properties of macroscopic objects) enjoy the following features:

- As properties, they are simultaneously meaningful in quantum mechanics and in classical physics.
- They satisfy classical logic, i.e., common sense. In particular, they obey classical determinism. An important aspect of determinism is the possibility of explaining the existence of records in the framework of quantum mechanics.
- Common sense and determinism are satisfied however only up to a small probability of error, which can be evaluated in principle in each specific case.
- Different properties are clearly separated events, i.e., there are no ambiguities originating from quantum interferences.
- For a system big enough (even if it is rather small in practice, as compared with our own scale), the classical features can never be shown to be erroneous by any experiment, however subtle and costly.

In a more straightforward way, one might just say that classical physics can be applied in most cases without afterthoughts, except for the occurrence of possible errors originating from large and rare quantum fluctuations, so rare that they have no practical consequence.

SUMMARY AND OUTLOOK

These conclusions hold with full confidence under the following "conditions of classicality": (i) The initial state is a state of fact, (ii) The dynamics of the system is regular for the properties to be considered during the time interval when they are occurring, (iii) There is enough dissipation for decoherence to take place, and (iv) The total number of continuous degrees of freedom is large enough so that decoherence cannot be circumvented.

These conditions are sufficient for a classical behavior. It is not strictly necessary that all of them hold together but, when one of them is violated, a specific investigation must be made. They are satisfied by practically all ordinary objects, except for the condition of regularity (ii), which is violated rather soon in the case of chaotic systems. This case looks at first sight like a serious potential source of trouble, since it might be that most objects are in fact chaotic when the collective variables to be considered are refined (as, for instance, when they include all the vibrations of a solid object). This is only however an apparent difficulty because, after a finite time, the chaotic variables must be described anyway by their statistics, even classically, and it turns out that the classical and the quantum statistical properties are practically identical.

9. Measurement Theory

Measurement theory can proceed in a completely deductive way starting from the previous results.

The Conditions for a Measurement

A measurement is an interaction of a special kind between two physical systems, to be denoted by Q and M. The system Q is to be measured; it can either be microscopic or macroscopic. The system M is the measuring apparatus and is necessarily macroscopic. It can exhibit several different phenomena, clearly separated from each other, which constitute the various possible *data* occurring at the end of the measurement. Their separation, i.e., their clearcut classical distinction as alternative events, requires decoherence and this is why the apparatus should be macroscopic or in contact with a macroscopic environment. Decoherence requires dissipation. It was often noticed that all the devices to detect particles rely upon an irreversible process (they often start from a metastable state) and it was sometimes believed that this was the reason why irreversibility was a necessary physical condition for the existence of a measurement. Though this point of view is not objectionable, one can also say that irreversibility is a necessary condition for the existence of decoherence and therefore for the occurrence of well-defined classical phenomena as so many separate data.

The interaction between Q and M takes place between two times, t and $t + \Delta t$, marking the beginning and the end of the measurement process. It is convenient to assume that, before time t, the two systems did not interact and that the initial state of Q was not correlated with the collective state of M. An observable A associated with Q, the one that is measured, plays a privileged role in the interaction. In order to describe the relevant properties, we consider a family of projectors E_n, belonging to the system Q and relative to a property occurring at time t. Each one of them says that the value of A at the beginning of the interaction is in some real interval (when the spectrum of A has a discrete part, some of these intervals may even become a single eigenvalue). These quantum properties are called the *results* of the measurement.

The distinction between data and results is central in the formulation of measurement theory and one of its main achievements is to clarify completely their relationship. To be a measurement of A, the interaction between Q and M must satisfy a specific condition involving the evolution operator $S = U(t + \Delta t, t)$ for the global system $Q + M$ during the time interval when the two systems interact, until a datum is registered. This "collision" operator must be semidiagonal so as to correlate strongly the initial properties of Q to be measured with the data. Denote by the index 0 the initial neutral state of M, by the index n the various data, and by F_0 and F_n the corresponding quasi-projectors (expressing these classical properties as so many quantum properties). The apparatus M is said to be a *perfect measuring device* if one has

$$S E_n(t) F_0(0) = F_n(t') S E_n(t) F_0(0). \tag{12.6}$$

This equation means that, when the apparatus is initially in its neutral state and it interacts with a state of Q satisfying the quantum property n, the corresponding datum n is finally shown by the apparatus. It should be stressed that the operator S is computed by means of the Schrödinger equation for the system $Q + M$. Equation (12.6) is therefore a purely dynamical condition. The measurement is said to be of Type I (in Pauli's notation) if one has

$$S E_n(t) F_0(0) = F_n(t') E_n(t') S E_n(t) F_0(0). \tag{12.7}$$

This condition says that the interaction does not change the value of A. One may also define a measurement as being strongly of Type I if equation (12.7) holds for every one-dimensional projector E_n.

The Main Theorems of Measurement Theory

The main outcome of measurement theory states that *the data and the results are logically equivalent*. This is of course a theorem in logic. It follows

SUMMARY AND OUTLOOK

from the examination of the relevant consistency conditions and the existence of implications going in both directions between data and results. It is accordingly a direct consequence of the universal rule of interpretation. This logical equivalence justifies a posteriori the lack of distinction between the two notions of data and result in most versions of the Copenhagen interpretation, in spite of their obvious difference. A more directly useful theorem is that the probability distribution for the data depends only upon the properties of the measured system and its initial state but not upon the detailed working of the measuring apparatus. It is given by

$$p_n = \text{Tr}\{\rho_Q E_n(t)\}. \tag{12.8}$$

One can recognize in this the probability rule first given by Born. It should however be stressed that it has now become a prediction of the theory.

One may also consider a great number of copies of Q and M, Q being always initially in the same initial state and M in its neutral state. One is then able to compare the theoretical probabilities with the limit of the frequencies for a large number of trials, as is usually done in probability calculus. It is well known from a multitude of experiments that they agree with what is obtained in a series of actual measurements.

Up till the present point, the probabilities considered were only purely theoretical tools to construct a meaningful language. Now, at last, they acquire a practical meaning as empirical probabilities, in the physicist's sense of the word and no longer in the abstract sense of pure mathematics. Among the other results one can obtain, the most interesting concerns the description of two successive measurements. When a first measurement of an observable A has been done and the system Q is left free to be measured again, one can compute the probabilities for the various possible results of a second measurement for another observable. This can be obtained from the following recipe. Assuming that the first measurement has given the result n, the probabilities of a later result can be computed by using again equation (12.8). Here one uses a density operator, for the measured system Q at the time $t' = t + \Delta t$ marking the end of the first measurement, given by

$$\rho_f(t') = \frac{\text{Tr}_M [F_n(t') \rho_Q \otimes \rho_M F_n(t')]}{\text{Tr}_Q [E_n(t) \rho_Q E_n(t)]}. \tag{12.9}$$

In the special case of a measurement strongly of Type I, this gives Lüders's formula:

$$\rho_f(t') = \frac{E_n(t) \rho_Q E_n(t)}{\text{Tr}_Q[E_n(t) \rho_Q E_n(t)]}. \tag{12.10}$$

Finally, when the result of the first measurement (of Type I) is a nondegenerate eigenvalue of A, one gets

$$\rho_f(t') = |a\rangle\langle a|. \tag{12.11}$$

The Status of Wave Function Reduction

Equations (12.9–11), as they stand, are only convenient recipes for a simple calculation of the probabilities for a second measurement. They express the rule of *wave function reduction*, also called sometimes wave function collapse. They do not imply however any other physical effect than the ones already discussed. The main one is decoherence, as it occurs in the apparatus M after the interaction. Nothing else is happening that would directly affect the measured system Q. As a matter of fact, the rule of reduction is not even a necessary ingredient of the theory, but rather a convenience. It is a straightforward consequence of the logical considerations about histories. When one wants to consider a second measurement, the reduction rule avoids having to keep track of the history of the first measuring device M, before and after its interaction with Q. It also allows one to sometimes ignore the detailed history of the preparation of the system Q before the first measurement. As such, the rule is very convenient but its use is not compulsory. One might as well keep track of the past and present history of Q and of both instruments and, as a matter of fact, this is how the rule is established. Perhaps the main point of all this is that reduction is not a physical effect.

Notwithstanding, equation (12.9) giving the density operator of the measured system after a measurement, when it is again isolated, gives the state operator correctly as required by Gleason's theorem. This is because it gives in principle the probability of every property and, according to this theorem, the operator giving that information is unique. This means that there is no ambiguity concerning the density operator of a measured system after a measurement. When one wants however to make sure that this assertion corresponds to a true property, some unexpected difficulties arise, as were first noticed by Einstein, Podolsky, and Rosen.

10. Actual Facts

While wave function reduction is not by itself a physical effect, a comparison of measurement theory with experiments shows a drastic and obvious effect not contained in the theory. Whereas quantum mechanics can only envision all the various data that can occur in the experiment on the same footing, reality shows a unique real datum: an actual fact. When one repeats a measurement many times under the same conditions, one finds that the data are distributed randomly, their probabilities agreeing with the theory, but the status of actual facts in the theory remains nevertheless an open and troublesome question. Where does this uniqueness and even this existence of facts come from? This is undoubtedly the main remaining problem of quantum mechanics and the most formidable one.

There are essentially three kinds of tentative answers intended to fill this lack of a theoretical explanation for the most essential feature of reality. The first one considers the existence of the problem as an indication that the theory is still incomplete. One should accordingly look elsewhere for the existence of real causes for actual facts. This is essentially the point of view advocated by the supporters of hidden variables. The second answer, which was introduced by Everett, is essentially a representation of reality as it would come from a pure probabilist, a Platonist,or an absolute idealist: All that the theory says is real, because theory itself is the law of reality. Accordingly, reality is not unique, as we see it, but it is multiple. This has been called the many-world interpretation. "Real reality" has many branches or alternatives, a new one opening when a new event akin to a quantum measurement occurs. You and I happen to be events occurring in one of these worlds and it seems to us to be unique because we cannot communicate with the other branches because of decoherence. This point of view, though modernized by Gell-Mann and Hartle, remains, to say the least, difficult to accept. There is a third answer going back essentially to Bohr, though the recent progress in the theory of interpretation has made it somewhat clearer. This is the one advocated in the present book, and we now recall it.

Past and Future

In Everett's representation, the initial state of the universe is assumed to be given and everything follows from it as a consequence of the Schrödinger equation. One can also use another representation, which stresses the uniqueness of reality. Right now, the universe shows a multitude of facts, but each one of them is unique. Many of these facts bear the memory of other past facts. This is obvious from observation but it was also shown to be consistent with quantum mechanics, with the justification of determinism. One can then show that, except for the past facts for which every memory has been lost, it is possible to reconstruct logically the past from the present and this past is also unique. It must be contrasted with the future, which remains completely open, except for the probabilities one can attribute to its various possibilities. The uniqueness of facts at the present time implies therefore the uniqueness of the past, while the future must remain potential.

One can therefore add to the whole logical construction an assumption according to which present phenomena are unique (and they are facts). This assumption leads to another remarkable agreement with evidence, namely that it is logically necessary to distinguish the characters of past and future. This is not just a formal distinction, as was found in equation (12.2) for the logical time ordering, but a drastic opposition of their natures agreeing moreover with the direction of time in thermodynamics.

The difference between past and future is something so obvious that one always takes it more or less for granted. The same is true for the uniqueness of reality with its uniquely defined facts. What the theory says is not that it can predict it, even far from that since the uniqueness of reality is a problem for it. It says however something unexpected, which is that these two features are connected, the character of time following almost unavoidably from the uniqueness of facts or, at least, quantum mechanics cannot separate one from the other without inconsistency.

A Rule Differing from All Others

When pushing this point of view further, one can also make a virtue of necessity. One may consider that the inability of the quantum theory to offer an explanation, a mechanism, or a cause for actualization is in some sense a mark of its achievement. This is because it would otherwise reduce reality to bare mathematics and would correspondingly suppress the existence of time. These considerations are however more appropriate for epistemology than for physics and they will be reconsidereded accordingly in the second part of this chapter. In any case, one must take care with the everflowing character of present time. If one remembers that Bohr considered the rule of wave packet reduction to be of a different character than all the other laws of physics, one arrives at something more or less equivalent, though much clearer, which is the following rule.

Rule 5. *Physical reality is unique. It evolves in time in such a way that, when actual facts arise from identical antecedents, they occur randomly and their probabilities are the ones given by the theory.*

It may be noticed that, when formulated in this way, the rule also covers the case of deterministic events as a special case. It might also be said, when looking backward at the whole construction of interpretation, that it consisted entirely up to then in developments internal to the theory itself. The present rule is the first and the last one to refer directly to reality and it is therefore not surprising that it is so different from all others.

11. Truth Criteria

The essentials of the interpretation of quantum mechanics may be considered as given by the present summary, as they were stated up to now. One may also add a few complements that are sometimes useful, not so much for understanding quantum mechanics itself but rather for clarifying some sophisticated issues. These issues themselves certainly do not have a cogent interest for all physicists.

Truth Criteria

The first issue concerns the question: what criteria should one ask for truth in physics? It was initially introduced by Heisenberg and it becomes particularly significant in a theory where logic is given a central role.

In empirical physics, the reality of a fact is usually taken as a criterion for the truth of the proposition asserting it. This is the famous criterion: "the rose is red" is true when the rose is red. Other propositions may also be said to be true when they follow logically from an observed one in a logic that is in accordance with reality. They must be nevertheless subjected to a verification. There are at least two reasons why the notion of truth becomes more intricate in the case of quantum mechanics: (i) It is not possible to check every proposition by comparing it with the corresponding fact, because a fact is only an actualized phenomenon, i.e., a classically meaningful property of a macroscopic object. This is a very restrictive kind of proposition and, if only facts could be shown to be true, one would never be able to state anything sensible concerning a particle or an atom alone. This is by the way exactly what Bohr maintained. (ii) One must also be careful when logic is used to infer a new truth from a known truth, because complementarity, with its multiplicity of consistent logics, is much trickier than ordinary classical (common sense) logic.

One can accept nevertheless a simple criterion of truth for at least a category of propositions, which is that the facts are true. It agrees formally with what logicians ask of truth and it is of course in agreement with common sense. It says that the proposition stating the possibility of a phenomenon is true when this phenomenon is a fact. It should be stressed that the facts mentioned in this criterion include in principle all the facts existing in reality. They are not restricted in principle to the facts that happen to be known to an observer. This is worth mentioning since often there has been a confusion between the objective status of a quantum system and the subjective knowledge of an observer about it.

One can take care explicitly of this objective character of the theory by restricting attention to *sensible logics*, which are the consistent logics including all the presently existing facts relative to the system under consideration. One can then extend the notion of truth to some other propositions that are not necessarily plain facts. If a proposition does not simply express a fact, it can nevertheless be held to be true if it satisfies the following two conditions: (i) It can be added to the field of propositions of every sensible logic without breaking its consistency, and (ii) In the logics so obtained, it is logically equivalent to a fact. It can be shown that this definition satisfies the formal conditions for truth in logic. It takes care in a rather simple way of the two major difficulties that were encountered and extends the notion of truth from the category of facts to at least two other categories of properties: (i) the result of a measurement and (ii) a phenomenon in the past, which cannot be directly observed but which is known from a record.

Truth and States. The EPR Problem

The present notion of truth applies to properties. One may also wonder whether it sometimes applies to the state of a system, i.e., whether one can consider the knowledge of the state as being true. This question has two aspects, namely: (i) When does the knowledge of the state amount to the statement of a property and, in that case, when can this property be said to be true? (ii) What kind of knowledge can one obtain about the state in general? In most cases, the physical state of a system cannot be summarized by a property. This is obvious from a mathematical standpoint, since a density operator is very seldom proportional to a projector. There is therefore generally no obvious meaning in saying that one has a true knowledge of the state of a system. Conversely, when the density operator is proportional to a projector, the question of its true knowledge makes sense and it worried physicists for a long time. For instance, if a pure state is associated with a vector $|\psi\rangle$, the property having the projector $|\psi\rangle\langle\psi|$ is well defined. It is then meaningful to ask whether this property is true or not. There was much insistence upon pure states in the early days of quantum mechanics and this question came accordingly to the forefront.

One can ask what kinds of facts allow a true knowledge of the state of an isolated system as it stands just after a measurement. A partial but precise answer can be given when the measurement gives as a result a nondegenerate eigenvalue of the measured observable: the property asserting the state at the end of the measurement is then true, according to the above criteria. On the other hand, when the measured eigenvalue turns out to be degenerate, it may be that reduction gives a state operator that is proportional to a projector, but the corresponding property is not found to be true (which does not mean of course that it is false; it is undecided or, as was said in Chapter 11, it is only reliable). When a second measurement takes place afterwards, it may happen that different consistent logics attribute different properties to the system just after the first measurement. Because of the no-contradiction theorem, no logical inconsistency can follow from this, though the choice of one logic or the other remains arbitrary. This interesting though unessential feature of the theory is particularly striking when the two measured observables happen to be strongly correlated in the initial state and this interesting situation was first noticed by Einstein, Podolsky, and Rosen. It was analyzed in Chapter 9 and this will not be repeated here, except for the conclusion.

Einstein, Podolsky, and Rosen were essentially criticizing the usual interpretation of quantum mechanics from a logical standpoint and they wanted to find some inconsistency in it. They did not have however a clearcut logical foundation to build on, any more than Bohr himself had. They relied upon an educated version of common sense, or if one prefers some sort of a philosophical principle, as everybody did as long as the nature of common sense was not clearly understood. In their case, the principle was the existence of what they called "elements of reality", namely some properties having at least at first sight an objective background. They considered a case when the state after the first measurement, as given by reduction, is well described by a property and the elements of reality were expressed by this property. Their knowledge looked therefore rather obvious and they were naturally assumed to be true.

What was found in Chapter 9 is something rather different—the "elements of reality" are not true. They are arbitrary insofar as there exist other, complementary logics, where they have no place and no meaning. To assign them some kind of reality was therefore an error, even if this error looked tempting and the pitfall was far from obvious. One cannot however close the contest by a judgment saying that Bohr was right and Einstein was wrong because Bohr's answer shows that he did not fully appreciate the interest of the question, perhaps because his interpretation was already too dogmatic to be adapted. Bohr did not seize this occasion to more consistently question the notion of truth in quantum mechanics.

It may be mentioned that one frequently meets physicists who say that there is no point in the EPR question and they can see why it is trivial, though they prefer not to be quoted. It is clear that they probably had no time to consider seriously the question: what can be said to be true when one accepts quantum mechanics?

12. The State of a System

Let us now consider what kind of knowledge one can obtain about the state of a system. Usually, no measurement allows one to completely know the state of an isolated system and its abstract definition through Gleason's theorem is not constructive. It only asserts that a unique density operator must exist. Accordingly, a complete formulation of the theory must give, at least in principle, an explicit construction of the density operator, even if it may be difficult to put in practice in every case.

A complete answer must use the definition of a state of the universe (or at least of a sufficiently isolated region of the universe containing the system in which one is interested). The density operator for a subsystem can then be obtained from it by a partial trace over the external degrees of freedom. The state of the "universe" $\rho(t)$ at a time t is given by the following construction. Let $\{F_k(t_k)\}$ denote the quasi-projectors representing all the facts having occurred everywhere in the "universe" at some time t_k earlier than the time t. One can then consider that the state of the universe at time t is the resultant of all these facts combined with the assumed knowledge of the initial state of the "universe" at some initial time. This is expressed by the formula

$$\rho(t) = M(t)/\mathrm{Tr}\, M(t),$$

where $M(t) = G(t)\rho_0 G^\dagger(t)$. The operator $G(t)$ recapitulates all the facts occurring between an initial time t_0, where the density operator of the universe was ρ_0, and the time t. More explicitly, one has

$$G(t) = T\left\{\prod_{t_k=t_0}^{t} F_k(t_k)\right\}.$$

This is a time-ordered product, as indicated by the symbol T (i.e., a projector referring to some specific time always stands somewhere on the left of another projector referring to a previous time). The small errors, i.e., the multitudinous epsilons affecting the statements of all the facts when they are considered as classical, give rise to an uncertainty in the definition of the state. This question has been investigated in Chapter 8 and it was found how one can define a *physically meaningful* density operator and consider different state operators as being equivalent, as far as all possible experiments are concerned. Their differences are furthermore mostly washed out by later decoherence. It appears nevertheless that there is an intrinsic, inaccessible, and unavoidable uncertainty in the possible knowledge of the state of a large system. It is not yet clear whether this feature is important and whether it has interesting consequences, at least for epistemology.

13. The Relation with Bohr's Interpretation

The present theory might be considered as being simply a modernized version of the interpretation first proposed by Bohr in the early days of quantum mechanics. It happens in fact to be nearer to Bohr's own views than to any other belonging to the Copenhagen nebula, concerning which we refer to the books by Jammer for a precise description, a history, and a discussion of the problems to which they gave rise.[1] It will however be useful to make clear where and when the new interpretation and the older one part, not of course because of minor refinements due to some technical advance, but in their really significant features. This will be done here without recalling in detail Bohr's interpretation, which was already described in Chapter 2 and may be

SUMMARY AND OUTLOOK

considered as known. It will also be convenient to appreciate how the new interpretation avoids the main difficulties encountered by the Copenhagen approach, rather than investigating these difficulties for themselves as was done many times by many authors.

Let us first look at the basic axioms. The principles asserting the mathematical and the dynamical framework of the theory, as given by Rules 1 to 3, are essentially unchanged. This is wonderful, considering that these principles withstood one of the most startling periods of progress in the whole history of physics without a change. One can perhaps find a little difference in the definition of a state, which was systematically identified with a wave function in the early days of the theory. There is however a striking difference in the axiomatics of the theory with the fundamental role now given to logic as a necessary cement between the mathematical formulation of the theory and reality. Its main role is to avoid an excessive reliance upon the slippery use of familiar words and familiar notions imprinted in our minds, in a domain where they become often unreliable.

The logical approach unfortunately went along the wrong track with Birkhoff and von Neumann when it diverted itself into the study of unconventional logics, as if some sort of unclear thinking could explain what is not understood. The next fruitful progress came therefore much later from Griffiths, who found how to extend the elementary notion of predicates by time-dependent histories. He also discovered the consistency conditions, which restrict the families of histories allowing the existence of a probability. It became then easier to start from his work to obtain an explicit interpretation of quantum mechanics within the framework of the conventional rules of logic.

The main foundation of this new approach, which is the universal rule of interpretation, has no analog in Bohr's interpretation. Some glimpses of it may perhaps be seen from hindsight in a few side remarks occurring in the vast Copenhagen literature, but it never came into full view. It marks clearly an essential difference with Bohr's interpretation, at least as far as the basic principles of the theory are concerned. It is more economical insofar as a unique axiom now replaces the rules of measurement theory as well as the various more or less subtle comments needed for applying them. This economy of principles, even if it gives an impression of getting nearer to the essentials, is not however really important in itself. What is much more significant is the change they provide in the nature of interpretation itself. It now becomes entirely deductive and foolproof, no longer an act of confidence in the wiseness of a great thinker. Furthermore nothing different from quantum mechanics (for instance, classical physics) is required when interpreting it.

Behind the differences in matters of principles, there is something much more important, which is a difference in the manner of dealing with classical phenomena by the theory. This was one of the biggest problems in the Copenhagen approach, since the basic quantum theory and the facts coming from experiments looked so far apart that they had to be put side by side as two practically independent foundations. Bohr insisted on a treatment of the measuring instruments by strict classical physics with no reference to a quantum background. It seems clear from hindsight, as well as from some quotations, that he was mainly motivated by logical considerations, namely to find an anchor for an unquestionable truth in the facts and in their classical characteristics. Although he was learned in philosophy, his culture came mainly from a German source. The influence of Hegel, who had quite personal views on logic, can be felt in some of his considerations, particularly when he made comments on the complementarity principle. He does not seem to have been well aware of the revival of another tradition in logic, with the work by Russell, Whitehead, and others in England and, even more actively, by Frege and by Hilbert and his coworkers in Germany. In any case, he was not in resonance with it nor did he see the opportunities it offered, despite the advances made by von Neumann, who was on the contrary well aware of these questions.

Bohr's point of view was by necessity very subtle because he was treading near an abyss when he thought it possible to combine two different classes of physical principles, classical and quantum. One of the most important advances of the new interpretation is undoubtedly the avoidance of this dangerous game, which is moreover difficult to teach to somebody else. In the long run, the resulting effect was even becoming vicious since physicists had a tendency to hide the difficulties, to forget them or to obliterate them in their teaching. Simultaneously, philosophers delighted in these exquisite subtleties and they transmitted them in their own way to their students, so that they reappeared in stranger disguise in unexpected places like psychology or theology. A kind of esoterism was beginning to be considered as a normal character of science and its opponents were in no better situation when they tried to fight it by relying upon the authority of the Master. Confusion reigned.

The theory of phenomena, though it still needs to be improved and further investigated, avoids these pitfalls. It allows the theory to stand clearly on a unique class of principles. It confirms in practice most of what Bohr asserted positively, when saying that a measuring apparatus, an ordinary object, obeys classical physics. The point is however that this is not any more a decree but something relying upon first principles, with errors that are known and very small in general. There are even cases where one must correct Bohr's assertions and some beautiful experiments made with a SQUID can now be used to confirm the new theory, though they were first conceived as a criticism of the Copenhagen interpretation.

SUMMARY AND OUTLOOK

Another consequence is to get rid of the false problem of an apparently drastic opposition between the probabilistic character of quantum theory and classical determinism. This is quite fortunate one could also begin to discern a serious deviation here and there. Classical determinism has become lately rather undervalued and the beautiful discoveries by Kolmogorov, Arnold, and Moser concerning its loss in chaotic systems were sometimes heralded as marking its death. Too few people were still clearly aware of the essential role of determinism in the existence of records, memory, and of its necessity for understanding experiments. The recovery of determinism is of course limited by the recognition that there exists a probability of error in the use of ordinary common sense, as well as in the classical course of events, because of possibly large quantum fluctuations. These sources of error are however completely negligible for all practical purposes and their origin is well understood.

Along with the theory of phenomena, important progress has also been made in the solution of another outstanding problem. It was concerned with the linear superposition of macroscopically different states at the end of an experiment and the problem was to reconcile this with the evidence of facts and, accessorily, with common sense. One can see glimpses of its solution in various remarks dispersed in many comments or investigations of the Copenhagen interpretation. Then came a rather clear idea,[2] and finally a quantitative theory of the decoherence effect. As often happens when there is a breakthrough, the exact assessment of its importance is not yet perfectly clear. It is undoubtedly very important but it is probably not the whole answer in itself to all the problems, as sometimes envisioned. On the other hand, some criticisms, particularly by John Bell, had a tendency to undervalue decoherence and they suggested that it only displaced the basic problem, making it less cogent but still exactly the same, as far as matters of principle are concerned.

This question has also been investigated here and it was asserted that decoherence cannot be circumvented in most cases, even as a matter of principle. It would be excessive however to consider that this statement is completely proved. There are nevertheless strong indications that the previous objections do not hold so clearly as they seemed to, though getting a better proof would be worthwhile. This need is felt more keenly when one considers the impact of a fundamentally uncircumventable decoherence upon several unsolved issues, including perhaps the foundations of thermodynamics. In any case, to decide whether or not the measurement of every observable is consistent with all the laws of physics is no longer a matter of free axiomatics but something that must be investigated with the available tools. This is typical of the present state of the Copenhagen interpretation. What it asserts can now be proved, rejected, or admitted under specific conditions, then kept within the bounds of estimated errors or, at least, submitted to a well-defined investigation. This state of affairs should encourage the renewal of a few old problems, which were left aside because of an apparent arbitrariness in their formulation.

To conclude, the new interpretation is almost completely identical with the older one in its practical aspects. This is why it is appropriate to put them in the same category. On the other hand, as far as matters of logic, of consistency, or even the meaning of knowledge are concerned, they are significantly different. The new theory may be considered an advance as compared with the older one, since it got rid of apparently inherent difficulties and also solved some major problems. Both theories however fail when asked to provide an explanation for the existence of facts. Their answer is essentially the same, though the most recent theory comes nearer to the essence of the question because it does not take as before a technical form requiring the reduction postulate. It makes clear a possible ultimate difference between the mathematics of theoretical physics and reality.

14. Gell-Mann and Hartle's Approach

After comparing the present interpretation with the Copenhagen approach, we should also say a few words about the possible variants of it and, particularly, the one proposed by Gell-Mann and Hartle.[3] This book relies mainly upon the author's approach for the sake of unity but many ideas originating from Gell-Mann and Hartle have been incorporated in it. It was shown moreover that the basic ideas are common and they differ mainly by their organization into a whole and by a different emphasis on their relative importance. We shall therefore be very brief since a rather complete comparison can be found elsewhere[4] and it would be out of place in a book with a more pedagogical bent.

Gell-Mann and Hartle want from the start to put the interpretation of quantum mechanics in a very general frame that will allow them to consider all sorts of unfamiliar but important applications going far beyond the simple nonrelativistic considerations, that have been dominant in this book. This includes cosmology and the situation at the beginning of the universe where the existence of facts, as they were discussed here, becomes quite questionable. It should also include in principle the interpretation of quantum fields, in a flat spacetime and a curved one, as well as quarks and strings. They also wish to extend the theory to include an objective account of observers as an "information gathering and utilizing unit" making use of the best of information theory and algorithmic complexity. Their enterprise is looking therefore far beyond the present one and it must rely upon even more general basic rules.

SUMMARY AND OUTLOOK

One of their essential ideas is to attribute completely to decoherence the dynamical origin of phenomena among the multitudinous features of the quantum evolution of the universe, as well as the selection of the significant collective observables and, as a consequence, the occurrence of the histories making physical sense. This point of view might very well be deeper than the one we followed but it has a practical drawback, which is that the theory of decoherence is still far from complete and its technical application is difficult. This is why the Gell-Mann–Hartle theory is still partly a program, though it would be easy to substantiate it in the ordinary conditions of experiments by the direct semiclassical methods we used in Chapter 6.

15. Perspectives

Although some progress has been made within the last ten years or so in the interpretation of quantum mechanics, a few questions still remain unanswered or incompletely understood. We shall not try to discuss them now in detail, because there is always something awkward in writing many pages upon what is not understood, but we shall however mention the main ones. Basically, there are two categories with, on the one hand, some important questions that may be at least partly considered as of a technical nature so that they can be submitted to a systematic investigation. There is on the other hand the basic question of the uniqueness of reality, which perhaps does not belong to physics as such.

Among the items in the first category, we have:

1. A satisfactory theory of the distinction between the collective and the microscopic observables is still missing. It is most probably a quantitative question, the character of a hierarchy of observables depending presumably upon the frontier to be drawn between the ones to be kept as collective and the ones describing the environment, this being linked with the corresponding errors in a semiclassical approximation.

The difficulty of the problem is obvious, as soon as one realizes that it is tantamount to precisely defining the notion of object from the standpoint of quantum mechanics. This is sometimes taken as a pretext to reject the problem as meaningless and its solution as impossible on the basis of arguments coming from the older versions of the theory. One may take however a very different point of view. Not long ago, the mathematician Charles Feffermann investigated how to approximately diagonalize a rather arbitrary hamiltonian involving possibly a large number of degrees of freedom.[5] Using the techniques of microlocal analysis, he solved the problem in terms of suitable coordinates showing an increasingly marked quantum character, as seen by the size of the domain of phase space in which they could be defined. These coordinates and the corresponding observables are local, i.e., they depend upon the state of the system, as might be expected. They are obtained by nontrivial canonical transformations, which at the present time can be shown to exist but are not yet constructed explicitly. This highbrow approach hints at a simple answer, namely that the problem is probably already solved but its solution is not yet fully understood.

2. The theory of decoherence is presently formulated in terms of special models of the environment, which are much more determined by what can be solved explicitly by the present methods of calculation than by the nature of the physical problems themselves. In view of the importance of this effect, a better understanding would be imperative. In particular, the beautiful approach to interpretation put forward by Gell-Mann and Hartle relies heavily upon decoherence and it risks partly remaining a program as long as no progress is made in the corresponding mathematical methods.

The logical interpretation described in the present book also remains partly incomplete because of the unsatisfactory state of decoherence theory. The corrections that must be introduced in the logical treatment of classical physics when friction is taken into account remain for instance badly known.

3. What was said about the impossibility of circumventing decoherence is still nearer to a conjecture than a proof, despite the obvious importance of its possible consequences in quantum mechanics and thermodynamics.

4. One might add of course the problems arising from the extension of the interpretation to relativistic situations. As a matter of fact, the frontier between the technical problems and the fundamental ones becomes somewhat fuzzier in that case. As an example, one may ask what it means to say that a quark is a particle and that it is nevertheless always confined. In another direction stands the awesome problem of what can be said about space and time when quantum mechanics is assumed to be essentially true and one insists upon an overall logical consistency. The present-day approaches to a synthesis between quantum mechanics and the relativistic theory of gravitation bear the mark of great mathematical ingenuity and high technicality. They still look however rather frail as far as their logical foundations are concerned.

Finally, there remains *the* problem, which is the existence of facts. It was somewhat hidden behind wave packet reduction in the older interpretation but, now that most other problems have been solved or at least clarified, it stands pure and alone. Once again, we shall not commit the error of digressing at great length upon something that is not clearly understood and we shall only recall the few directions along which an answer has been sought:

1. It was suggested in Chapter 8 that the existence of so unique a problem was perhaps the sign that a boundary of physics had been reached. Theory

SUMMARY AND OUTLOOK

and reality break apart only at one point, which is inaccessible to experiment as far as one can see. This point is their pure and unique point of contact where the meaning of the mathematical theory is compared with reality itself. It was also suggested that one should perhaps think more about the other basic problem of physical knowledge, which is why a mathematical construction can fit reality. One should not remain blind to its similarities with the problem of existence.

2. Other people prefer to remain in the idealistic cocoon of the Everett representation, involving many coexisting and separated realities.

3. Some people are so strongly impressed by the existence of the problem that they prefer to consider quantum mechanics itself as a drastically incomplete description of reality. Most of them look for hidden variables. Bell's inequalities and Aspect's experiments have made their lives more difficult but this is not enough to discourage them, as it is also not a reason for completely disproving their claim. What is however perhaps missing most obviously in their position is an acute analysis of what they criticize, except for their uneasiness, and a precise statement of what they really want.

4. Other directions for further thinking have been suggested from time to time. Should one consider that the motion of a particle is actually "computed" somewhere by something? Is the computation made by the particle or by the vacuum? If you find this question, once put forward by Feynman, to be crazy, just ask yourself what you think about the nature of the laws of physics. How does a system know about them, feel them, or respect them? Or is it that when you know the law you don't yet know anything about reality?

5. Should the universe itself be considered as a numerical computer or as an analogical computer as most of us see it in a confused manner?[6] Has its finiteness an effect upon the nature of the laws of physics? Is there something to be learned from algorithmic theory and its notion of algorithmic complexity? It was found possible in this framework to define randomness, which was assumed previously to be a primary concept. Might this have something to do with the randomness of the quantum events and the problem of actuality? It was found possible to express that an individual message (always reducible to a specific number by using a binary alphabet) is random without comparing it with a set of similar messages and the trick is to characterize randomness by using algorithmic complexity. This is of course not the place to elaborate upon these fascinating or crazy ideas, but there is an attractive symmetry between the randomness of a unique number and the actuality of an event emerging from a theoretical set of multivalued possibilities.

This might be a good place to stop while leaving quite a few beautiful and fascinating problems for the times to come.

PHILOSOPHICAL ASPECTS

Since its beginnings, quantum mechanics has given rise to many questions of a philosophical nature and demanded a deep revision of scientific knowledge.[7] The advent of another interpretation, even if it is only partially new, may therefore also lead to a change in the status of these questions. This is what will be considered in the first place in the present part of this chapter. We will then turn particularly towards two important questions dealing respectively with the objective and the realistic characters of the theory. It will be suggested that they are related with other questions of a still more philosophical nature, about which a few personal viewpoints inspired by physics will be presented by the author, for whatever they may be worth.

16. Twenty-one Theses

We shall now summarize the main features of the present interpretation of quantum mechanics as seen from the standpoint of epistemology. It will be convenient for that purpose to go to the essentials and not to distinguish again in detail between what has been rigorously established and what remains partly conjectural, since these distinctions have been made clear previously and they would only obscure the overall pattern. As a matter of fact, the main aim of a review such as the one to be given here, which is going to differ sometimes plainly with what is customarily said on the subject, is mostly to provoke the reader to respond by his own opinion and to have his own thinking. Although not all the theses to be offered here are equally interesting, it was found that asserting their respective level of interest is mainly a matter of personal opinion, so that it is difficult to assign them a hierarchy. They will be therefore listed without a preferential order.

These theses, which are more or less already implied by what was said previously, can be stated as follows:

1. The structure of quantum mechanics consists of two parts, which are intimately related but deeply different. One is concerned with dynamics and the other with logic.
2. The logical structure, though it is entirely controlled by strict mathematical methods, can be used to construct with precision the whole language of physics, together with all the propositions expressing its practice and its conceptions.
3. The interpretation of quantum mechanics relies completely upon a unique rule. The interpretation can proceed furthermore in a completely deductive way on this basis. This rule selects, among all

the conceivable propositions describing a physical situation, the meaningful ones from the meaningless. It also selects the trustworthy reasonings from the erroneous ones. Many descriptions can be conceived by our imagination, generated by our mental representations or suggested by the possibilities of language. The ones remaining outside of the frame of the rule are deemed to be gratuitous and meaningless.
4. The existence of phenomena follows directly from the theory and they need not be forced into it on the basis of empirical constraints. The phenomena are defined as the classically meaningful properties of a macroscopic object. They can be recorded and then retrieved from their records. They are neatly separated (showing no quantum interferences) so that they can be considered as classically alternative events. Their separation cannot be questioned in view of the result of any refined experiment, at least as long as the object is big enough.
5. In particular, classical determinism is a direct and controlled consequence of quantum mechanics.
6. Common sense logic, when applied to our usual environment, is usually correct and its exact domain of validity can be determined in principle on the basis of the universal rule.
7. The logic of common sense as well as determinism itself have however some probability of error, which can be estimated in principle. These probabilities are most often extremely small and the corresponding errors have no practical consequence.
8. Contrary to dynamics, the logical structure of quantum mechanics must select a definite direction of time, which necessarily coincides with the one occurring in thermodynamics.
9. The theory is unable to give an account of the existence of facts, as opposed by their uniqueness to the multiplicity of possible phenomena. This impossibility could mean that quantum mechanics has reached an ultimate limit in the agreement between a mathematical theory and physical reality. It might also be the underlying reason for the probabilistic character of the theory.
10. The logical structure of the theory allows a great variety of consistent logics, which are usually foreign to each other. This complementarity cannot however generate a paradox nor a logical inconsistency.
11. Because of complementarity, the notion of truth in physics is somewhat more subtle than in any other field of knowledge. A true proposition can express as usual an actual fact, but it can also be a proposition consistent with every sensible logic (i.e., every logic including the facts) where it is equivalent to a fact.

12. One can also introduce other so-called "reliable" propositions having no analog in other fields of knowledge. They appear as the logical consequence of a fact in some logical frame though they cannot be stated in every such frame, in every sensible logic. Their formulation remains therefore the matter of an arbitrary choice, even though they are not self-contradictory. The "elements of reality", which were introduced by Einstein, Podolsky, and Rosen, belong to this category of arbitrary and untrue, though not false, propositions.
13. Nevertheless, the experimental conditions selected by Einstein, Podolsky, and Rosen constitute, together with the simplest kinds of measurements, the only two canonical cases of measurement theory from a logical standpoint.
14. It can be said that quantum mechanics is truly separable, if one defines a physical system to be truly separable (or localized) when no true property of a subsystem can be modified by an action made elsewhere upon another subsystem that is not interacting with the first one. Separability (localizability) does not hold however if one tries to extend it arbitrarily from the true properties to the reliable ones. This rather subtle difference had been missed in many previous discussions and this was responsible for overstating the apparent difficulties of quantum mechanics.
15. The existence of actual facts can be added to the theory from outside as a supplementary condition issued from empirical observation. The structure of time must then be modified accordingly. Time must be split into a past, a present, and a future having very different qualities. Present and past are uniquely defined while the future must remain potential and subject to probabilistic expectations. This structure of time, so obvious from the standpoint of observation, turns out to be necessary from a theoretical standpoint.
16. In particular, the past of the universe is unique.
17. The state of a big isolated system (the solar system, the universe, ...) is defined uniquely and objectively by the totality of the past and present actual facts in it. The state of a subsystem can be obtained from it by a partial trace, though this formal definition is most often rather impractical.
18. The only nonfactual datum entering in the definition of the present state of the universe is its initial state.
19. Many properties that can be formulated in a consistent logic cannot be checked experimentally, even in principle, and are therefore neither true nor false. Similarly, many observables cannot be measured, even as a matter of principle, without violating some law of physics.

20. One can always consider an observer as being simply a part of the universe and no observer plays a privileged role because of his perception of physical reality or of what his mind is thinking of it.
21. Perception is a special case of a measurement.[8]

In view of these statements, it is clear that the modern version of the Copenhagen interpretation, if empirically very similar to the older one, is quite different from the standpoint of epistemology. The present theses should perhaps be stated more carefully and more at length for a serious philosophical discussion, but this will not be attempted here. Their justification is provided in full by the previous chapters.

17. Is the Theory Objective?

Is quantum mechanics an objective theory? Is it realistic? The first question still belongs to physics, though with a touch of philosophy, while the second one is plainly philosophical. Both have given rise to so many discussions since the advent of quantum mechanics that it is impossible to remain silent about them, even if it means giving up the comfortable protection afforded by the technical apparel of physics to advance as a naked incompetent philosopher.

Let us first consider the question of objectivity. This notion is usually defined according to Kant, something being said to be objective when it exists outside of the mind, as an object independent of the mind. The question of objectivity in quantum mechanics has a long story, which goes back once again to some puzzling aspects of the old idea of wave function collapse. No other permanent or transient principle of physics has ever given rise to so many comments, criticisms, pleadings, deep remarks, and plain nonsense as the wave function collapse. In spite of the risk of running into the last category, I would like to express here a feeling both admiring and irreverent. Reduction was a pirouette of genius for escaping a contradiction which came from simultaneously admitting two different kinds of laws of physics, the quantum and the classical ones. It belongs in some way to the category of great somersaults of thought, in company with the proof of God's existence by Anselm or the immobility of Achilles by Zeno of Elea. In short, it is a high summit in the history of metaphysics.

It resisted analysis for a long time, because one could not find the logical means for justifying or criticizing it. It refers to a measurement, a datum expressed by classical physics, and it states a conclusion about a wave function, i.e., a quantum notion. As long as the relation between the two theories is not clarified, one does not know how to progress in its understanding, since one is walking over two different banks of a river at the same time.

The meaning of the words that are used remains imprecise and slippery. A logician would say that the reduction of the wave function cannot be considered as an axiom and it is not even a sensible statement, because a logician wants to have a unique and well-defined field of propositions before saying anything, a universe of discourse in which he can identify an axiom. Here, on the contrary, one is simultaneously using two apparently exclusive logical frames. Under these conditions, one can only get an empty proposition or a rule of thumb.

It should not be surprising that, in view of the impossibility of pushing further the analysis of reduction in a formal way, one tried to think of it as being a real physical process, something occurring when a measured and a measuring system interact. Although there never was any evidence for such an effect, the underlying idea of a real physical process was essential in making popular a conception where reduction has a concrete character, where it occurs suddenly at a specific time and in such a way that it can be felt simultaneously everywhere. These views encountered of course other difficulties with relativity.

These apparently insuperable difficulties played an important role in the proposal of a brilliant and maleficent solution by von Neumann, later expounded in more detail by London and Bauer.[9] According to them, reduction had nothing to do with a real physical process. It takes place in the representation of the world existing in the consciousness of an observer. For this to occur, it is necessary that the wave function be itself an element of this consciousness rather than a part of reality. The wave function is the representation of reality that an observer may conceive in view of the information he has at his disposal. But the notion of a wave function, or of a density operator, is necessary for a complete construction of quantum mechanics. And quantum mechanics is itself a foundation for all of physics and chemistry, which themselves enter as building blocks in all the other sciences. So all of science was thereby reduced in this approach to the status of what happens in the consciousness of one or several observers. No objectivity was left and there was at best an agreement among learned people. This conception of science could be considered as a solipsist regression, i.e., a representation where thought remains enclosed within its own space without ever making sure of the existence of reality. It came nevertheless unfortunately to be called the "standard interpretation"[10] and it played a nonnegligible role in shaking the confidence that many serious people have for science, while leading some others towards strange deviations.

Conversely, one should not forget how much Bohr himself stressed the objective character of his interpretation. Unfortunately, his valuable arguments were not sufficient to untie the knot of the problem. He expressed himself in a subtle language, which did not have the forceful character of

SUMMARY AND OUTLOOK

the simple-minded ideas put forward outside of physics by trivialized versions of the standard model. One should therefore not minimize the problems arising from reduction. The character of science itself is at stake, as well as the image of physics outside. One might perhaps regret that a legitimate questioning by some highly concerned physicists leaked outside of a circle of specialists to reach an impassioned public lacking the patience to wait for decades before another solution could appear. Such a regret would however be unreasonable, since science should not be reserved to a cast of specialists. This is true of its problems as well as its success.

In view of what is at stake, it is difficult to serenely appreciate what a new solution brings with it, since one may be led as well to underestimate it for being a return to the ordinary and commonplace or to overestimate it as a return to order. If one tries to take a sober point of view, one may simply notice that the lack of objectivity attributed by some people to the theory was never established. It rested upon the existence of an easy subjective solution, much easier to understand than the subtle objective solution offered by Bohr. It was also enforced by the fact that the objectivity claimed by Bohr was itself far from obvious and not so convincingly consistent. This is why the occurrence of new obviously objective interpretations is important, even if they still need to be improved. Of course, one can never convince somebody who wants to get subjectivity for its own sake, because his own philosophy asks for it. The point is to make somebody who believes in objectivity feel really confident about it. It is enough for that purpose that a consistent objective interpretation should exist since then subjectivity is not compulsory. It does not matter whether the objective formulation is unique and whether someone might prefer a better interpretation satisfying some pet convictions, because this can wait. The main point is that an objective interpretation exists. One might say that another objective solution already existed, namely the many-worlds interpretation, where universes roll along their parallel histories as Everett proposed it. But this alternative has also unfortunately its own scent of mystery, leaving one's mind undecided as to the reality of these possible universes. In any case, it is not fitted for bringing back common sense into quantum physics.

The direction of the present discussion is perhaps too polemical. It will be better to go back to the initial question: is quantum mechanics objective? From the theses put forward in the previous section, it is clear that the answer is positive. When it asserts a property or reaches a conclusion, the theory only makes use of well-established facts. It has accordingly no more difficulty with objectivity than classical physics, though of course its background is less obvious. When it comes to using common sense or the ordinary representation of the world as being legitimate, it must be accompanied by detailed proofs. It also requires much care when it comes to the basic question of understanding what are these facts which constitute the key to objectivity. One must work harder and recognize that classical physics was in fact rather naive when it considered its own objectivity as obvious, but these subtleties do not affect the basic question, which is whether the theory refers to anything having to do with the mind. The answer is no. The theory only refers to the facts, to all of them even, as a matter of principle, and not necessarily only to the known ones. The consequences it draws from these facts are completely free from any arbitrariness originating in the mind, at least as long as one sticks to what can be said to be true. So, to conclude, the theory is plainly objective.

18. Is the Theory Realistic?

One often wonders whether quantum mechanics is a realistic theory. This is a much trickier question than the one concerning objectivity, if only because it is much less clearly stated. Some of the best analyses on the subject have been made by Bernard d'Espagnat[11] and it is therefore quite natural to elect him as a reference on the subject. We shall accordingly follow him when trying first to define realism, with its various possible meanings. We shall then consider what arguments can be given in favor of the "veiled realism" d'Espagnat favors having. Finally, we shall begin a critique of realism itself, which will lead us to return to foundations in the next section.

Realism

Realism is not as clearly defined as objectivity. This may be because it originated much farther back in the past. It was discussed during several millennia from Plato to Abélard, William of Okham, and Thomas Aquinas. Somewhat nearer our times, it was also a matter of opposition between the philosophies of Descartes and Leibniz, not to forget its grand reformulation by Spinoza. There is however no modern definition of realism by a well-known philosopher, whom one might take as a reference on the subject. The statement of realism as a philosophical position, or even a general agreement about what is supposed to be real, is far from being clear.

The name of realism traditionally covers a doctrine according to which "ideas" are as real as or more real than the objects they represent. This was one of the main statements to be made by Plato. One of the modern variants of Platonism is mathematical realism, according to which the intrinsic content of mathematics has its own reality and it is not simply the arbitrary outcome of a human game. Another kind of realism, which is more or less in the same vein, is strongly linked with theology when it asserts that there exists something more real than physical reality, namely God, at least in most versions. One may notice in this respect that some of the best arguments for the existence of a veiled reality can be found in the *Docta Ignorantia* (1440) by Nicola da Cusa, whose arguments are remarkably similar to the more recent ones by d'Espagnat. This idea is not even specific to western thinking since Shankaracharya in India was probably the philosopher who pushed it to its most extreme conclusions by considering physical reality as a pure illusion.

What is usually called realism, when one is talking of physics, is more or less the exact opposite of traditional realism, which is of course guaranteeing us a lot of confusion. One can borrow the corresponding definitions from d'Espagnat, at least in general terms. A realist is first of all somebody for whom there exists "something", to be called Reality, which is independent of the cognitive faculties of mankind and also independent of what man decides to observe or to measure. "Physical realism" goes farther and assumes that "Reality can be known in principle by the means of physics." What one means by knowing in the present case is not always perfectly clear and one assumes more or less that one knows what it means to know. D'Espagnat does not rely upon a specific theory of knowledge, which would help him to be more precise about it. He also finds difficulties in making clear what is the "principle" allowing the possibility of knowledge. These shortcomings indicate that serious difficulties are probably hidden behind what might seem at first sight to be a rather clear definition.

D'Espagnat also says that physical realism may assume two different versions, according to whether it uses the language of mathematics or not, but this is not essential. What is probably more significant is a distinction he makes between a "strong realism", when Reality is known in itself, and a weak realism, when this knowledge is only partial. It should be recalled that physical realism was already explicitly rejected by Kant in his Critique of Pure Reason since, according to him, one can never accede to a knowledge of the so-called numena (i.e., intelligible reality) but only to the phenomena (i.e., perceptible reality). This shows again how tricky the problem is, since a physical realist is more or less doing philosophy without having the possibility of relying upon a well-recognized philosophy. Realism is also often contrasted with another kind of philosophy, so as to make it clearer by a comparison. The opposite of realism is deemed to be positivism, in its version due to Stuart Mill (not the one by Auguste Comte) where the key to knowledge is an agreement between the consciousness of men. We shall see however that even this counter-definition is questionable.

A Brief Discussion of Realism

One of the main points in the modern discussion of realism is that one cannot get a completely reliable knowledge of reality, because of quantum limitations. This limitation in our knowledge of reality is of course more or less obvious, as far as practice is concerned. The point is that it is also limited "as a matter of principle". The first arguments reaching this conclusion relied once again upon the analysis introduced by Einstein, Podolsky, and Rosen.[12] It said that the properties of a system we are interested in could be changed, without our knowing it, through an action that is made far away upon another system. Then, in such conditions, reality would not be known with any kind of certainty.

D'Espagnat considered that this argument does not mean that Reality does not exist nor that it is utterly impossible to know it. It means only that the knowledge of it is deemed to remain incomplete, or that reality remains "veiled", as he says in a poetic mood by coining the name of "Veiled Realism". This analysis obviously needs to be reconsidered in view of the progress realized recently in the understanding of the EPR properties and particularly the importance of the notion of truth concerning the kind of knowledge to which one is referring. Nevertheless, if all the arguments in favor of veiled realism cannot be maintained, its main conclusions probably still hold. This can be shown as follows.

It is convenient to consider an example. We consider the energy levels of the matter in the Eiffel Tower in the order of increasing energies. Let us consider the nth level, taking for instance $n = 537,489,323,067$. This level exists. To say that the matter of the Eiffel Tower is presently in that state is an example of a conceivable knowledge. According however to what was discussed in Chapter 7, there is every reason to believe that this statement can be neither true nor false: it can never receive a truth value insofar as no conceivable experiment (i.e., as a matter of principle) can make sure of the truth or falsity of this property. It is condemned to be forever untrue. Similarly, there is a margin of uncertainty in the knowledge of the state of the real universe. One can then say that there is some restriction in what can be known about reality, so that it would be excessive to assert that reality can be known completely.

It is rather easy to draw the list of what can be known. It includes the facts and a few properties of some microscopic systems, which are mostly the results of a measurement. This is a pretty large set, since it includes all the facts, but it is also somewhat restricted as far as the knowledge of microscopic properties is concerned. Does it mean that reality is "veiled"? Why not? This short discussion has taught us anyway that an investigation of realism must rely upon a specific theory of knowledge, and this involves in particular an explicit notion of truth. Otherwise one does not know what one is talking about.

The Historical Extent of Realism

One may also wonder whether realism is an important doctrine about which one should take issue. Although there is no doubt that most past physicists, if not all of them, wished to adhere to objectivity, their position with respect to realism was much more cautious. As a matter of fact, the importance of realism in the history of physics is somewhat questionable and one may wonder whether one may follow completely d'Espagnat when he assumes that realism was the dominant attitude at the time of classical physics. As a matter of fact, there is only one wide-ranging work on the subject of realism written before the advent of quantum mechanics. It is due to Pierre Duhem[13] and it does not confirm the supremacy of realism.

Before mentioning Duhem's conclusions, one should distinguish two aspects in the definition of realism given above: (i) It assumes the existence of an independent reality and there is no doubt that this attitude was representative of the whole collectivity of physicists, at least before the time when a few people began to question it because of quantum desperation; and (ii) The second aspect is the one that must be considered with more caution, namely that Reality can be known, at least in principle, by using the means of investigation of physics. This is precisely the question in which Duhem was interested and he tried to answer it by considering the history of physics. Following him, one might ask: did one previously get a direct knowledge of Reality by the means of physics and can one see the existence of some barrier forbidding us from increasing indefinitely this knowledge up to its completion?

After investigating classical physics, chemistry, and astronomy in a search for an answer, Duhem said that he could identify three clearcut realists: Kepler, Descartes, and Boscovitch. Copernicus and Galileo made some statements in favor of realism and some others against it. A typical conflict where realism enters and in which some of these people were involved is well-known: can one say that the Earth is turning around the Sun rather than the opposite. The first statement was found to give a simpler account of the observations of planets while the second one came from various authorities. The Jesuits at the time of Galileo tried to reconcile the two views by resorting to something that was then well known, namely a position going back to Hipparchus and expressed most clearly by Simplicius (around A.D. 500) in his commentaries of Aristotle. Astronomy does not tell us what the celestial bodies are and how they move, it tells us how to describe them in such a way as to preserve all appearances.

According to Duhem, this is the true opposite of realism (Duhem did not even mention Mill's positivism). He then proceeded to find the many variants of this position in the later works of many physicists and philosophers of science. According to him, Newton considered that the principles of natural science provide a summary of many empirical observations, organized in a simple order, but this is more a description than an explanation. Among the English school of physics in the nineteenth century, he pointed out Kelvin's and Maxwell's insistence on the character of science as providing a model, something that was also emphasized by Ampère on the other side of the Channel. Mach himself made it clear that science provides a representation of reality, not an unassailable and intimate knowledge of it but something that can stand the assaults of experiment while keeping all appearances. Of course, one can also quote Kant in favor of the same point of view.

Whether or not Duhem showed bias in his selection of examples is a matter to be left to specialists. There is at least something that remains convincing in what he said, which is that realism is after all an extreme position, in the usual sense of extremism: how can you say that you *know* what Reality is? His insistence upon keeping all appearances, as substantiated by the existence of models, leads one naturally to a more modest and more promising attitude, which is to recognize science as a *representation* of reality, i.e., a description, an image, a picture, a model. If one accepts this view, the question of realism becomes much less important and another interesting question appears, which is how to understand in a better way the features and the meaning of this representation. This is what will be now considered, first from a general standpoint and also more particularly in view of what we have learned from quantum mechanics.

19. About the Foundations of Science

Other Philosophical Questions

As has just been mentioned, there might be something more important than the question of realism, or at least a necessary preliminary to its proper formulation, which is to better understand what physics is, considered as a representation of reality. This might look like another one of these unanswerable questions though, maybe, there is something to be said about it when one considers a few hints suggested by the logical interpretation.

It is easy to give a list of these hints, even if one does not try to exploit them immediately:

1. Facts, which had always been up till then a primary notion, have now acquired a more elaborate status. They are still the building blocks of empirical knowledge, but they appear to be restricted to the macroscopic world, even more clearly than in the framework of the Copenhagen interpretation. Another feature of facts is still more striking. As long as one only thinks of them (rather than experiencing them), by envisioning their occurrence as so many possible phenomena, the representation given by the theory takes care of them perfectly well. This representation however breaks down when one comes to their actuality. This shows that there is undoubtedly a serious question with actuality or, if one prefers, a serious question with realism. Moreover,

one cannot simply consider reality as being identical with a complete state of facts, as was assumed in classical physics. One must distinguish between the facts, the microscopic properties that can be said to be true, and also the vast amount of microscopic properties that cannot even be said to be true or false. All of this in some sense constitutes reality. Or does it?
2. Common sense had always been the foundation of every philosophical argument, even if ennobled under the guise of general principles and learned words. Here again, a significant change of status has occurred and common sense is no longer as privileged as it used to be. In other words, reason is now something quite different from common sense, at least in principle. What should then be taken as a primary notion? This feature was already mentioned previously and, without repeating ourselves, there is no doubt that something must change in epistemology when the status of facts as well as the status of common sense are both significantly reconsidered.
3. A positive aspect of this revision of common sense is that it stands now upon its own foundations, showing more clearly the logical skeleton of science. It also becomes linked in an unexpected way with the mathematical structure of science. These are in some sense discoveries implying philosophical consequences, if one takes them at their face value. A perhaps extreme question could also emerge: if philosophy relied mainly up till then upon common sense and if one can now see common sense in a new light, is it not possible also to see philosophy in another light?
4. Finally, the status of truth, though it was not deeply modified, has changed anyway and, even if it remains a primary notion, it is no longer as simple as it used to be.

These are a few remarks showing that the interpretation of quantum mechanics, in its new version, may have some interesting epistemological consequences. These consequences, or others coming from further progress, will probably be investigated by specialists in due time. It might be worthwile however to indicate presently a few possible avenues, as they appear dimly to a nonspecialist in these philosophical matters.

About the Theory of Knowledge

It looks as if really big questions are facing us. Shall we then hasten to reduce them as rapidly as possible to a few familiar things we have learned at school or shall we play the amusing and risky game of thinking? Why not try? After spending some time at worrying about the role of logic in science, one cannot escape from wandering among other problems that are still more concerned with the foundations of science. A very simple question comes first: What is science? But to consider it as simple is misleading. One can see that there are wonders inside science and that it is also concerned with other wonders, so that there is both clarity and mystery in it. Science is our essential source of knowledge and of understanding. When Virgilius said that the joy of understanding is the only one to last the whole life long, many of us agree with him. But do we know what this knowledge is and do we understand what it means to understand? In other words, how is it that one can understand reality?

This looks like the kind of question about which a beginner in philosophy may write a few pages as an exercise. As a matter of fact, few of us, either as scientists or as philosophers, are very different from these students when we comment upon what the ancient authors said about similar subjects. These respected authors discovered the game when the joy of thinking was young enough to allow crazy ideas to be proposed. We know much more than they did but we are too afraid to be crazy, so that the progress of science adds little to philosophy. We do not enjoy ourselves, we are boring, and what we do never changes the setting of the play.

The kind of change in perspective we met a moment ago might perhaps lead us to ask more daring questions. We need a theory of knowledge offering another point of view about the notion of facts, about the use of the language of common sense, and also about truth. It might be recalled on this occasion that the theory of knowledge is the setting offered by philosophy for the construction of science and, obviously, the changes we have met can modify this philosophical nest, as well as the construction inside it. But the theory of knowledge is also the foundation of philosophy itself. It says how philosophy is possible and what it may be expected to attain. So the game we are playing is no less than a questioning of the common foundation of science and philosophy.

This is after all a reassuring remark, because the question of foundations, as far as philosophy is concerned, is well known. So one may finally call the good old authors for help. The foundations of the theory of knowledge are usually assumed to proceed along one of two main lines. The first one takes for granted the existence of some primary data, which is most often both total and unique. This is the Being with Parmenides or Heidegger, the Logos with Plato, the Spirit for Hegel, God for Thomas Aquinas, the substance or Nature with Spinoza, Matter with Marx and probably, deep inside, reality for many physicists. The second approach, inaugurated by Descartes, takes the mind of the philosopher, with its "cogito", its faculty of thinking, as a starting point. One may thus ask as a first question whether one of these two ways will be chosen and, if so, which one.

Many physicists refuse to consider these questions as even being meaningful. They call themselves empiricists. They ask questions neither about

SUMMARY AND OUTLOOK

the existence of facts nor about the nature of facts, and this certainly brings some fresh air. They also more or less consider the existence and the content of science as being another kind of fact, which does not need to be excessively questioned. If one of them read this book, he probably gave up on the present chapter before reading this, so that he unfortunately won't appreciate the pun when he may be said to belong plainly to the first category of philosophers, the ones assuming a unique and universal kind of primary data. In his case, Fact.

Notwithstanding, it may be worth showing that these two main approaches are not antiquated and they are still playing an important part in modern thinking. This will best be shown by giving two examples of modern philosophers representing each school. It is often agreed that the most penetrating analysis of the "cogito" approach, also called positivist in its modern versions, is the one by Ludwig Wittgenstein (1889–1951). He was conscious of an acute problem in the foundations of science. He also investigated in great detail the nature, the structure, and the construction of language, as well as its relation with logic. This he did by analyzing acutely the language man must forge for his thinking. He came essentially to the conclusion that, by strictly following the cogito approach, one can never know anything about reality or, more properly, the approach is inconsistent. Reality is for Wittgenstein only what can be indicated by pointing a finger while saying "that". When he considered how a construction relying only upon thought, logic, and language can reach "that", he concluded it was an impossibility. Quite a few people believe that he pronounced a death sentence against positivism, but nevertheless quite a few positivists are still unaware of it or they do not care.

The modernity of the other approach can also be shown by the example of Martin Heidegger (1889–1976). He also gave much thought to the question of the nature of science, which he called "techne". According to him, the primary data is to be called the Being, and it shows at least two different aspects. One of them, the so-called "Dasein", is rather similar to what Wittgenstein would have called "that" and others Reality. The second aspect is what he calls "logos", which might be schematically defined as the thinkable (intelligible) aspect of the Being.

Heidegger did not show as much regard for logic as Wittgenstein did. One may even be surprised by his somewhat peculiar mode of thinking where a Greek etymology can replace a definition and the verse of a poet can take the place of a demonstration. Nevertheless, his vision most often remains rather clear and always rich. He does not propose as far as I know an explicit answer for the problem of founding a theory of knowledge and he even slights the kind of knowledge one can get from science. In his opinion, one cannot reach the Being by experiment, because the "Being withdraws itself as soon as it is disclosed in the instant." Whatever the meaning of this poetic sentence, it strangely looks when reading it in its context like a transmutation from potentiality to actuality that is not so foreign to physics.

In any case, one can at least conclude that the problem of finding a foundation for a theory of knowledge is as acute nowadays as it ever was. One can also say that the two traditional approaches are still representative of modern thinking. Finally, one may also add that the "essentialist" approach might be less open to trouble than the positivist one. After Wittgenstein, and also after what could be learned from quantum mechanics, one may pass judgment on the nature of ordinary language, common sense, and even philosophical thinking: they are frail, questionable, and nevertheless unavoidable.

Every question, whether concerning philosophy or science, is expressed in a language. Every answer, even when based upon experiment, must use logic. To judge its validity, to adopt it or not, one must use reason. The final formulation of the answer, even if it is a scientific law, will be expressed in some language, often partly in the mathematical one. The fundamental questions tend therefore always to be hidden behind the veil of our ignorance concerning the nature of language, logic, and reason. This ignorance erects a screen behind which science seems unable to make explicit its own relation to Reality.

It might look as if one were condemned to turn indefinitely inside the same circle when trying to understand Reality as such. One can now however dispose of a few significant hints upon which one can rely. Modern science is suggestive of a totality, i.e., of an ability to approach all the aspects of reality, even if some of them remain untouched or doubtful. Much more strikingly and undoubtedly, modern science shows universality, i.e., some of its most basic principles have a very wide, even huge if not unlimited domain of validity.

These rather recent features have deeply changed the whole character of science. Not so long ago, the discovery of a new phenomenon was more or less at the origin of a new science, so that science itself was an outgrowing process. By now and more and more often, science also shows an ingrowing process where vast parts of one branch of science become integrated in the principles of another, or they are both brought together into a common synthesis. The first example of this was the synthesis of electricity and optics by Maxwell, which was realized around 1864 and became plainly manifest near the end of the century. Atomism and later on its rules, namely quantum mechanics, brought a synthesis between chemistry and physics together with so many mergings and reorderings that it would take too long to enumerate them. Quite recently, electrodynamics and the weak interaction were also united. Moreover, the evergrowing family of elementary particles was also reduced to a few quarks and leptons. There is a strong stability of the basic physical laws under the occurrence of important new discoveries. Relativity and quantum mechanics itself did not make classical mechanics obsolete but only rejected its outer semiphilosophical frame. As far as practical knowledge and a corpus of results were concerned, they remained valid and did not pass into oblivion.

One might also take into account what the conceptual progress in science suggests about conceptual progress in philosophy. The clarification of the notions of space and time, even if one recognizes that it still leaves some difficult problems, was also an occasion to purify our philosophical conceptions. It is remarkable, for instance, how a significant part of the beautiful construction in Kant's Critique of Pure Reason is now disproved because of changes of that kind, though nothing as ambitious has come to replace it. One may even contemplate a troublesome process where science is over and over again blooming into new syntheses while philosophy is more and more reduced to smaller pieces. One may wonder in this connection whether the difficulties with which physics was confronted when trying to understand quantum mechanics were not responsible in part for a lack of ambition in the modern attempts to build a new theory of knowledge. By now, even if one were not to agree that quantum mechanics can be understood, it offers anyway so many hints about logic, language, and the nature of the scientific representation that it seems worthwhile to try making use of it.

A Question of Method

It will be convenient to put some order among these ideas before proceeding further. We are entering ground where science and philosophy meet, and this zone of contact is unclear. The culture of our times has always too strongly enforced the gap between science and philosophy. One has to go rather far back in time to find an approach where this divorce did not yet exist and where one of the two partners was not dominant. This can be found for instance in the *Novum Organum* by Francis Bacon (1561–1626), where he said that science is a work to be realized, which cannot be achieved in the span of one generation but must be pursued by others. He also said that the most basic axioms of science will only be reached at the end though. When they are obtained, they will not be empty or vague notions, but something well defined such that Nature itself would recognize them as its first principles, lying in the heart and the marrow of things. According to Bacon, the aim of philosophy was to deeply understand the results of science after having digested and transformed them.

Bacon's program, which assumed that the foundations of science and of philosophy would not be apparent at the beginning of the enterprise but would be revealed simultaneously at the time when the fruits became ripe, might again become the program of our age, which is perhaps the time of picking the fruits. Whether this is true depends of course upon a preliminary question, which is whether or not science has reached this kind of maturity. Even if it were so, the old difficulties arising from the ambiguous roles of language and reason would be a serious obstacle and, in order to overcome them, some specific method should be found.

To get a hint, one can start from science as it now exists, namely as an empiricist would see it, together with the ground where it grew, the facts, the language, and the use of reason, which themselves remained unquestioned. What can be seen, from the relative summit where science is now standing, may be used to enlighten, if only a bit, these old preliminaries. In other words, science is no longer an enterprise as in Bacon's time but it has now reached a mature level with both universality and depth; its almost incredible success has become a phenomenon, which is itself asking for an explanation. Can this be a new starting point to better explain its own roots and to reconsider its philosophical status? Why should it be examined always within the same old categories?

This idea of a method is better described by an image. The growth of science and philosophy can be seen as going along an ascending helix, so that the obscure preliminaries involved in a previous stage might be enlightened from a newly reached point of view. This effect involves first of all the status of language and reason. They have to be taken for granted at the beginning of the enterprise, but they can later become the objects of a learned investigation. The principles (by which one does not mean here the principles found by science but the philosophical principles asserting the status of science and of philosophy itself) cannot stand at the beginning, but they might become revealed during the process of progress.

It must be recognized that this kind of approach is radically opposed to the current main trends of epistemology. It assumes that something objective concerning the status of language and reason will be obtained sooner or later, so that the common formulation of science and philosophy will become less and less dependent upon our humanity and more and more an expression of reality. Modern epistemology goes in the opposite direction. It is fascinated by the historical, the sociological, and the psychological aspects of the development of science, its erratic events, its changes of perspective, and its controversies. It may go as far as denying the possibility of a method. As a matter of fact, few scientists recognize in modern epistemology what they see and know in science. This is because they are keenly interested, sometimes even fascinated, by reality and this is precisely what is missing most in our contemporary epistemology. It has gone to such extreme limits in its relativization of science that one can bet that the pendulum will soon have to go back in the opposite direction.

SUMMARY AND OUTLOOK

Orientation

It will be convenient to pause at this point and state directly a few simple-minded questions, though not simple ones, in the order in which they will be considered:

1. What is science and what is its relation to its object, namely reality?
2. In view of the specific role of logic and mathematics in science, one may also ask what they are themselves.
3. Why is the present extent and universality of science possible? In other words, can this remarkable fecundity give suggestions about the ontological status of science?
4. How does it happen that language, whichever one may be at issue, is possible?

These questions, which are certainly too wide, would require long discussions as well as a serious argumentation. This would not be appropriate in the few final pages of a book mainly devoted to a more specialized subject. We will therefore only give a few indications. Whether this will convince the reader or not is not essential because what is important is to launch the discussion and to have it going on though, as a matter of fact, it is easy to convince oneself that most questions concerning the foundations of physics amount to the present ones in one form or another.

Science as a Representation

The nature of science has already been considered on various occasions and we shall maintain the point of view according to which science is a representation of reality. It means that this representation evolves according to the tempo of history, it is made by men in their own cultural and social environment but it also deals with reality, which is something intrinsically different from humanity, something in fact wider. Science is expressed by various kinds of propositions, among which a few are principles, some are laws that can be derived from the principles or which are not yet completely related to them, and others are only the records of many facts more or less well understood. The representation is not completely contained in any mind and it remains distributed among many minds, many books, many papers or samples, like a representation of the surface of the Earth is distributed among the many maps of an atlas. It can be active or dormant. It changes as time goes on, sometimes even in its architecture as Kuhn or Foucault would have it when they talk respectively of paradigms and *épistèmes*.

There are many other representations of reality—religious, philosophical, and artistic. What is enough to distinguish science uniquely among them is its method, as was described in Chapter 2. The arguments by Feyerabend and others against the existence of a method will be disregarded. Their assertions come from an undue extension of the unavoidably human character of a representation towards a questioning of its quality, as if it were necessarily representing something that is only human. One could also criticize it as leaving aside the notion of time. It neglects the order of magnitude of the time necessary to get difficult problems settled, as compared with the shorter time involved in the continuous generation of more or less trivial human ideas and in controversies, as well as the temporary confusion accompanying a period of intense research activity. Whatever it may be, the main features or steps of this method, as mentioned in Chapter 2, are the following:

1. An empirical exploration of reality. Facts are observed, properties are classified, some empirical "laws" can be found.
2. The invention of a conceptual scheme to encompass the known facts and to order them. It almost always has a mathematical expression in the case of physics.
3. A theoretical elaboration of all the conceivable consequences of this scheme.
4. A verification, i.e., the submission of these consequences to experiments or observation.

About the Nature of Mathematics and Logic

The scientific method therefore involves an interaction among three poles: experiment, imagination, and logic-mathematics. We shall leave aside the fascinating role of imagination, just as one may leave aside the biography of Stanley when one looks at a map of Africa. What remains essentially, once this human aspect has been removed, is a confrontation between reality and logic, if one avoids separating logic from mathematics.

The extent and the depth of the agreement between reality and a logico-mathematical theory, as it appears in contemporary physics, is extraordinary. It goes so far that mathematics now frequently replaces ordinary language. The recent progress in quantum mechanics discussed in the previous chapters has even shown that this replacement could be complete, at least in principle. This feature is not only a matter of convenience or precision, it is unavoidable and one cannot give any reliable idea of what quantum mechanics is without using at least partly the language of mathematics.

One thus arrives at the second question concerning the nature of logic and mathematics. Since one cannot avoid using them when representing reality by science, one cannot dig further into the question of reality without committing oneself about the nature of logic and mathematics. It is not sufficient to register with d'Espagnat that the knowledge of reality might be expressed either in common terms or in mathematical ones. As soon as it becomes clear that mathematics is necessary, it is no longer an accessory but a key to the theory of knowledge. One might wonder whether reality enlightens the nature of mathematics or whether the converse is more appropriate, but one cannot in any case separate them.

The nature of mathematics (and accessorily logic) has been a subject of discussion among mathematicians and philosophers for centuries. Brain physiologists and informaticians have also recently entered the field. As usual in these very old problems, there are many positions differing from each other by important divergences or subtle shades. One can try nevertheless to get to the essentials by considering that there are mainly two opposite poles—namely, realism and nominalism. In the first one, it is assumed that there exists something, independent from the human mind, which is explored by the mathematician and discovered by him rather than being invented by him. This is why the name realism is used for it, even if one is careful to distinguish the "mathematical reality" from physical reality. The main representatives of this opinion are practicing mathematicians, often among the most creative ones, who claim that they find a significant or a huge gap between the ingenuity they put in their mathematics and the results they obtain. They are also impressed by the increasing consistency occurring in mathematics, when the concepts coming from a remote field bloom in a very different one, as well as with the extraordinary fecundity of some ideas. The renewal of production one is now contemplating in mathematics after an ascetic period of axiomatization can only increase the strength of these convictions.

Opposed to realism, nominalism goes back essentially to Leibnitz. It considers mathematics as some sort of a game where the rules are essentially arbitrary, like the rules of a chess game. The whole system is said to be hypothetico-deductive, i.e., consisting in a game of hypotheses from which one deduces consequences. Nominalism has the favor of a majority of philosophers and epistemologists. This is easily understood, because it is much easier to sustain than realism, since it is always easier in philosophy to find faults in an asserting argument than in a negative one. The nominalist reduces mathematics to other unknowns he has already accepted, like those concerning the mind or language and he prefers of course not to bother with a third unknown whose existence is perhaps intuitively obvious to a few people but remains rather uncertain. He is however unable to account for the fecundity of mathematics or for the surge of results, in all domains, coming from this supposedly arbitrary product of the brain.

It is worthwhile introducing physics into the contest because it brings with it a few significant new aspects. Its main point is not so much the consistency of mathematics or its intrisic fecundity, but rather its extraordinary efficiency when formulating science. It is a necessity in that role and therefore mathematics appears to have an intimate link with physical reality. It might be that some philosophical framework other than realism or nominalism would take these features into account, but it is simpler to keep only the two of them. So when it comes to choosing among them, one does not see what would be gained from nominalism. What can the relation be between an arbitrary hypothetico-deductive system and physical reality? When mathematics is assumed to be a branch of language, as one also sometimes does, nominalism is again illusory. It cannot deny an intrinsic value to language, which cannot be purely arbitrary since it must represent reality, if it means anything. So nominalism cannot deny the possibility that mathematics, by this ununderstood adequacy, partakes of reality, even if it is only a part of language. Finally, nominalism remains under Wittgenstein's criticism concerning the impotency of pure language.

We may therefore try to adopt mathematical realism as a preliminary step in our investigation towards a theory of knowledge. Of course, we must avoid a trivial difficulty of vocabulary, which would be to give the same name, reality, to what is explored by the natural sciences and what is explored by mathematics. Since the name of *Logos* has often been used in this framework, it will be convenient to keep it, barring of course other metaphysical or religious connotations of this word. The logos could be an entity, characterized as being different from physical reality, something more abstract, which is represented by logic and mathematics as reality is represented by science (for the sake of clarity, the name of science is reserved for the representation of physical reality). Logos is supposed to be objective, i.e., existing by itself and independent of the human mind.

This notion is far from obvious and somewhat difficult to sustain against a nonmathematician.[14] Some opponents would say that mathematics is one of the best examples of human creativity and it is accordingly nonsense to objectify it. A musician might say that he is also exploring and discovering ... why not? The difficulty is due to a tremendous difference between the two kinds of reality one is envisioning. Physical reality is everywhere present to our senses and we believe we know it from perception and intuition, long before asking devious questions about the validity of this knowledge. Our language as well as our other mental tools such as mathematics were first directly patterned by this concrete reality, before one began to pay attention to their intrinsic nature. Mathematics does not suggest a different kind of reality at an elementary level, but rather something that is imitating it or following from it.

Moreover this kind of consideration seems to go back to an epoch in the history of philosophy one might have thought to have ended together with the controversy on universals. As a matter of fact, the philosophical categories most able to express the existence of two different kinds of reality are to be found among older scholastic notions, like the *modes* expressing the mutually irreducible aspects of *substance*. Maybe this is just what one should revive after all, assuming physical reality and logos to be two modes of what Heidegger would call Being.

Again in favor of this point of view, one might say that it would perhaps be more easily accepted if it came directly from philosophy. The pure philosopher may start from a postulated unity and call it Being. He may then concede the necessity of distinguishing two modes of being and call them reality and logos or whatever else. Heidegger did that recently with other names and using another argumentation and this was considered to be serious philosophy. A physicist groping with his science is after all following the same path in the opposite direction. He is first confronted with physical reality and he cannot escape the use of mathematics. He then makes a jump and assigns an objective reality to mathematics and finds his own little logos. "Well, my boy," the philosopher would say, "what you have just done is called an hypostasis. In case you don't understand, it means giving a status of reality and objectivity to something you first knew as a concept. This is one of our entrance tests, please feel at home with us."

It might very well be of course that both the philosopher and the physicist are wrong. The notion of logos is obviously insufficiently developed and it is rather questionable. We shall see however that it offers a possible way out of several acute problems. Paraphrasing a famous sentence, one might say that the reality of logos may not be the best solution, but the others are much worse.

20. Total Realism

On the basis of this discussion, it becomes possible to bring up another issue in the old controversy about the realistic character of quantum mechanics. Recall that the position of realism, so well described and analyzed by d'Espagnat, suffered from three weaknesses: Kant's critique against the possibility of knowing reality in itself, i.e., a doubt about the meaning of the basic notions of reality; the lack of a theory of knowledge where what is meant by "knowing" reality would be clearer; and the lack of a wider philosophical background where the meaning of being able to know, "in principle" would be explained. We now try to answer these three questions.

The corresponding philosophical position might appropriately be called a total realism. It is total in so far as it ascribes a status of reality to a supposedly objective logos, which is explored by human logic and mathematics. Although when starting from this position, one reaches the same practical conclusions as "veiled realism", they no longer appear as the consequence of some kind of impotence but as an intrinsic necessity, which is a better bed to lie on. It also inscribes the problem we started with in a much wider framework, which provides another view of science itself.

A Founding Principle for Science

We first consider this perspective as offering a philosophical background for the formulation of science and its epistemic problems. It may be recalled that the present reflections went through other intermediate questions: wondering first how science is possible, then considering its remarkable present extent and universality, then wondering why reality is understandable. Many scientists have a spontaneous answer to both questions. Before beginning to indulge in philosophical considerations, they are tempted to answer: science is possible and it meets such a great success because there is an order in the universe. They might as well say that science is possible and efficient because reality is ordered.

This answer is however empty for a nominalist. Reality being what it is, if my logic and my mathematics are only arbitrary constructions and if I judge the existence of order by using them, I am only seeing a correspondence between reality and arbitrariness and the order one is talking about has no proper meaning. The answer is on the contrary quite satisfactory from the standpoint of total realism. One recognizes the objective existence of a reality and a logos. The first one is directly experienced while the other is not but, when one tries to make sure of what is best known about both of them, one has a good representation for each of them. Science is our representation of reality whereas our logic and mathematics offer a representation of logos. Humankind has explored them for centuries, and knowledge is the result of that effort. One finds that these two representations are much more strongly linked than what might have been expected beforehand. Not only is mathematics a convenient and more economical language for the formulation of science but it is now unavoidable, even intrinsic to science in a way. Since everything that can be said about reality can be surmised from its representation, the intimate link between science and mathematics can only be the image of a higher correspondence between reality and logos. In these conditions, the apparently meaningless belief in an order in the universe becomes the expression of something that is actually seen: science cannot do without mathematics, nor of course without logic. Reality and Logos are in correspondence. This is what one means when saying that reality is ordered. The extent and the success of science are then confirmations of the intensity of this correspondence. So for what it is worth, one has obtained a founding principle for science, which is this correspondence. It belongs of course to metaphysics (i.e., "after" physics, a reflection on physics) but the whole question of realism has always been a matter of metaphysics. One should not be ashamed of it, as long as no confusion arises concerning when one is doing science and when not.

A Theory of Knowledge

We also sketch a theory of knowledge, i.e., a framework for the concept of knowledge and some criteria for it. The difficulties with such a proposal, as they arise from the uncertain status of language, have already been stressed. "What is language and how is it possible?" was one of the questions we met on our way. It was also suggested that one might perhaps answer it, not in an a priori manner but from a certain height reached in the course of the development of science, as some sort of a view from above along the spiralling path of science. It might also be mentioned that the problem of understanding the status of language is nothing but the old problem of universals, which is still after all quite modern, at least according to Bertrand Russell. However it curiously remains absent from most attempts at epistemology.

Without trying to develop the ideas in detail, the main trend of the answer given by total realism is rather obvious. Ordinary language was probably born, like our visual representation of reality, from contemplating the insistent regularity of the facts surrounding us, the verticality of trees and the horizontality of the sea surface, the fall of objects (just look at the pleasure of a baby when he discovers that), the sequence of causes and effects. In other words, language was possible and efficient because reality is ordered. This looks like Hume's statements, except that the question he pronounced as inaccessible (how the world is ordered) is now answered by the laws of science.

One can then assert that a convenient theory of knowledge consists in considering it as the acquisition of two representations, one provided by science and one given by logic-mathematics. It involves the study of their correlations, under the assumption that they provide a representation. In other words, knowledge is essentially the representation of the correspondence between reality and logos.

This correspondence is so intimate that one can find many instances where it never failed, i.e., where no violation of a law, either an empirical rule or a principle, was ever noticed. This repetition and this determinism are at the origin of our concept of truth. Truth has two criteria of a very different nature. What is seen with attention is true, which is nothing but the existence of reality. What always occurs is true, which is the everpresent manifestation of order. The criterion for knowledge is truth, the criteria for truth are the ones just given.

The Status of Realism

Although these ideas remain rather sketchy, as one might expect when finding them in a book mainly devoted to a specific branch of science and not to a discourse about it, the consequences of total realism for the questions initially raised by ordinary realism are easily drawn. The interpretation of quantum mechanics enters in large part in their formulation. These consequences turn out to be quite different from what one would obtain with ordinary realism. Physical reality and logos are intrinsically different. There is a strong correspondence between them but its exact extent cannot be asserted a priori by a philosophical postulate. What we can guess about it can only come from what science shows.

Experiments belong to pure reality. How we express and transmit them belongs directly to the representation of reality. On the opposite, theory is a part of logos, or at least a special part of its representation that is particularly concerned with its relation to reality. How far their correspondence can go cannot be asserted a priori but we must learn it from science.

This correspondence is found to be limited in several respects, at least according to what we presently know. One manifestation of this limitation is the inadequacy of the theory in its coverage of actual facts. It might have been expected that something like that would happen anyway at some later stage of the theory, since it is only a manifestation of the intrinsic difference in nature between reality and logos. One also finds that some conceivable properties have no definite correspondence with reality, as happens when one talks about the value of an unmeasurable observable. Together, these two findings show the limitation of the correspondence at both ends. Some aspects of reality are not covered by theory and some aspects of logos have no counterpart in reality.

A correspondence that is by essence different from an identification must bring a limitation in the scope of knowledge, as was first shown by the approximate character of truth. Whereas truth most often holds with a high degree of accuracy, it is nevertheless quantitatively limited in each specific instance. What one can say about reality should always be expressed, in principle, with an assignment of its probability for being erroneous.

Quantum mechanics is extremely successful. It goes as far as expressing general rules for the basic correspondence, i.e., the laws of physics, which have not yet been found to fail in any instance and which, furthermore, go as far as estimating the probabilities of errors in the expression and the consequences of the facts. The theory is basically probabilistic with a precise meaning, which is that its whole construction and its intrinsic language rely upon some mathematical measures having the formal properties of probabilities. This structure belongs to the formulation of the correspondence with reality and it is useful information concerning the language best suited

SUMMARY AND OUTLOOK

to this correspondence. It becomes actually manifested by the randomness of events when the measures acquire the empirical status of probabilities, which also makes us confident in the actuality of the correspondence. To assume, with ordinary realism, that reality can be known arbitrarily well in principle, is a petitio principii relying upon a lack of a proper theory of knowledge. It indicates a lack of attention to the nature of science, a prebaconian scholastic metaphysical assumption, and finally a lack of foundations concerning the nature of truth.

The extraordinary fit between quantum mechanics and experiments indicates an unprecedented accomplishment, with its tremendous range covering many different orders of magnitude, its explanation of so many striking effects, and the deep hints it gives us for asserting the foundations of knowledge. This suggests that, maybe for the first time in history, we may have reached in this century a threshold where Bacon's program becomes fully meaningful, where perhaps a reliable philosophy of knowledge is able to begin. Whether it will have the features we have assumed here is of course another question. What is really important is that a new way may perhaps open. Our conclusion will be therefore that quantum mechanics is probably as realistic as any theory of its scope and maturity ever will be.

Notes

PREFACE

1. M. Gell-Mann and J. B. Hartle, in *Proceedings of the 3rd International Symposium on the Foundations of Quantum Mechanics in the Light of New Technology*, edited by S. Kobayashi, H. Ezawa, Y. Murayama, and S. Nomura (Physical Society of Japan, Tokyo, 1990). Practically the same paper also appeared in *Complexity, Entropy and the Physics of Information*, W. H. Zurek, ed., Santa-Fe Institute Studies in the science of complexity, No. 8 (Addison-Wesley, Redwood City, Calif. 1991). R. Omnès, J. Stat. Phys. **53**, 893 (1988), Ann. Phys. (N.Y.) **201**, 354 (1990). The essential equivalence of the two proposals was shown in R. Omnès, Rev. Mod. Phys. **64**, 339 (1992).

CHAPTER 1

1. M. Jammer, *The Conceptual Development of Quantum Mechanics*, McGraw-Hill, New York (1956).
2. N. Bohr, *Essays 1958–1962 on atomic physics and human knowledge*. Wiley-Interscience, New York (1963).
3. A. Einstein, B. Podolsky, and N. Rosen, Phys. Rev. **47**, 777 (1935).
4. J. von Neumann, *Mathematische Grundlagen der Quantenmechanik*, Springer, Berlin (1932). English translation: "Mathematical foundations of quantum mechanics." Princeton University Press, Princeton, New Jersey (1955).
5. Useful references concerning Hilbert spaces are the following books: P. Halmos, *Introduction to Hilbert Spaces;* F. Riesz and B. Sz-Nagy, *Functional Analysis*, Ungar, New York (1955); M. Reed and B. Simon, *Methods of Modern Mathematical Physics*, Academic Press, New York (1972).
6. See the previously mentioned references by Riesz and Sz-Nagy (especially clear) or by Reed and Simon.
7. In general, the continuous spectrum can be divided into two parts, the completely continuous spectrum and the singular spectrum. The completely continuous spectrum is Lebesgue measurable whereas the singular spectrum has zero Lebesgue measure and one must use a Stieltjes-Lebesgue integral in equations (1.44) and (1.45). Although singular spectra are met in the case of random potentials giving rise to Anderson localization, they do not occur frequently and they will not be considered in this book. This omission would be easy to remedy, if necessary, for specific applications.

8. J. Dieudonné, *Eléments d'Analyse,* Vol. 2, Gauthier-Villars, Paris (1974), Chapter XIV.

9. R. P. Feynman and A. R. Hibbs, *Quantum Mechanics and Path Integrals,* McGraw Hill, New York (1965). For a more mathematical approach, see B. Simon, *Functional Integration and Quantum Mechanics,* Academic Press, New York (1979).

10. E. P. Wigner, Ann. Math., **40**, 149 (1939); V. Bargman and E. P. Wigner, Proc. Nat. Acad. Sc., **34**, 211 (1948); E. Inonü and E. P. Wigner, Nuovo Cimento **9**, 705 (1952).

11. The universal covering group G' of a given continuous group G is the set of equivalent paths going from the identity to an element g of G, a path being equivalent to zero when it can be shrunk continuously to a point. In practice, this amounts to replacing the rotation group $O(3)$ by the spin group $SU(2)$, with a similar change for the Poincaré group.

12. Inonü and Wigner, Nuovo Cimento **9**, 705 (1952), have shown how the same results can be obtained directly by using nonrelativistic galilean changes of reference frame instead of Poincaré transformations.

13. L. D. Landau and E. M. Lifschitz, *Quantum Mechanics,* Pergamon Press, Oxford (1958). Chapter VI.

14. W. Thirring, *Quantum Mechanics of Atoms and Molecules,* p. 209, in: *A Course in Theoretical Physics,* Vol. 3, Springer, Berlin (1979).

15. L. Cohen, Ann. New York Acad. Sc., **480**, 283 (1986).

CHAPTER 2

1. Discussions of measurement theory can be found in B. d'Espagnat, *Conceptual Foundations of Quantum Mechanics,* Benjamin, Reading, Mass. (1976); H. Primas, *Chemistry, Quantum Mechanics and Reductionism,* Springer, Berlin (1983). See also J. von Neumann, *Mathematical Foundations of Quantum Mechanics,* Princeton University Press, Princeton (New Jersey) and E. P. Wigner, American J. Phys. 31, (1963), p. 6. A useful collection of reprints, with comments, is provided in J. A. Wheeler and W. H. Zurek, *Quantum Theory and Measurement,* Princeton University Press, Princeton (New Jersey) (1983).

2. As defined here, a phenomenon is some fact occurring on a macroscopic scale. In later chapters, it will be shown that real facts are necessarily macroscopic. We shall then reserve the word "fact" for something actual and the word "phenomenon" for something of the same nature though only theoretically possible. This distinction is mentioned presently to avoid leaving unnoticed a slight change in language that might be otherwise troublesome. It is however somewhat premature in the present chapter.

3. It will be shown in Chapter 3 that a wave function defines a property of a system.

4. See T. Tani, Physics Today, **42**, 36 (September 1989); J. Belloni-Cofler, J. Amblard, J. L. Marignier, M. Mostafavi, *Endeavour,* **15** (1), 2 (1991).

5. E. P. Wigner, Am. J. Phys. **31**, 6 (1963).

NOTES TO CHAPTER 2

6. J. von Neumann, *Mathematical Foundations of Quantum Mechanics*, last chapter.

7. There is apparently no such drastic and uncompromising statement in Bohr's writings as the one written here, but careful words are often ambiguous as anybody who has tried to translate them into another language may have noticed. In *Atomic Physics and Human Knowledge* (Wiley-Interscience, New York, 1963), where Bohr elaborates upon this question, the statements are so carefully ambiguous that one has clearly the impression of reading a philosophical text and not one about physics mentioning rules of physics. A clearcut statement of what is given here as the Copenhagen doctrine can be found, for instance, in the book by Landau and Lifschitz.

8. The effect was hinted at and even discovered several times, particularly by R. P. Feynman and F. L. Vernon, Jr., Ann. Phys. (N.Y.) **24**, 118 (1963); K. Hepp and E. H. Lieb, Helv. Phys. Acta, **46**, 573 (1974); W. H. Zurek, Phys. Rev. D **24**, 1516 (1981), ibid. D **26**, 1862 (1982).

9. J. von Neumann, *Mathematical Foundations of Quantum Mechanics*, Princeton University Press (1955).

10. J. S. Bell, Rev. Mod. Phys. **38**, 447 (1966).

11. J. von Neumann, *Mathematical Foundations of Quantum Mechanics*.

12. F. London and E. Bauer, *La théorie de l'observation en mécanique quantique*, Hermann, Paris (1939).

13. E. P. Wigner, Amer. J. Phys. **31**, 5 (1963). See also his essay "Remarks on the mind-body problem" in *The Scientist Speculates*, I. J. Good, ed., London (1962), p. 284; reprinted in E. P. Wigner, *Symmetries and Reflections*, Indiana University Press, Bloomington (1967).

14. N. Bohr, *Atomtheorie und Naturbeschreibung* (1931), English translation: "Atomic theory and the description of nature," Cambridge University Press (1934); *Kausalität und Komplementarität*, Erkenntnis, **6**, 293 (1936), English translation in Philosophy of Science, **4**, 289 (1937); *On the notions of causality and complementarity*, Dialectica, **2**, 312 (1948), reprinted in Science, **111**, 31 (1950); *Quantum physics and philosophy, causality and complementarity*, in Philosophy in the Midcentury; a Survey, R. Klibansky, ed., La Nuova Italia Editrice, Florence (1958); *Essays 1958-1962 on Atomic Physics and Human Knowledge*, Wiley-Interscience, New York (1963).

15. H. Primas, *Chemistry, Quantum Mechanics and Reductionism*, Springer, Berlin, 1983. The rich bibliography in this book is also particularly useful.

16. W. Pauli, *Die allgemeinen Prinzipien der Wellenmechanik*, in *Handbuch der Physik*, Vol. 24/1, Springer, Berlin (1933).

17. G. Lüders, Annalen der Physik, **8**, 322 (1951).

18. V. A. Fock, Usp. Fiz. Nauk, **62**, No. 6, 67 (1957); as quoted by Primas, *Chemistry, Quantum Mechanics, and Reductionism*.

19. E. Schrödinger, *Naturwissenschaften*, **23**, 807 (1935).

20. E. P. Wigner, Remarks on the Mind-Body Question, in *The Scientist Speculates*, I. J. Good, ed., Heinemann, London (1961).

21. E. P. Wigner, Am. J. Phys., **31**, 8 (1963).

22. See for instance A. F. Chalmers, *What Is This Thing Called Science? An Assessment of the Nature and Status of Science and its Methods*, University of

Queensland Press, St. Lucia (1976) where this question is discussed according to the modern trends in epistemology. This reference should not however be taken for an agreement of the present author with these trends.

23. A letter dated May 7, 1952 in A. Einstein, *Lettres à Maurice Solovine*, Gauthier-Villars, Paris (1956). I wish to thank L. Verlet for bringing this reference to my attention.

24. A. J. Leggett, Progr. Theor. Phys. (Suppl.) **69**, 1 (1980).

25. See Chapter 11 for references.

26. Aharonov, Albert, and Vaidman, Proceedings of the New York Academy of Science, **480**, 417 (1986).

INTERLUDE

1. John von Neumann, *Mathematical Foundations of Quantum Mechanics*.
2. R. B. Griffiths, J. Stat. Phys. **36**, 219 (1984); Am. J. Phys. **55**, 1 (1987).
3. R. Omnès, J. Stat. Phys. **53**, 893 (1988).
4. R. Omnès, J. Stat. Phys. **57**, 357 (1989); Ann. Phys. (N.Y.) **201**, 354 (1990).
5. R. P. Feynman and F. L. Vernon, Jr., Ann. Phys. (N.Y.) **24**, 118 (1963); K. Hepp and E. H. Lieb, Helv. Phys. Acta **46**, 573 (1974); W. H. Zurek, Phys. Rev. **D 26**, 1862 (1982); A. O. Caldeira and A. J. Leggett, Physica A 121, 587 (1983).

CHAPTER 3

1. J. von Neumann, *Mathematical Foundations of Quantum Mechanics*; G.W. Mackey, *The Mathematical Foundations of Quantum Mechanics*, Benjamin, New York (1963).

2. F. Riesz and B. Sz. Nagy, *Functional Analysis*, Ungar, New York (1955); M. Reed and B. Simon, *Methods of Modern Mathematical Physics*, Vol. 1, Academic Press, New York (1972).

3. A. M. Gleason, J. Math. Mechanics, **6**, 895 (1953).

4. R. Jost, in Studies in Mathematical Physics, *"Essays in Honor of Valentine Bargmann,"* E. H. Lieb, B. Simon, and A. S. Wightman, eds., Princeton Series in Physics, Princeton University Press (1976).

5. S. Maeda, Reviews in Mathematical Physics, **1**, 235 (1990). I wish to thank K. Chadan for bringing this reference to my notice.

CHAPTER 4

1. R. B. Griffiths, J. Stat. Phys. **36**, 219 (1984); Am. J. Phys. **55**, 11 (1987).

2. M. Gell-Mann and J. B. Hartle, in *Complexity, Entropy and the Physics of Information*, W. H. Zurek, ed., Santa Fe Institute Studies in the Science of Complexity, No. 8, Addison-Wesley, Redwood City, Calif. (1990).

3. R. B. Griffiths, J. Stat. Phys. **36**, 219 (1984); R. Omnès, J. Stat. Phys. **53**, 893 (1988).

4. P. R. Halmos, Trans. American Mathematical Society, **144**, 381 (1969). I thank J. Ginibre for calling this result to my attention.

CHAPTER 5

1. Voltaire, *L'Ingénu*. Complete Works, Vol. 2, Gallimard, Paris, 1952.
2. There are many original texts, often clearer than more modern books, in J. van Heijenoort, *From Frege to Gödel, A Source Book in Mathematical Logic*, Harvard University Press, 1967.
3. G. Birkhoff and J. von Neumann, Annals of Math. **37**, 823 (1936).
4. See various contributions in *Entropy, Complexity and the Physics of Information*, W. H. Zurek, ed., Proceedings of the Santa Fe Institute in the Science of Complexity, No. 8, Addison-Wesley, Redwood City, Calif. (1990).
5. See G. W. Mackey, "The Mathematical Foundations of Quantum Mechanics," Benjamin, N. Y. (1963).
6. F. W. Mott, Proc. R. Soc. London, Ser. A **126**, 74 (1929).
7. See, for instance, M. L. Goldberger and K. M. Watson, *Collision Theory*, Wiley, New York (1964), chap. 8.
8. B. Misra and E. C. G. Sudarshan, J. Math. Phys. **18**, 756 (1977); C. B. Chiu, E. C. G. Sudarshan, D. Misra, Phys. Rev. **D 16**, 520 (1977).
9. V. Weisskopf and E. P. Wigner, Zeit. Phys. **63**, 54 (1930), see also M. L. Goldberger and K. M. Watson, *Collision Theory*, Wiley, New York, (1964), chap. 8.
10. M. Gell-Mann and J. B. Hartle, in *Complexity, Entropy and the Physics of Information*, W. Zurek, ed., Santa Fe Institute Studies in the Science of Complexity, No. 8, Addison-Wesley, Redwood City, Calif. (1990).
11. E. Wigner, Göttinger Nachrichten **31**, 546 (1932).

CHAPTER 6

1. A. Bohr and B. R. Mottleson, *Nuclear Structure*, Benjamin, Reading, Mass. (1975).
2. C. Fefferman, Bull. Amer. Math. Soc. **9**, 129 (1983).
3. L. Hörmander, Comm. Pure Appl. Math. **32**, 359 (1979).
4. E. Wigner, Phys. Rev. **40**, 749 (1932).
5. H. Weyl, Bull. Amer. Math. Soc. **56**, 115 (1950).
6. L. Hörmander, Comm. Pure Appl. Math. **32**, 359 (1979).
7. L. Hörmander, *The Analysis of Linear Partial Differential Operators*, 4 Volumes, Springer, Berlin (1985).
8. C. Fefferman, Bull. Amer. Math. Soc. **9**, 129 (1983).
9. See for instance L. D. Landau and E. M. Lifschitz, *Quantum Mechanics*, Chapter 7.
10. J. Ginibre and G. Velo, Comm. Math. Phys. **66**, 37 (1979); G. Hagedorn, Comm. Math. Phys. **77**, 1 (1980); Ann. Phys. (N.Y.), **135**, 56 (1981).

11. This theorem is due to L. Hörmander, Arkiv für Math. **17**, 297 (1979) where it occurs as a lemma. Its adaptation to the framework of quantum mechanics and a more explicit form of the errors was given by R. Omnès, J. Stat. Phys. **57**, 357 (1989). The proof given in Appendix B is new.

12. Gaussian states are known to be conserved in two simple cases in which the kinetic energy depends only upon momentum and is a second-order polynomial: (i) when the potential is a second-order polynomial in terms of the position coordinates, and (ii) for the Kepler problem.

13. G. A. Hagedorn, Comm. Math. Phys. **71**, 77 (1980), Ann. Phys. (N.Y.) **65**, 58 (1981).

14. See, e.g., V. I. Arnold, *Mathematical Methods of Classical Mechanics*, Springer, Berlin (1987); L. P. Cornfeld, S. V. Fonin, and I. Ya. Sinai, *Ergodic Theory*, Springer, Berlin (1982); D. Ruelle, *Chance and Chaos*, Princeton University Press (1991).

15. This theorem is in some sense the converse of a well-known theorem by Ju. V. Egorov, Uspehi Mat. Nauk **24**, 235 (1969) showing how an approximate unitary representation is associated with a classical canonical representation. It was first given in R. Omnès, J. Stat. Phys. **57**, 357 (1989). The proof given in Appendix B is more elaborate.

16. G. Hagedorn, Comm. Math. Phys. **71**, 77 (1980).

17. Hörmander's book does not give this estimate but only less precise ones. It can be found in another work: L. Hörmander, Ark. Math. **17**, 297 (1973). I thank Professor Hörmander for communicating this work.

18. Ju. V. Egorov, Uspehi Mat. Nauk **24** (5), 235 (1969).

19. The tensor conventions for contravariant and covariant coordinates are not used in these exercises. This should be taken into account when comparing the results with Section 3.

20. M. Nauenberg, Phys. Rev. **A 40**, 1133 (1989).

CHAPTER 7

1. G. C. Wick, A. S. Wightman, and E. P. Wigner, Phys. Rev. **88**, 101 (1952).
2. W. H. Zurek, Phys. Rev. **D 26**, 1862 (1982).
3. K. Hepp and E. H. Lieb, Helv. Phys. Acta **46**, 573 (1973).
4. V. Weisskopf and E. Wigner, Zeitschrift Phys. **63**, 54 (1930).
5. W. H. Zurek, Physics Today, **44**, 36 (1991).
6. R. P. Feynman and F. L. Vernon, Jr., Ann. Phys. (N.Y.) **24**, 118 (1963).
7. A. O. Caldeira and A. J. Leggett, Physica A **121**, 587 (1983).
8. A. O. Caldeira and A. J. Leggett, Ann. Phys. (N.Y.) **149**, 374 (1983).
9. M. Gell-Mann and J. B. Hartle, Phys. Rev. **D 47**, 3345 (1993).
10. J. Mourad and R. Omnès, unpublished.
11. E. Joos and H. D. Zeh, Z. Phys. B **59**, 223 (1985).
12. G. Jona-Lasinio and P. Claverie, Progr. Theor. Phys., Suppl. **86**, 54 (1986).
13. The fluctuations in the refractive index in a small space region are known to be proportional to the fluctuations ΔN in the number N of molecules in that region, ΔN being of the order of $N^{1/2}$ according to statistical mechanics. These fluctuations

are responsible for the diffusion of light and they can therefore be easily detected by looking at the light diffused in all directions when a beam of light crosses a gas tank. In an optically active medium, the index of refraction is not the same for right-polarized and left-polarized light.

14. K. Hepp, Helv. Phys. Acta, **48**, 80 (1975).

15. J. S. Bell, Helv. Phys. Acta, **48**, 93 (1975).

16. It is stated for instance as Rule 7 in the list of the axioms of quantum mechanics by B. d'Espagnat, *Conceptual Foundations of Quantum Mechanics,* 2d ed., Benjamin, Reading, Mass. (1976), Chapter 3.

17. A. O. Caldeira and A. J. Leggett, Ann. Phys., **149**, 374 (1983). These authors take into account the very small probability of occupation of an eigenstate in the environment to show that each eigenstate can be replaced mathematically by a harmonic oscillator since this model oscillator will never reach its second excited state or a higher one. The state where all the oscillators are in their ground state is identified with the ground state of the environment, which is therefore treated in a different way. This model is probably justified for a superconductor.

18. R. J. Solomonoff, *Information and control,* **7**, 1; 224 (1964). A. N. Kolmogorov, *Problemy Perdachi Informartsii,* **1**, 1 (1965). English translation: *Prob. Infom. Transmission,* **1**, 1 (1965). G. J. Chaitin, *J. Assoc. Computing Machinery,* **13**, 547 (1966).

19. C. M. Caves in *Physical Origins of Time Asymmetry,* J. J. Halliwell, J. Pérez-Mercader, and W. H. Zurek, eds. Cambridge University Press. To be published in 1993.

20. A. K. Zvonkin and L. A. Levin, Usp. Mat. Nauk, **25** (6), 85 (1970). English translation: Russ. Math. Surveys, **25** (6), 83 (1970). C. H. Bennett, Int. J. Theor. Phys. **21**, 905 (1982). W. H. Zurek, Phys. Rev. **A 40**, 4731 (1989). C. M. Caves in *Complexity, Entropy and the Physics of Information,* W. H. Zurek, ed. Addison-Wesley, Redwood City, Calif., p. 137 (1990).

21. L. Van Hove, Physica, **18**, 145 (1952).

22. E. Joos and Z. H. Zeh, Z. Phys. B **59**, 223 (1985).

CHAPTER 8

1. H. Everett III, Rev. Mod. Phys. **29**, 454 (1957). See B. S. De Witt and N. Graham, eds., *The Many Worlds Interpretation of Quantum Mechanics,* Princeton University Press (1973).

2. G. Lüders, Ann. Phys. **8**, 322 (1951).

3. M. Gell-Mann, J. B. Hartle, in *Complexity, Entropy and the Physics of Information,* W.H. Zurek, ed. Santa Fe Institute Studies in the Science of Complexity, Addison-Wesley (1990); also in *Proceedings of the 3rd International Symposium on Quantum Mechanics in the Light of New Technology,* ed. by S. Kobayashi, H. Ezawa, Y. Murayarna, and S. Nomura, Physical Society of Japan (1990).

4. G. C. Ghirardi, A. Rimini, and T. Weber, Phys. Rev. **D 34**, 470 (1986).

5. G. C. Ghirardi, private communication.

6. In his *Tractatus logico-philosophicus* (1922), Ludwig Wittgenstein constructed a logical description of reality as follows. He considered the world to be a

totality of "facts" consisting in the existence of "states of affairs", also translated into English as "atomic facts". A state of affairs is composed of simple objects, each of which can be named. These names can be combined to give elementary propositions so as to describe a possible state of affairs, the elementary propositions being themselves independent from each other. Reality is then said to consist in the existence of a state of affairs and the nonexistence of all others.

This approach is easily mimicked in the logical formulation of quantum mechanics. Wittgenstein's "facts" are potential, i.e., what we called phenomena. If one assumes the existence of objects, the "elementary propositions" can be assimilated to the classically meaningful propositions; these propositions are independent with a very good approximation (i.e., their projectors almost commute). Accordingly, the notion of "state of affairs" and Wittgenstein's propositions about reality can be used properly. One should however notice that the world also involves microscopic objects, which neither manifest themselves by phenomena nor are they reducible to phenomena. It is not therefore identical to a totality of "facts" but it contains such a totality as well as something more. (It should be noticed that Wittgenstein's "world" is not identical to reality but a complete collection of possible realities so that it belongs clearly to the domain of theory.)

One may also notice the existence in quantum mechanics of properties that cannot be "facts" and cannot even be checked as being existent or not. An example of this was given in Chapter 7 with the property stating that "the energy level k of a macroscopic system is occupied." This also fits with a remark by Wittgenstein (Section 36 in his *Philosophical Remarks*, which was completed in 1945), where he envisions the possibility of things that are neither existent nor nonexistent. This was a puzzle for philosophers.

Wittgenstein proceeds from there by showing how elementary propositions can be combined by logic. He assumes of course a unique form of logic, but this is perfectly consistent with our standpoint, as long as he deals only with propositions concerning phenomena. He does it in a rather technical and complete way because this is important for his purpose. He aims to show the consistency and completeness of a philosophy of knowledge relying upon the elementary propositions, as well as establishing an inconsistency when one tries to say anything more than that. This is where his analysis of truth is pursued in greater detail, but one can retain from it only that the above criterion of truth for the elementary propositions (as existing states of affairs) is consistent from a philosophical and a logical standpoint. This is enough for our own purpose, which was to make sure that our own approach to reality and truth is significant from the standpoint of a valuable theory of knowledge.

One cannot however fail to mention Wittgenstein's famous sentence in the *Remarks* where he states that "the table I see is not made of electrons." This agrees well with the impossibility of identifying in quantum logic a classically meaningful property with a complete set of more elementary properties referring to each constituent particle in the object.

7. It might be noticed that one stands upon an even firmer ground than Wittgenstein could use since he would have accepted all the conceivable phenomena as possible, though this is not well defined on a quantum basis. One must indeed avoid the difficulty arising from the existence of nonexclusive phenomena (whose projectors do not multiply to 0). The only serious difficulty might come from relativity, since it

gives rise to a possible ambiguity in the definition of "present time". This last point will not be considered here.

8. A. Tarski, *Logic, Semantics, Metamathematics: Papers from 1923 to 1938.* Translated by J. H. Woodger. Clarendon, Oxford (1956).

9. R. Omnès, J. Stat. Phys. **62**, 841 (1991).

10. B. d'Espagnat, J. Stat. Phys. **56**, 747 (1989).

11. A. Einstein, B. Podolsky, and N. Rosen, Phys. Rev. **47**, 777 (1935).

12. To use a language introduced by Karl Popper and familiar to the philosophers of science, the knowledge afforded by the kind of state we are going to introduce could be said not to be falsifiable, just as Everett's and Bohr's representations cannot be falsified by experiment.

13. M. Gell-Mann and J. B. Hartle, in *Complexity, Entropy and the Physics of Information*, W. H. Zurek, ed., Santa Fe Institute Studies in the Science of Complexity, Addison-Wesley (1990) and in *Proceedings of the 3rd International Symposium on Quantum Mechanics in the Light of New Technology*, ed. by S. Kobayashi, H. Ezawa, Y. Murayama, and S. Nomura, Physical Society of Japan (1990).

14. For more details, see S. Weinberg, *Gravitation and Cosmology*, Wiley, New York (1972).

CHAPTER 9

1. A. Einstein, B. Podolsky, and N. Rosen, Phys. Rev. **47**, 777 (1935). A reprint is given by Wheeler and Zurek, eds., *Quantum Theory and Measurement*. Princeton University Press (1983).

2. N. Bohr, Phys. Rev. **48**, 696 (1935).

3. It may be noticed that in the case where the wave function (9.2) is a very sharp gaussian rather than a delta function, it can be normalized. There is then an uncertainty about the value of p_2 but it can be made arbitrarily small.

4. "Albert Einstein, Philosopher-scientist", II, 666, A. Schulpp, ed., Cambridge University Press (1982).

5. B. d'Espagnat made this statement particularly clearly in Scientific American **241**, November 1979, 128.

6. B. d'Espagnat, *Conceptual Foundations of Quantum Mechanics*, 2nd edition (Chap. 11).

7. D. Bohm and J. Bub, Rev. Mod. Phys. **38**, 453 (1966).

8. E. Nelson, *Dynamical Theories of Brownian Motion*, Princeton University Press (1967).

9. Quite recently, D. Dürr, S. Goldstein, and N. Zanghi, J. Stat. Phys. **67**, 843 (1992) have proposed an improved version of Bohm's theory. They show that the position of the particles, though intrinsically well defined, necessarily behave in a random way consistent with quantum mechanics for any experiment. As already mentioned, this confirms that the interpretation of their theory would have to go along the same steps as ordinary quantum mechanics. Their approach can only change, in their own words, the ontological status of particles, which is mostly a matter of philosophy.

10. J. Bell, Physics, **1**, 195 (1964). See also J. Bell, Rev. Mod. Phys. **38**, 447 (1966). A constructive generalization of Bell's inequalities has been given by M. Froissart, Nuovo Cimento **64 B**, 241 (1981). For extensive discussions, see A. Shimony in *Proceedings of the International Symposium on the Foundations of Quantum Mechanics,* Physical Society of Japan, 1983; B. d'Espagnat, *Conceptual Foundations of Quantum Mechanics*, 2d ed., Addison-Wesley, Reading, Mass. (1976) and *Physics Reports* **110**, 202 (1984); J. S. Bell, *Speakable and Unspeakable in Quantum Mechanics*, Cambridge University Press, 1987.

11. J. F. Clauser, M. A. Horne, A. Shimony, and R. A. Holt, Phys. Rev. Letters **23**, 880 (1969).

12. A. Aspect, P. Grangier, G. Roger, Phys. Rev. Letters **47**, 460 (1981); A. Aspect, J. Dalbart, R. Roger, Phys. Rev. Letters, **49**, 1804 (1982).

13. M. A. Horne, A. Shimony, and A. Zeilinger, Phys. Rev. Lett. **62**, 2209 (1989); Nature **347**, 429 (1990). N. D. Mermin, Am. J. Phys. **58**, 731 (1990).

CHAPTER 10

1. Our discussion is inspired by the *Feynman Lectures on Physics,* Vol. 3, Addison-Wesley, Reading, Mass. (1963).

2. See, for instance, S. Weinberg, Supplement Progr. Theor. Phys. **86**, 43 (1989).

3. Y. Aharonov and D. Bohm, Phys. Rev. **115**, 485 (1959).

4. R. C. Jaklevic, J. Lambe, A. H. Silver, and J. E. Mercereau, Phys. Rev. Letters **12**, 159 (1964).

5. A. J. Leggett, Progr. Theor. Phys. Supplement **69**, 1 (1980); Phys. Rev. **B30**, 1208 (1984).

6. A. O. Caldeira, A. J. Leggett, Ann. Phys. **149**, 374 (1983); Erratum **153**, 445.

CHAPTER 11

1. V. F. Weisskopf and E. P. Wigner, Zeit. f. Phys. **63**, 54 (1930); **65**, 18 (1930); F. E. Low, Phys. Rev. **88**, 53 (1952). For another detailed analysis, see M. L. Goldberger and E. M. Watson, *Collision Theory,* Wiley, New York (1964), Chap. 8.

2. B. Misra and E. C. G. Sudarshan, J. Math. Phys. **18**, 756 (1977); E. Joos, Phys. Rev. **D 29**, 1626 (1984); A. C. B. Chiu, E. C. G. Sudarshan, and B. Misra, Phys. Rev. **D 16**, 520 (1977); A. Peres, in *Quantum Theory without Reduction,* J. M. Lévy-Leblond and M. Cini, eds., Adam Hilger, Bristol (1990), p. 122; I. Singh and M. A. B. Whitaker, Am. J. Phys. **50**, 882 (1982).

3. Of course, the present interpretation differs with Bohr's by adding the "not always", since other examples have been given to the contrary.

4. A. Abragam, *The Principles of Nuclear Magnetism,* Clarendon, Oxford (1964), Chap. 8.

5. W. Nagourney, J. Sandberg, and A. Dehmelt, Phys. Rev. Letters **51**, 384 (1986); J. C. Bergquist, R. B. Hulet, W. M. Itano, and D. J. Wineland, Phys. Rev. Letters **57**, 1699 (1986); Th. Slauter, W. Neuhauser, R. Blatt, P. E. Toschek, Phys. Rev. Letters **57**, 1696 (1986).

NOTES TO CHAPTER 12

6. See W. Paul, Rev. Mod. Phys. **62**, 531 (1990).
7. See C. Cohen-Tanoudji in *Fundamental Systems in Quantum Optics,* Les Houches Summer School Proceedings for 1990, to be published.
8. C. Cohen-Tanoudji, J. Dalibard, Europhysics Letters, **1**, 441 (1986); M. S. King, P. L. Knight, and K. Wodkiewicz, Opt. Comm. **62**, 385 (1987); P. Zoller, M. Marte, D. F. Walls, Phys. Rev. **A 35**, 198 (1987); S. Reynaud, J. Dalibard, C. Cohen-Tanoudji, IEEE J. Quant. Electr. **24**, 1395 (1988).
9. C. Cohen-Tanoudji in *Frontiers in Laser Spectroscopy,* Proceedings of Les Houches Summer School July 1975, eds. R. Balian, S. Haroche, and S. Liberman, North Holland, Amsterdam (1977). H. Carmichael and D. Walls, J. Phys. **B9**, 1199 (1976).
10. D. Walls, Nature **280**, 451 (1979); C. Cohen-Tanoudji and S. Reynaud, Phil. Trans. Roy. Soc. London **A 293**, 233 (1979).
11. S. Reynaud, Annales de Physique, Paris **8**, 315 (1982).
12. G. Badurek, H. Rauch, and D. Tuppinger, Phys. Rev. **A 34**, 2600 (1985).
13. C. Dewedney, P. Guéret, A. Kyprianidis, and J. P. Vigier, Phys. Lett. **102 A**, 291 (1984).
14. R. J. Glauber, Phys. Rev. **131**, 2766 (1963).
15. K. F. von Weizsäcker, Zeit. f. Phys. **70**, 114 (1931).
16. In *Albert Einstein: Philosopher Scientist,* P. A. Schillp, ed., Library of Living Philosophers, Evanston, Illinois.
17. J. A. Wheeler, in: *Mathematical Foundations of Quantum Theory,* A. R. Marlow, ed., Academic Press, New York (1978).
18. T. Hellmuth, A. G. Zajonc, and H. Walther, in *New Techique and Ideas in Quantum Measurement Theory,* D. M. Greenberger, ed., Proc. New York Acad. Sc. **480**, 108 (1986). C. O. Alley, O. G. Jakubowicz, and W. C. Wikes in *Proc. 2nd Int. Symp. on the Foundations of Quantum Mechanics,* M. Namiki, ed., Physical Society of Japan, Tokyo, 36 (1987).
19. It may be noticed that, from the standpoint of quantum logic, this is a family of histories belonging to the so-called special type, according to the terminology introduced in the Appendix to Chapter 4. It should be contrasted with the simpler Griffiths families. This is the only case to our knowledge where Griffiths families are inadequate for a complete logical discussion and where a special family, rather than being as usual only more convenient, turns out to be necessary.
20. C. Tesche, in Greenberger, Proc. New York Acad. Sc. **480**, 36 (1986); S. Chakrabarty, idem, 25; C. Tesche, Phys. Rev. Lett. **64**, 2358 (1990).
21. There are many experimental contributions on the subject. For a lucid review, see J. Clarke, A. N. Cleland, M. H. Devoret, D. Estève, and J. M. Martinis, Science **239**, 992 (1988).
22. Ann. Phys. **149**, 374 (1983).
23. J. M. Martinis, M. H. Devoret, and J. Clarke, Phys. Rev. **B 35**, 4682 (1987).

CHAPTER 12

1. See Max Jammer: *"The Conceptual Development of Quantum Mechanics"* (McGraw Hill, 1966) and *"The Philosophy of Quantum Mechanics"* (Wiley, New York, 1974).

2. An attempt towards an axiomatic treatment can be found in G. Ludwig, *Deutung des Begriff "Physikalische Theorie"*. Springer, Berlin (1970).

3. M. Gell-Mann and J. B. Hartle in *Complexity, Entropy and the Physics of Information*, W. H. Zurek, ed., Santa Fe Institute Studies in the Science of Complexity, No. 8, Addison-Wesley, Redwood City, Calif. (1990).

4. R. Omnès, Rev. Mod. Phys. **64**, 339 (1992).

5. Bull. Am. Math. Soc., **9**, 129 (1983).

6. See the many contributions on this subject in the following book: W. H. Zurek (ed.), *Complexity, Entropy and the Physics of Information*, Santa Fe Institute Studies in the Science of Complexity, Addison-Wesley (1990).

7. See M. Jammer, *The Philosophy of Quantum Mechanics*.

8. This last thesis comes from a simple exercise where the optical detection of an illuminated object by the retina is treated according to the results of measurement theory. It was not given explicitly in the text because of its length and the inessential technicalities that are needed to treat it properly.

9. F. London and E. Bauer, *La Théorie de l'observation en Mécanique Quantique*, Hermann, Paris (1939).

10. E. P. Wigner, Am. J. Phys. **31**, 6 (1963).

11. B. d'Espagnat, *An Uncertain Reality*, Cambridge University Press (1989).

12. One can see however a decreasing importance of the arguments directly based upon the EPR correlations in the writings of B. d'Espagnat from *Scientific American* **241**, 128 (1979) to *An Uncertain Reality*, Cambridge University Press (1989) until their criticism in Found. Phys. **20**, 1147 (1990), which was inspired partly by the theory of truth given in Chapter 8.

13. Pierre Duhem, *La Théorie Physique* (1914), reprinted, Paul Brouzeng, ed., Vrin, Paris (1989). *Ibidem*, $\Sigma\Omega ZEIN\ TA\ \Phi AINOMENA$ (1908), P. Brouzeng, ed., Vrin, Paris (1990).

14. An example of this difficulty and a vivid controversy on the subject can be found in a book by the biologist Jean-Pierre Changeux and the mathematician Alain Connes, *Matière à penser*. Editions Odile Jacob, Paris (1989).

Index

Abragam, A., 542
Aharonov, Y., 536
Aharonov–Bohm effect, 418–23
algorithmic complexity, 314
Alley, C. O., 459
Aspect, A., 407, 542

Badurek–Rauch–Tuppinger experiment, 448–56
Bargmann, V., 56
Bauer–London interpretation, 80, 510
Bell, J., 80, 90, 304, 539, 542
Bell inequalities, 403–7
Bennett, C. H., 539
Bergquist, J. C., 542
Bernstein: norm, 250; theorem, 246
Birkhoff, G. See von Neumann
Blatt, R., 542
Bohm, D., 348, 353, 359, 541; theory, 401–3. See also Einstein–Podolsky–Rosen experiment, Bohm's version of
Bohr, N., 82, 86, 87, 88, 201, 382, 438, 459, 463, 476, 498, 510; and Einstein controversy, 13–14, 384; model, 4; quantization rule, 217. See also Copenhagen interpretation
Born, M., 7, 9–10, 36. See also probabilities: Born's rule for
Bub, J., 541
bubble chamber, 67

C*-algebras, 468
Caldeira, A. O., 291, 292, 309
Carmichael, H., 543
Caves, C. M., 314, 539
cells. See phase-space cells
Chaitin, G. J., 539
Chalmers, A. F., 535
Changeux, J. P., 544

chaotic systems, 229–31, 481; and quantum statistics, 426–32
Chiu, C. B., 537, 542
Clarke, J., 543
classicality parameter, 221, 479; best, 252–53; dynamical, 231, 481; effective, 221, 480; estimates of, 232
classical logic, 151–52, 213–15, 482; as common sense, 61, 95; consistency of, 259–61; derivation of, 234–36; limits of, 236–37
classical physics: conditions of validity for, 409; derivation of, 107, 201–62; status according to Bohr, 87; truth criterion in, 152
Clauser, J. F., 405
Claverie, P., 299
Cleland, A. N., 543
coherent states, 282–85
collective coordinates, 268
collective observables, 107, 205–6, 477; and collective Hilbert space, 212; construction of, 205; criterion for, 206, 261–62
collective system, 268
common sense, 108, 148, 204, 517. See also classical logic, as common sense
complementarity: Bohr principle of, 88, 160; origin of, 159–62
configuration space, 207
Connes, A., 544
consistency conditions, 132–40, 474–75; and additive probabilities, 132–34; alternative forms for, 134; from decoherence, 136; general form of, 137–40; for spin systems, 143
Copenhagen interpretation: discussion of, 89–90, 99–100, 498–502; practical rules of, 79; prescriptions of, 81–85. See also classical physics
correlated observables, 362–64

correspondence, 478; between classical and quantum physics, 202; between classical and quantum variables, 209–10, 478–79

Dalibard, J., 542, 543
datum, 325, 329
Daubechies, I., 265
De Broglie, L., 6, 379
decay, 176–79; 434–36
decoherence, 80, 108–9, 268–323, 484–86; for chiral molecules, 299–302; as a condition for measurement, 328; criticism and justification of, 304–15, 486–87; definition of, 269; by external environment, 297–302; intuitive approach to, 273–79; by scattering, 297–99, 319–22; solvable models for, 279–302; in SQUIDs, 465–66; theory of, 291–97
degrees of freedom, 44
Dehmelt, A., 542
delayed-choice experiments, 338, 456–63
Denkbereich, 149
density operator, 38–40, 471; collective, 275; construction of, 364–69, 498; and EPR correlations, 362–64; up to equivalence, 365–66; existence of, 119; reduced, 275, 484; spontaneous diagonalization of, 277; and truth, 360–64; for universe, 366–68
D'Espagnat, B., 399, 513, 514, 515, 527, 534, 539, 541, 542, 544
detectors, 62–68; necessity of, 71
determinism, 107, 215, 236, 483; as logical equivalence, 152. *See also* Theorem C
Devoret, M. H., 543
Dewedney, C., 543
De Witt, B. S., 348
Dirac, P.A.M., 7, 15, 47, 85, 153, 209; notation, 30–31
Duhem, P., 515, 516
Dürr, D., 541
dynamics, 481, 482; classical, 203, 227–34; regular, 231

Egorov, Ju, V., 538
Ehrenfest theorem, 43–44, 481
Einstein, A., 85, 351, 379, 536. *See also* Bohr, N., and Einstein controversy

Einstein–Podolsky–Rosen experiment, 176, 379–400, 496–97; Bohm's version of, 387–88; effect of errors in, 392; and logic, 386–87, 391–97; for photons, 388–91; relativistic case of, 397–98
elements of reality, 380–83
environment, 268, 477
Estève, D., 543
Everett, H., 327, 493, 511; representation of reality, 345–48
evolution operator, 32, 41, 113, 468; and Feynman histories, 32–34

facts, 87, 325, 516. *See also* uniqueness of facts
Feffermann, C., 206
Feynman, R. P., 291, 535, 542; and consistency, 180–81; histories or paths, 31–35, 130–32, 180–82, versus Griffiths histories, 122
Fock, V. A., 535
Foucault, M., 523
Froissart, M., 542

Garding's sharp inequality, 244
Geiger counter, 66–67
Gell-Mann, M., xiv, 136, 154, 182, 298, 319, 345, 367, 493, 502–3
Ghirardi, G. C., 349–50
Ginibre, J., 537
Glauber, R. J., 451
Gleason theorem, 105, 119, 471
Goldberger, M. L., 537
Goldstein, S., 541
Grangier, P., 542
Griffiths, R., xiv, 106, 122, 133, 189, 196, 197, 198, 376, 437, 473. *See also* consistency conditions; histories, Griffiths families of
group, 48–56; generators, 49, Heisenberg, 209, Lorentz, 51–53; Poincaré, 53–56
Guéret, P., 543

Hagedorn, G., 233, 266
Halmos, P. R., 141, 143
Hamilton: equations, 212, 254; function, 210, 479
Hartle, J. B., xiv, 136, 154, 182, 296, 319, 345, 367, 493, 502–3
Heidegger, M., 519, 527

Heisenberg, W., 82; frontier, 86, 87, 159; matrix mechanics, 6–7; microscope, 11; representation, 86. *See also* time; uncertainty relations
Hellmut, T., 543
Hepp, K., 282, 304, 535
hidden variables, 80, 400–408; Bohm theory of, 401–3
Hilbert, D., 153
Hilbert–Schmidt operators, 22
Hilbert space, 16–18; basis, 23–24; construction, 46–56; as framework, 112; subspaces, 23–24
histories, 105, 106, 122–43; compound, 137; consistent families of, 105–6 (*see also* consistency conditions); definition of, 125; describing experiments, 123–24; families of, 134–36, and Feynman histories, 130–32; graph representation of, 127, 139; Griffiths families of, 135, 189–90, 192; intuitive nature of, 123–24; in quantum logic, 155; and reality, 447–48; special families of, 138, 190–92. *See also* Feynman, R. P., histories or paths; history probabilities
history probabilities, 105–6, 126–32, 473, 474; and consistency conditions, 132–34; formula for, 129; logical criteria for, 128–29; uniqueness of, 129, 140–43
Holt, R. A., 405
Hörmander, L., 537, 538
Horne, M. A., 405, 542
Hulet, R. B., 542
Hume, D., 87
Huron, 146–47

implication, 150; approximate, 158; from conditional probabilities, 157–58
individual system, 85
information theory, 369, 375–77
Inönü, E., 534
interferences, 165–71, 448–56; macroscopic, 77–81
interpretation: complete, 99, 469; consistent, 99, 469; definition of, 98–100, 469; necessity of, 96–97; new, 103–9; standard, 510. *See also* Copenhagen interpretation

ion trap, 444–45
irreversibility: of measurements, 87, 328
isolated systems, 111
Itano, W. M., 542

Jakubowitz, O. G., 459
Jammer, M., 533, 543, 544
joint probability distributions, 58–59
Jona-Lasinio, G., 299
Joos, E., 297, 319
Josephson junction, 415–17
Jost, R., 536

King, M. S., 543
Knight, P. L., 543
knowledge: theory of, 529–30. *See also under* Wittgenstein, L.
Kolmogorov, A. N., 539
Kuhn, T., 523
Kyprianidis, A., 543

Landau, L. D., 38, 57, 535
Leggett, A. J., 99, 291, 292, 309; experiments, 423–26; 463, 466
Levin, L. A., 539
Lieb, E. H., 282, 535
Lifshitz, E. M., 57, 535
Liouville equation, 254
logic, 472; approximations in, 179–83; axioms of, 150, 183–85; consistency of, 158, 186–88; constituents of, 149–51; interpretations of, 149; and logics, 149; necessity of, 145–49; operations of, 149–50; in quantum mechanics, 106, 145–200; relations in, 150; and time reversal, 193–98. *See also* classical logic; logics; truth
logics, 472–73; complementary, 162; consistent, 475–76; extension of, 161; mutually consistent, 162; operations on, 161–62; sensible, 365, 495; simplification of, 161
logos, 526
London relation, 413
Lüders, V. G., 83, 339
Ludwig, G., 544

Mackey, G. W., 537
Maeda, S., 536
Mark, M., 543
Martinis, J. M., 543

master inequality, 245
mathematics versus physics, 524–27
measurements: of an apparatus by a second one, 308–9; circumstantial, 61; conditions for, 489–90; data, 325, 329; duration of, 330; of the first kind, 83, 84; ideal, 82; imprecise, 100–102; individual, 61; with lost measured system, 77; results of, 325, 329; structural, 61; successive, 337–39; systematically repeated, 436–38; of Type I, 83, 84; of a unique atom, 443–48. *See also* measurement theory; Zeno effect
measurement theory, 60–102, 324–77; elementary, 72–81; main theorems of, 490–91; S-operator in, 333, 490; von Neumann's formal, 72–75. *See also* measurements
Mermin, N. D., 542
method in science, 96
metric in phase space, 248
microlocal analysis, 211, 238–47
microscopic coordinates, 268
Misra, R., 537, 542
modus ponens, 340–41
Mott, F. W., 172

Nagourney, W., 445, 542
Nauenberg, M., 538
Nelson, E., 541
Neuhoser, W., 542
neutral position, 331
neutron interferences, 166–67
no-contradiction theorem, 162, 188–93, 476
nominalism, 525
nonsense, 127–28
norm of an operator: Bernstein, 250, 252, 257; L_1 norm, 245, 257; trace, 245

objectivity, 89, 509–12
object projector, 205
objects, 107, 203, 204–6, 477–78; versus physical systems, 204–5
observables, 469; measurable 306, 487–88, 545n.5; and operators, 114; values of, 115
operator, 18–24; domain of, 25; function of, 30; self-adjoint, 19; spectral decomposition of, 24–31; spectrum of, 20–21, 114–15; time-dependent, 41; trace of, 21–22

paradoxes: impossibility of, 162. *See also* no-contradiction theorem
partitions of unity, 246
Pauli, W., 82, 83
Paul trap, 445
Peres, A., 542
phase-space cells, 214; distinct, 225; margin of, 248, and quasi-projectors, 224; regular, 224
phenomena, 99, 325; according to Bohr, 87; conventional definition of, 60; theory of, 488–89
photographic detection, 63–65
photomultipliers, 65
Planck, M., 4–5
Poincaré recurrences, 281
Popper, K., 541
predicates, 115
Primas, H., 82, 85, 534, 535
principles. *See* quantum mechanics
probabilities: Born's rule for, 35–37; empirical status of, 335–36; formulas for, 83–84, 470–71; mathematical axioms for, 118–19; as mathematical and logical tools, 104–5, 118, 336; as primary, 10–11, 86. *See also* Gleason theorem; history probabilities
projection operators. *See* projectors
projectors, 23–24; for properties, 115–17; procrustean, 27; time-dependent, 121
properties, 103–5, 114–21, 469–70; classical, 107, 203, 213–26, 479; definition of, 115; mutually exclusive, 116; physical meaning of, 365; and predicates, 115–16; and projectors, 115–17; at given time, 120–21; reliable, 356–58; and systematic errors, 358
pseudo-differential calculus. *See* microlocal analysis
pseudo-differential operators, 211; product of, 240–43; seminorms of, 240; symbols of, 238, 239

quantum, 4
quantum jumps, 88, 438

quantum mechanics: logical principle of, 163; origin of, 4–15; philosophical aspects of, 506–31; principles of, 163, 467–69; structures of, 164–65

quasi-projectors, 223, 480–81; associated with a cell, 224; construction by coherent states, 226; construction by pseudo-differential operators, 226; equivalent, 224, 480; rank and order of, 223

realism, 512–16; status of, 530–31; total, 527–28
reasoning, 145–46
records, 202
reduction. *See* wave function reduction
reliable properties, 356–58
result of a measurement, 325, 329
Reynaud, S., 543
Rimini, A., 349–50
Roger, G., 542

S-operator, 333, 490
Sandberg, J., 542
Schrödinger, E., 15, 79, 85, 208; cat, 91, 109, 303–4; equation, 8–9, 41, 113, 468
semiclassical approximation, 216–17
semiclassical theorems. *See* Theorem A; Theorem B; Theorem C
seminorms, 240
sensible logics, 365, 495
separability, 398; and truth, 400
Shimony, A., 405, 542
Sinai billiards, 229–30
Slauter, T., 542
Solomonoff, R. J., 539
spin and relativity, 55–56
spin echoes, 278–79
SQUID, 417–18
state, 117–21, 488, 497–98; definition of, 118; historical, 118; operator (*see* density operator), predictive, 118
state of fact, 235, 482–83
Stern–Gerlach experiment, 68–72
Stone theorem, 42
straight-line motion, 171–76, 199–200
Sudarshan, E.C.G., 537, 542
symbol. *See* pseudo-differential operators

Tarski, A., 541
tautology, 127–28
tensor product, 45–46, 468
Tesche, C., 543
Theorem A, 224; proof of, 248–53
Theorem B, 225, 253, 267 (problem 19)
Theorem C, 233; proof of, 253–58
theory of knowledge, 529–30. *See also under* Wittgenstein
Thirring, W., 58
time: direction of, 342–45, 315–18; and logic, 164, 194–98; past and future in the structure of, 344–45, 493; reversal of, 193–94
Toschek, P. E., 542
trace, 21–22; norm, 245
truth, 517; in classical physics, 152; in the Copenhagen interpretation, 88, 355; criterion, 149, 150–51, 353–56, 494–96; function, 150, 185; of measurement results, 356; in quantum mechanics, 353–56

uncertainty relations, 11, 38, 59, 102; for time and energy, 13–14, 56–58, 102
uniqueness of facts, 341–53, 492–94, 504–5
universal rule of interpretation, 106, 163, 476
universe of discourse, 149

Vaidman, L., 536
Van Heijenoort, J., 537
Van Hove, L., 539
Velo, G., 537
Vernon, F. L., 291, 535
Vigier, J. P., 448, 543
von Neumann, 16, 26, 38, 72, 80, 104, 114, 115, 117, 153, 306, 471, 510, 534
Von Weizsäcker, C. F., 459

Walls, D. F., 543
Walther, H., 543
wave function reduction, 8, 90, 326, 337–41, 491–92; Copenhagen rule for, 83, 84; meaning of, 339–41, 492. *See also* uniqueness of facts
Weber, T., 349–50
Weinberg, S., 541

Weisskopf–Wigner equations, 286
Weyl, H., 210, 478
Weyl calculus. *See* microlocal analysis
Wheeler, J. A., 459, 534, 543
Wick, G. C., 538
Wightman, A. S., 538
Wigner, E., 47, 49, 56, 71, 80, 478, 534, 535, 538; friend, 91–92, 303; probability density, 210
Wikes, W. C., 459
Wineland, D. J., 542

Wittgenstein, L., 351, 354, 519; theory of knowledge, 539–41
Wodkiewicz, K., 543

Zajonc, A. G., 543
Zanghi, N., 541
Zeh, H. D., 297, 319
Zeilinger, A., 542
Zeno effect, 438–43
Zoller, P., 543
Zurek, W. H., xiv, 279, 291, 310, 534, 539
Zvonkin, A. K., 539